Mutagenicity

New Horizons in Genetic Toxicology

This is a volume in
CELL BIOLOGY
A series of monographs

Editors: D. E. Buetow, I. L. Cameron, G. M. Padilla, and A. M. Zimmerman

A complete list of the books in this series appears at the end of the volume.

Mutagenicity

New Horizons in

Genetic Toxicology

Edited by

JOHN A. HEDDLE

Department of Natural Science
York University
Downsview, Ontario, Canada

1982

ACADEMIC PRESS

A Subsidiary of Harcourt Brace Jovanovich, Publishers

New York   London
Paris   San Diego   San Francisco   São Paulo   Sydney   Tokyo   Toronto

ACADEMIC PRESS, INC.
111 Fifth Avenue, New York, New York 10003

United Kingdom Edition published by
ACADEMIC PRESS, INC. (LONDON) LTD.
24/28 Oval Road, London NW1 7DX

Library of Congress Cataloging in Publication Data
Main entry under title:

Mutagenicity, new horizons in genetic toxicology.

    (Cell biology)
    Includes bibliographies and index.
    1. Mutagenicity testing. I. Heddle, John A.
II. Series. [DNLM: 1. Mutagenicity tests. QH
460 M991]
QH465.A1M887   616'.042   81-22940
ISBN 0-12-336180-X   AACR2

PRINTED IN THE UNITED STATES OF AMERICA

82 83 84 85   9 8 7 6 5 4 3 2 1

Contents

3   The Use of Mutagenicity to Evaluate Carcinogenic Hazards in Our Daily Lives

Takashi Sugimura and Minako Nagao

4   Mutagenicity and Lung Cancer in a Steel Foundry Environment

D. W. Bryant and D. R. McCalla

8 Measurement of Mutations in Somatic Cells in Culture

Veronica M. Maher and J. Justin McCormick

9 Chromosomal Aberrations Induced in Occupationally Exposed Persons

Maria Kučerová

13 Cytogenetic Events *in Vivo*

J. G. Brewen and D. G. Stetka

14 Dominant Skeletal Mutations: Applications in Mutagenicity Testing and Risk Estimation

Paul B. Selby

List of Contributors

Numbers in parentheses indicate the pages on which the authors' contributions begin.

Richard J. Albertini (305), Department of Surgery, University of Vermont College of Medicine, Burlington, Vermont 05405

Elizabeth F. Allen (305), Department of Pathology, University of Vermont College of Medicine, Burlington, Vermont 05405

John Ashby (1), Genetic Toxicology, Central Toxicology Laboratory, Imperial Chemical Industries Limited, Alderley Park, Macclesfield, Cheshire 5K10 4TJ, England

H. Bartsch (35), Division of Environmental Carcinogenesis, International Agency for Research on Cancer, World Health Organization, 69372 Lyons, France

J. G. Brewen (351), Corporate Medical Affairs, Allied Chemical Corporation, Morristown, New Jersey 07960

D. W. Bryant (89), Department of Biochemistry, Health Sciences Centre, McMaster University, Hamilton, Ontario L8N 3Z5, Canada

Anthony V. Carrano (267), Biomedical Sciences Division, Lawrence Livermore National Laboratory, University of California, Livermore, California 94550

William F. Grant (407), Genetics Laboratory, Department of Plant Science, Macdonald Campus of McGill University, Ste. Anne de Bellevue, Quebec H9X 1C0, Canada

A. D. Kligerman (435), Chemical Industry Institute of Toxicology, Research Triangle Park, North Carolina 27709

Maria Kučerová (241), Pediatric Department, Postgraduate Medical Institute, and Thomayer's Hospital, Prague 4, Czechoslovakia

D. R. McCalla (89), Department of Biochemistry, Health Sciences Centre, McMaster University, Hamilton, Ontario L8N 3Z5, Canada

J. Justin McCormick (215), Carcinogenesis Laboratory, Departments of Microbiology and Biochemistry, Michigan State University, East Lansing, Michigan 48824

Veronica M. Maher (215), Carcinogenesis Laboratory, Departments of Microbiology and Biochemistry, Michigan State University, East Lansing, Michigan 48824

C. *Malaveille* (35), Division of Environmental Carcinogenesis, International Agency for Research on Cancer, World Health Organization, 69372 Lyon, France

Dan H. Moore II (267), Biomedical Sciences Division, Lawrence Livermore Laboratory, University of California, Livermore, California 94550

Minako Nagao (73), National Cancer Center Research Institute, Tsukiji, Chuo-ku, Tokyo 104, Japan

A. T. *Natarajan* (171), Department of Radiation Genetics and Chemical Mutagenesis, Sylvius Laboratories, University of Leiden, 2333 Al Leiden, The Netherlands, and J. A. Cohen Institute, Interuniversity Department of Radiation Protection and Radiation Pathology, Leiden, The Netherlands

G. *Obe* (171), Institut für Genetik, Freie Universität Berlin, D-1000 Berlin 33, Federal Republic of Germany

W. D. *Powrie* (117), Department of Food Science, University of British Columbia, Vancouver, British Columbia V6T 1W5, Canada

M. P. *Rosin* (117), Environmental Carcinogenesis Unit, British Columbia Cancer Research Centre, Vancouver, British Columbia V5Z 1L3, Canada

Paul B. Selby (385), Biology Division, Oak Ridge National Laboratory, Oak Ridge, Tennessee 37830

Andrew Sivak (143), Biomedical Sciences Section, Arthur D. Little, Inc., Cambridge, Massachusetts 02140

D. G. *Stetka* (351), Corporate Medical Affairs, Allied Chemical Corporation, New Jersey 07960

H. F. *Stich* (117), Environmental Carcinogenesis Unit, British Columbia Cancer Research Centre, Vancouver, British Columbia V5Z 1L3, Canada

Takashi Sugimura (73), National Cancer Center Research Institute, Tsukiji, Chuo-ku, Tokyo 104, Japan

David L. Sylwester (305), Biometry Facility, University of Vermont College of Medicine, Burlington, Vermont 05405

L. *Tomatis* (35), Division of Environmental Carcinogenesis, International Agency for Research on Cancer, 69372 Lyon, France

Alice S. Tu (143), Biomedical Sciences Section, Arthur D. Little, Inc., Cambridge, Massachusetts 02140

C. H. *Wu* (117), Department of Food Science, University of British Columbia, Vancouver, British Columbia V6T 1W5, Canada

Andrew J. Wyrobek (337), Biomedical Sciences Division, Lawrence Livermore National Laboratory, University of California, Livermore, California 94550

K. D. *Zura* (407), Genetics Laboratory, Department of Plant Sciences, Macdonald Campus of McGill University, Ste. Anne de Bellevue, Quebec H9X 1C0, Canada

Preface

The most astonishing aspect of environmental mutagenesis, aside from the field's explosive growth, is the diversity of applications of the new assays for mutagenicity. Originally developed with the idea of screening food additives and drugs, these essays are now being used to investigate human disease, to monitor the natural environment, to identify hazards in the workplace, and even to evaluate the quality of food and the impact on it of cooking. These and other applications that are the "new horizons" in genetic toxicology and environmental mutagenesis form the theme of this volume. Those who seek details of particular assays or a compilation of test results will not find them here. Rather, this volume provides an overview of the field, particularly of the exciting frontier areas, which is not provided by any other volume. It should be useful to anyone interested in environmental mutagenesis, genetic toxicology, occupational health, or the regulation of toxic substances.

Although the testing of chemicals in many assays is now routine, the first chapter of this volume (Ashby) makes it clear that the interpretation of the results is not. It should be required reading for anyone developing a new assay, using multiple assays, or regulating the use of a chemical. In Chapter 2, Bartsch *et al.* deal with the problem of potency—whether a positive in the Ames test is likely to be a potent human carcinogen or mutagen. This is an area of intense controversy and investigation. In practice, scientists are forced to make de *facto* extrapolations of potency almost daily as will be seen on reading Chapter 3 by Sugimura and Nagao on naturally occurring mutagens, Chapter 4 by Bryant and McCalla on a search for an occupational carcinogen, and Chapter 5 by Stich and his collegues on the evaluation of food and cooking methods.

In the initial chapters of the book, the emphasis is primarily on the use of the bacterial assays. Subsequent chapters deal with the use of mammalian cells in culture and in whole animals. Some of the mammalian cell assays are now almost as fast, cheap, and successful as the best of the microbial assays. They have the advantage that events such as transformation (Sivak and Tu, Chapter 6) and chromosomal events (Natarajan and Obe, Chapter 7) that have no direct counterpart in bacteria but are of great concern in relation to human disease can be measured. Of course mutations can be measured in mammalian cells too (Maher and McCormack, Chapter 8). Indeed, studies on human cells, particularly cells deficient in DNA repair, can be used to test models of the mechanism by which diseases such as cancer arise. Furthermore, measurements on cells cultured from occupationally exposed persons can be used to investigate *in vivo* exposures. It is noteworthy, for example, that chromosomal aberrations were detectable in blood cells cultured from vinyl chloride workers exposed to conditions that were giving rise to occupational cancers, but were not detected in workers who had only been exposed since the new, lower permissible levels were established. The armamentarium of the industrial toxicologist has been considerably augmented by the new techniques developed in environmental mutagenesis as shown in four chapters in this volume (Kučerová, Chapter 9, on chromosomal aberrations; Carrano and Moore, Chapter 10, on sister chromatid exchanges; Albertini *et al.*, Chapter 11, on mutations; and Wyrobek, Chapter 12, on sperm abnormalities). The use of whole animals is, of course, the best existing model for human exposure and thus is vital in risk evaluation. Readers will find the chapters by Brewen and Stetka (Chapter 13) and by Selby (Chapter 14) very useful in recognizing both the advances that have been made and the uncertainties that remain.

The final two chapters of the volume concern environmental sentinals—organisms in which genetic events can easily be measured which can thus be used as environmental monitors. This is an area in which I anticipate dramatic progress. As will be seen from Chapter 15 by Grant and Zura (on plants) and from Chapter 16 by Kligerman (on fish), powerful tools already exist.

The diversity of the chapters of this volume shows the vitality and excitement that characterize the application of the new assays for genetic events and the concepts these assays have engendered. I hope you will find this volume useful, interesting, and stimulating.

John A. Heddle

Chapter 1

Screening Chemicals for Mutagenicity: Practices and Pitfalls

John Ashby

MUTAGENICITY:
NEW HORIZONS IN GENETIC TOXICOLOGY

I. INTRODUCTION

If a chemical is naturally reactive to nuclear DNA, or can be transformed to such a species by mammalian enzymes, it is usually also mutagenic, clastogenic, and carcinogenic to appropriately exposed organisms. The fact that this consistency of activity is not always observed is worthy of study, and this forms the basis of the present chapter.

Vesicants such as sulfur mustard (I) and the nitrogen mustards bis (2-chloroethyl) methylamine (NH2) (II) and HN3 (III) received sporadic attention between 1930 and 1950 as possible genotoxic agents. As early as 1931, Berenblum had shown that sulfur mustard exerted antitumor effects, and in 1941 (published 1944, 1946), Auerbach and Robson demonstrated that several of these chemicals [(I), (II), (III), and other alkylating agents] were capable of simulating the mutagenic properties of X-rays in *Drosophila*. With these observations the modern science of chemical mutagenesis was born.

The mutagenic properties* of these mustards were rapidly confirmed in other organisms including *E. coli* and *Neurospora* by Horowitz *et al.* (1946) and Tatum (1947).

$$S \begin{cases} CH_2CH_2Cl \\ CH_2CH_2Cl \end{cases} \qquad H_3CN \begin{cases} CH_2CH_2Cl \\ CH_2CH_2Cl \end{cases} \qquad N \begin{cases} CH_2CH_2Cl \\ -CH_2CH_2Cl \\ CH_2CH_2Cl \end{cases}$$

I   II   III

* *Definition of Terms:* A *mutagen* is considered herein to be an agent capable of disturbing the integrity of the hereditary mechanism of the cell or organism. However, such interferences can be mediated by several mechanisms, each of which has acquired separate names. Experiments that demonstrate a chemical to be capable of altering the base sequence of a gene define a gene or point mutagen. Those that illustrate the ability to induce structural changes in eukaryotic chromosomes define a *clastogen*. The chemical induction of micronuclei is regarded as a clastogenic event. Changes in the chromosome complement of a cell (apart from changes in ploidy) are referred to as *aneuploidy*. Chemically induced mutations may occur in either somatic cells or in germ cells, the former primarily having potential carcinogenic significance, the latter heritable mutagenic significance.

The term *genotoxic* implies that a test chemical is capable of damaging DNA in a chemical sense, an effect usually monitored by the induction of point mutations or chromosomal aberrations. A carcinogen is defined as a chemical capable of increasing the incidence of cancer in any species of mammal when administered by any route. In this discussion the conclusions of individual investigators are usually regarded as definitive.

These several demonstrations of mutagenicity prompted Boyland and Horning to evaluate HN2 for carcinogenicity, and in a study conducted in 1949 they demonstrated a systemic carcinogenic effect for this chemical in mice, and thereby provided the first evidence that chemical mutagens in general might also be mammalian carcinogens. The paper that described these findings provides one of the earliest, and certainly one of the most succinct, discussions of the several genotoxic properties to be expected of an overt electrophile such as HN2, and for these reasons, part of its introduction section is reproduced below as a historical perspective for the chapters that follow.

> Recent experience has shown that many of the growth-inhibiting agents used in the palliative treatment of cancer are carcinogenic. They also cause specific damage to cell nuclei and chromosomes and are able to induce mutations. The association of these biological effects of (1) growth inhibition, (2) chromosome damage, (3) production of mutations and (4) induction of cancer suggests that they may have a common fundamental biochemical mechanism. If the induction of cancer is indeed a somatic mutation, then cancer induction might be included as a special mutation . . . The present authors consider that the initial irreversible conversion to latent tumour cells is the change which is associated with chromosome damage and mutations (Boyland and Horning, 1949).

This quotation is enhanced by the subsequent involvement in the discussion of the then recently defined phenomena of carcinogenic promotion (Berenblum and Shubik, 1947) and Haddow's paradox (anticancer agents may also be carcinogenic; Haddow et al., 1948). Thus, in a single paper published in 1949, were discussed the various genotoxic phenomena that have so influenced the development and deployment of environmental chemicals during the past few years.

The fact that it took 20 more years to exploit these findings is due in large part to the fact that most carcinogens and mutagens are not naturally reactive to DNA (electrophilic), but rather need to be transformed to such species by metabolic enzymes. This fact was formally recognized by Miller and Miller (1969) and translated to the field of in vitro mutagenicity testing by Malling in 1971. Here again, it was a carcinogen (DMN) that was chosen to be activated to a mutagen in vitro.

The present interest in environmental carcinogens, and subsequently environmental mutagens, was inspired by the demonstration by Ames et al. (1975) and McCann et al. (1975) that the majority of mammalian carcinogens are also mutagenic to bacteria, a finding that complements the early conclusions discussed above. These several considerations indicate that any attempt to view mutagenicity in complete isolation from carcinogenicity, and vice versa, would be unjustified, and for this reason this has not been done here.

II. TEST BATTERIES

Within the above context it would seem logical to assume that a chemical capable of inducing gene mutations in bacteria would also be a clastogen and mammalian carcinogen, and vice versa, but such has not always proved to be the case. Consequently, although the several genotoxic properties of compounds such as ethyl methanesulfonate (EMS) and HN2 appear to be causally related, the genetic/carcinogenic effects produced by other chemicals are increasingly regarded as separate and perhaps mechanistically unrelated phenomena. The trend to isolate various chemically induced genetic/carcinogenic effects is based on a few classical exceptions to the "unified" theory of genotoxic expression, and the majority of these are shown in Table I. This table is selective in terms of the assay responses referred to and the differences in biological activity emphasized. The assay responses omitted were either undetermined or considered not to be important to the present discussion. Table I is discussed below.

It is now clear that the early workers in this field, by concentrating their studies on ionizing radiation and *direct-acting* alkylating agents such as ethyl methanesulfonate (EMS) and HN2 (II), temporarily obscured the possibility that not every chemical carcinogen/mutagen would give a positive response in every mutation assay. However, there exists now a sufficient number of exceptions to justify not regarding the activity of a chemical in any particular assay as automatically predictive of its activity in another; and from this realization a logical basis for the design of a battery of test systems might have been derived. In reality, however, the several current test batteries were designed more in an attempt to represent adequately a range of different genetic end points than to detect as positive all established mammalian mutagens and carcinogens. Nonetheless, the disparate activities of chemicals, such as those shown in the lower part of Table I, may still be used to refine or extend the usefulness of existing batteries. Some objective criteria for the selection and validation of a battery of tests for the detection of potential mammalian carcinogens and mutagens are discussed later in this chapter.

III. POSSIBLE BASIS OF DIVERGENT ASSAY RESPONSES OBSERVED BETWEEN ASSAY SYSTEMS

There are four possible origins for qualitative differences between assay responses observed for the same chemical in different assay

TABLE I

Responses of Ten Selected Chemicals in Seven Classes of Genotoxicity Assay[a]

Chemical	Point mutation (prokaryotes)	Dominant lethal (mammal)	Heritable translocation (mammal)	Aneuploidy (yeast)	Clastogen (in vitro)	Clastogen (in vivo)	Carcinogenicity
EMS	+	+	+	+	+	+	+
HN2	+	+			+	+	+
DES	−			+	+	+	+
HMPA	−	−				(micronucleus)	+
Benzene	−	−			+	+	+
EO	+	+	+		+	−/+ (?) see text	+/− ?
DMF	−	+			−	−	−
DMN	+	−					+
4CMB	+	−		−	+	(micronucleus)	− ?
IsoPC	−			+	−	(micronucleus)	− (see text)

[a] EMS (ethyl methanesulfonate) is a universally employed positive control chemical. The several activities of HN2 [bis(2-chloroethyl)methylamine] are described in the text and in a recent detailed review by Fox and Scott (1980). The data quoted for diethylstilbestrol (DES), hexamethylphosphoramide (HMPA), dimethylformamide (DMF), and isopropyl N-(3-chlorophenyl)carbamate (IsoPC) were in the main derived from the "International Study" (de Serres and Ashby, 1981) with the following additions: the negative mouse dominant lethal assay response for HMPA is from Von Schefler (1980), and the epidemiological linking of occupational exposure to DMF with an increased incidence of cytogenetic damage in the peripheral lymphocytes of exposed workers is reported by Kondela in an article by Sram and Kuleshov (1980) (see text). The several genotoxic properties of 4-chloromethylbiphenyl (4CMB) are discussed in the Proceedings of the United Kingdom Environmental Mutagen Society Collaborative Study (UKEMS, 1982). The negative dominant lethal assay response observed for the carcinogenic bacterial mutagen dimethylnitrosamine (DMN) has been discussed by Bateman and Epstein (1971). The genotoxic profile of benzene has been reviewed by Dean (1978) and the clastogenicity in vitro of several of its metabolites was recently described by Morimoto and Wolff (1980). The genotoxicity of ethylene oxide (EO) was recently reviewed by Generoso et al. (1980) and Snellings et al. (1981).

systems, an appreciation of which will aid interpretation and extrapolation of assay data derived for new chemicals in the future.

A. Genetic End-Point Specificity of Action

Benzene and DES appear to affect adversely the structure of mammalian chromosomes, yet they are inactive as point mutagens in bacteria. Likewise, IsoPC (see Table I) specifically induces aneuploidy in yeast, but is inactive as a point mutagen, clastogen, or carcinogen. These three examples provide probably the clearest evidence of a genuine specificity in the mutagenic action of some chemical mutagens, and thereby provide a logical basis for the inclusion of at least one cytogenetic assay in any battery of test systems.

B. Metabolic Differences between Assays

Probably the major factor contributing to qualitative differences in assay responses arises from differences in the metabolic competence of individual assay marker organisms to activate nonelectrophilic test chemicals to a DNA-reactive species. These differences are often exacerbated *in vitro* by variations in the source and nature of the enzyme preparations usually employed as an adjunct to these tests (referred to hereafter as S9 mix), and *in vivo*, by species, strain, or sex differences of the test animals. For example, the inactivity of HMPA as a point mutagen in bacteria is probably associated with their inability (± S9 mix) to activate it appropriately to an electrophilic species. The activity of this chemical as a point mutagen in several mammalian cell *in vitro* assays supports this explanation. Similarly, the fact that the bacterial mutagen DMN gives a negative response in the mouse dominant lethal mutation assay is probably associated with its failure either to reach the testes or to be appropriately metabolized therein.

In order to emphasize the possible importance of the *genetic* versus the *metabolic* explanation for assay response differences a section of Table I is reproduced in Table II. The examples selected demonstrate that qualitatively similar changes in test responses observed in separate assays for different chemicals may arise for mechanistically unrelated reasons.

C. False-Positive and False-Negative Results

When attempting to rationalize or understand qualitative differences in a series of assay responses, it is important to consider the credibility

TABLE II

Comparison of the Bacterial Mutagenicity and Cytogenetic Activity *in Vivo* of Diethylstilbestrol (DES) and Hexamethylphosphoramide (HMPA)[a]

Chemical	Point mutation in bacteria	Micronucleus formation *in vivo*	Suggested reason for changed response
DES	−	+	DES is specifically a clastogen?
HMPA	−	+	HMPA is a promutagen that is not effectively activated in bacteria?

[a] Although these chemicals present a similar genotoxic profile, it is suggested that they do so for different reasons. These data are abstracted from the results of the "International Study" (de Serres and Ashby, 1981).

that can be accorded to individual test responses. For example, the isolated positive response observed for IsoPC in the yeast aneuploidy assay (Table I) is known to be reproducible, and therefore worthy of consideration. In contrast, had this result represented an isolated observation made with an impure sample of the test chemical, then its significance would have been reduced. The point is made particularly important by the experience gained in the recent "International Evaluation of Short-Term Tests for Carcinogens" [referred to hereafter as the "International Study" (1981)] which indicated that irreproducible assay responses occur, and with some regularity. Another aspect of this concern is that a specific genotoxic activity may be associated inadvertently with the wrong chemical. Apart from the possibility that a mutagenic impurity might influence the assay response observed for an otherwise inactive test chemical, there remains the chance that positive epidemiological observations may sometimes be associated with the most easily detected chemical, rather than with the *causative* genotoxic agent present in the environment sampled. For example, it was recently reported that dimethylformamide (DMF, see Table I) was responsible for an increase in the incidence of cytogenetic damage in the peripheral lymphocytes of a group of "DMF-exposed" workers (Sram and Kuleslov, 1980). This finding is in sharp contrast to the general lack of genetic activity shown by this chemical in the "International Study" (1981), an evaluation that included four independently conducted *in vivo* cytogenetic bioassays of DMF. It therefore seems justifiable to regard these recent epidemiological findings with some caution: a biological effect has been observed, but the claim that it was specifically mediated by DMF, although possible, appears unproved. The importance of these considerations is that had DMF been defined previously as giving a positive response in a range of *in vitro* and *in vivo* assays, then these epidemiological findings may have been accepted without question as confirming a mutagenic trend already well established in experimental systems.

D. Unexplained Differences in Assay Responses

Although certain of the differences in assay responses observed for some chemical mutagens may be tentatively rationalized (Section III, (A–C), the differences observed for others remain obscure. The most challenging current examples of this are presented by the several genotoxic properties of ethylene oxide (EO, see Table I) and 4-chloromethylbiphenyl (4CMB, see Table I).

1. Ethylene Oxide

Ethylene oxide (EO) is a potent *direct-acting* mutagen and clastogen *in vitro*. Further, it gives a clear positive response in both the dominant lethal and heritable translocation mutation assays conducted in rats. It was therefore somewhat surprising when a recent inhalation carcinogenicity bioassay of this chemical revealed that the test animals failed to develop an increased incidence of cancer in the tissues apparently most at risk, the nasal epithelium and the respiratory tract.

A slight enhancement of the spontaneous incidence of leukemia was recorded, but of greater interest, a statistically and biologically significant increase in the incidence of peritoneal mesotheliomas was also evident (Snellings *et al.*, 1981). These tumors originated in the tunica vaginalis surrounding the testes; thus, EO elicits both mutagenic and carcinogenic effects in the relatively "remote" testicular tissues, while in the respiratory tract, where its alkylating properties should be most evident, and where its *potential* carcinogenicity was considered most likely to be expressed, no effect was observed.

These activities may be due to the rapid "detoxification" of EO in the respiratory tract [perhaps to yield a glutathione adduct ($GS—CH_2—CH_2—OH$)] and the subsequent *reactivation* of one of these conjugates to a secondary mutagen in the testes [perhaps, the corresponding O-acetate derivative ($GS—CH_2—CH_2—OAc$)]. Whatever the explanation for these several activities of EO, they indicate that superficial correlations of activities observed for a chemical in different assays may occasionally prove to be misleading. In particular, it is suggested that the bacterial mutagenicity of EO is associated with the parent molecule, while the mutagenic and carcinogenic effects observed in mammals may be related to the DNA reactivity of a secondary metabolite.

2. 4-Chloromethylbiphenyl

4-Chloromethylbiphenyl (4CMB) was evaluated recently by members of the United Kingdom Environmental Mutagen Society (UKEMS, 1982). The potent responses observed *in vitro* for 4CMB confirmed earlier reports, but were in stark contrast to the uninterrupted sequence of negative responses observed in the assays conducted *in vivo*. These included eight independently conducted bone marrow micronucleus assays, the interim results of three skin-painting carcinogenicity bioassays, and a dominant lethal mutation assay, each conducted in mice. 4-Chloromethylbiphenyl entered the UKEMS study as a potent genotoxic agent *in vitro*, and exited as a very weak or perhaps completely inactive genotoxic agent *in vivo*. These marked disparities in assay responses have yet to be fully rationalized, but they already warn of the dangers inherent in attempts to relate quantitatively the magnitude of positive assay responses observed *in vitro* to the likely carcinogenic/mutagenic hazard of exposing mammals to the same chemical *in vivo*.

In order to emphasize the challenge inherent in the genotoxic profile of 4CMB, part of it has been reproduced below (Table III) and contrasted with the corresponding assay responses observed for the potent animal carcinogens dimethylnitrosamine (DMN, see also Table I) and

TABLE III

Comparison of the Responses Given by 4-Chloromethylbiphenyl (4CMB), Dimethylnitrosamine (DMN), and Hexamethylphosphoramide (HMPA) in Four Groups of Genotoxicity Assay[a]

Chemical	Bacterial point mutation	Mammalian cell point mutation	Rodent dominant lethal mutation	Rodent carcinogenicity	Suggested reason for assay disparities
4CMB	+ + (± S9)	+	−	− (?)	Rapidly hydrolyzed or detoxified *in vivo*?
DMN	+ (+ S9)	+	−	+ ⎫	Activated *in vivo* but
HMPA	− (± S9)	+	−	+ + ⎭	not in the testes?

[a] The data presented for 4CMB were abstracted from the results of the United Kingdom Environmental Mutagen Society (UKEMS) Study of 1980 (published 1982). The data for DMN and HMPA are referenced fully in the legend to Table I. "S9" in the table refers to the use of a rat liver S9 mix as a source of auxiliary metabolism.

hexamethylphosphoramide (HMPA, see also Table I). The diametrically opposed responses of these *selected* chemicals in the *in vitro* and *in vivo* mutation assays is of obvious interest, as is the reverse correlation evident between their bacterial mutagenicity and mammalian carcinogenicity. It must be emphasized, however, that these examples were *selected* from a much larger data base in order to emphasize a principle.

The four principles discussed above are probably sufficient to explain why not all chemical mutagens produce a positive response in each of the currently used assay systems. However, a constructive approach would be to use these principles to design a battery of assays that *would* be capable of detecting genotoxic agents already present in the environment together with those inadvertently under development. How this might be achieved is discussed in Section IV.

IV. DESIGN, VALIDATION, AND USE OF TEST BATTERIES

Many countries already have, or are in the process of drafting, legislative guidelines that will aid the detection of new mammalian carcinogens and mutagens. These guidelines list a range of established assays from which a battery (usually of three tests) may be selected by the individual investigator. It is also usual to gather these assays into groups or phyla sharing a similar genetic end point, but whether the assay is conducted *in vitro* or *in vivo* sometimes remains optional. For example, emphasis is not usually placed on whether cytogenetic assays are conducted *in vivo* or *in vitro*; the unique position of the *Drosophila* assays in this debate are discussed later. It is suggested in the discussion that follows that the results of mammalian *in vivo* assays may have a unique role to play in the assessment of effects observed *in vitro*. Consequently, *in vitro* and *in vivo* assays are discussed separately.

A. *In Vitro* Assays

The *Salmonella* mutation assay of Ames is so widely used and validated that it is generally regarded as the primary *in vitro* assay for the detection of potential mammalian carcinogens and mutagens. The real need is, therefore, to define assays that can complement the *Sal-*

monella test, i.e., be capable of detecting those carcinogens, mutagens, and clastogens that it fails to identify. With this object in mind, eleven carcinogens that are difficult or impossible to detect as positive in the Salmonella assay were included in the recent "International Study" (de Serres and Ashby, 1981). These carcinogens are shown in the bottom half of Fig. 1, and it can be seen that most of the seventeen independent investigators using bacterial mutation assays failed to detect the majority of them. Nonetheless, these compounds gave a positive response in several eukaryotic assays. Thus, it is feasible to

Compound	← STANDARD SALMONELLA TEST →	NOR	HEP	8AZA	FLUC	FLUC
β-Propiolactone	• • • • • • • • • • •			•	•	•
4-Nitroquinoline-N-oxide	• • • • • • • • • • • •			•	•	•
Benzo[a]pyrene	• • • • • • • • • • •			•	•	•
2-Acetylaminofluorene	• • • • • • • • • • • •			•	•	•
Epichlorhydrin	• • • • • ? • • • • •			•	•	•
Benzidine	• • • • • • • • • • • • •			−	•	•
2-Naphthylamine	• • • • • • • • • • •			−	•	•
9,10-Dimethylanthracene	• − • • • • − • • • •			•	•	•
Cyclophosphamide	• • • • • • • • • •	−	−	•	•	•
Hydrazine sulphate	• • − • • • • • • • •	•	−	•	•	•
N-Nitrosomorpholine	• • − − • • • • • •	−	−	•	•	?
4,4'-Methylenebis(2-chloroaniline) (MOCA)	• • • • • ? • − • • • •			−	•	•
Dimethylcarbamoyl chloride	? • ? • • − • • • • • •			−	•	?	•
Methylazoxymethanolacetate	− • • − • • • • ? − •				•	•
Auramine (technical grade)	• ? − − • − • • ? • − − •			•	−	?
4-Dimethylaminoazobenzene (butter yellow)	• − − − − − • − − − − •		•	•	−	?
O-Toluidine hydrochloride	− − − − − • − − − − • •		•	•	−	
Hexamethylphosphoramide (HMPA)	− − − − − − − − − − − − •	−		−	−	
Safrole	− − − − − − − • • − − −		•	−		
Urethane	− − • − − − • − − − • −		•	−	•	−
Chloroform	− − • − − − − − − − − −					?
DL-Ethionine	− − − − − ? − − − − • −			−		
Ethylenethiourea	− − • − • − − − − − − −	−	−			
Diethylstilboestrol	− − − − − ? − − − − • −	−	−			
3-Aminotriazole	− − − − − − − − − − − −					?

Fig. 1. Responses observed in the 17 independent Salmonella mutation assays included in the recent "International Study" (de Serres and Ashby, 1981). The 14 carcinogens presented at the top were previously established as being readily detected in this assay, while those below were known to be either difficult to detect, or to be undetectable by this assay. All of the 39 positive or questionable responses recorded in the lower part of the figure were weak, and in several cases, unreproducible by the same investigators (● = positive, − = negative, ? = undecided response). NOR represents addition of norharman; HEP, the use of hepatocyte activation; 8AZA, a forward mutation assay; FLUC, the use of the fluctuation protocol.

detect the vast majority of known carcinogens and mutagens by using a simple combination of the *Salmonella* assay and a second eukaryotic point mutation *in vitro* assay.

It is suggested here that the "complementary" properties of the eukaryotic assays are associated primarily with their superior ability (usually with the auxiliary assistance of S9) to activate promutagens to mutagens, rather than with some peculiarity of their chromosomes, or of their constituent DNA. The fact that *direct-acting* mutagens such as EMS and HN2 (see Table I) are positive in all *in vitro* assays is consistent with this suggestion. In contrast, the fact that several *in vitro* cytogenetic assays are successful in detecting the carcinogens benzene and DES (Table I) is clearly due to the change in genetic end point, and these two chemicals, in fact, provide the majority of the limited evidence available to justify using a third class of *in vitro* assay (for clastogenesis) when screening chemicals for potential carcinogens and mutagens.

The rationale for the design and use of a battery of three *in vitro* tests is shown schematically in Fig. 2. Below each test system is a list of some of the carcinogens and mutagens detected as positive therein, but not by the previous class of assay. The majority of the examples listed under the "complementary eukaryotic assay" may be detected as positive in the *Salmonella* assay if extensive changes are made to the auxiliary source of enzymes employed (S9 mix), and this raises the question of whether it would be better to conduct the *Salmonella* assay using a wide range of test protocols, and completely omit the use of a complementary assay. Few investigators would support this viewpoint at present, but this, in turn, requires that when employing

---

Fig. 2. An example of how three distinct classes of assay are required to detect as positive the 17 carcinogens selected. Compounds active in the primary test usually give a positive response in the complementary test, but the reverse is not always true. Similar restraints apply to third tier cytogenetic assays. It is suggested in the text that the differential sensitivities to carcinogens and mutagens of the primary versus the complementary assays is probably due mainly to differences in the metabolic competance of the different assays. In contrast, those compounds uniquely active in cytogenetic assays may be specific clastogens. The test data implied in this figure were derived mainly from the Proceedings of the International Study (de Serres and Ashby, 1981) although most are well established in the literature. The genotoxic profile of formaldehyde has been discussed in detail elsewhere (see Table IV). Acrylonitrile gives a contested positive response in the *Salmonella* assay (Venitt, 1978) but transforms BHK cells *in vitro* (J. A. Styles, 1980, unpublished data) and mutates mouse lymphoma cells (P388) at the thymidine kinase locus (Cross and Anderson, 1981, unpublished data). The clastogenic properties of benzene metabolites *in vitro* have been described by Morimoto and Wolff (1980) and its genotoxic profile in general by Dean (1978).

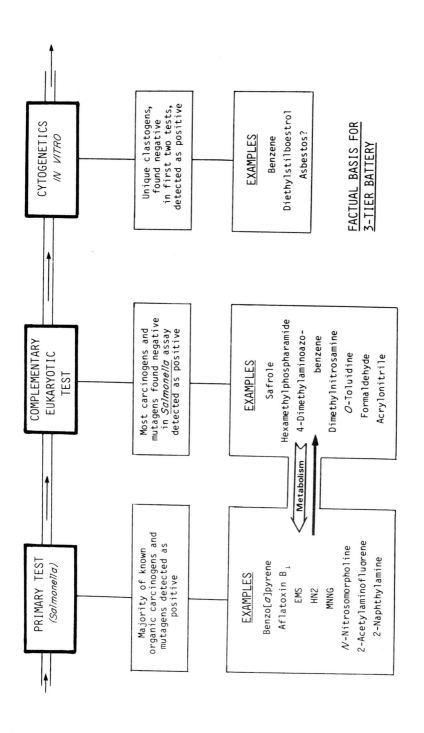

a complementary assay it should be demonstrated to be sensitive to chemicals such as safrole and HMPA, as well as to "primary" test positive control chemicals such as EMS and benzo[a]pyrene.

It was implied above that the major function of a complementary assay (Fig. 2) is to provide a second metabolic environment, and thus, a second opportunity for the test chemical to be activated to a DNA-reactive species. This possibility, as distinct from the genetic differences between assays, should be considered when a range of different assay responses for a chemical is being compared. For example, the statement "safrole is nonmutagenic to bacteria yet initiates unscheduled DNA synthesis in mammalian cells" should perhaps read "safrole is activated to an electrophilic species more readily under the prevailing metabolic environment of mammalian cell assays than it is in that of the bacterial assays."

It follows from the above discussion that once a chemical has shown a reproducible and statistically significant positive response in an established in vitro assay there is little to be gained from supplementing this with responses observed in other in vitro assays.

B. Short-Term *in Vivo* Assays

These assays are more time and resource consuming than in vitro tests; thus, their deployment should be considered, rather than used routinely. Perhaps the most important point is that few, if any, organic mammalian carcinogens, germ cell mutagens, or clastogens are uniquely active in vivo. This means, in effect, that were the definition in vitro of potential mammalian genotoxicity to be generally regarded as adequate there would be no need to conduct any lifetime studies in vivo. Thus, it is suggested that a major role for short-term assays conducted in vivo is to assess the likely significance to mammals of genotoxic effects observed in vitro, i.e., to distinguish potent agents from weak or inactive ones. However, all short-term in vivo assays, at least as they are currently constituted, cannot be relied on to define absolute inactivity, mainly because the limited numbers of animals usually employed for such tests inevitably precludes the detection of very rare events. Nevertheless, the majority of mammalian carcinogens and mutagens are readily detected by such assays.

The recent "International Study" (de Serres and Ashby, 1981) included three independently conducted micronucleus assays, one SCE assay, and a sperm abnormality test, each conducted in mice. In that study, in vivo tests with end points other than carcinogenicity were evaluated specifically for their ability to detect carcinogens, but this does not lessen the significance of the trends and conclusions discerned:

1. Of 63 determinations made on noncarcinogens, only five positive responses were recorded, and four of these were associated with compounds whose noncarcinogenicity has seriously been called into question as a result of the total findings of the study. A major conclusion that could be drawn from these findings is that if a positive effect is seen for a bacterial mutagen in a well established *in vivo* mammalian assay, then it is likely to represent a greater potential hazard to appropriately exposed humans than is one that is inactive. Such a conclusion places a duty on the investigator to establish with confidence the limits of routine variation in the genetic parameter of the test.

2. Some carcinogens that are mutagenic and/or clastogenic *in vitro* gave a negative response in some of the *in vivo* assays. For example, the liver carcinogen and mammalian cell mutagen, safrole, gave a negative response in the four independently conducted mouse cytogenetic assays, and several other carcinogens such as *o*-toluidine and hydrazine gave a positive response in only a proportion of these tests. This weakness of *in vivo* assays is of particular importance when assaying potential carcinogens, because many established carcinogens are organ specific in their effect. Thus, it is perhaps not too surprising that the weak liver carcinogen safrole failed to damage mouse bone marrow cell chromosomes *in vivo*, etc. In that case, the assay gave a correct response *vis-à-vis* its own end point, cytogenetic damage in the bone marrow, but this inactivity failed to correlate with the ability of safrole to induce liver cancer in mice. Such instances probably reflect the organotropy of the test chemical, and are therefore not always as clear-cut as observed for safrole. For example, several carcinogens that are carcinogenic to both the liver and other tissues (e.g., 2-acetylaminofluorene and 4-dimethylaminoazobenzene) gave a positive response in the micronucleus assay.

1. Heritable Mutation Assays in Mammals

In contrast to the difficulties implicit in designing an *in vivo* test capable of detecting all mammalian carcinogens, the design of *in vivo* mutagenicity assays appears to be relatively straightforward: males are exposed to the test chemical and their progeny assayed for heritable, chemically induced changes in phenotype. The earliest mammalian chemical mutation studies were conducted by Strong (1945, 1946) who fed the polycyclic aromatic hydrocarbon carcinogen 3-methylcholanthrene to mice in conjunction with selective breeding of the treated animals. This was shown to induce heritable biochemical, coat color, and carcinogen sensitivity mutations in the mice.

The most commonly used *in vivo* mammalian mutation assay is the dominant lethal test in which increases in lethal mutations are assessed by scoring the number of nonviable embryonic implants in the untreated females crossed with the treated males. This assay finds a logical, albeit expensive, extention in the more elaborate F_1/F_2-heritable translocation assay. In addition to these "lethal" mutation assays, several "viable" ones have been described such as the mouse coat color spot test. These and related tests are expensive and time consuming to conduct, due mainly to the need to use large numbers of animals in order to credit observations with statistical significance. The latter problem, and the inevitable compromise that has to be reached concerning the study size, may contribute to these assays appearing "insensitive." It was mentioned earlier that certain chemicals may either fail to reach the gonads or fail to become activated therein; thus, any potential mutagenic activity that they may possess cannot be expressed. These compounds will therefore give a negative response in these assays, and this will represent significant and real data. In contrast, other bacterial mutagens may reach the gonads but fail to induce a mutagenic change therein, and it is only the latter results that should be considered when assessing the "insensitivity" of this and related assays.

A problem common to all *in vivo* mutation assays is that some chemicals may induce heritable mutations in the germ line that either occur at an undetectable level, are recessive in nature and may not, therefore, become evident in the F_1/F_2 generations, or finally, are not being monitored at the time (e.g., some subtle, but potentially significant, biochemical change). It will be impossible to resolve all these uncertainties in the near future; thus, a compromise will have to be reached, the exact nature of which will vary according to individual circumstances.

2. Heritable Mutation Assays in Drosophila

The sex-linked recessive lethal (SLRL) assay in *Drosophila* is widely used, and three were included in the recent "International Study." This assay has been treated separately from mammalian mutation assays herein, mainly for the following reasons. Although exposed to the test chemical *in vivo*, *Drosophila* are probably so remote in physiology and anatomy from mammals that potentially important limiting factors in the latter, such as the absorption, distribution, detoxification, and excretion of the test chemical in relation to the half-life of the ultimate reactive species, are unlikely to be represented adequately. This potential weakness is important within the context of using data generated "in vivo" to assess the likely significance to mammals of effects seen in vitro.

The Drosophila assays could perhaps best be regarded as bridging the gap between assays conducted in vitro and those conducted in mammals in vivo. They are more expensive to conduct than most in vitro assays but possess some of the advantages of mammalian in vivo assays, e.g., the ability to activate effectively certain promutagens such as HMPA (Table IV). In contrast, they are probably too remote from the mammalian situation to be regarded as useful aids to risk estimations; in fact, they may share some of the problems of in vitro assays in tending to detect the *potential* rather than the *ability* of a chemical to produce genotoxic effects in mammals [see the specific case of dinitrosopentamethylenetetramine (DNPT) Table IV].

A particular point in favor of these assays is the wide range of cytogenetic and mutagenic end points that can be assessed, a topic recently reviewed in detail by Committee 1 of ICPEMC Newsletter (1980). Nonetheless, that review concluded that assays in Drosophila cannot at present be regarded as capable of replacing mammalian cytogenetic assays.

TABLE IV

Comparative Biological Activities of Hexamethylphosphoramide (HMPA) and Dinitrosopentamethylenetetramine (DNPT) in the assays shown[a]

Chemical	Bacterial mutation	Drosophila SLRL mutation	Micronucleus (mouse)	Carcinogenicity (rodent)
HMPA	−	+	+	+
	(16 assays)	(3 assays)	(3 assays)	(1 assay)
DNPT	−	+	−	−
	(17 assays)	(1 assay)	(3 assays)	(3 assays)

[a] The data were abstracted from the "International Study" (de Serres and Ashby, 1981). The carcinogenicity data are included as an alternative and complementary phenotype to cytogenetic disturbance in the bone marrow. In the case of HMPA, two very weak responses were observed out of 18 determinations made in bacterial mutation assays, likewise, DNPT gave one weak and one questionable response in 19 determinations. The overall conclusions of the working groups were that no mutagenic effects had been clearly demonstrated in either case in any of the bacterial assays. The possible significance of these data have been discussed in more detail by Ashby and Lefevre (1982).

3. In Vivo Cytogenetic Assays

Somatic cell chromosome aberration assays are often conducted in the belief that any chemical capable of affecting the structural integrity of condensed somatic cell chromosomes must also be both potentially carcinogenic and likely to be a germ cell mutagen. These assumptions have not been rigorously tested, although their emotive appeal and apparent logic are evident. The discussion that follows attempts to place positive cytogenetic assay findings in perspective, i.e., as events that may herald significant changes in phenotype, but of themselves provide nothing more than evidence of exposure of the test animals to the test chemical.

The chemical induction of chromosomal aberrations or changes in number (aneuploidy) can be assayed in vitro using a variety of mammalian cells or yeasts, but the present discussion is concerned solely with the phenotypic significance of cytogenetic damage induced either in bone marrow cells or in peripheral blood lymphocytes following exposure of mammals to the test chemical in vivo. For the present purposes, the mouse micronucleus assay (assessed in mature erythrocytes) is regarded as a cytogenetic assay.

As early as 1947, Tatum had implicated chromosome damage as an early and probably critical event in the generation of abnormal phenotypes such as cancer. Again, a limited quotation from an early paper will serve to dispel any illusions that the fundamental tenets of the science of chemical carcinogenesis and mutagenesis were recently established:

> Since a mutation, germinal or somatic, gene or chromosomal, determines the potentiality of a cell and not necessarily the expression of the character, it may be suggested that environmental factors, nutrition, irritation, hormonal levels, or biochemical changes resulting from further somatic mutation or virus infection may determine the expression of the character, in this case neoplastic growth...(Tatum, 1947).

Despite this early optimism, the possible unique role of cytogenetic analysis, i.e., when regarded in isolation from the parallel induction of either cancer or heritable germ cell mutagenic effects, remains uncertain. Perhaps the main reason for this uncertainty is that the major classes of damage detectable by chromosomal analyses of somatic cells are usually also lethal; thus, the stability (heritable nature) of the induced damage is usually inferred but seldom, if ever, established. Similarly, unless a cell with a chemically induced mutation can propagate at a rate equal to, or greater than that of the normal population, it will be unlikely to gain phenotypic importance.

It is suggested herein that the most important aspect of an *in vivo* cytogenetic assay is that it has been conducted *in vivo*, i.e., that the data generated can be used to assess the significance to mammals of mutagenic or clastogenic effects previously observed only *in vitro*. Three main uses for such assays can be defined, each predictive, rather than diagnostic, of a possible change in the phenotype of exposed mammals:

1. An agent shown to be either mutagenic or clastogenic *in vitro* is also shown to induce cytogenetic damage in mammals following exposure to the chemical *in vivo*. From this it can be inferred that the agent remains active *in vivo*, and, thus, may also be capable of inducing cancer or heritable mutations in mammals. It follows, equally, that if such an agent fails to induce cytogenetic effects *in vivo*, then the chances of it being either a potent carcinogen or a heritable mutagen in mammals are reduced.

2. In proved cases of human exposure to a known carcinogen or other genotoxic agent, it may be possible to monitor exposure by conducting cytogenetic analyses of the peripheral lymphocytes of exposed individuals. In such cases, a positive effect will be indicative of exposure to a known genotoxic agent, but cannot, at present, be considered predictive of the future occurrence of any particular change in phenotype, such as cancer.

3. Probably the most challenging situation would be the induction in mammals of a clastogenic effect (either in bone marrow cells or in peripheral lymphocytes) by an agent, both well established as a noncarcinogen and demonstrably incapable of reaching the gonads, i.e., the definition of a clastogen that is both noncarcinogenic and nonmutagenic to germ cells. To the author's knowledge, this situation has not yet arisen, but were it to, the most immediate question would be whether the clastogenic effects were induced as a secondary response to an initial toxic reaction in the cell, in which case, their genetic significance would be reduced. Perhaps the simplest method of guarding against this eventuality would be to ensure that cytogenetic observations are only regarded as significant when achieved using dose levels that do not cause overt toxic effects in the cells being assayed.

The above discussion suggests that cytogenetic analyses may fulfill two separate roles, roles that are often confused. First, some genotoxic agents are uniquely detected in these assays (Table I), and for these purposes an *in vitro* assay (± liver S9 mix) will probably suffice. Second, cytogenetic assays conducted *in vivo* represent a useful short-term method of assessing the likely significance to mammals of genotoxic effects previously observed for a chemical *in vitro*. Nonetheless, it must be concluded that the results of these assays, at least as they are currently conducted, are not, in isolation, predictive of any adverse change in the phenotype of the mammal.

V. PITFALLS IN THE USE OF MUTATION ASSAYS

A. False-Negative Responses

The response observed for a chemical in a well established *in vitro* mutation assay is obviously used to define it as a mutagen or a non-mutagen. The only restraints are that the response should be reproducible, and, if positive, statistically significant and dose related. It was suggested earlier that some negative responses may be due to the inability of the host organism to metabolize the test chemical to an electrophile, or, less likely, to an inappropriate genetic end point being monitored. Thus, the possibility that a negative response may be "false" is the first serious pitfall to avoid, and this can be achieved by conducting more than one assay *in vitro*, as discussed above (Section II). It is usually possible to be alerted at an early point to a false negative response by consideration of the chemical structure of the test compound. This inevitably introduces a subjective element, and this is illustrated by the suggestion that an initial negative response observed for compound (IV) would probably be genuine, while a negative response observed for compound (V) would probably be worth confirming in a second test system.

IV

No potential or overt electrophilic centers apparent

V

Both the —NH group and the isolated C=C double bond are potentially electrophilic after metabolism

B. False-Positive Responses

Prior to the early 1970s, the detection of new mammalian carcinogens and mutagens relied solely on conducting the appropriate animal experiments, and was therefore a very slow and expensive process. Now, however, *potential* mammalian carcinogens and mutagens can be detected in a matter of days, and at relatively little expense. This acceleration in the detection process, coupled with the large number of laboratories conducting such tests, has revealed a much greater number of genotoxic chemicals than had been originally anticipated. Unfortunately, the increase in the number of agents shown to be active *in vitro* has not been paralleled by studies to *confirm* their activities *in vivo*. The possibility remains, therefore, that some bacterial mutagens (or yeast recombinogens, etc.) may be neither carcinogenic nor mutagenic *in vivo*. Even if, as seems probable, the incidence of such "false positive" predictions of activity *in vivo* were to be low, individual instances could prove important. These concerns are reinforced by the inevitable tendency to abandon the development of a chemical if it is active in an *in vitro* assay, simply because this activity, irrespective of its significance to mammals, would probably inhibit the widespread deployment, and thus the ultimate profitability, of the candidate drug, pesticide, dyestuff, etc. Unless this tendency is resisted, *in vitro* assays may ultimately replace mammalian *in vivo* assays, simply by default, i.e., activity *in vitro* might become an end in itself. A discussion of the reasons why this change in emphasis should not be consolidated forms the final section of this chapter.

Any discussion of "false-positive" *in vitro* responses is hampered by the difficulty of proving a negative response *in vivo*, no matter how convincing the available observations might appear to be. For example, a recent NCI bioassay found 4-nitro-*o*-phenylenediamine to be noncarcinogenic to both male and female B6C3F1 mice, and male and female Fischer rats (Dunkel and Simmon, 1980). This level of certainty about the noncarcinogenicity of a chemical is unusual, but the question remains as to whether the positive response observed for this material in the *Salmonella* mutation assay (Dunkel and Simmon, 1980) is really "false." For example, it may be argued that had a larger number of test animals been employed, or had hamsters (for example) been chosen as the test species, then a carcinogenic effect may have been observed. Likewise, the possibility that this chemical could induce dominant lethal mutations in primates (for example) has not been eliminated. Clearly, such searches for perfection are not consistent with the initially implied goal of preventing human exposure to significant mutagenic or carcinogenic hazards, and thus, a pragmatic compromise to the above open-ended concerns must be devised.

The compromise referred to above might be based on an acknowledgment that certain chemicals active in one or more *in vitro* assays are able to elicit a wide range of genotoxic effects, including cancer, in a range of species, at low dose levels and with, in the case of cancer, a short latent (or expression) period. These compounds could perhaps be called "potent" genotoxic agents, and human exposure to them should be limited if not completely avoided. In contrast, other chemicals active in vitro either fail to elicit genotoxic effects in vivo, or do so only after extensive testing. It is suggested here, and elsewhere herein, that limited assays conducted in vivo may prove useful in separating agents active in vitro into one of these two groups, but this would inevitably result in the definition of a series of inadequately defined "false-positive" predictions in vitro of genotoxic activity in vivo.

Several examples that illustrate this use of short-term in vivo assays were observed in the recent "International Study" (de Serres and Ashby, 1981), and these have been exemplified below under three separate headings, each of which are suggested to illustrate the several underlying reasons why false-positive predictions of activity in vivo should be expected of any assay conducted in vitro.

1. Inadequate "Detoxification" of the Test Chemical in Vitro

Both the carcinogen 2-acetylaminofluorene (2AAF) and its noncarcinogenic analogue, 4AAF, gave a positive response in the majority of the bacterial mutation assays conducted. The Salmonella data submitted by Trueman (International Study, 1981) for these two chemicals is shown in Fig. 3, together with the micronucleus assay responses reported by Salamone et al. and Tsuchimoto et al. In addition, the in vivo peritoneal macrophage assay of Nashed and Chandra (1980) have been included. From these data it is clear that the bacterial mutagenicity of 2AAF is predictive of potent genotoxicity in vivo, while that of 4AAF is only indicative, at the best, of weak genotoxic activity in vivo. The most likely explanation for these divergent biological profiles is that 4AAF is preferentially detoxified in vivo by conjugating enzymes represented inadequately in the liver activation system employed in vitro.

2. "Unrepresentative" Metabolism in Vitro

The carcinogen benzo[a]pyrene and its noncarcinogenic analogue, pryene, each gave a positive response in the majority of the in vitro assays represented in the "International Study" (de Serres and Ashby, 1981). While the incidence and magnitude of the positive responses observed for benzo[a]pyrene were greater than those seen for pyrene, each chemical was clearly active in vitro. In contrast, only the carcinogen gave a positive response in the five in vivo assays conducted in mice on both chemicals. This divergence of activities in vivo is

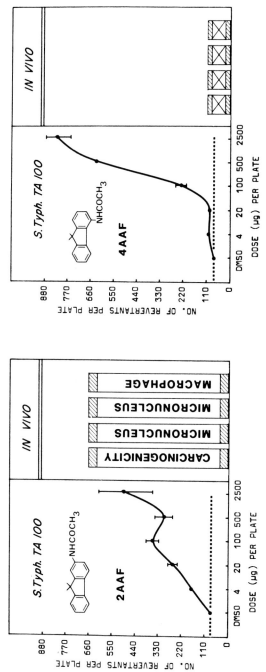

Fig. 3. Comparison of the gross genotoxicity profiles of 1-acetylaminofluorene (2AAF) and its 4-acetylamino isomer (4AAF). The mutagenic dose–response curve observed for each chemical (+S9) in *Salmonella* is derived from the data submitted by Trueman of these laboratories to the "International Study" (de Serres and Ashby, 1981). The mouse micronucleus *in vivo* assay data were similarly submitted to that study by Salamone *et al.* and Tschuchimoto *et al.* These two chemicals were recently evaluated independently by Nashed *et al.* (1980) in their rat peritoneal macrophage assay, a positive and negative response being observed, respectively, as shown. The carcinogenicity classifications are as used in the "International Study" (de Serres and Ashby, 1981) and are discussed in more detail therein by Purchase *et al.* (Chapter 4 thereof); the other papers cited above also appear in this volume.

suggested to be associated with the formation, from both compounds *in vitro*, of metabolites unassociated with the carcinogenicity of benzo[a]pyrene *in vivo* (namely, a mutagenic K-region epoxide from each compound *in vitro* and a carcinogenic diol-epoxide derivative from only benzo[a]pyrene *in vivo*). This example suggests that the activity of some chemicals *in vitro* may be mediated by metabolites unconnected with those that might cause cancer *in vivo*. It should be noted that this possibility only became apparent after conducting relatively limited studies on these two bacterial mutagens *in vivo*.

3. *Excretion Pathways Unique to Assays Conducted in Vivo*

An approximately equal incidence of positive *in vitro* responses was observed in the above study for the noncarcinogen methyl orange (VI) and for the structurally related carcinogen, 4-dimethylaminoazobenzene (DAB, VII). Nonetheless, methyl orange gave a negative response in the micronucleus assay of Salamone and in the sperm morphology assay of Wyrobek, while DAB gave a positive response in each of these assays. It is suggested that the polar sulfonic acid group ($-SO_3H$) present in methyl orange influences its distribution and excretion *in vivo* such that it is prevented from reaching nuclear DNA, and that the genotoxicity evident *in vitro* is therefore not expressed *in vivo*. Were this effect to prove general, many polar/hydrophilic chemicals might be associated incorrectly with potential genotoxicity *in vivo*, based on tentative observations made *in vitro*.

VI   VII

In concluding this discussion of the findings of the "International Study" (de Serres and Ashby, 1981), it must be emphasized that in the *majority* of the assays conducted, and with the *majority* of the chemicals evaluated, qualitative agreement was observed between assay responses, including the available carcinogenicity bioassay data. While this confirms the current stability of the science, the exceptions noted provide a valuable means of anticipating future areas of disagreement in assay responses, and of underlining the potential pitfalls of interassay extrapolations of data, including the most relevant, those concerning man.

C. The Illusion of Effects Observed *in Vivo*

It was suggested earlier that observations made *in vivo* may contribute to an assessment of the likely "potency" *in vivo* of a genotoxic agent as well as providing specific data relating to the genetic end point being monitored. Within this context it is important to establish that any genetic changes observed were actually produced *in vivo*, rather than *in vitro*, during sample preparation. This potential pitfall was first recognized by Dufrain *et al.* (1980) who demonstrated that administration of streptonigrin to rats *in vivo* produced cytogenetic damage to their peripheral lymphocytes when assayed *in vitro*.

These authors, however, also observed that if the "treated" lymphocytes were carefully washed after harvesting, and before treatment with the mitogen (PHA), no damage was observed. These findings are open to at least two explanations. First, it is possible that the drug was separately transposed in the blood with the target cells and only gained access to them during the division stimulated *in vitro* by the PHA, in which case the cytogenetic damage observed was induced uniquely *in vitro*. Alternatively, the drug may have entered the cells *in vivo*, but have been unable to exert any cytogenetic effects while the lymphocytes remained in the normal resting phase of their cell cycle (G_0). However, when stimulated into division *in vitro* by PHA, the drug may have had an effect during, for example, the S phase (DNA synthesis) of the cell cycle which is rarely encountered *in vivo*. Both of these explanations are consistent with removal of the active agent by washing, and with its probable mode of genotoxic action— intercalation in DNA as opposed to covalent addition to it. If the latter explanation is correct, then it follows that the presence of streptonigrin in, for example, normally dividing bone marrow cells might lead to cytogenetic damage therein that had been genuinely mediated *in vivo*. In contrast, were the active agent to have entered the lymphocytes for the first time *in vitro*, and if these cells were considered representative of other cells, then the significance of the observations of Dufrain *et al.* (1980) would be as they had concluded, a pitfall to be avoided in future.

The possibility that chemicals administered to rodents *in vivo* may uniquely enter the assay target cells *in vitro* during their preparation for analysis is perhaps greatest for water-soluble, ionic (charged) chemicals. An example from our own researches (J. Ashby, C. Richardson, and J. Styles, unpublished) illustrates this potential source of confusion. We have observed that treatment *in vitro* of the bone marrow cells of rats with either quaternary ammonium salts such as $(CH_3)_4N^+Cl^-$ or curare, guanidinium salts such as arginine, guanidine, or creatine, or sulfonium salts such as $S(CH_3)_3{}^+ I^{2-}$, leads to the production of G-banded metaphase figures upon staining the chromosomes with Giemsa. These observations led us to expose Chinese hamster ovary cells in culture to either guanidine hydrochloride (VIII) or trimethylsulfonium iodide (IX), followed by metaphase analysis. The extent of the cytogenetic damage observed was greater than that seen with "classical" clastogens such as EMS or HN2 (Fig. 4). A possible explanation for these effects is that the cationic test chemicals

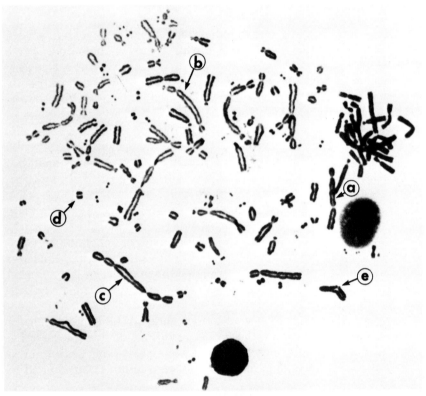

Fig. 4. Chinese hamster ovary cells were cultured for 24 hours in the presence of 10^{-3} M guanidine hydrochloride. The cells were prepared for metaphase analysis by standard methods and stained with Giemsa. The spread illustrated is representative of the damage observed in ~7% of the cells examined. Of particular interest is the incidence within the same nucleus of quadricentric (c), tricentric (b), and dicentric figures (a), together with a high incidence of minutes (d). The isolated chromatid break (e) is unusual in a spread that otherwise only contains evidence of chromosomal damage (Ashby et al., 1980, unpublished data).

are interrupting the histone–phosphate–magnesium ion interactions that condense chromatin during the approach to metaphase. Whatever their basis, the fact that such effects can be elicited *in vitro* by mammalian cell constituents is significant within the context of the observations of Dufrain *et al.* (1980). In particular, it is now possible to conceive of a study in which large doses of arginine, for example, are administered to rodents *in vivo* and in which cytogenetic damage is induced uniquely *in vitro* via carryover of the test chemical with the cytogenetic target cells.

$$\underset{\substack{\text{H}_2\text{N}^{\diagup}\text{C}^{\diagdown}\text{NH}_2}}{\overset{\overset{\text{NH}}{\|}}{}} \cdot \text{HCl} \quad = \quad \left[\underset{\text{H}_2\text{N}^{\cdots}\text{C}^{\cdots}\text{NH}_2}{\overset{\overset{\text{NH}_2}{\cdot|\cdot}}{}} \right]^{\oplus} \text{Cl}^{\ominus} \qquad\qquad \text{Me}_3\text{S}^{\oplus}\,\text{I}^{\ominus}$$

VII   **IX**

VI. MUTAGENIC SYNERGISM AND PROTECTION

In most human exposure situations a single xenobiotic chemical dominates the environment, but others are always present, if only in low concentrations. This is in contrast to the previous discussions which were concerned solely with the genotoxic properties of individual and pure chemicals when assayed in isolation. The possibility must be considered, therefore, that subsidiary chemicals present in a given environment may influence the expression of the mutagenic properties of a primary mutagenic test chemical.

Few observations have been made concerning interactions between mutagenic and nonmutagenic chemicals *in vivo*, except in the areas of carcinogenic synergism and carcinogenic protection, but there seems little reason to doubt that a range of mutagenic interactions could be demonstrated if mammals were exposed to appropriate mixtures of chemicals. In the immediate absence of the necessary experiments it seems appropriate to review some of the general principles established in the fields of carcinogenic synergism and protection, as they may relate directly to possible future cases of mutagenic/clastogenic interactions between chemicals *in vivo*.

The majority of the established examples of carcinogenic synergism or carcinogenic protection are suggested herein to be mediated *via* interference by a secondary chemical with the metabolic activation (or deactivation) of the primary genotoxic agent. Such interactions would be expected to apply equally to synergistic (or protective) clastogenic or mutagenic effects in mammals; and thus, two extreme examples from the field of chemical carcinogenesis are described below. The critical message intended is that such effects are generally exclusive to the situation *in vivo*, i.e., they cannot usually be monitored or predicted by experiments conducted *in vitro*.

A. Carcinogenic Synergism and Its Reflection *in Vitro*

This subject was reviewed recently by Ashby and Styles (1980), and one recent example of this phenomenon has been selected to represent the principal arguments presented in that article. The carcinogenicity of the bacterial mutagen and rodent carcinogen ethylene dibromide (EDB) is markedly enhanced when administered to animals concomitantly exposed to the noncarcinogen, and bacterial nonmutagen, disulfiram. However, the results shown in Fig. 5 illustrate that this synergistic interaction was not reflected in the *Salmonella* in vitro mutation assay. In contrast, limited experiments conducted *in vivo* provided evidence that enzyme changes induced exclusively *in vivo* by disulfiram were probably sufficient to explain the observed synergistic response (Elliott and Ashby, 1981).

B. Carcinogenic Protection and Its Reflection *in Vitro*

This subject has been reviewed in detail by Wattenberg (1978), and was alluded to in the article by Ashby and Styles (1980). One example from our own researches is illustrated in Fig. 6 in an attempt to emphasize the uncertainty implicit in any quantitative conclusions derived from effects observed *in vitro*. The data shown in this figure illustrates that 3′-methyl-4-dimethylaminoazobenzene is mutagenic to *Salmonella* only when activated with a liver S9 mix derived from PCB-treated rats. In contrast, Makiura *et al.* (1974) has shown that concomitant administration of PCB to rats *in vivo* inhibits the liver carcinogenicity of the same azo compound.

VII. GENOTOXIC THRESHOLD DOSE LEVELS

Man is rarely, if ever, exposed to high concentrations of a known carcinogen, and a chemical capable of inducing heritable mutations

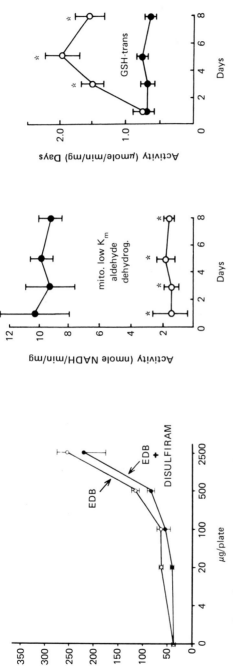

Fig. 5. The rodent carcinogen ethylene dibromide (EDB) is potentiated as a rodent carcinogen by the rodent noncarcinogen disulfiram (Plotnick, 1978). The data in the left of the figure demonstrate that the mutagenicity of EDB to *Salmonella* (±S9) is not potentiated by disulfiram when added to the assay incubation medium. In contrast, oral administration of disulfiram to rats induces a marked lowering of the liver mitochondrial (low K_m) aldehyde dehydrogenase enzyme activity (○-○, treated; ●-● control) and an elevation of the glutathione transferase activities (○-○, treated; ●-●, control). These enzyme changes have been suggested to be consistent with both the carcinogenic synergism seen *in vivo* and the absence of an enhancing effect *in vitro* (Elliott and Ashby, 1981).

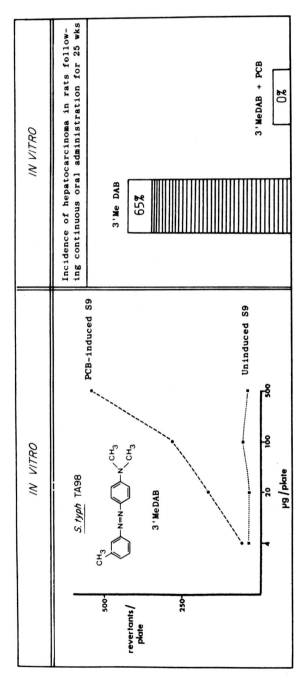

Fig. 6. Response observed for 3'-methyl-4-dimethylaminoazobenzene (3'MeDAB) in the *Salmonella* mutation assay when tested in the presence of an Aroclor (PCB) induced rat liver S9 mix (---), and in the presence of uninduced S9 mix (···) (Lefevre and Ashby, 1981). In contrast are the findings in rats of Makiura *et al.* (1974) who demonstrated that the incorporation of low levels of PCBs in the diet abolished (or at least attenuated to below the sensitivity of the bioassay) the liver carcinogenicity of 3'MeDAB.

in humans has yet to be identified. As a consequence, most genotoxicity data generated in laboratory assays are destined for use as an aid to risk estimations in situations where humans are likely to be exposed to only low doses of the chemical in question. In the average situation, laboratory data are used to assess the potential risk posed to populations intermittently exposed to ppm levels of the test chemical for, at the most, a few hours each week.

Although not scientifically sustainable at present, it seems probable that organic genotoxic chemicals will have a dose level below which the risk implicit in exposure becomes negligibly small, or even zero. Clearly, if threshold values do exist, albeit they be different for individual chemicals, and variable among a given human population for each chemical, they should form an integral part of the risk estimation process. It is probably this belief that has given rise to the concept of genetically significant dose (GSD) referred to in a recent ICPEMC Newsletter (1980). Although ill defined at present, such concepts may have an important role to play in the future.

VIII. CONCLUSIONS

The vast majority of known mutagenic, clastogenic, or carcinogenic chemicals are detected by the standard *Salmonella* mutation assay of Ames *et al.* (1975). Some agents require multistage metabolic activation to an ultimate electrophilic intermediate, and these are sometimes optimally detected by assays that employ a eukaryotic marker organism. A small minority of genotoxic chemicals appear to be uniquely active as clastogens. There is, therefore, clearly a need for a minimum battery of three independent assays if *all* new mammalian genotoxic agents are to be detected *in vitro*.

The available short-term *in vivo* assays are expensive to conduct and are potentially insensitive; nonetheless, they provide a means of assessing the probable mutagenic/carcinogenic potency to mammals of bacterial mutagens. Further, they also provide probably the only method of assaying the synergistic/protective, mutagenic/carcinogenic effects that may operate between chemicals *in vivo*.

Cytogenetic analysis *in vitro* of the peripheral lymphocytes or bone marrow cells of rodents exposed to a test chemical *in vivo* may have a useful role to play in any assessment of the biological significance of exposure of mammals to a genotoxic chemical. Further, analysis of the peripheral lymphocytes of humans exposed to a genotoxic chemical may contribute to an assessment of industrial hygiene measures.

However, the results of all such cytogenetic analyses, when viewed in isolation, are of uncertain phenotypic significance.

Investigators who routinely conduct a variety of tests on a variety of chemicals should accept that they are probably also about to advance the frontiers of their science. In particular, it is unlikely that current hypotheses and theories will be capable of absorbing each and every future observation made with previously untested chemicals.

The science of genetic toxicology is developing rapidly, thus its general deployment should be carefully considered in view of the difficulties inherent in using any biological model to predict the response of man to xenobiotic chemicals.

ACKNOWLEDGMENTS

I wish to thank Chris Richardson, Jerry Styles, and Eric Longstaff of these laboratories for the many helpful discussions and comments they have contributed during the development of this chapter. The secretarial assistance of Wendy Singleton is also gratefully acknowledged. Thanks are also due to Hilary, Louise, and Gregory for their patience and support.

REFERENCES

Ames, B. N., McCann, J., and Yamasaki, E. (1975). *Mutat. Res.* **31,** 347.

Ashby, J., and Lefevre, P. A. (1981). *In* "Proceeding of the CIIT Symposium on Formaldehyde Toxicity," Raleigh, N.C., Nov. 1980 (in press).

Ashby, J., and Styles, J. A. (1980). *Brit. Med. Bull.* **36,** 33.

Auerback, C., and Robson, J. M. (1944). *Proc. R. Soc. Edinburgh* **62,** 284.

Auerbach, C., and Robson, J. M. (1946). *Nature (London)* **157,** 302.

Bateman, A. J., and Epstein, S. S. (1971). "*Chem. Mutagens*" Vol. 2.

Berenblum, I. (1931). *J. Pathol. Bacteriol.* **34,** 731.

Berenblum, I., and Shubik, P. (1947). *Br. J. Cancer* **1,** 383.

Boyland, E., and Horning, E. S. (1949). *Br. J. Cancer* **3,** 118.

Dean, B. J. (1978). *Mutat. Res.* **47,** 75.

Dufrain, R. J. *et al.* (1980). *Mutat. Res.* **69,** 101.

Dunkel, V. C., and Simmon, V. F. (1980). *In* "Molecular and Cellular Aspects of Carcinogen Screening Tests" (R. Montesano, H. Bartoch, and L. Tomatis, eds.), p. 283. IARC, Lyon, France.

Elliott, B. M., and Ashby, J. (1981). *Carcinogenisis* **2,** 1049.

Fox, M., and Scott, D. (1980). *Mutat. Res.* **75,** 131.

Generoso, W. M., Cain, K. T., Krishna, M., Sheu, C. W., and Gryder, R. M. (1980). *Mutat. Res.* **73,** 133.

Haddow, A., Harris, R. J. C., Kon, G. A. R., and Roe, E. M. F. (1948). *Philos. Trans. R. Soc. London, Ser. A* **241,** 147.

Horowitz, N. H., Houlahan, M. B., Hungate, M. G., and Wright, B. (1946). *Science* **104,** 233.

ICPEMC Newsletter (1980). No. 3, Committee 1, B. J. Kilby, Chairman. *Mutat. Res.* **76,** 217.

"International Evaluation of Short-term Tests for Carcinogens" (1981). Report of the International Collaborative Program, (F. J. de Serres and J. Ashby, eds.), Elsevier, North-Holland, Amsterdam.

Lefevre, P. A., and Ashby, J. (1981). *Carcinogenesis* **2,** 927.

McCann, J., Choi, E., Yamasaki, E., and Ames, B. N. (1975). *Proc. Natl. Acad. Sci. U.S.A.* **72,** 5135.

Makuira, S., Aoe, H., Sugihara, S., Hirao, K., Arai, M., and Sto, N. (1974). *J. Natl. Cancer Inst.* **53,** 1253.

Malling, H. V. (1971). *Mutat. Res.* **13,** 425.

Miller, J. A., and Miller, E. C. (1969). Physiochemical Mechanisms of Carcinogenesis. *Jerusalem Symp. Quantum Chem. Biochem.* **1,** 237–261.

Morimoto, K., and Wolff, S. (1980). *Proc. 11th Ann. Meet. Am. Environ. Mutagen Soc., Nashville* (Abstr. Ea –6).

Nashed, N., and Chandra, P. (1980). *Cancer Lett. (Shannon, Irel.)* **10,** 95.

Plotnick, H. B. (1978). *J. Am. Med. Assoc.* **239,** 1609.

Snellings, W. M., Weil, C. S., and Maronpot, R. R. (1981). "Ethylene Oxide Two Year Inhalation Study in Rats." Bushy Run Research Centre, Pittsburgh, Pennsylvania.

Sram, R. J., and Kuleslov, N. P. (1980). *Arch. Toxicol. Suppl.* **4,** 11.

Strong, L. C. (1945). *Proc. Natl. Acad. Sci. U. S. A.* **31,** 290.

Strong, L. C. (1946). *Yale J. Biol. Med.* **18,** 359.

Tatum, E. L. (1947). *Ann. N. Y. Acad. Sci.* **49,** 87.

United Kingdom Environmental Mutagen Society (UKEMS) (1982). Collaborative Genotoxicity Trial. *Mutat. Res.* (1982) **100.**

Venitt, S. (1978). *Mutat. Res.* **57,** 107.

Von Scheufler, H. (1980). *Biol. Rdsch.* **18,** 94.

Wattenberg, L. W. (1978). *Adv. Cancer Res.* **26,** 197.

Chapter 2

Qualitative and Quantitative Comparisons between Mutagenic and Carcinogenic Activities of Chemicals

H. Bartsch, L. Tomatis, and C. Malaveille

35

MUTAGENICITY:
NEW HORIZONS IN GENETIC TOXICOLOGY

I. INTRODUCTION: ENVIRONMENTAL CHEMICALS IN HUMAN CARCINOGENESIS AND THE ROLE OF EXPERIMENTAL DATA IN THE ASSESSMENT OF RISK

Over the last 15 years, the percentage of cases of cancer that has been attributed to environmental causes has varied, according to different authors, from 70 to 90% (WHO, 1964; Haddow, 1967; Higginson, 1968; Boyland, 1969). These estimates were based on two main considerations: (1) a certain number of environmental factors have been identified in the etiology of human cancer, and (2) considerable variations in the incidence of cancer in different population groups in different countries suggest that environmental factors play a prevailing role.

Discussion about the percentage of cancer cases that may be attributable to environmental factors continues. Although a recent publication has authoritatively surveyed the evidence for estimating the proportions of human cancer cases which could be avoidable (Doll and Peto, 1981), the following considerations should contribute to that debate:

1. A certain proportion of human cancers have been shown incontrovertibly to be due to exposure to identified chemicals or complex mixtures. Such cause–effect relationships have been established in relatively small, easily identifiable groups exposed intensively to either occupational or iatrogenic carcinogens. These situations were therefore similar to those imposed in experimental carcinogenesis, where limited numbers of animals are exposed to relatively high dose levels in order to maximize the sensitivity of the experimental models.

2. A large number of environmental chemicals have been identified as possible hazards to humans, although proof of their carcinogenicity relies only on experimental evidence.

3. At present, the computerized registry of the Chemical Abstracts Service (CAS) of the American Chemical Society contains over 4,000,000 organic and inorganic chemicals, of which 3,400,000 have been fully structurally defined. Although the number of chemicals in the register increases at the rate of 6000 per week, only a minority are used widely: the CAS has estimated that about 63,000 are in common use (Maugh, 1978). Acute and chronic toxicity data are available for only a minority of these chemicals. There is therefore a backlog accumulated over many decades of chemicals which remain to be tested.

Because a greater effort to investigate the possible long-term toxicity of environmental chemicals was initiated only in recent years, and because that effort was obviously addressed first to chemicals suspected of being carcinogenic, it is not surprising that an increasing number of reports have recently indicated the carcinogenicity of many of the chemicals tested (Tomatis, 1979).

Although the recognition of a substance as a human carcinogen has relied until now basically on epidemiological methods, this approach suffers from two main, unavoidable limitations: (1) it cannot, in most cases, control individual exposures, as it relies almost entirely on observations in human populations; and (2) because most people are exposed to such low levels that risk cannot be measured directly, this approach can only rarely, if ever, detect slight (<10%) increases in the incidence of commonly occurring tumors (Peto, 1978).

The identification of chemicals as possibly carcinogenic to humans will therefore still rely largely on results from experimental systems. Of these, long-term animal tests are still the only ones capable, in the absence of adequate human data, of providing conclusive evidence of the carcinogenic effect of a chemical. However, short-term tests that provide evidence that a chemical can react with DNA and cause mutations indicate that it induces a response known to be produced by most carcinogens, and that it should, therefore, be considered a potential carcinogen. The key problem is still, therefore, to assess the significance in predicting hazards to humans from results obtained in experimental systems. Although correlations can be made between data obtained *in vitro* and those obtained *in vivo*, none of the currently available mutagenicity (short-term) tests have yet been shown to be totally satisfactory for predicting the carcinogenic potential of all the carcinogens tested. However, mutagenicity assays (in particular, prokaryotic test systems) offer a rapid and inexpensive means of acquiring information that is useful for selecting and ranking chemicals to be submitted to long-term carcinogen bioassays.

Whether mutagenicity and other short-term assays (singly or in combination) will eventually acquire the status of long-term animal tests in predicting human hazards depends entirely on further demonstrations of their consistency with results obtained in adequately conducted animal tests and human epidemiological studies.

This chapter presents an analysis of the qualitative relationship between carcinogenicity and mutagenicity (DNA-damaging activity), based on chemicals which are known to be or suspected of being carcinogenic to humans and/or to experimental animals. In view of the increasing demand for quantitative data for the purpose of risk assessment, the question of whether there are quantitative relationships between the potency of carcinogens in animals and in humans and their mutagenic activity is also briefly examined.

II. COMPARISON BETWEEN DATA ON CARCINOGENICITY IN HUMANS AND RESULTS OF MUTAGENICITY AND OTHER SHORT-TERM TESTS

There is increasing evidence to suggest that DNA damage (expressed mainly as mutations) is involved in the induction of many cancers; however, the relevance of the various biological end points used in short-term assays (e.g., those listed in Table III) to mechanisms of tumor induction is not known precisely. All test procedures must therefore be validated before they can be used to predict the carcinogenicity of chemicals. Ideally, such validations would be based on correlations between responses in short-term tests and data from epidemiological studies in humans.

Chemicals evaluated in Volumes 1–25 of the *IARC Monographs on the Evaluation of the Carcinogenic Risk of Chemicals to Humans* (IARC, 1972, 1973a,b, 1974a,b,c,d, 1975a,b, 1976a,b,c, 1977a,b,c, 1978a,b,c, 1979a,b,c,d, 1980a,b,c, 1981) as either definitely or probably carcinogenic to humans, and which have been tested in various mutagenic and other short-term assays, offer a basis for such an analysis. For the purpose of this discussion, therefore, we used information available through the IARC program on the evaluation of the carcinogenic risk of chemicals to humans, in which monographs are prepared on individual chemicals, groups of chemicals, or industrial processes (Tomatis *et al.*, 1978). A total of 532 compounds have been evaluated in that program.

Epidemiological studies and/or case reports were available for only about 60 chemicals, groups of chemicals, industries, or industrial processes; for 22 of these the available evidence was sufficient to support a causal relationship with the occurrence of cancers in humans (Table I). Among the latter are seven industrial processes: the manufacture of auramine, chromate-producing industries, hematite mining, the manufacture of isopropyl alcohol, nickel refining, boot and shoe manufacture and repair, and the furniture/cabinet-making industry. For these processes, no direct correlation can be made between data for humans and for experimental animals, because the identity of the agent(s) responsible for the carcinogenic effect in humans is unknown. The remaining 15 compounds were also found to be carcinogenic in one or (mostly) several experimental animal species. A recent carcinogenicity test of benzene has provided, for the first time, some experimental evidence of its capacity to induce tumors in rats (Maltoni and Scarnado, 1979). Results of carcinogenicity tests on arsenic were

TABLE I

Chemicals, Groups of Chemicals, Industries, or Industrial Processes Associated with the Induction of Cancer in Humans[a]

1. 4-Aminobiphenyl	11. Cyclophosphamide[b]
2. Arsenic and arsenic compounds	12. Diethylstilbestrol
3. Asbestos	13. Furniture and cabinet-making
4. Auramine (manufacture of)	industry (certain industries)[b]
5. Benzene	14. Hematite mining (radon?)
6. Benzidine	15. Isopropyl alcohol (manufacture of,
7. N,N-Bis(2-chloroethyl)-2-	using the strong-acid process)
naphthylamine	16. Melphalan
8. Bis(chloromethyl)ether and	17. Mustard gas
technical-grade chloromethyl	18. 2-Naphthylamine
methyl ether	19. Nickel refining
9. Boot and shoe manufacture and	20. Soots, tars, and mineral oils[c]
repair (certain industries)[b]	21. Vinyl chloride
10. Chromium and certain chromium	22. Conjugated estrogens[b]
compounds	

[a] Compiled from Volumes 1–25 of the *IARC Monographs on the Evaluation of the Carcinogenic Risk of Chemicals to Humans.*

[b] Added to IARC (1976c) by subsequent working groups at IARC.

[c] Mineral oils may vary in their composition, in particular in their content of carcinogenic polycyclic aromatic hydrocarbons.

negative, although there is sufficient evidence that arsenic compounds induce skin and lung cancer in humans.

An additional 18 compounds are probably carcinogenic to humans (Table II). While the carcinogenicity to humans of the previous group of chemicals and industrial processes could be assessed exclusively on the basis of epidemiological data that provided sufficient evidence of a causal relationship, the carcinogenic risk of this second group of chemicals was evaluated by taking into consideration evidence from studies in both humans and experimental animals (IARC, 1979c). The evidence that chemicals in this group are carcinogenic to humans varies from being almost sufficient (subgroup A) to suggestive (subgroup B).

The remaining compounds for which epidemiological data were available are chloramphenicol, chlordane/heptachlor, chloroprene, DDT, dieldrin, epichlorohydrin, hematite, hexachlorocyclohexane (BHC and lindane), N-phenyl-2-naphthylamine, phenytoin, reserpine, styrene, trichloroethylene, triaziquone, o-dichlorobenzene, dichlorobenzidine, phenylbutazone, 2,3,7,8-tetrachlorodibenzo-p-dioxin, o- and p-toluidine, and vinylidene chloride. These could not be classified as to their carcinogenicity to humans, due to limitations to the available epidemiological data and/or to the fact that only limited evidence of

TABLE II

Chemicals, Groups of Chemicals, Industries, or Industrial Processes Strongly Suspected of Being Associated with the Induction of Cancer in Humans[a]

Subgroup A: Higher degree of human evidence
   1. Aflatoxins
   2. Cadmium and certain cadmium compounds
   3. Chlorambucil
   4. Nickel and certain nickel compounds
   5. Tris(1-aziridinyl)phosphine sulfide (thiotepa)
Subgroup B: Lower degree of human evidence
   1. Acrylonitrile
   2. Amitrole
   3. Auramine
   4. Azathioprine[b]
   5. Beryllium and certain beryllium compounds
   6. Carbon tetrachloride
   7. Dimethylcarbamoyl chloride
   8. Dimethyl sulfate
   9. Ethylene oxide
  10. Iron dextran
  11. Oxymetholone
  12. Phenacetin
  13. Polychlorinated biphenyls (PCBs)

[a] Compiled from Volumes 1–25 of the *IARC Monographs on the Carcinogenic Risk of Chemicals to Humans.*
[b] Added to IARC (1979c) by a subsequent working group at IARC.

carcinogenicity was provided by data from experimental animals. For those compounds, therefore, no comparison can be made between epidemiological and experimental data.

For a comparison between the carcinogenicity of a chemical in humans and its behavior in mutagenicity and other short-term tests, a number of systems were selected on the basis of data in the literature that indicate their value for predicting carcinogenicity or their ability to detect specific classes of carcinogens (Table III). The list is not exhaustive, because many assays are still being evaluated in terms of their usefulness, reproducibility, and comparability with carcinogenicity data obtained *in vivo* (for reviews, see Hollstein *et al.*, 1979; IARC, 1980d). The tests considered were divided arbitrarily into six categories on the basis of their end points: (1) mutagenicity in the *Salmonella typhimurium* microsome assay; (2) mutagenicity in other submammalian systems, including *Escherichia coli*, *Saccharomyces cerevisiae*, *Neurospora crassa*, and *Drosophila melanogaster*; (3) mutagenicity in cultured mammalian cells; (4) chromosomal abnormalities in mammalian cells; (5) DNA damage and repair in mammalian cells; (6) cell transformation (or altered growth properties) in vitro.

The test systems considered either incorporate some aspects of mammalian metabolism, e.g., by adding a microsomal fraction of rodent or human liver in vitro or by using metabolically competent rodent cells, or involve activation in vivo, as in the host-mediated assay in intact mammalian organisms and in the test in Drosophila melanogaster (Table III, column 3). Because of its efficiency, low cost, and rapidity, the Salmonella/microsome test has been used most extensively; it therefore also has been most extensively validated, and 30 identified and suspected human carcinogens have been assayed (Tables IV and V). Of these, 21 (70%) were detected as mutagens. Of the known human carcinogens (Table IV), arsenic compounds (arsenite, As III), asbestos, benzene, and diethylstilbestrol were not mutagenic in this test. Sodium arsenite induces point mutations in E. coli WP2 strain and caused chromosomal aberrations in cultured human peripheral lymphocytes. Metal carcinogens are normally not mutagenic in the Salmonella test when it is carried out by the standard procedure, although certain metal salts, such as hexavalent chromium compounds, are genotoxic in bacterial and mammalian systems. Several metal carcinogens also decrease the fidelity of DNA polymerase in vitro and are active in the cell transformation test. Diethylstilbestrol weakly stimulates unscheduled DNA synthesis in HeLa cells (Martin et al., 1978), induces mutations in the mouse lymphoma L5178 ($TK^{+/-}/TK^{-/-}$) system (Clive et al., 1979) but not in Chinese hamster V79 cells in the presence of rat hepatocytes (Drevon et al., 1981), and transforms early-passage Syrian hamster embryo cells but not BHK-21 cells (Purchase et al., 1978; Pienta, 1979).

Of the possible human carcinogens (Table V), amitrole, carbon tetrachloride, and polychlorinated biphenyls (PCBs), for which there is sufficient evidence of carcinogenicity in experimental animals, were not mutagenic in the Salmonella test. Polychlorinated biphenyls were negative in all other short-term assays. Phenacetin can be detected as a bacterial mutagen in S. typhimurium if hamster liver fractions are used instead of the rat liver preparations generally added in routine testing (Bartsch et al., 1980; Matsushima et al., 1980). Aflatoxin B_1 and cyclophosphamide gave uniformly positive results in all six test systems. Because of the limitations of individual systems, confidence in positive results obtained with new compounds is increased when the results are confirmed in other short-term tests, using either non-repetitive end points (e.g., those mentioned in Table III) or different activation systems. When results obtained in several test systems

TABLE III

Selected Short-Term Tests for the Detection of Chemical Carcinogens or Promoting Agents[a]

System[b]	Genetic/biochemical end point monitored	Metabolic activation system
A. Mutagenesis in submammalian indicator organisms		
Salmonella typhimurium (1–3)	Histidine auxotrophs	Postmitochondrial rodent (or human) liver fractions
B. *Escherichia coli* (4–7)	Arginine and tryptophan auxotrophs Prophage induction, growth inhibition (repair-deficient strains)	Host-mediated assays (urine and feces analysis) *in vivo*
Saccharomyces cerevisiae (8)	Mutations, gene conversion, and mitotic recombinations	
Neurospora crassa (9)	Adenine auxotrophs	
Drosophila melanogaster (10)	Recessive lethal mutations	
C. Mutagenesis in cultured mammalian cells		
Chinese hamster ovary (CHO) and lung (V79) (11–15)	Mutations at *HGPRT* locus	Postmitochondrial rodent liver fractions
Mouse lymphoma (L 5178Y) (16)	$TK^{+/-}/TK^{-/-}$ mutations	Cell-mediated assays (cocultivation of lethally irradiated rat embryo cells or hepatocytes)
Rat liver epithelial cells (17)	8-Azaguanine resistance	Postmitochondrial rodent liver fraction, *in vivo*
D. Chromosome analysis		
Chinese hamster cells and human fibroblasts; human peripheral blood lymphocytes (18–20)	Sister chromatid exchanges, chromosal aberrations	

E. DNA Damage and repair

Test system	Endpoint	Activation
Chinese hamster lung (V79) (21)	Single-strand breaks in DNA (alkaline elution)	Postmitochondrial rodent liver fraction, *in vivo*
Various rodent tissues (treatment *in vivo*) (22)	Unscheduled DNA repair	
HeLa cells, rat hepatocytes, human skin fibroblasts (23–25)		
DNA synthesis *in vitro* (26)	Decreased fidelity	

F. *In vitro* cell transformation (altered growth properties)

Test system	Endpoint	Activation
Early-passage Syrian hamster embryo (27)	Morphological transformations	None
Mouse embryo C3H 10T$\frac{1}{2}$ (28–29)	Morphological transformations	None
Newborn Syrian hamster kidney (BHK21) (3,30)	Growth in agar	Postmitochondrial rodent liver fraction

[a] These tests were selected on the basis of data that indicate (1) their sensitivity in detecting several classes of carcinogens and of discriminating between carcinogens and noncarcinogens, or (2) their unique capability to detect particular classes of carcinogen or promoting agent. This list is not exhaustive and the degree to which these tests have been validated varies widely.

[b] Key to references: (1)McCann et al. (1975); (2)Sugimura et al. (1976); (3)Purchase et al. (1978); (4)Mohn et al. (1974); (5)Green and Muriel (1976); (6)Moreau and Devoret (1977); (7)Rosenkranz and Poirier (1979); (8)Zimmerman (1975); (9)Ong and de Serres (1972); (10)Vogel (1977); (11)O'Neill et al. (1977); (12)Clive et al. (1979); (13)Kuroki et al. (1977); (14)Huberman and Sachs (1974); (15)Langenbach et al. (1978); (16)Clive et al. (1979); (17)San and Williams (1977); (18)Wolff (1977); (19)Perry and Evans (1975); (20)Abe and Sasaki (1977); (21)Swenberg et al. (1976); (22)Petzold and Swenberg (1978); (23)Martin et al. (1978); (24)Williams (1977); (25)San and Stich (1975); (26)Sirover and Loeb (1976); (27)Pienta (1979); (28)Reznikoff et al. (1973); (29)Benedict et al. (1977); (30)Bouck and diMayorca (1976).

TABLE IV

Identified Human Carcinogens and Their Effects in Some Short-Term Assays[a,b]

Human carcinogen	Mutagenicity in Salmonella (A)	Mutagenicity in other submammalian assays (B)	Mutagenicity in mammalian cells (C)	Chromosome analysis (D)	DNA damage and repair (E)	Cell transformation (F)
4-Aminobiphenyl	+	+			+	+
Arsenic compounds	−	+		+		
Asbestos	−			+		
Auramine (dye mixture)	+			+		
Benzene	−				+	+
Benzidine	+			+		
N,N-Bis(2-chloroethyl)-2-naphthylamine	+					+
Bis(chloromethyl)ether	+	+	+	+	+	+
Chromium compounds	+	+	+	+	+	+
Cyclophosphamide[c]	+		±		+	±
Diethylstilbestrol			±			+
Melphalan	+			+	+	+
Mustard gas	+	+		+		
2-Naphthylamine	+					±
Soot	+					
Vinyl chloride	+	+	+	+		+

[a] +, correctly identified carcinogen; −, false-negative response. From *IARC Monographs* 1–25 and from references in Table III; classification (A)–(F) refers to test systems grouped in Table III.

[b] Conjugated estrogens and industrial processes, i.e., hematite mining, manufacture of isopropyl alcohol, nickel refining, boot and shoe manufacture and repair, and furniture and cabinet-making industries, have been omitted, since no results from short-term tests were available.

[c] Classified as a human carcinogen by a working group at IARC, Lyon, October 1980.

TABLE V

Possible Human Carcinogens and Their Effects in Some Short-Term Assays[a]

Possible human carcinogen	Mutagenicity in *Salmonella* (A)	Mutagenicity in other submammalian systems (B)	Mutagenicity in mammalian cells (C)	Chromosome analysis (D)	DNA damage and repair (E)	Cell transformation (F)
Acrylonitrile	+	+				+
Aflatoxins[b]	+	+	+	+	+	+ +
Amitrole	−	−	+		+	−
Auramine (pure)	−	+				−
Azathioprine[c]	+	+				
Beryllium compounds	−	−	+	+	+	+
Cadmium compounds				+	+	
Carbon tetrachloride	−	−				
Chlorambucil	+	+		+		+ +
Dimethylcarbamoyl chloride	+	+				+ +
Dimethyl sulfate	+					
Ethylene oxide	+	+	−	+		
Iron dextran						
Nickel compounds			+		+	+
Oxymetholone						
Phenacetin	+	−		+		
Polychlorinated biphenyls (PCBs)	−	−		−		
Tris(1-aziridinyl)phosphine sulfide (thiotepa)	+	+		+		−

[a] +, correctly identified carcinogen; −, false-negative response. From *IARC Monographs* 1–25 and from references in Table III; classification (A)–(F) refers to test systems grouped in Table III.

[b] Results in short-term tests refer to aflatoxin B_1 only.

[c] Evaluated and classified as possible human carcinogen by a working group at IARC, Lyon, October 1980.

(Tables IV and V) are combined, it can be seen that 19 out of 34 known or possible human carcinogens were tested in both systems A and B; while 13 and 14 of the 19 were positive in both A and B, respectively, 15 were positive in at least one of the two assays.

Negative results obtained in a battery of short-term tests in the absence of animal data are certainly reassuring; however, given the present limitations, it is still necessary to await the results of long-term tests in animals to confirm the absence of a carcinogenic effect, as illustrated by the example of PCBs (Table V). Cancer induction may occur in multiple steps; some compounds may act not as complete carcinogens or initiating agents but as promoters, and are therefore not detectable as electrophilic mutagens. It is therefore essential that assays be developed to detect agents that do not appear to act via electrophilic intermediates, but enhance or initiate carcinogenesis by other mechanisms, which today would be missed even in a comprehensive screening program. The possibly multifactorial origin of certain human cancers indicates the need for assays to study the interactions between viruses, carcinogens, and tumor-promoting agents (Fisher et al., 1978).

III. COMPARISONS BETWEEN DATA FROM LONG-TERM ANIMAL CARCINOGENICITY TESTS AND RESULTS OF MUTAGENICITY (SHORT-TERM) TESTS

It is generally agreed that exposure to chemicals causally associated (or strongly suspected of being associated) with the occurrence of cancer in humans must be avoided. A different and major problem is the evaluation of the possible carcinogenic hazard to humans of chemicals which have not been studied epidemiologically or noted in case reports. In an attempt to provide better assistance to regulatory bodies, the IARC revised the criteria used with the IARC Monographs Programme for assessing the significance of experimental animal data for predicting the possible carcinogenic risk of chemicals to humans (IARC, 1979a,b,c). According to these criteria, "sufficient evidence" of carcinogenesis is provided by experimental studies that show an increased incidence of malignant tumors: (1) in multiple species and strains; and/or (2) in multiple experiments (routes or doses); and/or (3) to an unusual degree (with regard to incidence, site, type and/or precocity of onset).

"Limited evidence" of carcinogenicity is provided by experimental data that suffer from certain drawbacks: (1) they were obtained in a single animal species, strain or experiment, or in experiments that were restricted by inadequate dosage levels, by inadequate duration of exposure or of period of follow-up, or by poor survival; (2) the neoplasms seen occur spontaneously or are difficult to classify as malignant by histological criteria alone; and (3) there is uncertainty about whether the incidence of tumors in test animals was increased in comparison with that in control animals. ("Sufficient evidence" of carcinogenicity and "limited evidence" of carcinogenicity do not represent categories of chemicals, but indicate varying degrees of experimental evidence and do not refer to the potency of the compound as a carcinogen.)

Of the chemicals evaluated in the first 25 volumes of the *IARC Monographs*, 130 had "sufficient evidence" of carcinogenicity in experimental animals (Table VI). According to the criteria, in the absence of adequate human data, chemicals for which there is "sufficient evidence" of carcinogenicity in laboratory animals should be regarded, for practical purposes, as if they presented a carcinogenic risk to humans. The use of the expressions "for practical purposes" and "as if they presented a carcinogenic risk" indicates that the correlation between the experimental data and possible human risk was not made on a purely scientific basis, but rather in an attempt to provide regulatory bodies with one of the elements on which priorities in the formulation of preventive measures can be based.

As shown above, there is a good empirical correlation between epidemiological and experimental data, and experimental data may predict a qualitatively similar response in humans; however, correlation cannot be used to predict quantitative variations in the responses of different species. We are still a long way from the possibility of making a scientifically acceptable, direct extrapolation from experimental data to the human situation.

We are even further from an extrapolation to human risk from experimental situations, such as those occurring with short-term tests, which do not have the production of tumors as their end point. The number of chemicals which have been definitely recognized or are suspected of being carcinogenic to humans is too small (Tables IV and V) to provide a basis for validation of short-term tests. At present, objective judgment of the value of mutagenicity tests for predicting the carcinogenicity of chemicals must perforce be based on comparison with the much larger number of chemicals shown to be carcinogenic (or noncarcinogenic) in experimental animals.

The selection of such chemicals (or classes of chemicals) for validation studies is biased by the fact that it is limited to those for which carcinogenicity data are available. Moreover, the number of chemicals for which there is adequate evidence of noncarcinogenicity is very

TABLE VI

Chemicals Evaluated in Volumes 1–25 of the *IARC Monographs* for Which There Is
Sufficient Evidence of Carcinogenicity in Experimental Animals[a]

Compound	IARC Monograph volume and page number
1. Actinomycins	10,29
2. o-Aminoazotoluene	8,61
3. 2-Amino-5-(5-nitro-2-furyl)-1,3,4-thiadiazole	7,143
4. Aramite	5,39
5. Azaserine	10,73
6. Benz[a]anthracene	3,45
7. Benzo[b]fluoranthene	3,69
8. Benzo[a]pyrene	3,91
9. Benzyl violet 4B	16,153
10. Beryllium oxide	1,17; 23,143
11. Beryllium phosphate	1,17; 23,146
12. Beryllium sulfate	1,17; 23,146
13. β-Butyrolactone	11,225
14. Cadmium chloride	2,74; 11,39
15. Cadmium oxide	2,74; 11,39
16. Cadmium sulfate	2,74; 11,39
17. Cadmium sulfide	2,74; 11,39
18. Calcium chromate	2,100; 23,212
19. Chlordecone (Kepone)	20,67
20. Chloroform	20,401
21. Citrus red No. 2	8,101
22. Cycasin	1,157; 10,121
23. Daunomycin	10,145
24. N,N'-Diacetylbenzidine	16,293
25. 4,4'-Diaminodiphenyl ether	16,301
26. 2,4-Diaminotoluene	16,83
27. Dibenz[a,h]acridine	3,247
28. Dibenz[a,i]acridine	3,254
29. Dibenz[a,h]anthracene	3,178
30. 7H-Dibenzo[c,g]carbazole	3,260
31. Dibenzo[a,e]pyrene	3,207
32. Dibenzo[a,h]pyrene	3,207
33. Dibenzo[a,i]pyrene	3,215
34. 1,2-Dibromo-3-chloropropane	15,139; 20,83
35. 3,3'-Dichlorobenzidine	4,49
36. 3,3'-Dichloro-4,4'-diaminodiphenyl ether	16,309
37. 1,2-Dichloroethane	20,429
38. Diepoxybutane	11,115
39. 1,2-Diethylhydrazine	4,153
40. Diethyl sulfate	4,277
41. Dihydrosafrole	1,170; 10,233
42. 3,3'-Dimethoxybenzidine (o-Dianisidine)	4,41
43. p-Dimethylaminoazobenzene	8,125
44. trans-2[(Dimethylamino)methylimino]-5-[2-(5-nitro-2-furyl)vinyl]-1,3,4-oxadiazole	7,147
45. 3,3'-Dimethylbenzidine (o-Tolidine)	1,87

TABLE VI (continued)

Compound	IARC Monograph volume and page number
46. 1,1-Dimethylhydrazine	4,137
47. 1,2-Dimethylhydrazine	4,145
48. 1,4-Dioxane	11,247
49. Ethinylestradiol	6,77; 21,233
50. Ethylene dibromide	15,195
51. Ethylenethiourea	7,45
52. Ethyl methanesulphonate	7,245
53. 2-(2-Formylhydrazino)-4-(5-nitro-2-furyl)thiazole	7,151
54. Glycidaldehyde	11,175
55. Hexachlorobenzene	20,155
56. Hexamethylphosphoramide	15,211
57. Hydrazine	4,127
58. Ideno[1,2,3-*cd*]pyrene	3,229
59. Isosafrole	1,169; 10,232
60. Lasiocarpine	10,281
61. Lead acetate	1,40; 23,327
62. Lead chromate	23,208
63. Lead phosphate	1,40; 23,327
64. Lead subacetate	1,40; 23,327
65. Merphalan	9,167
66. Mestranol	6,87; 21,257
67. Methoxsalen + ultraviolet light	24,101
68. 2-Methylaziridine	9,61
69. Methylazoxymethanol and its acetate	1,164; 10,131
70. 4,4′-Methylene bis(2-chloroaniline)	4,65
71. 4,4′-Methylene bis(2-methylaniline)	4,73
72. Methyl iodide	15,245
73. Methyl methanesulfonate	7,253
74. *N*-Methyl-*N*′-nitro-*N*-nitrosoguanidine	4,183
75. Methylthiouracil	7,53
76. Mirex	5,203; 20,283
77. Mitomycin C	10,171
78. Monocrotaline	10,291
79. 5-(Morpholinomethyl)-3[(5-nitrofurfurylidene)-amino]-2-oxazolidinone	7,161
80. Nafenopin	24,125
81. Nickel subsulfide	2,126; 11,75
82. Niridazole	13,123
83. 5-Nitroacenaphthene	16,319
84. 1-[(5-Nitrofurfurylidene)amino]-2-imidazolidinone	7,181
85. *N*-[4-(5-Nitro-2-furyl)-2-thiazolyl]acetamide	1,181; 7,185
86. Nitrogen mustard and its hydrochloride	9,193
87. Nitrogen mustard *N*-oxide and its hydrochloride	9,209
88. *N*-Nitrosodi-*n*-butylamine	4,197; 17,51
89. *N*-Nitrosodiethanolamine	17,77
90. *N*-Nitrosodiethylamine	1,107; 17,83
91. *N*-Nitrosodimethylamine	1,95; 17,125
92. *N*-Nitrosodi-*n*-propylamine	17,177

(*Continued*)

TABLE VI (continued)

Compound	IARC Monograph volume and page number
93. N-Nitroso-N-ethylurea	1,135; 17,191
94. N-Nitrosomethylethylamine	17,221
95. N-Nitroso-N-methylurea	1,125; 17,227
96. N-Nitroso-N-methylurethane	4,211
97. N-Nitrosomethylvinylamine	17,257
98. N-Nitrosomorpholine	17,263
99. N'-Nitrosonornicotine	17,281
100. N-Nitrosopiperidine	17,287
101. N-Nitrosopyrrolidine	17,313
102. N-Nitrososarcosine	17,327
103. Estradiol-17β and its esters	6,99; 21,279
104. Estrone and its esters	6,123; 21,343
105. Oil orange SS	8,165
106. Panfuran-S	24,77
107. Phenazopyridine and its hydrochloride	24,163
108. p-Phenoxybenzamine and its hydrochloride	24,185
109. Ponceau MX	8,189
110. Ponceau 3R	8,199
111. 1,3-Propane sultone	4,253
112. β-Propiolactone	4,259
113. Propylthiouracil	7,67
114. Safrole	1,69; 10,231
115. Sintered calcium chromate	23,302
116. Sintered chromium trioxide	23,302
117. Sodium saccharin	22,113
118. Sterigmatocystin	1,175; 10,245
119. Streptozotocin	4,221; 17,337
120. Strontium chromate	23,215
121. Testosterone and its esters	6,209; 21,519
122. Thioacetamide	7,77
123. Thiourea	7,95
124. Toxaphene (polychlorinated camphenes)	20,327
125. Tris(2,3-dibromopropyl)phosphate	20,575
126. Trypan blue (commercial grade)	8,267
127. Uracil mustard	9,235
128. Urethane	7,111
129. Zinc beryllium silicate	23,146
130. Zinc chromate	23,215

[a] Excluding those chemicals associated with cancer induction in humans listed in Tables I and II.

small. Thus, the empirically established predictive value of short-term tests (Purchase *et al.*, 1978; Cooper *et al.*, 1979; Bartsch *et al.*, 1980) is clearly influenced by the quality of the animal data used as a standard for the validation. The level of correlation between results from mutagenicity or other screening tests and those from animal bioassays can thus most reliably be examined by testing chemicals for which there is sufficient evidence of carcinogenicity in animals (Table VI, excluding those listed in Tables I and II). Of these, two-thirds (85) have been tested in the *Salmonella*/microsome mutagenicity test, and 79% (67/85) were found to be mutagenic. Those which were not mutagenic in the *Salmonella*/microsome plate test were actinomycins, benzyl violet 4B, beryllium sulfate, chloroform, 1,2-dimethylhydrazine, 1,4-dioxane, ethinylestradiol, lead acetate, hexamethylphosphoramide, nafenopin, N-nitrososarcosine (tested in the host-mediated assay using *Salmonella*), 17β-estradiol, estrone, safrole, sodium saccharin, thioacetamide, thiourea, and urethane.

Some of the reasons why certain carcinogens may produce false-negative results in bacterial mutagenicity tests have been discussed in detail (McCann and Ames, 1977). In the case of 1,2-dimethylhydrazine, such results may be attributable to inadequacies in the *in vitro* metabolic activation system currently used, because this compound was mutagenic in the host-mediated assay (Moriya *et al.*, 1978). Similarly, those chemicals such as mitomycin C which produce mutations in eukaryotic organisms only, e.g., by interference with functions that are not present in prokaryotes, would also be missed in bacterial mutagenicity tests (cf. Ashby, Chapter 1). Certain classes of compounds, however, may not be detectable as mutagens, even with improvements in *in vitro* activation systems or increased sensitivity of genetic indicator organisms; these appear to include sex hormones, thyroid-active compounds, tumor promoters, and physically acting agents. As emphasized above (Section II), it is essential to develop short-term assays that can detect agents that are definitely carcinogenic in animals but that probably do not act via electrophilic intermediates.

IV. QUANTITATIVE CORRELATIONS BETWEEN THE CARCINOGENIC POTENCY OF CHEMICALS AND THEIR MUTAGENIC ACTIVITY

The mutagenic and carcinogenic activity of chemicals can vary over a range of 1,000,000 (McCann and Ames, 1977; Nagao *et al.*, 1978;

Bartsch et al., 1980). Estimates of the carcinogenic risk of chemicals or complex mixtures (in the absence of sufficient experimental or epidemiological data) would therefore be greatly facilitated if a quantitative relationship could be established between carcinogenic activity and mutagenic activity (Ames and Hooper, 1978). Such a correlation would also aid in the design of carcinogenicity tests, e.g., in deciding the number of animals necessary to detect weakly active carcinogens. The following is a brief summary of published experimental data on certain major classes of carcinogens, which investigate the possibility of a correlation between the potency of a chemical as an animal carcinogen and its activity as a mutagen in the Salmonella/microsome or other short-term tests.

The published data are insufficient to establish whether or not there is such a correlation for a number of reasons. First, a universally accepted index for the carcinogenic potency of chemicals has still to be defined. Second, too few correlation studies have been published, and most are based on a relatively small number of carcinogens/mutagens. Third, the carcinogenic potency indexes for laboratory animals have rarely been calculated and never for a sufficient number of chemicals to cover representative classes of environmental agents to which humans are exposed. A further difficulty in establishing a quantitative relationship between epidemiological and experimental data is that observations in humans rarely yield the precise dose–response relationships needed for an exact estimation of potency, whereas these can be obtained more easily in controlled laboratory experiments.

Published studies that have examined a possible quantitative correlation between carcinogenicity in vivo and mutagenicity in vitro include those of Meselson and Russell (1977), who calculated the carcinogenic potency of fourteen chemicals as the TD_{50} (the daily dose of a carcinogen which gives a 50% incidence of cancer in rodents after two years' exposure). Mutagenic activity was determined from results in the Salmonella/microsome test, using the most sensitive bacterial strain. In a double logarithmic plot of mutagenic and carcinogenic activity, most of the compounds showed a linear correlation, with the notable exception of several N-nitroso compounds. Clive et al. (1979) reported correlation studies on 25 chemicals. Carcinogenic activity in rats and mice was expressed as the frequency of tumor-bearing animals per μmole of compound administered per kilogram body weight. This was compared with mutagenic activity in the L5178Y $TK^{+/-} \rightarrow TK^{+/-}$ mouse lymphoma system in the presence of rat liver fractions, expressed as number of $TK^{-/-}$ mutants per cell per μmole-hour per milliliter. An approximately linear relationship was obtained over a 10^5-fold range in activity.

Hsieh *et al.* (1977) compared the rat liver microsome-mediated mutagenicities of aflatoxin B_1 and several structural analogues with their potency as hepatocarcinogens in several animal species. A good parallel was found, although the carcinogenicity indexes were not calculated. Nagao *et al.* (1977a) tested 31 N-nitrosamines, either structurally or metabolically related to N-n-butyl-N-(4-hydroxy-n-butyl) nitrosamine or to N,N-di-n-butylnitrosamine, in the *Salmonella*/microsome mutagenicity assay, using a testing procedure whereby the compound and a 9000 g supernatant from PCB-treated rats were preincubated 20 minutes in the presence of *S. typhimurium* strains TA100 and TA1535 and then plated. The authors concluded that the mutagenicities of these compounds were not related quantitatively to their potencies as carcinogens.

Langenbach *et al.* (1980) assayed a series of β-oxidized derivatives of N-nitrosodi-n-propylamine for mutagenicity in two systems: (1) liquid incubation assays in the presence of *S. typhimurium* TA1535 and hamster liver homogenate, and (2) Chinese V79 hamster cells cocultivated with freshly isolated hamster hepatocytes. The mutagenic activity of the four nitroso compounds correlated better with their carcinogenic activity in the hamster in assay (2) than in assay (1). In another study, several hydrazine derivatives were tested both in the *Salmonella*/microsome assay in the presence of rat liver fractions, and for the induction of DNA damage in liver or lung tissue *in vivo* using an alkaline elution assay (Parodi *et al.*, 1981). The authors concluded that the ability of the 12 compounds to induce lung tumors in mice was better reflected by the assay for DNA damage.

Coombs *et al.* (1976) measured the liver microsome-mediated mutagenicity of 35 polycyclic hydrocarbons (derivatives of cyclopentaphenanthrene and chrysene) using Aroclor-pretreated rats and *S. typhimurium* TA100 strain. These results were compared with data on carcinogenicity obtained from skin-painting experiments in mice and expressed as Iball index: (percent tumor incidence × 100)/mean latent period in days. The authors reported little quantitative correspondence between carcinogenic potency and mutagenic activity. Huberman and Sachs (1976), however, using a cell-mediated mutagenicity assay with Chinese hamster V79 cells cocultivated with lethally irradiated rat embryo cells for metabolic activation, found that the carcinogenicity of 10 polycyclic hydrocarbons paralleled their mutagenicity, as measured by 8-azaguanine or ouabain resistance.

The discrepancies observed between studies in which metabolic activation was provided by cell-free systems and those in which cellular metabolic activation systems were used may be due in part to the fact that certain ultimate reactive mutagenic metabolites produced by rat liver microsomal systems *in vitro* may be different from those which are generated in cells (Selkirk, 1977; Malaveille *et al.*, 1977a; Bigger *et al.*, 1978). This observation may explain the lack of correlation between the mutagenicities of five hydrocarbons assayed in the presence of a rat liver microsomal system and their carcinogenicities (expressed as Iball indexes) on mouse skin (Table VII) (Bartsch *et al.*, 1979), which was particularly evident for the benz[a]anthracene (BA) series. Mutagenic activity decreased in the order BA > 7-methyl-BA > 7,12-dimethyl-BA, while carcinogenicity increased in that order.

Wislocki *et al.* (1980) also reported no quantitative agreement between the mutagenicity in *S. typhimurium* TA100 in the presence of activating systems of hydroxymethyl and other derivatives of 7,12-dimethylbenz[a]anthracene and their tumor-initiating activity in the mouse skin (two-stage tumorigenesis model).

However, Bartsch *et al.* (1979) found a very close positive association between the liver microsome-mediated mutagenicities of dihydrodiols that can yield bay-region diol-epoxides and the carcino-potencies of the parent hydrocarbons. These data are consistent with the assumption that, under the assay conditions utilized, liver microsomes *in vitro* predominantly produce simple, mutagenic oxides, whereas cultured cells or cells *in vivo* can carry out a three-step activation process involving the sequential formation of epoxides, diols, and diol-epoxides. The latter are now assumed to be the ultimate carcinogenic metabolites of polycyclic hydrocarbons (Grover and Sims, 1978; Jerina *et al.*, 1979). However, liver microsomes incubated with the appropriate diol precursor catalyze the formation of vicinal diol-epoxides.

Differences in the pathways leading to intermediates that are mutagenic to *S. typhimurium in vitro* and the electrophilic metabolites known to bind to cellular macromolecules *in vivo* have also become apparent for certain aromatic amines, e.g., 2-acetylaminofluorene (AAF). Reactive esters like AAF-N-sulfate and N-acetoxy-2-aminofluorene, which are formed *in vivo* and *in vitro*, appear not to be involved in bacterial mutagenesis when N-hydroxy-AAF is incubated with rat liver postmitochondrial supernatant and *S. typhimurium* strains (Weeks *et al.*, 1980). Such differences could profoundly influence any quantitative correlation between the bacterial mutagenicity and the carcinogenicity of certain aromatic amines.

In order to eliminate the vagaries of metabolic activation, ultimate reactive compounds that do not require enzymatic activation and which are structurally related were compared qualitatively and quantitatively in several short-term tests (Bartsch *et al.*, 1977; Kuroki and Bartsch, 1979). Reactive esters derived from N-hydroxy-2-aminofluorene were assayed for electrophilicity by reaction with methionine,

TABLE VII

Relationship between the Mutagenicity of Polycyclic Aromatic Hydrocarbons and of Certain Related Dihydrodiols in Microsome-Mediated Assays with *Salmonella typhimurium* TA100 and the Extents of Reaction with DNA and of Tumor Initiation and Carcinogenesis in Mouse Skin Treated with Polycyclic Hydrocarbons[a]

Polycyclic hydrocarbon	Mutagenicity[c] (his+ revertants/ nmole) (A)	Extent of reaction with DNA in mouse skin[d] (pmoles/mg DNA) (B)	Tumor initiation on mouse skin (tumors/μmole) (C)	Carcinogenicity[e] (D)	Related dihydrodiol[b] mutagenicity[c] (his+ revertants/ nmole) (E)
Benz[a]anthracene	6	2	0.9	5	8.5
7-Methylbenz[a]-anthracene	5	25	1.7	45	33
7,12-Dimethylbenz[a]anthracene	2.4	42	819	95	80
3-Methylcholanthrene	17	26	102	90	35
Benzo[a]pyrene	29	25	25	70	101

[a] From Bartsch et al. (1979).

[b] The trans-dihydrodiols expected to be the metabolic precursors of "bay-region" vicinal diol-epoxides were used in each case. These were the 3,4-diols derived from benz[a]anthracene and 7-methylbenz[a]anthracene and 7,12-dimethylbenz[a]anthracene, the 9,10-diol derived from 3-methylcholanthrene, and the 7,8-diol derived from benzo[a]pyrene.

[c] Mutations to his+ were estimated in *Salmonella typhimurium* TA100 and the values have been taken from the ascending linear portion of the dose–response curves.

[d] Estimated from Sephadex LH20 column elution profiles of hydrolysates of DNA obtained from the skin of C57BL mice treated *in vivo* with a ³H-labeled polycyclic hydrocarbon (1 μmole/mouse) for 19 hours.

[e] Iball indexes for skin tumor formation in mice.

for mutagenicity in S. typhimurium strains and in Chinese V79 hamster cells, or for the induction of unscheduled DNA repair in cultured human fibroblasts (measured by incorporation of [³H]thymidine, followed by autoradiography). Overall, the data showed a general, qualitative correlation between induction of DNA repair, electrophilicity, and carcinogenic activity of these esters. However, quantitative correlations among these activities were poor: the large difference observed in the carcinogenic potency of N-myristoyloxy-2-acetylaminofluorene (the most active carcinogen) and that of N-acetoxy-2-acetylaminofluorene (the least active carcinogen) was not reflected by the biological parameters measured in the in vitro systems.

In another study of direct-acting carcinogens (Bartsch et al., 1980), 10 monofunctional alkylating agents (including carcinogenic N-nitrosamides, alkylmethane sulfonates, epoxides, β-propiolactone, and 1,3-propane sultone) were assayed for mutagenicity in two S. typhimurium strains, TA1535 and TA100, and in two test procedures, plate and liquid assays. The mutagenic activities in TA100 and TA1535 strains

TABLE VIII

Comparison of Carcinogenic Activity (TD_{50}) and Mutagenicity in *Salmonella typhimurium* TA100 and TA1535 of 10 Direct-Acting Alkylating Agents[a]

Compound	Mutagenicity in S. typhimurium[b]		Range of TD_{50}[c] in rodents
	TA 1535	TA 100	
N-Nitrosoethylurea	7450	2790	<0.3 – 35
N-Nitrosomethylurea	660	550	<5.4 – 155
N-Methyl-N-nitroso-N'-nitroguanidine	5.7	4.2	1.1 – 179
N-Nitrosomethylurethane	12	9	6.9 – <119
Ethylmethane sulfonate	19,100	14,200	[d]
1,3-Propane sultone	40	40	<3.5 – 1345
β-Propiolactone	310	250	104 – 619
Methylmethane sulfonate	n.d.[e]	680	1082 – 1399
Epichlorohydrin	1130	1130	13,718
Glycidaldehyde	19	15	1422 – 16,865

[a] From Bartsch et al. (1980; 1982.)

[b] Expressed as micromolar concentration of the test compound to produce 500 revertants/plate.

[c] Total dose of carcinogen in milligrams/kilogram body weight required to reduce by one-half the probability of the animals being tumor-free throughout a standard lifetime (calculated according to Hooper et al., 1977); ranges of TD_{50} values in different rodent species and after different modes of administration.

[d] TD_{50}>175, noncarcinogenic.

[e] Not detected.

(plate assays) were then compared with the carcinogenic activities of these alkylating agents, expressed as TD_{50} values (Table VIII). Although the TD_{50} values for the 10 compounds varied with the mode of administration and animal species, there was no obvious proportionality between carcinogenicity in rodents and mutagenicity in either *Salmonella* strain. For example, on the basis of the TD_{50} values, N-nitrosoethylurea was the most potent carcinogen studied, but it was only weakly active as a mutagen; glycidaldehyde was one of the most mutagenic compounds, but it was only weakly carcinogenic. These data on a limited number of compounds indicate that a quantitative relationship between the carcinogenesis and mutagenesis of these direct-acting carcinogens in the two *Salmonella* strains tested cannot be established with enough precision to allow a confident prediction of the carcinogenic potency of new compounds of this class.

V. COMPARISON OF THE *IN VITRO* MUTAGENICITY AND THE *IN VIVO* COVALENT BINDING INDEX OF CHEMICAL CARCINOGENS

Since the data that have been published on the carcinogenic potency of chemicals do not adequately cover representative classes of environmental agents, we have attempted to replace the TD_{50} values by the Covalent Binding Index (CBI) of chemicals in rat liver DNA. Lutz (1979, 1981) proposed this index to estimate the carcinogenic potency of chemicals, but on the basis of a very small number of carcinogens. We therefore compared microsome-mediated mutagenicity in *Salmonella* with the CBIs of chemicals in rat liver *in vivo*, as summarized by Lutz in his review article (1979). Covalent Binding Indexes (defined in footnote a to Table IX) and mutagenicity data were available for 44 compounds, about one-half of which were carcinogens with the liver as the principal target organ in the rat. As the data on mutagenicity were taken from the literature, possible limitations of this compilation are that different assay procedures were used to determine mutagenicity, i.e., plate, liquid incubation, or preincubation assays, and different enzyme inducers were used to pretreat rats before preparation of the liver homogenates used to activate the test chemicals. Similarly, different S. typhimurium strains were tested; only those that gave the highest responses are listed.

The ranges for the CBI values for each compound are given in Table IX; the arithmetic mean and the extremes of the ranges are plotted in Fig. 1. The CBIs for the 44 compounds varied over a range of five

TABLE IX

Comparison of Mutagenic Activity and Covalent Binding Index of Chemical Carcinogens *in Vivo*

No.	Compound	CBI[a] (range)	Revertants/μmole	Reference
			Mutagenicity in *S. typhimurium* strains[b]	
1	2-Acetylaminofluorene	59–560	5.6×10^3	1
			1.08×10^5	2
			1.38×10^4	3
2	2-Acetylaminofluorene, N-acetoxy	198	1.5×10^4	1
			5×10^4	2
3	2-Acetylaminofluorene, l-hydroxy	18	0	2
4	2-Acetylaminofluorene, N-hydroxy	225–949	1.2×10^3	1
			4.8×10^4	2
5	Acrolein	360	0	1
6	Aflatoxin B$_1$	10,300–31,000	1.36×10^7	1
			7.06×10^6	2
			2.66×10^6	4
7	Aflatoxin B$_2$	560	2.1×10^3	2
			5.66×10^3	4
8	Aflatoxin G$_1$	680	1.16×10^5	2
			9.35×10^4	4
9	Aflatoxin M$_1$	1600	1.12×10^5	2
			9.02×10^4	4
10	p-Aminoazobenzene	2.4	2.9×10^2	2
			4.7×10^2	3
11	o-Aminoazotoluene	59–230	1.5×10^4	2
			5.9×10^3	3
12	2-Aminofluorene	260	2.05×10^5	2
13	Benz[a]anthracene	68	6×10^3	1
14	Benzene	1.7	0	1
15	Benzo[a]pyrene	4–20	2.9×10^4	1
			1.21×10^5	2
			4.3×10^4	3
16	Carbon tetrachloride	51	0	2
17	Cyclophosphamide	62	1.4×10^3	2
18	1,2-Dibromoethane	180	15[c]	5
			28	6
19	Diethylstilbestrol	0.4–0.6	0	1
20	4-Dimethylaminoazo-benzene	4.5–10	725	7
21	N,N-Dimethyl-4-aminoazobenzene, 2-methyl	11–19	6×10^{2d}	2
			6.2×10^2	3
22	N,N-Dimethyl-4-aminoazobenzene, 3'-methyl	66	3.4×10^2	2
			4.3×10^3	3

Table IX (continued)

No.	Compound	CBI[a] (range)	Mutagenicity in S. typhimurium strains[b]	
			Revertants/μmole	Reference
23	trans-N,N-Dimethyl-4-aminostilbene	130	2.2×10^4	2
24	7,12-Dimethylbenz[a]anthracene	4–37	2.5×10^3	1
			1.9×10^4	2
			1.12×10^4	3
25	1,2-Dimethylhydrazine	1570–1730	0	1
26	3,3-Dimethyl-l-phenyltriazene	49	6.6×10^{2d}	1
27	Ethionine	0.2–0.7	0	2
28	Ethylmethane sulfonate	62	8–14	1
			1.6×10^2	2
29	N-Methyl-4-aminoazobenzene	8	1.4×10^{2d}	2
30	Methylazoxymethanol acetate	4400	21.6	3
31	3-Methylcholanthrene	<1.1	2×10^4	1
			5.8×10^4	2
32	Methylmethane sulfonate	360–556	3.5×10^2	1
			6.3×10^2	2
33	N-Methyl-N'-nitro-N-nitrosoguanidine	~1000	$(0.4–1.5) \times 10^5$	1
			1.38×10^6	2
34	2-Naphthylamine	~2.7	8.5×10^3	2
			8.3×10^3	3
35	Nitrogen mustard	83	1.3×10^3	2
36	N-Nitrosodiethylamine	42–430	0.45	1
			10^d	2
37	N-Nitrosodimethylamine	516–11200	15–48	1
			20^d	2
			22.2	
38	N-Nitrosoethylurea	9	50–120	1
			1.1×10^3	2
39	N-Nitrosomethylurea	400–640	470–760	1
			4.4×10^3	2
			1.6×10^3	3
40	N-Nitrosomorpholine	44	2.4×10^2	1
			60	2
41	N-Nitrosopiperidine	118	80	1
			10	2
42	N-Nitrosopyrrolidine	176	80	1
			20	2
43	Safrole, 1'-hydroxy	280	0	2
44	Vinyl chloride	525	18^e	8

(continued)

Table IX (continued)

---

[a] Covalent binding index (CBI) in rat liver from Lutz (1979). The CBI is calculated as: (µmoles of carcinogen bound per mole nucleotides)/(µmole of carcinogen administered per kilogram body weight).

[b] Mutagenicity of chemical carcinogens in *Salmonella typhimurium his*⁻ strains in the presence (for carcinogens requiring activation) or in the absence (for direct-acting agents) of rat liver 9000 g supernatant (S9). Data were collated from the following references: (1) Bartsch *et al.*, 1980; (2) McCann *et al.*, 1975; (3) Kawachi *et al.*, 1980; (4) Wong and Hsieh, 1976; (5) Rannug *et al.*, 1978; (6) van Bladeren *et al.*, 1980; (7) Nagao *et al.*, 1977b; (8) Bartsch *et al.*, 1975. Data from references (1), (2), (4), (5), (6), and (8) were obtained using the plate incorporation assay, except those values marked [d]. Results from references (3) and (7) were obtained from preincubation assays. The figures listed refer to the most sensitive strain tested following incubation with liver S9 from rats treated with inducers (PCBs, phenobarbitone, or 3-methylcholanthrene).

[c] This value corresponds to the S9-mediated mutagenicity, from which the direct action of the test compound has been subtracted.

[d] Mutagenicity was measured in liquid incubation assays.

[e] Rat liver S9-mediated mutagenicity in TA1530 after 9 hours exposure to 20% vinyl chloride in air; the direct action of the test compound has been subtracted.

orders of magnitude; ethionine had the lowest and aflatoxin B_1 the highest value. For those compounds that were mutagenic to *S. typhimurium*, the CBIs varied over a range of seven orders of magnitude; the lowest values being found for N-nitrosodiethylamine and the highest for aflatoxin B_1.

Eight chemicals that have been observed to bind covalently to rat liver DNA *in vivo* did not produce microsome-mediated mutagenicity in *Salmonella* strains, although their CBIs varied over four orders of magnitude. Six of these, diethylstilbestrol, benzene, carbon tetrachloride, 1′-hydroxysafrole, ethionine, and 1,2-dimethylhydrazine, have been shown to be carcinogenic in experimental animals. For acrolein, also nonmutagenic, there were no experimental data to evaluate its carcinogenic effects (IARC, 1979a). The eighth chemical, 1-hydroxy-2-acetylaminofluorene, reported as being noncarcinogenic and also nonmutagenic (McCann *et al.*, 1975), is an exception in that it is the only compound for which a CBI value was found that has been reported to be noncarcinogenic. Apart from these eight nonmutagenic compounds, all the remaining 36 chemicals showed a positive correlation between CBI and mutagenicity *in vitro* ($r = 0.847$; $p < 0.001$; $n = 36$) in a double logarithmic plot.

The 11 compounds that deviated from the approximately linear relationship can be grouped into two classes: one, for which the CBI values are too high or the mutagenicity too low, or both, includes N-nitrosopyrrolidine, N-nitrosopiperidine, methane azoxymethanolace-

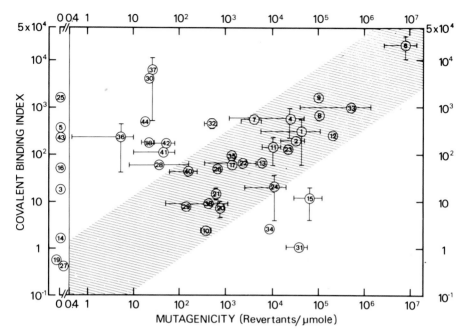

Fig. 1. A double logarithmic plot of the Covalent Binding Index (CBI) of 44 chemicals *versus* their mutagenicity in *Salmonella typhimurium* strains *in vitro*. The values are taken from Table IX, and the numbers in the circles correspond to those of the chemicals listed in the same table. When available, arithmetic means and ranges of CBI and mutagenicity values (indicated by bars) are indicated. Chemicals that did not fall within the shaded area are discussed in Section V.

tate, methylmethane sulfonate, 1,2-bromoethane, and vinyl chloride. The second group, for which the reported mutagenicity is either too high or the CBI too low, or both, includes 3-methylcholanthrene, benzo[*a*]pyrene, and 2-naphthylamine.

It is difficult to decide which of the two variables over- or underestimates the carcinogenicity of these 11 compounds, since the CBI in the liver and mutagenicity *in vitro* both have limitations as predictors of carcinogenic activity. It is reasonable that the CBI for the simple alkylating agents is too high, since both miscoding and nonmiscoding DNA adducts are measured whereas only certain *O*-alkylated products are believed to have promutagenic potential (Lawley, 1976). Accordingly, compounds that have a CBI value that is too high were found to belong to this class of precarcinogens from which (monofunctional) alkylating agents are released.

By contrast, nitroso compounds have been reported to show too little mutagenic activity in the *Salmonella*/microsome assay, partic-

ularly in the plate assay (Bartsch *et al.*, 1976). They were also among
the exceptions noted by Meselson and Russell (1977) which do not
show the proportionality between carcinogenic potency and mutagenicity seen with certain other carcinogens/mutagens. The reasons for
this low mutagenic activity may be related to metabolic parameters
of the *in vitro* activation system used and to details of the assay
procedure, as discussed below.

It has been shown that for some members of the class of compounds
with too high mutagenicity or too low a CBI (2-naphthylamine,
benzo[*a*]pyrene, and 3-methylcholanthrene) the mutagenic metabolites
produced *in vitro*, and their relative concentrations, are not identical
to those known to be relevant to carcinogenic processes. This dis-
crepancy, as well as the fact that these three carcinogens do not have
the liver as their principal target, may account for the deviation from
linearity observed in Fig. 1.

VI. FACTORS THAT MAY AFFECT THE ACTIVITY OF CARCINOGENIC CHEMICALS IN SHORT-TERM TESTS AND IN EXPERIMENTAL ANIMALS

As an expansion of a previous discussion (Ashby and Styles,
1978a,b; Ames and Hooper, 1978) and of why certain compounds in
our experimental systems do not show a proportionality between their
activity *in vitro* and that *in vivo* as carcinogens, a number of factors
may be taken into account which can be expected to influence the
numerical results obtained *in vitro* and *in vivo*.

A. Discrepancies between Cell-Free and Cellular Metabolic Activation and Mutability of the Genetic Indicator

A good correlation between carcinogenicity and mutagenic activity
is more likely when the pathways involved in the activation of the
carcinogens into electrophilic metabolites (and thus the molecular
binding products) are qualitatively and quantitatively similar *in vivo*
and *in vitro*. This important aspect has not yet been investigated thoroughly because in only a few instances has the structure of the metabolites generated from carcinogens *in vivo* and the nature of their
macromolecular binding products (assumed to be critical for carcinogenesis) been elucidated. Furthermore, no systematic studies to iden-

tify bacterial DNA adducts produced by mutagenic metabolites of car-cinogens in the *Salmonella*/microsome test have yet been carried out, except for studies on benzo[a]pyrene (Santella *et al.*, 1979). Thus, a comparison of mutagenicity *in vitro* and carcinogenic activity relies on the assumption that the pattern of mutagenic metabolites produced and their relative concentrations *in vitro* are identical to those relevant to carcinogenic processes *in vivo*. Recent studies with aromatic amines (Weeks *et al.*, 1980) and polycyclic hydrocarbons indicate that the mutagens produced by a rat liver microsomal system are different from those which are generated by enzymes in intact cells (Selkirk, 1977; Malaveille *et al.*, 1977b; Bigger *et al.*, 1978). This explanation may help to understand the controversial findings of Coombs *et al.* (1976) in the *Salmonella*/microsome test and of Huberman and Sachs (1976) in a cell-mediated assay.

In mutagenicity tests *in vitro*, in which metabolic activation is pro-vided by subcellular or cellular systems, the mutagenic activity of the premutagen can be assumed to be proportional to the product of the mutagenic potential of the DNA adducts with the time integral of the concentration of mutagenic metabolite(s) available for reaction with the DNA of the genetic indicator [Eq. (1)]:

$$\text{Mutagenicity} \sim \sum_i MP_i \text{x} \int_0^{t_i} C_i \, dt \qquad (1)$$

Where C_i is the concentration of mutagenic metabolite i available for reaction with DNA of the indicator organism during time t_i (C_i is dependent on the rate of metabolic activation/deactivation of the pre-mutagenic compound and on the electrophilicity of the mutagenic metabolite i.) The MP_i is the mutagenic potential of the DNA adducts generated. It is dependent on the structure of the DNA adduct(s) formed and their genetic consequences in the indicator cell, which depend on the DNA repair system(s) in operation. The time interval t_i is that during which mutagenic metabolite i is generated. Thus, the muta-genicity of a chemical in an *in vitro* assay is determined by these variables, and what is actually measured as the number of mutants per concentration of parent compound could be a reflection of the rate-limiting formation or deactivation of mutagenic metabolites during the incubation period. For certain N-nitrosodialkylamines, Negishi and Hayatsu (1980) found that if the pH of the incubation medium con-taining the bacteria (*S. typhimurium his*⁻ strains) and a liver subcel-lular activation system was acidic, mutagenicity was increased, pos-sibly by stabilizing the presumed α-hydroxyalkyl intermediates (Mochizuki *et al.*, 1980).

Equation (1) shows clearly that results obtained in mutagenicity test systems *in vitro* depend on the time integral of the concentration of mutagenic metabolites generated and available for reaction with the DNA of the genetic indicator organisms. Thus, assay conditions, e.g., pH, whether a liquid or plate assay, type of metabolic activation system, and time span during which mutagenic metabolites are generated, influence the mutagenic activity of a given premutagen.

The mutability of the genetic indicator may also affect the level of mutagenicity shown by a chemical: the ranking of mutagenic activity of a series of structurally related *N,N*-dialkyl and heterocyclic nitrosamines was different in *S. typhimurium* and in Chinese V79 hamster cells scoring for 8-azaguanine resistance, although very similar conditions of incubation and metabolic activation were used (Bartsch *et al.*, 1976; Kuroki *et al.*, 1977). Similarly, the mutagenic activities of the syn and anti isomers of the 7,8-dihydrodiol-9,10-oxide of benzo[a]pyrene in *S. typhimurium* strains showed the reverse order from that in Chinese hamster V79 cells (Newbold and Brookes, 1976, Wood *et al.*, 1976; Malaveille *et al.*, 1977b).

B. Influence of Treatment of Rodents with Enzyme Inducers on Their Liver Microsome-Mediated Mutagenicity of Carcinogens

Liver fractions from rodents pretreated with inducers of drug-metabolizing enzymes are frequently used in mutagenicity assays to enhance the sensitivity, by increasing the metabolic activation capacity. Thus, the liver microsome-mediated mutagenicity of most carcinogens is increased many times when liver fractions from rodents treated with, e.g., phenobarbitone, 3-methylcholanthrene, or PCBs, are used, in comparison with preparations from untreated animals. However, this enhancing action *in vitro* is not always reflected in the effect on carcinogenicity of many chemicals *in vivo*, since coadministration of such enzyme inducers leads in many cases to a reduction in carcinogenicity (for a review, see Wattenberg, 1975).

C. Activation of Carcinogens by Enzymes of the Genetic Indicator

Indicator organisms commonly used to detect genotoxic agents may also be capable of metabolizing the test compound or its metabolites, regardless of the source of the exogenous mammalian metabolic activating system. For example, bacteria (*E. coli, S. typhimurium*) contain

nitroreductases which can efficiently activate aromatic nitroso compounds to DNA-binding species (McCalla *et al.*, 1976; Blumer *et al.*,
1980; Rosenkranz and Mermelstein, 1980).

D. Host Factors That Determine the Carcinogenic Activity of Chemicals in Animals

Circumstantial evidence suggests that the induction of tumors in a
given organ or in a given species is, at least in part, related to the dose
of the ultimate reactive carcinogenic metabolites that reaches the tissues and is available for reaction with cellular macromolecules. Numerous environmental and genetic factors do influence the metabolism
of foreign compounds (Vesell and Passananti, 1977) and thus mediate
the biological response to drugs (carcinogens) in experimental animal
species, and probably in humans. This provides an added reason for
exercising great caution in transferring results obtained *in vitro* to the
in vivo situation. The question is left open as to which animal model
should be used if not only a qualitative but also a quantitative estimation of the risk of a particular compound to humans, individually
or collectively, is to be made. The importance of such data for human
risk assessment indicates that more studies should be made to increase
knowledge of possible pharmacological and toxicological differences
between test animals and human individuals.

Factors that have been shown to affect carcinogenesis include, apart
from binding reactions of carcinogenic metabolites to cellular macromolecules following their generation in the body, differences in the
persistence of miscoding alkylated DNA bases in organs and cells
(Goth and Rajewsky, 1974), cellular replication, or other promoting
stimuli on the growth of tumors in a given organ, e.g., the enhancement
of tumors by partial hepatectomy prior to a single dose of a N-nitroso
compound (Craddock, 1976). The existence of multiple factors in determining tumor formation may further explain some of the conflicting
results reported in Section IV with regard to a proportionality between
carcinogenic potency *in vivo* and biological effects *in vitro*.

VII. CONCLUSIONS

Although in the absence of adequate studies in humans, long-term
animal tests are still today the only ones capable of providing conclusive evidence of the carcinogenic effect of a chemical, the developement and application of an appropriate combination of mutagen-

icity or other short-term tests to screen the human environment, in order to identify both man-made and naturally occurring carcinogens or mutagens, and to quantify their adverse biological effects, is of particular importance. The achievement of this goal will depend heavily on progress made in elucidating the mechanisms of carcinogenesis. Increasing demand for quantitative carcinogenicity data should stimulate further examination of whether there is a quantitative relationship between the potency of a carcinogen in experimental animals and in humans, and its genotoxic activity in short-term tests. Because mutagenic and carcinogenic activities vary over a range of 1,000,000, it has been argued that even if a rough correlation could be established between these two biological activities, it would aid in the assessment of risk of chemicals. However, a recent study of 101 chemicals (Bartsch et al., 1980) revealed that about 90% of the chemicals showed mutagenic activity ranging over only four orders of magnitude. Thus, an approximate correlation would be of limited practical value. The conflicting results of experimental data published so far with regard to a possible quantitative correlation between the potency of a chemical carcinogen in animals and its activity in short-term mutagenicity tests do not yet sufficiently establish such a relationship for all classes of carcinogens to allow its general use for the confident prediction of carcinogen potency of new compounds.

ACKNOWLEDGMENTS

The authors would like to thank Mrs. Sheila Stallard for valuable secretarial assistance and Mrs. E. Heseltine for editorial aid. Financial support for the authors' research activities in this area was provided partly by contract No. NO1-CP-55630 with the National Cancer Institute, NIH, USA and partly by contract No. 190-77-ENVF with the Commission of the European Communities.

REFERENCES

Abe, S., and Sasaki, M. (1977). Chromosome aberrations and sister chromatid exchanges in Chinese hamster cells exposed to various chemicals. *J. Natl. Cancer Inst.* **58**, 1635–1641.

Ames, B. N., and Hooper, K. (1978). Does carcinogenic potency correlate with mutagenic potency in the Ames assay? A reply. *Nature (London)* **274**, 19–20.

Ashby, J., and Styles, J. A. (1978a). Does carcinogenic potency correlate with mutagenic potency in the Ames assay? *Nature (London)* **271**, 452–455.

Ashby, J., and Styles, J. A. (1978b). Factors influencing mutagenic potency *in vitro*. *Nature (London)* **274**, 20–22.

Bartsch, H., Malaveille, C., and Montesano, R. (1975). Human, rat and mouse liver-mediated mutagenicity of vinyl chloride in S. typhimurium strains. *Int. J. Cancer* **15**, 429–437.

Bartsch, H., Camus, A.-M., and Malaveille, C. (1976). Comparative mutagenicity of N-nitrosamines in a semisolid and liquid incubation system in the presence of rat or human tissue fractions. *Mutat. Res.* **37**, 149–162.

Bartsch, H., Malaveille, M., Stich, H. F., Miller, E. C., and Miller, J. A. (1977). Comparative electrophilicity, mutagenicity, DNA repair induction activity and carcinogenicity of some N- and O-acyl derivatives of N-hydroxy-2-aminofluorene. *Cancer Res.* **37**, 1461–1467.

Bartsch, H., Malaveille, C., Tierney, B., Grover, P. L., and Sims, P. (1979). The association of bacterial mutagenicity of hydrocarbon-derived 'bay region' dihydrodiols with the Iball indices for carcinogenicity and with the extents of DNA-binding on mouse skin of the parent hydrocarbons. *Chem. Biol. Interact.* **26**, 185–196.

Bartsch, H., Malaveille, C., Camus, A.-M., Martel-Planche, G., Brun, G., Hautefeuille, A., Sabadie, N., and Barbin, A. (1980). Validation and comparative studies on 180 chemicals with S. typhimurium strains and V79 Chinese hamster cells in the presence of various metabolizing systems. *Mutat. Res.* **76**, 1–50.

Bartsch, H., Malaveille, C., Terracini, B., Tomatis, L., Brun, G., and Dodet, B. (1982). Quantitative comparisons between carcinogenicity mutagenicity and electrophilicity of direct-acting N-nitroso compounds and other alkylating agents. In "N-Nitroso Compounds: Occurrence and Biological Effects" (H. Bartsch, I. K. O'Neill, M. Castegnaro and M. Okada, eds.), Lyon, International Agency for Research on Cancer, IARC Scientific Publications No. 41 (in press).

Benedict, W. F., Banarjee, A., Gardner, A., and Jones, P. A. (1977). Induction of morphological transformation in mouse C3H/10T½ clone 8 cells and chromosomal damage in hamster A (T_1)Cl Cl/3 cells by cancer chemotherapeutic agents. *Cancer Res.* **37**, 2202–2208.

Bigger, C. A. H., Tomaszewski, J. E., and Dipple, A. (1978). Differences between products of binding of 7,12-dimethylbenz[a]anthracene to DNA in mouse skin and in a rat liver microsomal system. *Biochem. Biophys. Res. Commun.* **80**, 229–235.

Blumer, J. L., Friedman, A. L. W., Fairchild, E., Webster, L. T., Jr., and Speck, W. T. (1980). Relative importance of bacterial and mammalian nitroreductases for niridazole mutagenesis. *Cancer Res.* **40**, 4599–4605.

Bouck, N., and diMayorca, G. (1976). Somatic mutation as the basis for malignant transformation of BHK cells by chemical carcinogens. *Nature (London)* **264**, 722–727.

Boyland, E. (1969). The correlation of experimental carcinogenesis and cancer in man. *Prog. Exp. Tumor Res.* **11**, 222–234.

Clive, D., Johnson, K. O., Spector, F. S., Batson, A. G., and Brown, M. M. M. (1979). Validation and characterization of the L5178Y/TK mouse lymphoma mutagen assay system. *Mutat. Res.* **59**, 61–108.

Coombs, M. M., Dixon, C., and Kissonerghis, A.-M. (1976). Evaluation of the mutagenicity, belonging to the benz[a]anthracene, chrysene and cyclopenta[a]phenanthrene series, using Ames's test. *Cancer Res.* **36**, 4525–4529.

Cooper, J. A., Saracci, R., and Cole, P. (1979). Describing the validity of carcinogen screening tests. *Br. J. Cancer* **39**, 87–89.

Craddock, V. M. (1976). Replication and repair of DNA in liver of rats treated with dimethylintrosamine and with methyl methanesulfonate. In "Fundamentals in Cancer Prevention" (P. N. Magee, S. Takayama, T. Sugimura, and T. Matsushima, eds.), 293–311. Univ. of Tokyo Press, Tokyo.

Doll, R., and Peto, R. (1981). The causes of cancer: quantitative estimates of avoidable risks of cancer in the United States today. *J. Natl. Cancer Inst.* **66**, 1193–1308.

Drevon, C., Piccoli, C., and Montesano, R. (1981). Mutagenicity assays of estrogenic hormones in mammalian cells. *Mutat. Res.* **89**, 83–90.

Fisher, P. B., Weinstein, I. B., Eisenberg, D., and Ginsberg, H. S. (1978). Interactions between adenovirus, a tumor promotor, and chemical carcinogens in transformation of rat embryo cell cultures. *Proc. Natl. Acad. Sci. U. S. A.* **75**, 2311–2314.

Goth, R., and Rajewsky, M. F. (1974). Molecular and cellular mechanisms associated with pulse-carcinogenesis in the rat nervous system by ethylnitrosourea: Ethylation of nucleic acids and elimination rates of ethylated bases from the DNA of different tissues. *Z. Krebsforsch.* **82**, 37–64.

Green, M. L., and Muriel, W. J. (1976). Mutagen testing using *trp* reversion in *Escherichia coli*. *Mutat. Res.* **38**, 3–32.

Grover, P. L., and Sims, P. (1978). DNA binding and polycyclic hydrocarbon carcinogenesis. *Adv. Pharmacol. Ther.* **9**, 13–27.

Haddow, A. (1967). Speeches of the Opening Ceremony of the Ninth International Cancer Congress. *UICC Monogr. Ser.* **9**, 111–116.

Higginson, J. (1968). Present trends in cancer epidemiology. *Can. Cancer Conf. Proc.* **8**, 40–75.

Hollstein, M., McCann, J., Angelosanto, F. A., and Nichols, W. W. (1979). Short-term tests for carcinogens and mutagens. *Mutat. Res.* **69**, 133–226.

Hooper, N. K., Friedman, A. D., Sawyer, C. B. and Ames, B. N. (1977). Carcinogenic potency: analysis, utility for human risk assessment and relation to mutagenic potency in Salmonella (Progress Report for IARC/WHO meeting, Lyon, October 1977). International Agency for Research on Cancer, Lyon.

Hsieh, D. P. H., Wong, J. J., Wong, S. A., Michas, C., and Ruebner, B. H. (1977). Hepatic transformation of aflatoxin and its carcinogenicity. *In* "Origins of Human Cancer" (H. H. Hiatt, J. D. Watson, and J. A. Winsten, eds.), pp. 697–707. Cold Spring Harbor Lab., Cold Spring Harbor, New York.

Huberman, E., and Sachs, L. (1974). Cell-mediated mutagenesis of mammalian cells with chemical carcinogens. *Int. J. Cancer* **13**, 326–333.

Huberman, E., and Sachs, L. (1976). Mutability of different genetic loci in mammalian cells by metabolically activated carcinogenic polycyclic hydrocarbons. *Proc. Natl. Acad. Sci. U. S. A.* **73**, 188–192.

IARC (1972). *IARC Monog. Eval. Carcinog. Risk Chem. Man* **1**.

IARC (1973a). *IARC Monog. Eval. Carcinog. Risk Chem. Man* **2**.

IARC (1973b). *IARC Monogr. Eval. Carcinog. Risk Chem. Man* **3**.

IARC (1974a). *IARC Monogr. Eval. Carcinog. Risk Chem. Man* Vol. **4**.

IARC (1974b). *IARC Monogr. Eval. Carcinog. Risk Chem. Man* **5**.

IARC (1974c). *IARC Monogr. Eval. Carcinog. Risk Chem. Man* **6**.

IARC (1974d). *IARC Monogr. Eval. Carcinog. Risk Chem. Man* **7**.

IARC (1975a). *IARC Monogr. Eval. Carcinog. Risk Chem. Man* **8**.

IARC (1975b). *IARC Monogr. Eval. Carcinog. Risk Chem. Man* **9**.

IARC (1976a). *IARC Monogr. Eval. Carcinog. Risk Chem. Man* **10**.

IARC (1976b). *IARC Monogr. Eval. Carcinog. Risk Chem. Man* **11**.

IARC (1976c). *IARC Monogr. Eval. Carcinog. Risk Chem. Man* **12**.

IARC (1977a). *IARC Monogr. Eval. Carcinog. Risk Chem. Man* **13**.

IARC (1977b). *IARC Monogr. Eval. Carcinog. Risk Chem. Man* **14**.

IARC (1977c). *IARC Monogr. Eval. Carcinog. Risk Chem. Man* **15**.

IARC (1978a). *IARC Monogr. Eval. Carcinog. Risk Chem. Man* **16**.

IARC (1978b). *IARC Monogr. Eval. Carcinog. Risk Chem. Humans* **17**.

IARC (1978c). *IARC Monogr. Eval. Carcinog. Risk Chem. Humans* **18**.

IARC (1979a). *IARC Monogr. Eval. Carcinog. Risk Chem. Humans* **19**.

IARC (1979b). *IARC Monogr. Eval. Carcinog. Risk Chem. Humans* **20**.

IARC (1979c). *IARC Monogr. Eval. Carcinog. Risk Chem. Humans, Suppl. 1*.

IARC (1979d). *IARC Monogr. Eval. Carcinog. Risk Chem. Humans* **21**.

IARC (1980a). *IARC Monogr. Eval. Carcinog. Risk Chem. Humans* **22**.

IARC (1980b). *IARC Monogr. Eval. Carcinog. Risk Chem. Humans* **23**.

IARC (1980c). *IARC Monogr. Eval. Carcinog. Risk Chem. Humans* **24**.

IARC (1980d). *IARC Monogr. Eval.` Carcinog. Risk Chem. Humans, Suppl. 2*.

IARC (1981). *IARC Monogr. Eval. Carcinog. Risk Chem. Humans* **25**.

Jerina, D. M., Yagi, H., Thakker, D. R., Karle, J. M., Mah, H. D., Boyd, D. R., Gadaginamath, G., Wood, A. W., Buening, M., Chang, R. L., Levin, W., and Conney, A. H. (1979). Stereoselective metabolic activation of polycyclic aromatic hydrocarbons. *Adv. Pharmocol. Ther.* **9**, 53–61.

Kawachi, T., Komatsu, T., Kada, T., Ishidate, M., Sasaki, M., Sugiyama, T., and Tazima, T. (1980). Results of recent studies on the relevance of various short-term screening tests in Japan. *In* "The Predictive Value of Short-Term Screening Tests in Carcinogenicity Evaluation" (G. M. Williams, R. Kroes, H. W. Waaijers, and K. W. van de Poll, eds.), pp. 253–267. Elsevier/North-Holland Biomedical Press, Amsterdam.

Krahn, D. F., and Heidelberger, C. (1977). Liver homogenate-mediated mutagenesis in Chinese hamster V79 cells by polycyclic aromatic hydrocarbons and aflatoxins. *Mutat. Res.* **46**, 27–44.

Kuroki, T., and Bartsch H. (1979). Mutagenicity of some *N*- and *O*-acyl-derivatives of *N*-hydroxy-2-aminofluorene in V79 Chinese hamster cells. *Cancer Lett. (Shannon, Irel.)* **6**, 67–72.

Kuroki, T., Drevon, C., and Montesano, R. (1977). Microsome mediated mutagenesis in V79 Chinese hamster cells by various nitrosamines. *Cancer Res.* **37**, 1044–1050.

Langenbach, R., Freed, H., Ravek, D., and Huberman, E. (1978). Cell specificity in metabolic activation of aflatoxin B_1 and benzo[a]pyrene to mutagens for mammalian cells. *Nature (London)* **276**, 277–280.

Langenbach, R., Gingell, R., Kuszynski, C., Walker, B., Nagel, D., and Pour, P. (1980). Mutagenic activities of oxidized derivatives of *N*-nitrosodipropylamine in the liver cell-mediated and *Salmonella typhimurium* assays. *Cancer Res.* **40**, 3463–3467.

Lawley, P. D. (1976). Carcinogenesis by alkylating agents. *In* "Chemical Carcinogens" (C. E. Searle, ed.), ACS Monograph 173, pp. 83–244. Amer. Chem. Soc., Washington, D.C.

Lutz, W. K. (1979). *In vivo* covalent binding of organic chemicals to DNA as a quantitative indicator in the process of chemical carcinogenesis. *Mutat. Res.* **65**, 289–356.

Lutz, W. K. (1981). Constitutive and carcinogen-derived DNA binding as a basis for the assessment of potency of chemical carcinogens. *In* "Biological Reactive Intermediates II" (R. Snyder, D. V. Parke, J. J. Kocsis, and D. J. Jollow, eds.), Plenum, New York (in press).

McCalla, D. R., Olive, P. L., and Yu, Tu. (1976). Damage to DNA activated nitrofurans. *In* "Fundamentals in Cancer Prevention" (P. N. Magee, S. Takayama, T. Sugimura, and T. Matsushima, eds.), pp. 229–249. Univ. of Tokyo Press, Tokyo/Univ. Park Press, Baltimore.

McCann, J., and Ames, B. N. (1977). The *Salmonella*/microsome mutagenicity test: Predictive value for animal carcinogenicity. *In* "Origins of Human Cancer" (H. H.

Hiatt, J. D. Watson, and J. A. Winsten, eds.), pp. 1431–1450. Cold Spring Harbor Lab., Cold Spring Harbor, New York.

McCann, J., Choi, E., Yamasaki, E., and Ames, B. N. (1975). Detection of carcinogens as mutagens in the *Salmonella*/microsome test: Assay of 300 chemicals. *Proc. Natl. Acad. Sci. U. S. A.* **72,** 5135–5139.

Malaveille, C., Tierney, B., Grover, P. L., Sims, P. and Bartsch, H. (1977a). High microsome-mediated mutagenicity of the 3,4-dihydrodiol of 7-methylbenz[a]anthracene in S. typhimurium TA98. *Biochem. Biophys. Res. Commun.* **75,** 427–433.

Malaveille, C., Kuroki, T., Sims, P., Grover, P., and Bartsch, H. (1977b). Mutagenicity of isomeric diol-expoxides of benzo[a]pyrene and benz[a]anthracene in S. typhimurium TA98 and TA100 and V79 Chinese hamster cells. *Mutat. Res.* **44,** 313–326.

Maltoni, C., and Scarnado, C. (1979). First experimental demonstration of the carcinogenic effects of benzene: Long-term bioassays on Sprague-Dawley rats by oral administration. *Med. Lav.* **5,** 352–357.

Martin, C. N., McDermid, A. C., and Garner, R. C. (1978). Testing of known carcinogens and noncarcinogens for their ability to induce unscheduled DNA synthesis in HeLa cells. *Cancer Res.* **38,** 2621–2627.

Matsushima, T., Yahagi, T., Takamoto, Y., Nagao, M., and Sugimara, T. (1980). Species differences in microsomal activation of mutagens and carcinogens, with special reference to new potent mutagens from pyrolysates of amino acids and proteins. *In* "Microsomes, Drug Oxidations and Chemical Carcinogenesis" (M. J. Coon, A. H. Conney, R. W. Estabrook, H. V. Gelboin, J. R. Gillette and P. J. O'Brien, eds.), Vol. II, Academic Press, New York, pp. 1093–1102.

Maugh, T. H. (1978). Chemicals: How many are there? *Science* **199,** 162.

Meselson, M., and Russell, K. (1977). Comparisons of carcinogenic and mutagenic potency. *In* "Origins of Human Cancer" (H. H. Hiatt, J. D. Watson, and J. A. Winsten, eds.), pp. 1473–1481. Cold Spring Harbor Lab., Cold Spring Harbor, New York.

Mochizuki, M., Anjo, T., and Okada, M. Isolation and characterization of N-alkyl-N-(hydroxymethyl)nitrosamines from N-alkyl-N-(hydroperoxymethyl)nitrosamines by deoxygenation. *Tetrahedron Lett.* **21:** 3693–3696, 1980.

Mohn, G., Ellenberger, J., and McGregor, D. (1974). Development of mutagenicity tests using *Escherichia coli* K-12 as an indicator organism. *Mutat. Res.* **23,** 187–196.

Moreau, P., and Devoret, R. (1977). Potential carcinogens tested by induction and mutagenesis of prophage λ in *Escherichia coli* K-12. *In* "Origins of Human Cancer" (H. H. Hiatt, J. D. Watson, and J. A. Winsten, eds.), pp. 1451–1472. Cold Spring Harbor Lab., Cold Spring Harbor, New York.

Moriya, M., Kato, K., Ohta, T., Watanbe, K., Watanbe, Y., and Shirasu, Y. (1978). Detection of mutagenicity of the powerful colon carcinogen 1,2-dimethylhydrazine (DMH) by host-mediated assay and its inhibition by disulfiram. *Mutat. Res.* **54,** 244–245.

Nagao, M., Yahagi, T., Seino, U., Sugimura, T., and Ito, N. (1977a). Mutagenicity of quinoline and its derivatives. *Mutat. Res.* **42,** 335–342.

Nagao, M., Yahagi, T., Honda, M., Seino, Y., Kawachi, T., and Sugimura, T. (1977b). Comutagenic actions of norharman derivatives with 4-dimethylaminoazobenzene and related compounds. *Cancer Lett.* (*Shannon, Irel.*) **3,** 339–346.

Nagao, M., Sugimura, T., and Matsushima, T. (1978). Environmental mutagens and carcinogens. *Annu. Rev. Genet.* **12,** 117–159.

Negishi, T., and Hayatsu, H. (1980). The pH-dependent response of *Salmonella typhimurium* TA100 to mutagenic N-nitrosamines. *Mutat. Res.* **79,** 223–230.

Newbold, F. R., and Brookes, P. (1976). Exceptional mutagenicity of benzo[a]pyrene diol epoxide in cultured mammalian cells. *Nature* (*London*) **261,** 52–54.

O'Neill, J. P., Couch, D. B., Machanoff, R., San Sebastien, J. R., Brimer, P. A., and Hsie, A. W. (1977). A quantitative assay of mutation induction at the hypoxanthine-guanine phosphoribosyl transferase locus in Chinese hamster ovary cells (CHO/ HGPRT system): Utilization with a variety of mutagenic agents. *Mutat. Res.* **45**, 103–109.

Ong, T., and de Serres, F. J. (1972). Mutagenicity of chemical carcinogens in *Neurospora crassa. Cancer Res.* **32**, 1890–1893.

Parodi, S., de Flora, S., Cavanna, M., Pino, A., Robbiano, L., Bennicelli, C., and Brambilla, G. (1981). DNA-damaging activity *in vivo* and bacterial mutagenicity of sixteen hydrazine derivatives as related quantitatively to their carcinogenicity. *Cancer Res.* **41**, 1469–1482.

Perry, P., and Evans, J. J. (1975). Cytological detection of mutagen-carcinogen exposure by sister chromatid exchange. *Nature (London)* **258**, 121–125.

Peto, R. (1978). Carcinogenic effects of chronic exposure to very low levels of toxic substances. *Environ. Health Perspect.* **22**, 155–159.

Petzold, G. L., and Swenberg, J. A. (1978). Detection of DNA damage induced *in vivo* following exposure of rats to carcinogens. *Cancer Res.* **38**, 1589–1594.

Pienta, R. J. (1979). *In vitro* transformation of cultured cells. In "Proc. XII Int. Cancer Congress, Buenos Aires, Vol. 1, Carcinogenesis" (J. P. Margison, ed.), pp. 1–317. Pergamon Press, Oxford.

Purchase, I. F. H., Longstaff, E., Ashby, J., Styles, J. A., Anderson, D., Lefevre, P. A., and Westwood, F. R. (1978). Evaluation of six short term tests for detecting organic chemical carcinogens. *Br. J. Cancer* **37**, 873–959.

Rannug, U., Sundvall, A., and Ramel, C. (1978). The mutagenic effect of 1,2-dichloro-ethane on *Salmonella typhimurium*. I. Activation through conjugation with glutathione *in vitro. Chem. Biol. Interact.* **20**, 1–16.

Reznikoff, C. A., Bertram, J. S., Brankow, D. W., and Heidelberger, C. (1973). Quantitative and qualitative studies of chemical transformation of cloned C3H mouse embryo cells sensitive to postconfluence inhibition of cell division. *Cancer Res.* **33**, 3239–3249.

Rosenkranz, H. S., and Mermelstein, R. (1980). The *Salmonella* mutagenicity and *E. coli* Pol A$^+$/Pol A$_1^-$ repair assays: Evaluation of relevance to carcinogenesis. In "The Predictive Value of Short-Term Screening Tests in Carcinogenicity Evaluation" (G. M. Williams, R. Kroes, H. W. Waaijers, and K. W. van de Poll, eds.), pp. 5–26. Elsevier/North-Holland Biomedical Press, Amsterdam.

Rosenkranz, H. S., and Poirier, L. A. (1979). Evaluation of the mutagenicity and DNA-modifying activity of carcinogens and non-carcinogens in microbial systems. *J. Natl. Cancer Inst.* **62**, 873–892.

San, R. H. C., and Stich, H. F. (1975). DNA repair synthesis of cultured human cells as a rapid bioassay for chemical carcinogens. *Int. J. Cancer* **16**, 284–291.

San, R. H. C., and Williams, G. M. (1977). Rat hepatocyte primary cell culture-mediated mutagenesis of adult rat liver epithelial cells by procarcinogens. *Proc. Soc. Exp. Biol. Med.* **156**, 534–538.

Santella, R. M., Grunberger, D., and Weinstein, I. B. (1979). DNA-benzo[a]pyrene adducts formed in a *Salmonella typhimurium* mutagenesis assay system. *Mutat. Res.* **61**, 181–189.

Selkirk, J. K. (1977). Divergence of metabolic activation systems for short-term mutagenesis assays. *Nature (London)* **270**, 604–607.

Sirover, M. A., and Loeb, L. A. (1976). Infidelity of DNA synthesis in *in vitro* screening for potential metal mutagens or carcinogens. *Science* **194**, 1434–1436.

Sugimura, T., Sato, S., Nagao, M., Yahagi, T., Matsushima, T., Takeuchi, M. and Kawachi,

T. (1976). Overlappings of carcinogens and mutagens. In "Fundamentals in Cancer Prevention" (P. N. Magee, ed.), pp. 191–215. Univ. of Tokyo Press, Tokyo/Univ. Park Press, Baltimore.

Swenberg, J. A., Petzold, G. L., and Harback, P. R. (1976). In vitro DNA damage alkaline elution assay for predicting carcinogenic potential. Biochem. Biophys. Res. Commun. 72, 732–738.

Tomatis, L. (1979). The predictive value of rodent carcinogenicity tests in the evaluation of human risks. Annu. Rev. Pharmacol. Toxicol. 19, 511–530.

Tomatis, L., Agthe, C., Bartsch, H., Huff, J., Montesano, R., Saracci, R., Walker, E., and Wilbourn, J. (1978). A review of the monograph programme of the International Agency for Research on Cancer (1971–1977). Cancer Res. 38, 877–885.

van Bladeren, P. J., Breimer, D. D., Rotteveels-Smijs, G. M. T., and Mohn, G. R. (1980). Mutagenic activation of dibromoethane and diiodomethane by mammalian microsomes and glutathione-S-transferases. Mutat. Res. 74, 341–346.

Vesell, E. S., and Passananti, G. T. (1977). Genetic and environmental factors affecting host response to drugs and other chemical compounds in our environment. Environ. Health Perspect. 20, 161–184.

Vogel, E. (1977). Identification of carcinogens by mutagen testing in Drosophila: The relative reliability for the kinds of genetic damage measured. In "Origins of Human Cancer" (H. H. Hiatt, J. D. Watson, and J. A. Winsten, eds.), pp. 1483–1497. Cold Spring Harbor Lab., Cold Spring Harbor, New York.

Wattenberg, L. (1975). Effects of dietary constituents on the metabolism of chemical carcinogens. Cancer Res. 35, 3326–3331.

Weeks, C. E., Allaben, W. T., Tresp, N. M., Lowie, S. C., Lazear, E. J. and King, C. M. (1980). Effects of structure of N-acyl-N-2-fluorenylhydroxylamines on arylhydroxamic acid transferase, sulfotransferase and deacylase and on mutations in Salmonella typhimurium. Cancer Res. 40, 1204–1211.

WHO (W. H. O. Tech. Rep. Ser.) (1964). "Prevention of Cancer," No. 276, Geneva.

Williams, G. M. (1977) Detection of chemical carcinogens by unscheduled DNA synthesis in rat liver primary cell cultures. Cancer Res. 37, 1845–1851.

Wislocki, P. G., Gadek, K. M., Chou, M. W., Yang, S. K., and Lu, A. Y. H. (1980). Carcinogenicity and mutagenicity of the 3,4-dihydrodiols and other metabolites of 7,12-dimethylbenz[a]anthracene and its hydroxymethyl derivatives. Cancer Res. 40, 3661–3664.

Wolff, S. (1977). Sister chromatid exchange. Annu. Rev. Genet. 11, 183–201.

Wong, J. J., and Hsieh, D. P. H. (1976). Mutagenicity of aflatoxins related to their metabolism and carcinogenic potential. Proc. Natl. Acad. Sci. U. S. A. 73, 2241–2244.

Wood, W. A., Wislocki, P. G., Chang, R. L., Levin, W., Lu, A. Y. H., Yagi, H., Hernandez, U., Jerina, D. M., and Conney, A. H. (1976). Mutagenicity and cytotoxicity of benzo[a]pyrene benzo-ring epoxides. Cancer Res. 36, 3358–3366.

Zimmerman, F. K. (1975). Procedures used in the induction of mitotic recombination and mutation in the yeast Saccharomyces cerevisiae. Mutat. Res. 31, 71–86.

Chapter 3

The Use of Mutagenicity to Evaluate Carcinogenic Hazards in Our Daily Lives

Takashi Sugimura and Minako Nagao

I. INTRODUCTION

Cancer is thought to be related to environmental factors (Doll, 1977; Wynder and Gori, 1977). Among such environmental factors, the clearest causative relation has been observed in the case of occupational cancers. However, the number of cases of occupational cancer is small compared with the total number of cases of cancer. Studies on the incidence of cancer in immigrants has also indicated that environmental factors expressed in the lifestyle are important in development of the most common types of cancer in various countries (Higginson and Muir, 1979). As an example, the high incidence of stomach cancer in Japan may be cited. Autopsy records in Japan about 100 years ago, before modern industries developed, showed the occurrence of many cases of stomach cancer (Sugimura, 1979). Thus, apparently the high incidence of stomach cancer is due to the traditional Japanese lifestyle, as represented by traditional Japanese foods. With gradual West-

73

MUTAGENICITY:
NEW HORIZONS IN GENETIC TOXICOLOGY

ernization of the lifestyle, the incidence of colon cancer is increasing while that of stomach cancer has started to decline (Hirayama, 1979). The lifestyle means the way of ordinary, everyday life of people in a country. The high incidence of hepatomas in some areas of the world may be due to mycotoxin contamination of the food and/or hepatitis B virus infection. The high incidence of cancer in some parts of the alimentary tract may be due to the composition of the food and its contents of carcinogenic factors. The high incidence of cancer in endocrine-related organs, including the breast, may be related to fat intake. The high incidence of pulmonary cancers in the general population may be related to smoking and also to inhalation of air containing carcinogenic factors produced by fumes and smoke of factories, cars, and houses.

Cancer is produced through very complicated and various mechanisms as indicated below. Genetic background should certainly not be neglected (Knudson, 1973). There are, however, also at least two environmental carcinogenic factors, an initiating factor and a promoting factor (Berenblum, 1941), and most typical carcinogens have both initiating activity and promoting activity. Initiating activity can be determined by measuring activity to produce DNA damage or activity to induce mutation in prokaryotes and eukaryotes. However, promoting activity cannot be determined as activity to induce mutations. Typical tumor promoters such as tetradecanoylphorbol acetate (TPA) (Hecker, 1967; Van Duuren, 1969) and teleocidin (Fujiki *et al.*, 1981; Sugimura *et al.*, 1981) can induce many cancers when applied to mouse skin treated with a limited amount of a carcinogen, such as dimethylbenzanthracene, that is insufficient to produce cancer during the life-span of the mice. Thus, carcinogenic hazards are due to both promotion and initiation, and, unfortunately, the former cannot be detected by mutagenicity tests: the use of mutagenicity tests to evaluate carcinogenic hazards in our daily life is limited to detection of hazards of initiators. Since most typical carcinogens may have initiating and promoting activity, mutagenicity tests can be used to detect substances that may cause human cancers, but there is no evidence that mutagens detected in microbial tests have carcinogenic activity as a result of simultaneous tumor-promoting activity. Moreover, it is very possible that the bottleneck for human cancer development is the step of tumor promotion, not tumor initiation, since so many mutagens have been found in our environment. Epidemiological studies have shown that a high fat diet results in high incidences of breast and colon cancers (Wynder and Reddy, 1975; Hirayama, 1978), but this phenomenon does not necessarily mean that an initiator(s) in a high fat diet or derived from a high fat diet, which could be detected by mutagenicity tests, is involved in development of these cancers. On the contrary, it may mean that promoters, such as hormones for breast cancer and bile acids for colon cancer, which could not be detected in mutagenicity tests are involved in cancer development as secondary effects of the high fat diet (Reddy et al., 1980). It is important to realize that nobody yet knows whether the step of initiation or of promotion is the more important in evaluating environmental carcinogenic hazards. However, nevertheless, it is still important to determine initiator-type carcinogens, which can be demonstrated in mutagenicity tests.

This chapter gives several examples of carcinogenic hazards and possibly carcinogenic hazardous substances present in the normal daily environment. This chapter is not intended to include all the available information, but to indicate current problems, illustrated by findings in the authors' laboratory.

II. EXPERIMENTAL STUDIES

Carcinogenic Hazardous Substances in Foods

1. Ordinary Cooked Foods

Ordinary broiled beefsteak, hamburger, and fish have mutagenic materials on their charred surface (Nagao et al., 1977b; Commoner et al., 1978; Sugimura, 1979). This mutagenicity was demonstrated with a metabolic activation system, and it was too great to be explained by the content of benzo[a]pyrene, which had long been thought to be the most likely hazardous carcinogen. Foods are mainly composed of protein, carbohydrate, fat, and nucleic acid, and of these, only protein yielded potent mutagenicity upon pyrolysis (Nagao et al., 1977a). Pyrolysates of individual amino acids also yielded mutagens (Sugimura, 1979). A series of new potent mutagens was isolated and identified from amino acid pyrolysates, and some of these compounds were found to exist in protein pyrolysates and cooked foods. Some mutagens were first isolated from protein pyrolysates (Yoshida et al., 1978). New potent mutagens were also isolated and identified from broiled sardines (Kasai et al., 1980) and broiled beef (Kasai et al., 1981); their structures are given in Fig. 1, and their mutagenic potentials on *Salmonella typhimurium* TA98 and TA100 with an optimum amount of S9 are given in Table I. Table I also includes distributions of these mutagens and references. It is clear that most of these mutagens are

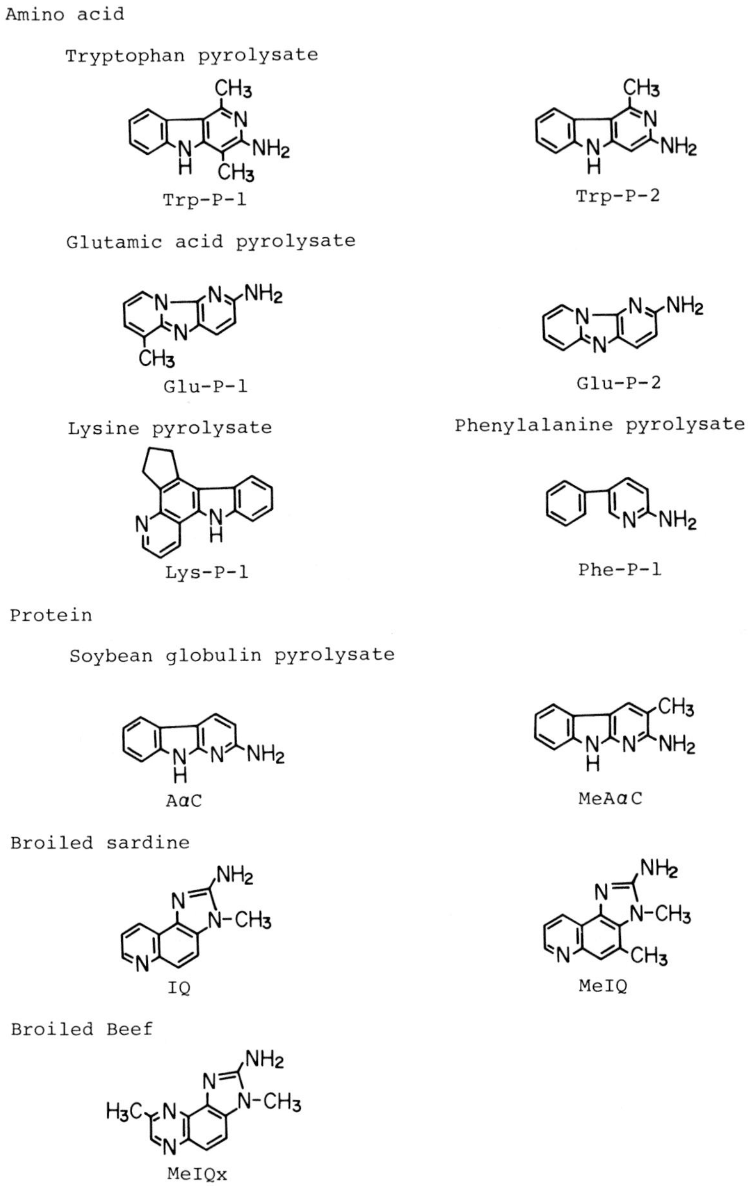

Amino acid

    Tryptophan pyrolysate

        Trp-P-1   Trp-P-2

    Glutamic acid pyrolysate

        Glu-P-1   Glu-P-2

    Lysine pyrolysate   Phenylalanine pyrolysate

        Lys-P-1   Phe-P-1

Protein

    Soybean globulin pyrolysate

        AαC   MeAαC

Broiled sardine

        IQ   MeIQ

Broiled Beef

        MeIQx

Fig. 1. Mutagens isolated from pyrolysates.

TABLE I

Mutagenic Activities of the Mutagens Isolated from Pyrolysate of Amino Acids and a Protein and Broiled Fish and Meat

Abbreviation	Chemical name	Mutagenicity (revertants/μg)		Identified in	ng/gm of heated material	Reference
		TA98	TA100			
Trp-P-1	3-Amino-1,4-dimethyl-5H-pyrido[4,3-b]indole	39,000	1,650	Broiled sun-dried sardine	13.3	Yamaizumi et al., 1980
				Broiled beef	53 ng/gm of raw beef	Yamaguchi et al., 1980a
Trp-P-2	3-Amino-1-methyl-5H-pyrido[4,3-b]indole	104,200	1,750	Broiled sun-dried sardine	13.1	Yamaizumi et al., 1980
Glu-P-1	2-Amino-6-methyldipyrido-[1,2-a:3',2'-d]imidazole	49,000	3,200			
Glu-P-2	2-Aminodipyrido[1,2-a:3',2'-d]imidazole	1,900	1,200	Broiled sun-dried cuttlefish	280	Yamaguchi et al., 1980b
Lys-P-1	3,4-Cyclopentenopyrido[3,2-a]carbazole	86	100			
Phe-P-1	2-Amino-5-phenylpyridine	41	23			
AαC	2-Amino-9H-pyrido[2,3-b]indole	300	20	Tryptophan pyrolysate	250,000 ng/gm of tryptophan	Yoshida and Matsumoto, 1979
				Pyrolysed chicken meat	34,000	Yoshida et al., 1979
				Albumin pyrolysate	43,000	Yoshida et al., 1979
MeAαC	2-Amino-3-methyl-9H-pyrido[2,3-b]indole	200	120	Tryptophan pyrolysate	20,000 ng/gm of tryptophan	Yoshida and Matsumoto, 1979
IQ	2-Amino-3-methylimidazo[4,5-f]quinoline	433,000	7,000	Broiled sun-dried sardine	158	Yamaizumi et al., unpublished data
				Broiled beef	0.59	Wakabayashi et al., unpublished data
MeIQ	2-Amino-3,4-dimethylimidazo[4,5-f]quinoline	661,000	30,000	Broiled sun-dried sardine	72	Yamaizumi et al., unpublished data
MeIQx	2-Amino-3,8-dimethylimidazo[4,5-f]quinoxaline	145,000	14,000	Broiled beef		

TABLE II

Incidence of Hepatic Tumors in Mice Fed Trp-P-1 and Trp-P-2[a]

Treatment	Sex	No. of effective mice[b]	No. of mice with hepatocellular carcinoma
None	M	25	0
	F	24	0
Trp-P-1	M	24	4
	F	26	14
Trp-P-2	M	25	3
	F	24	22[2][c]

[a] Mice were fed on basal diet with or without 200 ppm Trp-P-1 or 200 ppm Trp-P-2 for up to day 621.

[b] Number of mice surviving on Day 402, when the first hepatic tumor was found.

[c] Numbers in parentheses are numbers of mice with pulmonary metastases of hepatocellular carcinomas.

heterocyclic amines, and that the specific mutagenic activities of some of them exceed that of aflatoxin B_1.

Since these mutagens are found in cooked foods, their carcinogenic potentials *in vivo* must be elucidated. So far only *in vivo* studies on Trp-P-1 and Trp-P-2 are complete (Matsukura *et al.*, 1981). Table II shows data on the *in vivo* carcinogenicities of synthetic Trp-P-1 and Trp-P-2 in mice; these compounds were added to pellet diet and supplied to the mice *ad libitum*. Many hepatomas developed in mice of both sexes given Trp-P-1 or Trp-P-2, but it is interesting that females were more susceptible than males.

These heterocyclic amines are converted to the N-hydroxy derivatives by cytochrome *P*-448, induced in rat liver by polychlorinated biphenyl treatment (Yamazoe *et al.*, 1980). The Trp-P-2 and Glu-P-1 form adducts with guanine base (Hashimoto *et al.*, 1979, 1980). These heterocyclic amines are inactivated by peroxidase with hydrogen peroxide (Yamada *et al.*, 1979), by nitrite under acidic conditions (Tsuda *et al.*, 1980), and by chlorine ion in tap water (M. Tsuda, personal communication). Although Trp-P-1 and Trp-P-2 showed definite carcinogenicity in mice (Matsukura *et al.*, 1981) and Trp-P-1 was carcinogenic in rats (Ishikawa *et al.*, 1979a), their carcinogenic potencies were much lower than expected from their mutagenic potencies. This may partly be due to the inactivation mechanism described above. It is also likely that these heterocyclic amines act preferentially as tumor initiators rather than as complete carcinogens with both initiating and promoting activities. It was found that oral administrations of Trp-P-1 and Trp-P-2 followed by phenobarbital produced numerous enzyme-altered foci in rat liver (Ishikawa *et al.*, 1979b). Little is known about promoters in development of human cancer, and these heterocyclic amines with highly potent mutagenicity may be of importance. This example illustrates how difficult it is to evaluate carcinogenic hazard on the basis of mutation tests only.

2. A Food Additive, AF-2

Nitrofuran derivatives were once used as convenient and practical food and feed preservatives. Nitrofurazone was used as a food additive from 1952 to 1965 and nitrofurylacrylamide from 1954 to 1965. In 1965 these nitrofurans were replaced by AF-2, 2-(2-furyl)-3-(5-nitro-2-furyl)acrylamide. It was used as a food preservative until 1974. It is now realized that all nitrofuran derivatives are mutagenic in microbes (Yahagi *et al.*, 1976). The discovery of mutagenicity of AF-2 on *Escherichia coli* WP-2 in 1973 (Kondo and Ichikawa–Ryo, 1973; Kada, 1973) attracted the attention of both scientists and nonscientists, because this compound was being added to normal foods. Moreover, it was suspected that because AF-2 was mutagenic, it would also be carcinogenic. The doses of AF-2 permitted as food additives were 5 ppm in fish sausage, 20 ppm in fish-meal paste, and 5 ppm in soybean curd and red bean paste. The AF-2 was so effective that it preserved fish-meat sausage for at least a year at room temperature; when AF-2 was added, no refrigeration was required. Curiously, AF-2 and related nitrofuran derivatives were not mutagenic toward *Salmonella typhimurium* TA1535 or TA1538, but strain TA100 derived from TA1535 and strain TA98 derived from TA1538 by introduction of plasmid pKM101 were good indicator strains for detecting the mutagenicity of nitrofuran derivatives, including AF-2 (Yahagi *et al.*, 1974, 1976). The mutagenicity of AF-2 is fairly high with *Salmonella typhimurium* TA100 and TA98, as shown in Table III. Mice fed on diet containing 0.05, 0.15, and 0.45% AF-2 developed squamous cell carcinomas of

TABLE III

Mutagenicity of AF-2 toward *E. coli* WP2 *uvrA* and *S. typhimurium* TA1535, TA1538, TA100, and TA98[a]

WP2 uvrA	TA1535	TA1538	TA100	TA98
56,000	0	0	65,000	5,600

[a] Revertants/10^8 cells/μg.

the forestomach, the incidence depending on the AF-2 concentration in the diet (Ikeda et al., 1974). Rats fed on diet containing 0.08 and 0.4% AF-2 (Takayama and Kuwabara, 1977) or 0.2% AF-2 (Cohen et al., 1977) developed mammary carcinomas. Hamsters fed on diet containing 0.08 and 0.16% AF-2 developed squamous cell carcinoma of the forestomach (Kinebuchi et al., 1979). The case of AF-2 indicates clearly that scientists can predict the carcinogenicity of a chemical from data on its mutagenicity.

The total annual production of AF-2 was reported to be 3 tons in 1973. The TD_{50}, defined as the dosage producing tumor in 50% of the animals when fed continuously throughout life, was 24 mg/kg/day for AF-2. Thus, a simple calculation from the data on hamsters indicates that the AF-2 intake per head of Japanese was 0.0005 of the TD_{50} (Kinebuchi et al., 1979). When the mutagenicity of AF-2 was first observed, its carcinogenic risk was overemphasized, since AF-2 showed such high specific mutagenicity. Data from in vivo long-term animals experiments indicated that in fact AF-2 is not a very potent carcinogen. Thus, the carcinogenicity of AF-2 could be predicted from its mutagenicity, but its relatively low carcinogenic potential could not be predicted from its high mutagenic potential. (This issue is discussed fully in Chapter 2.)

3. Are Naturally Occurring Mutagenic Flavonoids Carcinogenic?

Pyrrolizidine alkaloids which are found in edible vegetables are reported to be carcinogenic and also mutagenic (Hirono et al., 1979; Yamanaka et al., 1979), as shown in Table IV. The preincubation method (Sugimura and Nagao, 1980) is essential for demonstrating the mutagenicities of pyrrolizidine alkaloids. Further investigations are required to determine the mutagenicities of a few carcinogenic pyrrolizidines such as monocrotaline. However, it can be said that in general there is a good correlation between the carcinogenicities and mutagenicities of many pyrrolizidine alkaloids.

Flavonoids are more widely distributed than pyrrolizidine alkaloids in plants, and their mutagenicities have been reported by several groups (Bjeldernes and Chang, 1977; Sugimura et al., 1977; Hardigree and Epler, 1978; MacGregor and Jurd, 1978; Brown, 1980; Nagao et al., 1981a). The flavonoids quercetin and kaempferol are moderately mutagenic, their specific activities on TA98 being about 62 and 32 revertants/μg, respectively, which are in the same order as those of o-aminoazotoluene and 4-aminobiphenyl. The glycoside of quercetin, rutin, and the glucoside of kaempferol, astragalin, are cleaved by var-

TABLE IV

Relationship between Mutagenicity and Carcinogenicity of Pyrrolizidine Alkaloids

Name of alkaloid	Source	Muta-genicity	Carcino-genicity
Clivorine	Ligularia dentata	+	?
Fukinotoxin (Petasitenine)	Petasites japonicus	+	+
Heliotrine	Heliotropium europaeum	+	+
Lasiocarpine		+	+
Ligularidine	Ligularia dentata	+	?
Lindelofine	Eupatorium stoechadosmum	−	?
LX-201	Ligularia dentata	+	?
Lycopsamine	Messerschmidia sibirica	−	+
Monocrotaline	Crotalaria spectabilis	−	+
Retronecine	Syneilesis palmata	−	?
Senecionine	Syneilesis palmata	−	?
Seneciphylline	Senecio cannabifolius	−	+
Senkirkine	Farfugium japonicum	+	+

ious glycosidases, such as hesperidinase from *Aspergillus*, to yield aglycones (Nagao *et al.*, 1981a) (Fig. 2). Intestinal flora contain glycosidase which forms quercetin from rutin because "fecalase" can split this glycoside (Brown and Dietrich, 1979; Tamura *et al.*, 1980).

Quercetin, which was shown to be mutagenic, is present in vegetables and fruits, and if it is shown to be carcinogenic will provide us with much information for evaluating carcinogenic hazard in ordinary life. Since quercetin is present in many foods, the flavonoid intake per person may be as much as 10–100 mg/day. If flavonoids are shown to be carcinogenic, they will also provide an example of the value of mutagenesis tests in evaluating carcinogenic hazard in our daily life.

In 1952, Ambrose *et al.* reported that addition of 0.25, 0.5, and 1% quercetin or quercitrin, quercetin-3-O-rhamnose, to the diet for 410 days did not induce tumors in rats. Moreover, in 1964 Nagase *et al.* reported that quercetin feeding suppressed the carcinogenicity of 3′-methyl-4-dimethylaminoazobenzene. However, recent results by Pamukcu and associates (1980) demonstrated the carcinogenicity of quercetin in Norwegian rats: 80% of their rats fed on 0.1% quercetin diet for 406 days developed many tumors in the lower part of the ileum and urinary bladder. In contrast more recent results again showed that quercetin and rutin were not carcinogenic: Saito *et al.* (1980) could not detect any increase in tumor incidence in ddY mice fed on 2%

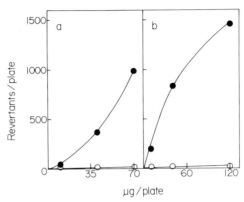

Fig. 2. Mutagenicity of rutin (a) and astragalin (b) after pretreatment with hesperidinase. The test compounds with (●-●) or without (○-○) hesperidinase pretreatment were incubated with *S. typhimurium* TA98 in the presence of S9 mix.

quercetin diet throughout life; Hirono *et al.* (1981) could not demonstrate the carcinogenicity of quercetin by feeding ACI rats on 10% quercetin diet throughout life, and Morino *et al.* (1982) observed no increase in the incidence of tumors in golden hamsters fed 10% quercetin diet throughout life.

The mutagenicity of plant foods may mostly be due to that of flavonoids. For instance, kaempferol was found to be a major mutagen in Japanese pickles (Takahashi *et al.*, 1979). A methanol extract of dill weed showed fairly strong mutagenicity, and isorhamnetin 3-sulfate and quercetin 3-sulfate were identified as its major mutagens (Fukouka *et al.*, 1980). All the mutagenic activity in sumac, the red seeds of a plant of the genus *Rhus*, was found to be due to quercetin (Seino *et al.*, 1978). Various flavonoids were identified from black tea, green tea, and red wines as mutagens after treatment with a glycosidase, such as hesperidinase or fecalase (Mazaki *et al.*, 1979; Nagao *et al.*, 1979; Tamura *et al.*, 1980). The mutagenicity of quercetin was detected not only in a *Salmonella* mutagenesis test, but also with mammalian cells by assays on mouse lymphoma with L5178Y TK$^{+/-}$ mutation (Amacher *et al.*, 1980; Meltz and MacGregor, 1981), and Chinese hamster lung fibroblasts with a diphtheria toxin resistance mutation (Nakayasu, M. personal communication). Quercetin also transformed Syrian golden hamster cells (Umezawa *et al.*, 1977) and BALB/c 3T3 cells (Meltz and MacGregor, 1981) *in vitro*.

If quercetin is actually carcinogenic, its hazard to humans should be very serious, whereas if it is not carcinogenic, short-term tests for detection of environmental carcinogens must be reevaluated.

4. How Important Are Direct-Acting Mutagens Which Are Ubiquitously Distributed in Various Foods and Beverages?

We usually drink several cups of coffee, tea, or green tea every day. Coffee, tea, and green tea were demonstrated to have direct-acting mutagens to TA100 (Nagao et al., 1979); 200 ml of coffee, tea, or green tea contain a substance(s) that can induce $2 \times 10^4 \sim 10^5$ revertants of TA100, and the mutagenic activity is not correlated with the caffeine content. Instant coffee, with or without caffeine, and regularly brewed coffee have the same mutagenic activity and they also have phage-inducing activity without metabolic activation (Kosugi et al., 1980). Recently, an epidemiological study showed that there is a correlation between pancreatic tumors and coffee intake (McMahon et al., 1981). However, there is no direct evidence for a carcinogenic effect of coffee on experimental animals.

Alcoholic beverages have been considered to be causative agents of esophageal cancer in humans (Tuyns, 1979). Alcohol itself may be a promoter and it is possible that it is also an initiator, because one of its metabolites, acetaldehyde, induces sister chromatid exchanges in Chinese hamster cells in vitro (Obe and Ristow, 1977). In addition, nonvolatile residues of whiskeys, brandies, and an apple brandy were found to contain a mutagen(s) that was active without metabolic activation on S. typhimurium TA100 (Nagao et al., 1981b). Homemade apple brandies were also found to contain a direct acting mutagen(s) on TA98 and TA100 (Loquet et al., 1981). The intake of 100 ml of a certain brand of whiskey was found to correspond to the intake of mutagens inducing 3.8×10^4 revertants of TA100 under defined conditions (Nagao et al., 1981b).

The direct-acting mutagenic activity present in coffee, tea, green tea, and alcoholic beverages was abolished or markedly reduced by addition of rat liver S9 mix, except in the cases of some brands of Chinese alcoholic spirits, the mutagenicities of which were expressed only in the presence of S9 mix (Lee and Fong, 1979).

The direct-acting mutagens in coffee and whiskey were formed during roasting of coffee beans and heating of the barrel in which whiskeys were stored and matured. This type of mutagen may be widely distributed in our environment. Another example is the mutagen in caramel. The specific mutagenicity of caramel varies greatly and is sometimes very low, but caramel-type mutagens are widely distributed in our environment. Unfortunately, it is not known definitely whether this type of mutagen is carcinogenic; in some long-term animal tests caramel was not carcinogenic (Ito, 1981; Kurokawa, 1981).

III. CONCLUSIONS AND FUTURE STUDIES

There is no doubt that lifestyle has an influence on the incidence of cancer in various organs of humans. Mutagenesis tests using microbes, which are now widely used, are very useful for rapid detection of mutagens. In the past 10 years many new mutagens have been found in microbial tests and some of these compounds have been shown to be carcinogenic *in vivo*. For instance, the new mutagens found in grilled foods would have been very difficult to examine in long-term animal tests and only the availability of microbial tests has made this study possible.

However, it is now time to make a quantitative estimation of the risk of these mutagens to humans. The mutagens discovered in microbial tests may be important as tumor initiators, but the presence of a second factor in carcinogenesis, namely, promoters, is required to make these mutagens complete carcinogens. Further studies are definitely required on this problem, and at present it is still premature to decide which mutagens in our normal environment are important in developement of cancer.

However, based on recent information it seems wise to reduce exposure to these mutagens as much as possible. This would probably be possible without disturbing the quality of life. For instance, it would be simple to avoid charring food. Contamination of food by fungi could also be avoided by appropriate methods of food storage. Housewives could stop using plants containing pyrrolizidine alkaloids in the kitchen. Moreover, various methods for nonmetabolic and metabolic inactivation of mutagens could be developed; antioxidants and appropriate amounts of vitamin E and vitamin C should be useful for this purpose.

From the results of microbial tests, scientists tend to emphasize the presence of mutagens in the environment as tumor initiators. However, it should not be forgotten that our lifestyle, including our feeding habits, may have an influence, though indirect, on cancer development. A high salt intake may cause chronic gastritis and a high fat intake may result in more bile acid formation, which may act as a tumor promoter in the large intestine. Furthermore the composition of our food may change the intestinal flora which is responsible for the formation of mutagens/tumor initiators as well as of promoters.

Thus, an open-minded attitude is required in considering the integrated nature of human cancer risk in our environment.

REFERENCES

Amacher, D., Paillet, S. C., Turner, G. N., Ray, V. A., and Salsburg, D. S. (1980). Point mutations at the thymidine kinase locus in L5178Y mouse lymphoma cells. II. Test validation and interpretation. *Mutat. Res.* **72**, 447–474.

Ambrose, A. M., Robbins, D. J., and DeEds, F. (1952). Comparative toxicities of quercetin and quercitrin. *J. Am. Pharm. Assoc.* **41**, 119–122.

Berenblum, I. (1941). The cocarcinogenic action of croton resin. *Cancer Res.* **14**, 44–47.

Bjeldernes, L. F., and Chang, G. W. (1977). Mutagenic activity of quercetin and related compounds. *Science* **197**, 577–578.

Brown, J. P. (1980). A review of the genetic effects of naturally occurring flavonoids, anthraquinones and related compounds. *Mutat. Res.* **75**, 243–277.

Brown, J. P., and Dietrich, P. S. (1979). Mutagenicity of plant flavonols in the Salmonella/mammalian microsome test. Activation of flavonol glycosides by mixed glycosidases from rat cecal bacteria and other sources. *Mutat. Res.* **66**, 223–240.

Cohen, S. M., Ichikawa, M., and Bryan, G. T. (1977). Carcinogenicity of 2-(2-furyl)-3-(5-nitro-2-furyl)acrylamide (AF-2) fed to female Sprague-Dawley rats. *Gann* **68**, 473–476.

Commoner, B., Vithayathil, A. J., Dolora, P., Nair, S., Madyastha, P., and Cuca, G. C. (1978). Formation of mutagens in beef and beef extract during cooking. *Science* **201**, 913–916.

Doll, R. (1977). Strategy for detection of cancer hazards to man. *Nature (London)* **265**, 589–596.

Fujiki, H., Mori, M., Nakayasu, M., Terada, M., Sugimura, T., and Moore, R. E. (1981). Indole alkaloids; dihydroteleocidin B, teleocidin and lyngbyatoxin A, as a new class of tumor promoters. *Proc. Natl. Acad. Sci. U.S.A.* **78**, 3872–3876.

Fukuoka, M., Yoshihira, K., Natori, S., Sakamoto, K., Iwahara, S., Hosaka, S., and Hirono, I. (1980). Characterization of mutagenic principles and carcinogenicity test of dill weed and seeds. *J. Pharmacobio. Dyn.* **3**, 236–244.

Hardigree, A. A., and Epler, J. L. (1978). Comparative mutagenesis of plant flavonoids in microbial system. *Mutat. Res.* **58**, 231–239.

Hashimoto, Y., Shudo, K., and Okamoto, T. (1979). Structural identification of a modified base in DNA covalently bound with mutagenic 3-amino-1-methyl-5*H*-pyrido[4,3-b]-indole. *Chem. Pharm. Bull.* **27**, 1058–1060.

Hashimoto, Y., Shudo, K., and Okamoto, T. (1980). Metabolic activation of a mutagen, 2-amino-6-methyldipyrido[1,2-a:3',2'-d]imidazole: Identification of 2-hydroxyamino-6-methyldipyrido[1,2-a:3',2'-d]imidazole and its reaction with DNA. *Biochem. Biophys. Res. Commun.* **92**, 971–976.

Hecker, E. (1967). Phorbol esters from croton oil, chemical nature and biological activities. *Naturwissenschaften* **54**, 282–284.

Higginson, J., and Muir, C. S. (1979). Environmental carcinogenesis: Misconceptions and limitations to cancer control. *J. Natl. Cancer Inst.* **63**, 1291–1298.

Hirayama, T. (1978). Epidemiology of breast cancer with special reference to the role of diet. *Prev. Med.* **7**, 173–195.

Hirayama, T. (1979). The epidemiology of gastric cancer in Japan. In "Gastric Cancer" (C. J. Pfeiffer, ed.), pp. 60–82. Gerhard Witzstrock Publishing House, Inc. New York.

Hirono, I., Ueno, I., Hosaka, S., Takanashi, H., Matsushima, T., Sugimura, T. and Natori, S. (1981). Carcinogenicity examination of quercetin and rutin in ACI rats. *Cancer Lett.* (Shannon, Irel.) **13**, 15–21.

Hirono, I., Mori, H., Haga, M., Fuji, M., Yamada, K., Hirata, Y., Takanashi, H., Uchida, E., Hosaka, S., Ueno, I., Matsushima, T., Umezawa, K., and Shirai, A. (1979). Edible plants containing carcinogenic pyrrolizidine alkaloids in Japan. In "Naturally Occurring Carcinogens—Mutagens and Modulators of Carcinogenesis" (E. C. Miller, J. A. Miller, I. Hirono, T. Sugimura, and S. Takayama, eds.), pp. 79–87. Univ. Park Press, Baltimore, Maryland.

Ikeda, Y., Horiuchi, S., Furuya, T., Uchida, O., Suzuki, K., and Azegami, J. (1974). Induction of gastric tumors in mice by feeding of furylfuramide. Food Sanitation Study Council, Ministry of Health and Welfare, Japan.

Ishikawa, T., Takayama, S., Kitagawa, T., Kawachi, T., Kinebuchi, M., Matsukura, N., Uchida, E., and Sugimura, T. (1979a). In vivo experiments on tryptophan pyrolysis products. In "Naturally Occurring Carcinogens—Mutagens and Modulators of Carcinogenesis" (E. C. Miller, J. A. Miller, I. Hirono, T. Sugimura, and S. Takayama, eds.), pp. 159–167. Univ. Park Press, Baltimore, Maryland.

Ishikawa, T., Takayama, S., Kitagawa, T., Kawachi, T., and Sugimura, T. (1979b). Induction of enzyme-altered islands in rat liver by tryptophan pyrolysis products. J. Cancer Res. Clin. Oncol. **95**, 221–224.

Ito, N. (1982). Annual Report of the Cancer Research, Ministry of Health and Welfare of Japan (in press).

Kada, T. (1973). Escherichia coli mutagenicity of furylfuramide. Jpn. J. Genet. **48**, 301–305.

Kasai, H., Yamaizumi, Z., Wakabayashi, K., Nagao, M., Sugimura, T., Yokoyama, S., Miyazawa, T., Spingarn, N. E., Weisburger, J. H., and Nishimura, S. (1980). Potent novel mutagens produced by broiling fish under normal conditions. Proc. Jpn. Acad. **56B**, 278–283.

Kasai, H., Yamaizumi, Z., Shiomi, T., Yokoyama, S., Miyazawa, T., Wakabayashi, K., Nagao, M., Sugimura, T., and Nishimura, S. (1981). Structure of a potent mutagen isolated from fried beef. Chem. Lett. 485–488.

Kinebuchi, M., Kawachi, T., Matsukura, N., and Sugimura, T. (1979). Further studies on the carcinogenicity of a food additive, AF-2, in hamsters. Food Cosmet. Toxicol. **17**, 339–341.

Knudson, A. G., Jr. (1973). Mutation and human cancer. Adv. Cancer Res. **17**, 317–352.

Kondo, S., and Ichikawa-Ryo, H. (1973). Testing and classification of mutagenicity of furylfuramide in Escherichia coli. Jpn. J. Genet. **48**, 295–300.

Kosugi, A., Nagao, M., Wakabayashi, K., Kawachi, T., and Sugimura, T. (1980). Mutagenicity of coffee. II. Proc. 9th Jpn. Environ. Mutat. Soc. (Okayama), p. 39.

Kurokawa, Y. (1982). Annual Report of the Cancer Research, Ministry of Health and Welfare of Japan (in press).

Lee, J. S. K., and Fong, L. Y. Y. (1979). Mutagenicity of Chinese alcoholic spirits. Food Cosmet. Toxicol. **17**, 575–578.

Loquet, C., Toussaint, G., and Le Talaer, J. Y. (1981). Studies on mutagenic constituents of apple brandy and various alcoholic beverages collected in Western France, a high incidence area for oesophageal cancer. Mutat. Res. **88**, 155–164.

MacGregor, J. T., and Jurd, L. (1978). Mutagenicity of plant flavonoids: Structural requirements for mutagenic activity in Salmonella typhimurium. Mutat. Res. **54**, 297–309.

MacMahon, B., Yen, S., Trichopoulos, D., Warren, K., and Nardi, G. (1981). Coffee and cancer of the pancreas. N. Eng. J. Med. **304**, 630–633.

Matsukura, N., Kawachi, T., Morino, K., Ohgaki, H., Sugimura, T., and Takayama, S. (1981). Carcinogenicity in mice of mutagenic compounds from a tryptophan pyrolysate. Science **213**, 346–347.

Mazaki, M., Taue, S., and Ueta, M. (1979). Mutagenicity of green-tea extract. *Proc. 8th Jpn. Environ. Mutat. Soc. (Hakone)*, p. 57.

Meltz, M. L., and MacGregor, J. T. (1981). Activity of the plant flavonol quercetin in the mouse lymphoma L5178Y TK$^{+/-}$ mutation, DNA single-strand break, and BALB/c 3T3 chemical transformation assays. *Mutat. Res.* **88**, 317–324.

Morino, K., Matsukura, N., Kawachi, T., Ohgaki, H., Sugimura, T. and Hirono, I. (1982). Carcinogenicity test of quercetin and rutin in golden hamsters by oral administration. *Carcinogenesis* (in press).

Nagao, M., Honda, M., Seino, Y., Yahagi, T., Kawachi, T., and Sugimura, T. (1977a). Mutagenicities of protein pyrolysates. *Cancer Lett. (Shannon, Irel.)* **2**, 335–340.

Nagao, M., Yahagi, T., Kawachi, T., Seino, Y., Honda, M., Matsukura, N., Sugimura, T., Wakabayashi, K., Tsuji, K., and Kosuge, T. (1977b). Mutagenesis in foods, and especially pyrolysis products of protein. *In* "Progress in Genetic Toxicology" (D. Scott, B. A. Bridges, and F. H. Sobels, eds.), pp. 259–264. Elsevier/North-Holland Biomedical Press, Amsterdam.

Nagao, M., Takahashi, Y., Yamanaka, H., and Sugimura, T. (1979). Mutagens in coffee and tea. *Mutat. Res.* **68**, 101–106.

Nagao, M., Morita, N., Yahagi, T., Shimizu, M., Kuroyanagi, M., Fukuoka, M., Yoshihira, K., Natori, S., Fujino, T., and Sugimura, T. (1981a). Mutagenicities of 61 flavonoids and 11 related compounds. *Environ. Mutagen.* **3**, 401–419.

Nagao, M., Takahashi, Y., Wakabayashi, K., and Sugimura, T. (1981b). Mutagenicity of alcoholic beverages. *Mutat. Res.* **88**, 147–154.

Nagase, S., Fujimaki, C., and Isaka, H. (1964). Effect of administration of quercetin on the production of experimental liver cancers in rats fed p-dimethylaminoazobenzene. *Proc. Jpn. Cancer Assoc., 23rd Meet.* pp. 26–27.

Obe, G., and Ristow, H. (1977). Acetaldehyde, but not ethanol, induces sister chromatid exchanges in Chinese hamster cells *in vitro. Mutat. Res.* **56**, 211–213.

Pamukcu, A. M., Yalciner, S., Hatcher, J. F., and Bryan, G. T. (1980). Quercetin, a rat intestinal and bladder carcinogen present in bracken fern (*Pteridium aquilinum*). *Cancer Res.* **40**, 3468–3472.

Reddy, B. S., Cohen, L. A., McCoy, G. D., Hill, P., Weisburger, J. H., and Wynder, E. L. (1980). Nutrition and its relationship to cancer. *Adv. Cancer Res.* **32**, 237–345.

Saito, D., Shirai, A., Matsushima, T., Sugimura, T., and Hirono, I. (1980). Test of carcinogenicity of quercetin, a widely distributed mutagen in food. *Teratog. Carcinog. Mutagen.* **1**, 213–221.

Seino, Y., Nagao, M., Yahagi, T., Sugimura, T., Yasuda, T., and Nishimura, S. (1978). Identification of a mutagenic substance in a spice, sumac, as quercetin. *Mutat. Res.* **58**, 225–229.

Sugimura, T. (1979). Naturally occurring genotoxic carcinogens. *In* "Naturally Occurring Carcinogens—Mutagens and Modulators of Carcinogenesis" (E. C. Miller, J. A. Miller, I. Hirono, T. Sugimura, and S. Takayama, eds.), pp. 241–261. Univ. Park Press, Baltimore, Maryland.

Sugimura, T., and Nagao, M. (1980). Modification of mutagenic activity. *Chem. Mutagens* **6**, 41–60.

Sugimura, T., Nagao, M., Matsushima, T., Yahagi, T., Seino, Y., Shirai, A., Sawamura, M., Natori, S., Yoshihira, K., Fukuoka, M., and Kuroyanagi, M. (1977). Mutagenicity of flavone derivatives. *Proc. Jpn. Acad.* **53B**, pp. 194–197.

Sugimura, T., Fujiki, H., Mori, M., Nakayasu, M., Terada, M., Umezawa, K., and Moore, R. E. (1982). New naturally occurring tumor promoter. *In* "Carcinogenesis" (E. Hecker, N. Fusening, F. Marks, and W. Kunz, eds.), pp. 69–73. Raven Press, New York.

Takahashi, Y., Nagao, M., Fujino, T., Yamaizumi, Z., and Sugimura, T. (1979). Mutagens in Japanese pickle identified as flavonoids. *Mutat. Res.* **68**, 117–123.

Takayama, S., and Kuwabara, N. (1977). Carcinogenic activity of 2-(2-furyl)-3-(5-nitro-2-furyl)acrylamide, a food additive, in mice and rats. *Cancer Lett.* (*Shannon, Irel.*) **3**, 115–120.

Tamura, G., Gold, C., Ferro-Luzzi, A., and Ames, B. N. (1980). Fecalase: A model for activation of dietary glycosides to mutagens by intestinal flora. *Proc. Natl. Acad. Sci. U. S. A.* **77**, 4961–4965.

Tsuda, M., Takahashi, Y., Nagao, M., Hirayama, T., and Sugimura, T. (1980). Inactivation of mutagens from pyrolysates of tryptophan and glutamic acid by nitrite in acidic solution. *Mutat. Res.* **78**, 331–339.

Tuyns, A. J. (1979). Epidemiology of alcohol and cancer. *Cancer Res.* **39**, 2840–2843.

Umezawa, K., Matsushima, T., Sugimura, T., Hirokawa, T., Tanaka, M., Katoh, Y., and Takayama, S. (1977). *In vitro* transformation of hamster embryo cells by quercetin. *Toxicol. Lett.* **1**, 175–178.

Van Duuren, B. L. (1969). Tumor-promoting agents in two-stage carcinogenesis. *Proc. Exp. Tumor Res.* **11**, 31–68.

Wynder, E. L., and Gori, G. B. (1977). Contribution of the environment to cancer incidence: An epidemiologic exercise. *J. Natl. Cancer Inst.* **58**, 825–832.

Wynder, E. L., and Reddy, B. S. (1975). Dietary fat and colon cancer. *J. Natl. Cancer Inst.* **54**, 7–10.

Yahagi, T., Nagao, M., Hara, K., Matsushima, T., Sugimura, T., and Bryan, G. T. (1974). Relationships between the carcinogenic and mutagenic or DNA-modifying effects of nitrofuran derivatives, including 2-(2-furyl)-3-(5-nitro-2-furyl)acrylamide, a food additive. *Cancer Res.* **34**, 2266–2273.

Yahagi, T., Matsushima, T., Nagao, M., Seino, Y., Sugimura, T., and Bryan, G. T. (1976). Mutagenicity of nitrofuran derivatives on a bacterial tester strain with an R factor plasmid. *Mutat. Res.* **40**, 9–14.

Yamada, Y., Tsuda, M., Nagao, M., Mori, M., and Sugimura, T. (1979). Degradation of mutagens from pyrolysates of tryptophan, glutamic acid and globulin by myeloperoxidase. *Biochem. Biophys. Res. Commun.* **90**, 769–776.

Yamaguchi, K., Shudo, K., Okamoto, T., Sugimura, T., and Kosuge, T. (1980a). Presence of 3-amino-1,4-dimethyl-5H-pyrido[4,3-b]indole in broiled beef. *Gann* **71**, 745–746.

Yamaguchi, K., Shudo, K., Okamoto, T., Sugimura, T., and Kosuge, T. (1980b). Presence of 2-aminodipyrido[1,2-a:3′,2′-d]imidazole in broiled cuttlefish. *Gann* **71**, 743–744.

Yamaizumi, Z., Shiomi, T., Kasai, H., Nishimura, S., Takahashi, Y., Nagao, M., and Sugimura, T. (1980). Detection of potent mutagens, Trp-P-1 and Trp-P-2, in broiled fish. *Cancer Lett.* (*Shannon, Irel.*) **9**, 75–83.

Yamanaka, H., Nagao, M., Sugimura, T., Furuya, T., Shirai, A., and Matsushima, T. (1979). Mutagenicity of pyrrolizidine alkaloids in the Salmonella/mammalian-microsome test. *Mutat. Res.* **68**, 211–216.

Yamazoe, Y., Ishii, K., Kamataki, T., Kato, R., and Sugimura, T. (1980). Isolation and characterization of active metabolites of tryptophan-pyrolysate mutagen, Trp-P-2, formed by rat liver microsomes. *Chem. Biol. Interact.* **30**, 125–138.

Yoshida, D., Matsumoto, T., Yoshimura, R., and Matsuzaki, T. (1978). Mutagenicity of amino-α-carbolines in pyrolysis products of soybean globulin. *Biochem. Biophys. Res. Commun.* **83**, 915–920.

Yoshida, D., and Matsumoto, T. (1979). Isolation of 2-amino-9H-pyrido[2,3-b]indole and 2-amino-3-methyl-9H-pyrido[2,3-b]indole as mutagens from pyrolysis product of tryptophan. *Agric. Biol. Chem.* **43**, 1155–1156.

Yoshida, D., Nishigata, H., and Matsumoto, T. (1979). Pyrolytic yields of 2-amino-9H-pyrido[2,3-b]indole and 3-amino-1-methyl-5H-pyrido[4,3-b]indole as mutagens from protein. *Agric. Biol. Chem.* **43**, 1769–1770.

Chapter 4

Mutagenicity and Lung Cancer in a Steel Foundry Environment

D. W. Bryant and D. R. McCalla

I. INTRODUCTION

The current scientific and public concern about toxic pollutants in the environment and workplace is the product of a number of factors. Among areas of primary concern are the handling, storage, and accidental release of large amounts of environmentally toxic materials; occupational exposure and related health hazards of such materials as asbestos and vinyl chloride; and excess exposure to ionizing and other types of radiation. In a distressing number of cases, humans have become the experimental animals in which the chronic and tumorogenic potential of substances has been recognized. Given the latent periods required for development of chronic illness (especially

89

cancer) large numbers of individuals frequently suffer significant exposure before ill effects are brought to public attention. Thus, in spite of preventative measures taken where a hazard is recognized, there remains the problem of those who carry a significant burden of damage from prior exposure to the toxic environment. Classically, it has been epidemiology that was responsible for recognition and characterization of health related hazards. However, were a chemical agent to induce only a modest increase in the incidence of a tumor that is already common in the population, such a carcinogen might well go unrecognized.

The notion of establishing procedures to monitor the workplace for known toxic substances is not new. Such practices, however, usually concentrate on identified hazards (e.g., heavy metals, dust, or radioactive materials), leaving a majority of substances, some of whose health effects may be equally devastating, undetected. More ambitious programs of chemical analysis directed at identifying and characterizing components of a particular environment have been described. Modern chromatographic techniques such as gas or high-performance liquid chromatography (GC and HPLC) coupled with mass spectrometry (MS) and associated computer data systems offer the means to detect and quantitate large numbers of substances collected from an environment.

A related approach utilizes bioassays to detect compounds of interest in the basis of biological activity. To some extent this type of research obviates any requirement for assumptions with respect to the chemical nature of a particular environment and concentrates on "most relevant" compounds. Bioassays have a long and honorable history. Polycyclic aromatic hydrocarbons (PAH) isolated from coal tar were first recognized as potentially carcinogenic as a result of whole animal studies. Because of the expense and time-consuming nature of such assays, the data on animal carcinogenicity investigations are limited and the biological analysis of environmentally detected substances fragmentary. Development of short-term tests that are both inexpensive and provide a reasonable indication of carcinogenic/mutagenic potential of chemicals has resulted in an explosive increase in our knowledge of such materials. One might expect, therefore, that monitoring and characterization of local environments for carcinogenic hazards should be enormously simplified. The very sensitivity of our chemical and biological analytical techniques has created unforeseen problems; we have found that small amounts of harmful substances are almost ubiquitous. This discovery raises the question of level of exposure, whether such substances pose a real threat to human health, and more particularly how we should respond to regulate the level of such carcinogens.

One theme of this chapter will be the need to make intelligent use of the information gained from short-term tests coupled with chemical analysis to probe hazards that have already been identified by epidemiological studies. This information should lead not only to causal factors, but should be balanced by insight into the degree of risk that the level of particular pollutants detected pose for human health. This approach is illustrated with an ongoing study of a lung cancer hazard in a steel foundry. Obviously this approach is not limited to the examination of agents found in the external environment, but has also to be applied to examination of tissues and body fluids for endogenous mutagens where epidemiological results suggest such agents might be of significance (Falck et al., 1980; Hopkin and Evans, 1980; Møller and Dybing, 1980).

The second question addressed by this chapter illustrates the use of short-term assays to examine urban air and the impact of the anticipated increase in diesel- rather than gasoline-fueled engines.

II. FOUNDRY MUTAGENICITY STUDY

A. The Epidemiological Background

Some years ago, Gibson and associates (1977) prompted by results of lung function tests of steel foundry workers reported an analysis of foundry workers at DOFASCO, Inc., in Hamilton, Ontario, Canada. In a retrospective study a cohort of 1542 individuals was examined over a 10-year period from 1967 to 1977. All members of the cohort were employees 45 years of age or older in 1967. These were divided into two main groups: workers who had at least 5 years' experience in the foundry prior to 1967 (foundry workers); and those workers who had been employed for a similar period but had never worked in the foundry. Standardized mortality ratios (SMR), based on data for the city of Toronto, calculated for each group indicated that the foundry workers had a slightly elevated risk for total cancers (138 foundry, 92 nonfoundry; where SMR = 100 indicates a risk equal to that of the control population). When lung cancer risk was assessed, the SMR for foundry workers was 250 while that of the remainder of the cohort was only 66. Subclassification of foundry workers by job description (e.g., crane operators, finishers, molders, coremakers, and electric furnace/open hearth operators) showed that the lung cancer risk was not shared equally among all workers. Although the number of cases of lung cancer in each job class group was small, crane operators appeared six times more likely to develop lung cancer than the general population ($p = 0.01$). In contrast, electric furnace operators were the lowest risk group with an SMR of 1.14. Finishers were nearly three times as likely to develop lung cancer, while molders and coremakers shared risks only twice that of furnace operators but these differences were not statistically significant. Later, Tola and associates (1979) reported that workers in an iron foundry in Finland also showed an increased cancer risk and that casters and floor molders were most severely affected.

In the steel casting process, molten metal at 1400°–1500°C is poured into sand molds which contain organic binders to provide rigidity and other additives to ensure favorable casting conditions. The hot metal pyrolyses some of the organic material, a process accompanied by the generation of PAH. Gibson and co-workers (1977) found a number of carcinogenic PAH in environmental particulate samples collected in the foundry. No correlation was found between the type PAH observed and the cancer risk at the collection site. They did, however, point out that some of the crane runways are directly over the pouring floors, where most of the fumes and particulates originate (Gibson et al., 1977).

The DOFASCO study raised a number of important questions. First, what is the nature of the carcinogenic material present in the foundry environment? Second, do these substances vary in concentration from one site to another? Third, what is the source of the hazard? Fourth, can the risk be eliminated or ameliorated in some way? Answers to these questions require a means of quantitating the carcinogenicity of materials present in the environment. A prerequisite to such an investigation was the availability of cheap and rapid biological assays. Furthermore, chemical and biological analysis of isolated substances required reliable and rapid procedures for testing chemically fractionated environmental samples. Clearly conventional animal bioassays did not meet our criteria. We chose the Salmonella typhimurium S9 assay developed and described by Ames and co-workers (1975). This assay, which discriminates among chemicals for mutagenic/carcinogenic potential, has been the subject of extensive validation tests (Ames et al., 1975; Yamasaki and Ames, 1977; Ames, 1979; Rinkus and Legator, 1979). Although it is generally accepted that the Salmonella S9 assay gives a reliable indication of the mutagenic and potential carcinogenic activity of many classes of compounds, it appears to work better with more powerful organic carcinogens than with weaker ones (Rinkus and Legator, 1979). In spite of its shortcomings, the *Salmonella* S9 assay was the one of choice at initial stages of development of this project.

In this assay, easily recognized reverse mutations of defective histidine genes are scored. Positive results (i.e., induction of more revertants than are obtained in control experiments) in the presence of a substance alone indicate that the material tested was "direct acting." When positive results are obtained only after addition of S9 (9000 g supernatant fraction of homogenized liver from Aroclor 1254 induced rats) it implies that the test substance itself was inactive but was able to interact with bacterial DNA after metabolism by the mammalian enzymes. Many PAH and aromatic amines belong to this latter class of "indirect-acting" mutagens, while many nitrocompounds, quinones, and epoxides are "direct acting."

The choice of this, or any short-term assay represents a compromise, since the use of a battery of tests could provide greater assurance that a majority of mutagens were detected. However, the cost of expanding the test battery is a decrease in the number of samples that can be processed. Because we found a good deal of mutagenic activity in various kinds of foundry air particulates, we have concentrated on these materials rather than expanding the number of short-term assays used.

B. Sampling

The composition of local environments is extremely complex, and efforts to analyze them have generally centered about procedures which collect organic material bound to a solid substrate. A number of studies (Commoner et al., 1978; Huisingh et al., 1978; Hughes et al., 1980), although recognizing the existence of a volatile fraction, have largely ignored these environmental components because collection requires sophisticated trapping devices and the procedures for bioassay of vapor phase organics collected on site and analyzed in the laboratory *in vitro* have not yet been developed (Hughes et al., 1980). Furthermore, it may be noted that the residence time of the volatile fraction in, for example, the lung is considerably shorter than material bound to inhaled particulates (Natusch and Wallace, 1974). Most of the knowledge of the composition and mutagenicity of air particulates is based on the study of samples collected on glass fiber filters (Commoner et al., 1978; Huisingh et al., 1978; Hughes et al., 1980; Ohnishi et al., 1980). In addition to the loss of the more volatile constituents such as naphthalene and anthracene when air is drawn through glass fiber filters, there is the problem of artifacts which arise from alteration of particulate bound organics by interaction with oxygen, oxides of nitrogen, and SO_3 (Pitts et al., 1979). Peters and Seifert (1980) have carried out a careful analysis of the stability of [^{14}C]benzo[a]pyrene (B[a]P) on glass fiber filters over a 24-hour sampling period in a high volume collection device (96 m^3/hour). Although sublimation accounted for a loss of only 10% or less of the radioactivity after 24 hours, an average of 75% of the B[a]P was converted to other products. Among the oxidation products were 3 isomeric quinones (1,6', 3,6', and 6,12-quinone) and 7H-benz[d,e]anthracene-7-one-3,4-dicarboxy-cyclic acid (B[a]P acid). The rate of degradation of B[a]P was accelerated by sunlight and was closely related to ozone concentration in the atmosphere. Pitts and co-workers (1979) also found that the half-life of B[a]P in the presence of ozone at a concentration of 0.1 ppm in the atmosphere was less than 1 hour. The formation of nitro-PAH on filters exposed to NO or NO_2 in acidic environments has also been documented (Pitts et al., 1978). Although less detailed studies were carried out in our laboratory, we have found similar results. We found that part of a sample of B[a]P spotted onto a high volume filter was lost in 24 hours and a portion of that which remained was no longer B[a]P. Though nitrobenzo[a]pyrene was not observed on TLC plates, a significant amount of the extracted material no longer migrated with the B[a]P standard (results not shown).

The collection procedures used in the DOFASCO study have attempted specifically to address problems of changes in particulate-bound organics during the sampling period. Particulates were trapped using three methods: low volume (0.18 m^3/hour), personal cassettes containing 0.8 μm glass fiber–silver membrane filters, as well as medium (1.26 m^3/hour) and high (96 m^3/hour) volume apparatus which collect particulates on standard glass fiber filters (8 × 10 GELMAN/Type A/E glass fiber). The personal monitors were useful for detection of a variety of polycyclic aromatic hydrocarbons adsorbed to particulates (Gibson et al., 1977) but proved inadequate for use in mutagenicity assays. The quantity of material provided by these low volume samples was not sufficient for fractionation or assay by the standard Ames test. Preliminary analysis of medium and high volume sampling on glass fiber filters showed that no advantage was gained by extending the sampling time. Figure 1 shows that with strain TA98, revertants induced per microgram of particulates collected was approximately the same for high, medium, and low volume samplers over a wide range of particulate concentration. Most of the activity observed in

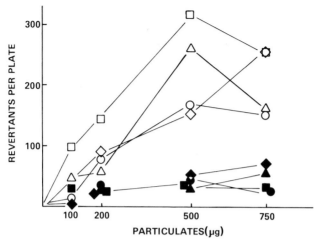

Fig. 1. Dose–response curves for particulate extracts. Three types of air samplers were used to determine particulate mutagenicity. Results are shown for two separate high volume samples (open circles, open squares), a medium volume (open triangles), and a composite of particulates pooled from several low volume samples (diamonds). Particulates on filters collected at different face velocities (see Table I) were weighed, extracted with methanol, and analyzed for mutagenicity with S. *typhimurium* TA98. The open symbols represent revertants induced after activation by S9 (6%); while the filled symbols show direct-acting mutagenic activity in the extracts.

crude methanol extracts of the filters required microsomal (S9) activation. Table I compares the yield of material deposited on filters using the different types of samplers. Although different collection rates and periods were used, the actual amount of material deposited per cubic meter of air were similar. Thus, when we consider the results of Fig. 1 and Table 1, it would appear that no significant losses of mutagenic material occurred when high-velocity collection methods were used.

C. Extraction

Different workers have used a variety of solvents for extraction of air particulates including a mix of methanol, benzene, and dichloromethane (Pitts *et al.*, 1977) or benzene, hexane, and isopropanol (Commoner *et al.*, 1978), benzene (Huisingh *et al.*, 1980), acetone (Talcott and Harger, 1980), or methanol (Dehnen *et al.*, 1977). We chose to use methanol (16 hours in a Soxhlet apparatus) since this recovers polar as well as nonpolar compounds. The solvent is easily removed and residue is not toxic to bacteria. No additional mutagenic activity was

TABLE I

Coke Oven and Foundry Environmental Air Particulate Samples

Date	Filter	Sample weight (mg)	Volume air (m^3)	Concentration (mg/m^3)
High volume (1.60 m^3/minute)				
Jan. 3	36	590	194	3.041
Jan. 10	37	220	139	1.583
Jan. 11	38	170	240	0.708
Jan. 30	39	230	197	1.167
Medium volume (0.021 m^3/minute)				
Jan. 3	C	19.80	11.32	1.749
Jan. 10	D	3.50	2.09	1.675
Jan. 11	E	6.00	3.47	1.515
Jan. 30	F	5.40	3.91	1.381
Low volume (.003 m^3/minute)				
Jan. 3	646	3.012	1.435	2.098
	647	2.763	1.483	1.863
	648	2.431	1.328	1.830
	649	2.851	1.534	1.858
	596	1.956	1.112	1.758
Jan. 10	24	0.742	1.121	0.662
	47	0.634	1.121	0.565
	111	1.492	0.963	1.549
	128	0.742	0.977	0.760
	138	0.936	0.937	0.998
Jan. 11	2	0.795	1.010	0.791
	4	0.831	0.880	0.944
	6	0.826	1.020	0.810
	41	0.954	1.154	0.827
	45	0.704	0.800	0.879
Jan. 30	68	0.648	0.973	0.666
	76	1.318	0.934	1.411
	77	1.202	0.910	1.321
	123	0.478	1.159	0.412

recovered from filters previously extracted with methanol by further treatment with benzene or a mix of benzene, hexane, and isopropanol (unpublished observations).

D. Fractionation

A number of methods have been described for the fractionation of the organic compounds in the particulate extract into various subfractions (Commoner et al., 1978; Hughes et al., 1980). Figure 2 shows a protocol we have developed using mild conditions for separating

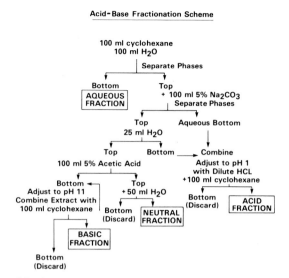

Fig. 2. Protocol for fractionation of crude methanol extracts of collected particulates.

such extracts by liquid–liquid partitioning. The technique produces four fractions: water soluble compounds plus acidic, basic, and neutral substances which partition into the organic solvent. When a clean filter, spiked with [^{3}H]B[a]P was extracted using the procedure outlined in Fig. 2, 95% of the activity was found in the neutral fraction as would be expected (see Table VI).

An alternative approach to liquid–liquid partitioning is the use of preparative reverse-phase high-performance liquid chromatography (HPLC) to separate complex mixtures. Injection of crude methanol extracts onto a preparative column (Whatman RP18 Magnum 9) and elution with a methanol gradient results in successive elution of polar, intermediate, and nonpolar constituents. Extracts eluted with a water:50–100% methanol gradient were fractionated, evaporated to dryness, and then redissolved in DMSO. The concentrated eluates were subsequently tested for their ability to revert TA98 to histidine prototrophy with or without S9 activation.

III. RESULTS

Preliminary experiments indicated that foundry air particulates were a rich source of mutagens. Our primary concern with respect to the particulates collected was the type of mutagenic activity which could

TABLE II

Comparison of Response of TA98 and TA100 to Extracts of Foundry Air Particulate Samples

Sample no.	TA98		TA100	
	− S9	+ S9	− S9	+ S9
70	212[a]	477	118	125
71	181	358	95	120
72	206	192	105	103
73	313	196	100	106
74	181	525	103	115
75	750	681	125	148
76	369	265	105	106
79	350	581	107	138
82	219	250	93	110
83	769	665	109	130
85	219	423	97	111
86	163	250	90	94
87	163	150	88	79
88	263	219	100	108
91	169	227	107	113

[a] Because the spontaneous reversion rate of strain TA98 is about 15–20 per plate and that of TA100 is about 150 per plate, it is not appropriate to compare the response of these strains to chemicals by simply examining the absolute number of chemically induced revertants. A much more appropriate basis for comparison is the percentage increase in revertants over the background controls. (Thus, the result for a sample that caused a fourfold increase in revertant frequency is 400%.)

be detected. Table II shows accumulated data from a number of filters which represent independent foundry air samples from various locations. All extracts were from particulates collected on glass fiber filters after a 2-hour period in a high volume sampler. The important points shown in Table II are, first, that strain TA100 is much less sensitive to reversion by particulate extracts compared with TA98. Even microsomal activation (+ S9) did not detect significant levels of promutagens of the class that revert TA100 (single base substitutions). However, TA98 detected considerable mutagenic activity in nearly all extracts with or without activation. Second, Table II shows wide variation in mutagenic activity between samples collected on separate occasions and at different locations. This point will be dealt with in detail later. These observations led to a general concentration on the activities of extracts with strain TA98, although the use of TA100, especially at later stages of fractionation was not ignored.

A second concern originally raised by Gibson and co-workers (1977) was the possibility that some of the mutagens could be generated from pyrolysis of organic binders used in the mold making process. To see whether the mold constituents themselves contained mutagens before exposure to the hot metal, we performed assays on these materials using strains TA98 and TA1535. The materials tested in Table III gave no significant mutagenicity in any of the starting components. Recycled molding sand (recovered from used molds) showed a level of mutagenicity corresponding to only 1% (weight/weight) of that recovered from foundry air particulates. Thus it appears that a substantial portion of the mutagenic material generated comes from the direct contact of the molten metal and molding sand during the casting process and is largely contained in fumes released. Some pyrolysis products (fine particles) that remain on the casting and which are removed during shake-out and finishing are not recycled and have not yet been tested for mutagenicity.

It was previously noted that air samples collected at different locations showed unequal mutagenicity (Table II). Figure 3 is a map showing a number of collecting stations located at points throughout

TABLE III

A Partial Listing of Binder Materials Tested for Mutagenicity with TA98 and TA1535

System sand	
before cleaning[a]	0.1–150 mg/plate[b]
after cleaning[a]	0.1–40 mg/plate[b]
Face tex mold binder (pure grain product)[a]	1–10 mg/plate[b]
Cell flow (cellulose: ground oat hulls)[a]	1–10 mg[b]
Presstite Loron (46% aromatic, 53% asbestos)	[c]
Mika coat parting agent (petroleum naphtha, mica)	[c]
Glutrin core binder (leached cellulose resin; a pulp and paper by-product)	[c]
Lignosol calcium lignosulfonate	[c]
Kold Set Oil (vegetable oil and petroleum distillate)	[c]
Thiem 2372 core oil (95% aliphatic hydrocarbon, 5% aromatic hydrocarbon, 0.018% iron)	[c]
Resin BR 58782 shell core (phenol, formaldehyde, denatured alcohol)	[c]

[a] 305 gm of raw material was extracted 16 hours with methanol in a Soxhlet apparatus.

[b] Expressed in equivalents of material originally extracted.

[c] These materials were tested directly without being previously extracted with methanol. Some toxicity but no mutagenesis was found.

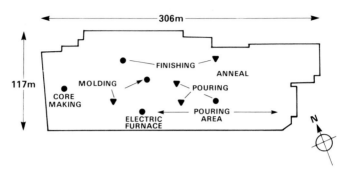

Fig. 3. A map of the DOFASCO steel foundry showing either crane (filled triangles) or floor (filled circles) locations of high volume samplers used for collection of particulates. The dimensions shown are approximate. The average height of the building is 12.75 m.

the foundry. It is important to note that the various work sites are not separated by walls so data gathered from any one station could represent activities within a fairly wide radius. The locations were chosen to correspond generally to the work areas identified by Gibson and co-workers (1977) in their epidemiological study. Several samples were collected at each location and particulate extracts were tested for mutagenicity in TA98 with and without activation by an S9 preparation. The results given in Table IV show a general parallel between areas of high lung cancer risk and areas of high mutagenic activity. The samples which produced the most revertants per cubic meter of air were from crane areas, particularly those over the core-making and molding sites. Samples collected at similar locations but from ground floor level rather than on crane runways showed less mutagenic activity. The mutagenic activity in the finishing and furnace areas was elevated over that found just outside the building but they were considerably below other areas within the foundry.

The day to day variations in mutagenicity of samples collected at the same site are impressive. Table V shows the total number of revertants observed after extraction of filters from the sites indicated. It is clear that not only do the sites vary qualitatively but also that from one day to the next the quantity of material extracted from the same site fluctuates considerably. Perhaps the relatively short length of the sampling period (2–3 hours) could account for some of this variation. Because the particulates collected represent only a segment of those present on a given day, one must assume that the differences are largely due to the discontinuous batch type nature of the steel casting

TABLE IV

Average Mutagenicity Values for Various Foundry Sites

Site	Weight (mg/m³)	Revertants/m³	
		(−)S9	(+)S9
Crane molding	4.5	99	157
	4.8	40	156
	6.9	9.4	48
Floor molding	2.3	38	87
Crane core making	1.4	73	205
Floor core making	4.9	17	55
Crane pouring No. 1	3.7	34	138
	3.6	27	95
Crane pouring No. 2	2.4	7.1	71.5
	3.0	5	26
Crane pouring No. 3	2.2	13.5	55
Crane finishing No. 1	2.5	9	33
Crane finishing No. 2	3.6	6.6	43
	7.0	1.3	30
Floor finishing	4.5	21	22
Floor electric furnace	1.5	7.3	27
	1.2	8.8	28
	2.7	8.7	45
	2.4	3.2	25
Outside[a]	0.076	5.0	7.0

[a] Environmental sample 24 hours.

process. Although we have noted some differences in particulate concentration and mutagenic activity on days when atmospheric conditions might be expected to cause significant effect (i.e., thermal inversions), no clear pattern has yet emerged. Any analysis of particulate associated organics must take these sources of variation into account, because it is through repeated monitoring that significant trends become clear. An analysis of variance shows that differences between the sites described in Table IV are statistically significant (details to be published elsewhere).

The most encouraging point to be made from the results described above is that even at this relatively crude level of analysis it is possible to discriminate among local environments. We believe the *Salmonella* assays can in this, and similar industrial settings, be used to determine which work sites pose the greatest risk. We have noted that the samples assayed (Table V) contain both direct- and indirect-acting mutagens

TABLE V

Variability of Mutagenicity Found in Particulate Extracts from Samples Collected over a 1-Week Period (TA98)

Date	mg[a]	Activation −	Activation +	mg	Activation −	Activation +
Crane (No. 3 sand system)				Bench molding		
26/11/79	1040	30160	43888	590	0	2360
27/11	980	33908	80360	480	9408	26016
28/11	1500	22100	25750	390	9060	19055
29/11	1020	17340	20400	520	12500	22000
30/11	650	21400	29400	470	7400	16900
3/12	630	9009	17199	300	2200	6600
West core room				Chip shop		
26/11/79	1400	0	3360	1320	400	0
27/11	1990	8458	19700	1360	11424	11424
28/11	1490	4605	19875	1140	7752	10944
29/11	1310	4600	22200	740	3922	4400
30/11	300	5200	6200	860	2000	2600
3/12	510	1800	7000	1370	3014	1600
Electric furnace				Environmental[b]		
26/11/79	170	0	0	110	660	660
27/11	450	2700	7605	220	22600	56400
28/11	280	2600	11600	170	8600	8600
29/11	380	1406	3610	210	3800	7800
30/11	330	2000	7800	50	2400	2600
3/12	240	1008	3000	180	9800	7000

[a] Collection times varied between 2 and 3 hours.

[b] Environmental sample collection times: 24 hours except 30/11 (8 hours).

and that the proportions of each varies from day to day and perhaps from site to site. Another significant point is that the levels of mutagenicity found throughout the foundry are high not only when compared to urban air samples from the Hamilton area (Kaiser *et al.*, 1980, 1981) but also compared to those from other urban centers (Dehnen *et al.*, 1977; Møller and Alfheim, 1980).

The observation that different ratios of direct- and indirect-acting mutagens were present in various samples provided the first indication that particulate extracts varied in composition. Table VI confirms this by showing examples of two fractionated particulate extracts from samples collected on successive days. A majority of the mutagenic material from filter No. 56 was found in the neutral organic portion with the remaining activity being equally distributed between aqueous and acidic fractions. By contrast, the material collected on filter No.

TABLE VI

Liquid–Liquid Fractionation of Foundry Air Particulates Collected on Successive Days

Fraction	Total revertants				[^{3}H]B[a]P (cpm)
	Filter No. 56		Filter No. 58		
	− S9	+ S9	− S9	+ S9	
Aqueous	485	660	1,380	7,490	1,260
Acidic	635	500	440	590	1,540
Basic	0	80	80	5,960	1,720
Neutral	1,560	10,900	600	3,680	96,400
Total	2,680	12,140	2,500	17,720	
Recovery (%)	30	100	40	25	102

58 when extracted in the same way, had a majority of its mutagenic activity in the aqueous and basic fractions. The last column in Table VI simply illustrates that [^{3}H]B[a]P can be quantitatively recovered from a filter by the fractionation processes.

The results of Table VI are both interesting and somewhat disquieting. First, they show clearly that there are great variations in the types of mutagenic material associated with the foundry environment. Second, they also show that recovery of mutagenic activity in the fractionation process is often less than 100%. We have fractionated a total of 13 filters collected on separate occasions at different locations. The top panel of Fig. 4 shows the mutagenicity of crude extracts both with and without activation by S9. The lower panel of Fig. 4 shows the percentage recovery of this activity when the total mutagenic activity of the various fractions are summed and compared to the original mutagenicity of the crude particulate extract. With two exceptions, the recovery of direct acting material was less than 75%, and varied between 10 and 74%. Recoveries of mutagenicity seen after S9 activation were generally better, with five filters yielding 100% or greater recovery. Variability, however, remained a problem as reflected in recoveries of, in one case, only 10%. There is a possibility that synergistic effects among both direct acting and promutagens found in crude extracts are eliminated by fractionation. Alternatively, these altered activities may be due to selective losses of material in the acid/base treatment used in the fractionation protocol. Attempts to reconstitute the crude extracts from their subcomponents have given equiv-

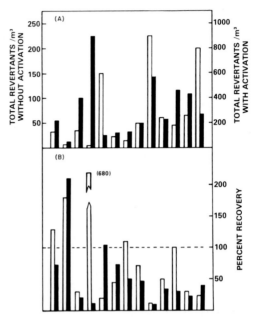

Fig. 4. Mutagenic activity recovered from separate air samples before (a) and after (b) fractionation according to the scheme shown in Fig. 2. (a) The mutagenicity of extracts before fractionation. Solid bars show revertants induced in TA98 by particulate extracts with no activation; open bars show mutagenicity after activation by S9. (Note the differences in scale for revertants induced with or without activation.) (b) The percent recovery of original mutagenicity when the sum of the total revertants induced by each portion of the fractionated extract (acid, basic, and neutral solvent and aqueous soluble) is compared to the original mutagenicity shown in (a).

ocal results (usually less than 100% of organic activity; results to be published elsewhere).

An alternative approach to liquid–liquid partitioning of extracts was the use of preparative reverse-phase HPLC. Figure 5 shows a pair of mutachromatograms generated from the fractionation of two separate crude extracts. The top panel of the figure shows results of a molding area crane sample, while the middle panel shows mutagens found in a dust sample collected from a catwalk in the same vicinity. For reference, the bottom panel shows the elution pattern of some PAH standards. It is immediately apparent that in neither sample is activity restricted to any particular area of the chromatogram. In the crane air sample (Fig. 5a) a large portion of the more polar (early eluting) mutagenic material was direct acting, while the less polar material gave significantly more revertants when S9 was included in the incubation mix. A broadly similar pattern was observed with the dust sample (Fig. 5b) but this material also contained a significant amount of very

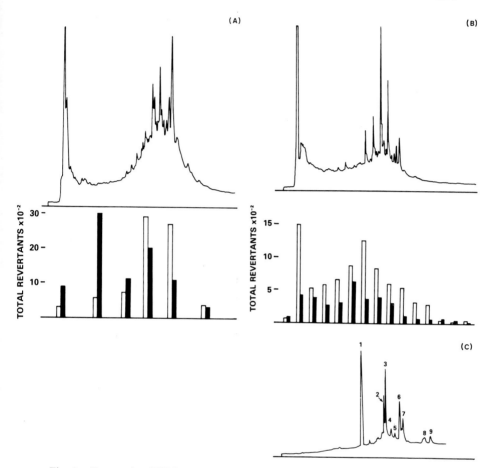

Fig. 5. Preparative HPLC separation of particulate extracts to produce mutachromatograms. Extract (250 μl) was injected onto a preparative column (Whatman M9, RP18) equilibrated with 50% methanol in water, and connected to a Beckman Model 332 liquid chromatograph system (Beckman Instruments, Berkeley, Calif.), a spectrophotometer (Hitachi Model 100-10), and a plotter (Altex C-RIA). A gradient of 50–100% methanol in water pumped at 4 ml/minute over a period of 50 minutes fractionated the extract. The upper portions of (a) and (b) show the absorbance profiles (254 nm) while the histograms indicate mutagenicity with TA 98 recovered from fractions evaporated and dissolved in DMSO. Panel (a) represents particulates collected at a crane located in the molding area. Each fraction was tested with (open bars) or without (solid bars) activation by S9. Panel (b) represents mutagenicity of a dust sample collected in the same area. Panel (c) records the elution pattern of a set of standard compounds: (1) toluene, (2) phenanthrene, (3) anthracene, (4) fluoranthene, (5) pyrene, (6) triphenylene, (7) benzanthracene, (8) perylene, and (9) benzo[a]pyrene.

polar mutagens which required activation. With the exception of the very early component, the dust sample extract contained direct-acting material which eluted early in the water:methanol gradient, while the later material eluted (nonpolar) was more mutagenic after treatment with S9. In both extracts, the profiles of absorbance at 254 nm give some indication of the complexity of the mixtures. There is reason to hope that repeated fractionation procedures using some combination of liquid–liquid partitioning and HPLC, followed by GC and MS, will lead to identification of some of the many components present. Clearly PAH can account for only a part of the total mutagenicity; such compounds are neutral organics requiring activation. The characterization of such mutagenic compounds as are found in the aqueous, basic, and acidic fractions shown in Table VI should provide clues to their origin. Such identification could lead to changes in the materials used in the foundry process and thus eliminate the most hazardous components. The work of Hites *et al.* (1981), who identified a large number of the compounds found in the extracts of diesel soot, provides a good example of what can be achieved. Their results show that a preponderance of the aromatic compounds isolated could be derived from a single source: phenanthrene. Schimberg (1981) found that the removal of pitch used as a binder in molds in an iron foundry resulted in an appreciable drop in B[a]P associated with collected particulates. The correlation of B[a]P concentration in the air samples collected at two Finnish iron foundries (Skytta *et al.*, 1980) and mutagenicity of particulate extracts toward TA98 and TA100 were somewhat disappointing. Of two sites studied, one gave good results while the samples collected at a second foundry showed a poor correlation between B[a]P concentration and mutagenicity of extracts.

IV. CONCLUSIONS AND DISCUSSION

A. Mutagenicity in the Steel Foundry Environment

The steel foundry study which our laboratory has embarked on has the merit of having a well-defined epidemiological basis. This has enabled us to evaluate the locations associated with highest risk for development of lung cancer among foundry workers for biological activity using the *Salmonella* test for mutagenicity. The correlation between high risk areas and mutagenic potency of particulate extracts collected at these locations is reasonably good. It is important to note limitations inherent in this type of study when considering risk. First, although foundry activities at DOFASCO have been located in the same area for many years, it is possible that conditions and/or processes have evolved or changed so that measurements made today do not necessarily reflect conditions which previously induced lung cancer. A continuation of the epidemiology is essential. A second point that merits consideration is the problem of mixtures. From what is known from work with pure compounds there is no reason to assume that mutagenic potency will always or even generally correlate with carcinogenic potency (Rinkus and Legator, 1979). The myriad of individual mutagens detectable in such a mixed environment as a steel foundry present considerable difficulties for assessing the genotoxic burden of respirable particulates. Finally, distribution of mutagens on particles of differing size may influence risk analysis. Talcott and Harger (1980) have shown that particles in the respirable fraction (less than 1.7 μm) to be consistently higher in mutagenic activity than fractions containing larger particulates. In our study we have observed that extracts of particulates of inhalable size (under 15 μm) were generally more active in the *Salmonella* test (weight for weight of particulates collected) than extracts of total particles.

We have taken considerable care, particularly with respect to collection methods and recovery of material, to ensure our procedures are valid. We have carried out rudimentary fractionation of particulate crude extracts into neutral, polar organic and aqueous moieties. The results of our mutagenicity assays are not unlike those reported by workers in air pollution and diesel particulate analyses. Much of the activity observed is direct acting and found among the more polar compounds (aqueous and basic fractions). More sophisticated fractionation of particulate extracts with reverse-phase HPLC has further distinguished mutagenic activity of foundry samples on the basis of polarity. The mutagens collected on the filters are clearly the result of the foundry activity, since we have been unable to detect significant mutagenic activity in the untreated or recycled starting materials involved in the manufacture of castings. The combustion products are obviously complex as revealed by the HPLC tracings shown in Fig. 5.

Two major obstacles encountered in the course of this study are shared to some degree with others involved in air sampling. The first is day to day variation which we presume to be related to the type of activity at the time of collection and possibly the weather, although the effect of the latter is not yet clear. The second is the great complexity of the organics bound to particulates as revealed by chromatography. These complex mixtures are difficult to analyze by conventional fractionation since reconstitution assays do not regenerate the original mutagenic activity. Perhaps synergistic activation present in crude extract is lost during the fractionation procedure. Kaden and co-workers (1979) reconstructed mutagenic activities of components in PAH isolated from kerosene soot extracts and were able to show no synergistic activity. However Commoner *et al.* (1978) showed increased activity following fractionation. It is possible that "losses" of mutagenic activity which we observed were related to components other than those found in the neutral fractions. The regeneration of original mutagenicity from the fractionated materials should be an important goal in the characterization of environmental carcinogens isolated from the foundry air samples.

Preparative HPLC provides a useful tool for partially purifying particulate extracts for more sensitive gas chromatography (GC) and mass spectrometry (MS). The correlation of high mutagenic activity detected in fractions collected from chromatography with major components identified by GC/MS techniques will be a major thrust of future investigations. Once the substances which pose the greatest hazard are identified it may be possible to develop alternative foundry procedures which could at least reduce the genotoxic effects in local environments.

B. Mutagenicity of Urban Air Pollutants

The possibility of characterizing complex environmental mixtures has been recognized by a number of investigators and applied to a variety of situations. While some of these have led to the successful identification of the components responsible for mutagenic activity, a majority of investigators have been satisfied with characterization of classes of mutagenic compounds.

The most ambitious efforts have been directed at the urban environment. Commoner *et al.* (1978) attempted to analyze the local environment at the Washington School located in a highly industrialized neighborhood of Chicago. They fractionated air particulate samples and showed that much of the neutral organic extracted material required microsomal activation. Furthermore, the mutagenic material isolated from thin-layer chromatograms of benzene, hexane, isopropanol extracts had UV and mass spectra which were similar to B[a]P. They concluded that the existence of coke ovens and steel works as well as other heavy industrial activity in the vicinity of the collection location contributed a major portion of the extracts' mutagenicity. Dehnen and co-workers (1977), on the other hand, performed a similar analysis on air particulates collected in a heavily industrialized West German town and found little correlation between mutagenic activity and PAH content. The most active fractions contained polar compounds which partitioned into methanol but not cyclohexane. Much of this material was direct acting. It has been pointed out already that, over long sampling times, B[a]P is oxidized to other products, particularly quinones and acids (Peters and Seifert, 1980). Löforth (1978) also found that combustion of wood and peat yielded particulates rich in polycyclic organic material (POM) but not PAH. These nitrated, sulfonated, and oxygenated polycyclic compounds did not require activation to induce revertants in the Ames test.

Hughes and co-workers (1980) have taken great care to establish procedures for collection, extraction, and fractionation of air pollution samples. They reported activities in acidic and basic as well as PAH fractions of total air extracts. Among the PAH identified by gas chromatography and by mass spectra were isomers of anthracenes, benzo[a]pyrene, benz[a]anthracene, perylenes, as well as fluorene, pyrene, phenylnaphthalamine, and naphthalene. Acidic compounds identified were phenol, creosol, and benzenes; while bases included quinolines and acridines. The nonpolar neutral and polar neutral organics were more mutagenic with microsomal activation, while the acidic and basic fractions were direct acting.

Møller and Alfheim (1980) have reported that air samples collected in Oslo over a 3-month period varied in PAH content. This variability was ascribed partially to the meteorological conditions but also to changes in heating requirements as winter weather ended. They found considerable direct-acting material, though mutagens requiring activation were significantly predominant. No conclusions could be drawn regarding the relation between PAH content and variation in total mutagenic activity. Wang and associated researchers (1980), on the other hand, have found that air extracts collected in and around the city of Detroit in the United States contained significant mutagenicity unrelated to neutral organic material. They showed further that nitroaromatic derivatives probably account for part of the direct-acting material. The rate of reversion to histidine prototrophy induced by particulate extracts in a nitroreductase-deficient strain of TA98 was reduced compared to the parent. These authors suggested that this reduction of mutagenic activity could be attributed to the absence of bacterial nitroreductase needed for activation. They also postulated that the majority of direct-acting nitrocompounds detected were derivatives of B[a]P or perylene formed directly by exposure of the PAH to atmospheric concentrations of nitrogen oxides or nitric acid (Pitts et al., 1978). Recently, McCoy and Rosenkranz (1980) suggested that additional direct-acting or ultimate mutagens found in atmospheric pollutants could arise from photoconversion of promutagens by reaction of a triplet state PAH with singlet oxygen. In the laboratory these investigators were able to convert 3-methylcholanthrene and chrysene to direct-acting mutagens by treatment with light in the presence of a photodynamically activated singlet oxygen generating system. The increase in mutagenicity of these two promutagens, which normally require activation by S9, was proportional to the amount of illumination of the generating system.

C. Mutagenicity of Diesel Exhaust Particulates

The production of activated species in the environment might arise from the interaction of light with increasing levels of oxides of nitrogen in the atmosphere produced by gasoline and diesel exhausts (Pitts, 1979). In addition to industrial pollutants in urban environments, this second large source of potentially hazardous emissions has been the focus of several recent investigations. The internal combustion engine and particularly the diesel have been found to produce large amounts of incomplete combustion products and particulates. Secondary processes in the environment such as those reported by Pitts (1979) and McCoy and Rosenkranz (1980) may indeed be important. Lee and co-workers (1980) have found that direct inhalation of diesel emissions resulted in poor induction of arylhydrogen hydroxylase (AAH) or expoxide hydroxylase (EH) in rats. Animals exposed up to 42 days (20 hours/day) to diesel emissions showed only moderate induction of AAH in liver, lung, and prostate, and no significant induction of EH.

There can be no doubt that a considerable decrease in mutagenic emissions from gasoline engines has occurred in recent years. Ohnishi and co-workers (1980) compared particulates from engines with or without emission controls. They found that compared with uncontrolled engines, exhaust particulates from regulated engines had a ninefold reduction in mutagenicity with TA98 but only a threefold reduction with TA100. The mutagenicity of particulates collected from diesel engine exhaust (commercial and automobile) is considerably higher than those of gasoline powered vehicles (Ohnishi et al., 1980) in terms of revertants/m^3. Air samples collected in a highway tunnel during periods of heavy truck traffic showed increased mutagenicity of extracted particulates compared with rural samples. Fractionated extracts of such particulates showed that most of the activity requiring activation resided in the neutral fraction, but the largest yield of mutants was found in the polar organics (ether soluble acids and bases) and required no activation. Huisingh and others (1978) also showed that source collected particulates of diesel engines yielded a majority of mutagenic activity in the polar organics (substituted PAH). This material largely did not require activation by microsomes to induce revertants in either TA98 or TA100. Curiously, this situation was reversed with TA1535 and TA1538, the less sensitive *Salmonella* strains for detection of substitution and frame-shift mutations. Huisingh and associates (1978) explain this result as being the consequence of the greater sensitivity of the strains carrying the plasmid to direct-acting mutagens in general.

Hites *et al.* (1981) have examined organic compounds associated with diesel soot collected on glass fiber filters and then extracted with methylene chloride. Most mutagenic activity was eluted from silicic acid chromatographic columns with either hexane/toluene (1:1) or toluene. The PAH fraction (hexane/toluene) contained a large number of compounds, 62 of which were identified by gas chromatography. Only low levels of B[a]P were detected, but significant amounts of alkylphenanthrenes and alkylfluorenes were observed. Although generally less potent mutagens than B[a]P, these compounds probably represent the major biologically active components among the PAH detected. Oxygenated aromatics were partitioned from the silicic acid by elution with toluene. The authors observed some 80 different species in this fraction, including substantial amounts of aromatic ketones and aldehydes. Generally the biological activity of these products of combustion include toxicity as well as weak mutagenicity. Hites *et al.* (1981) have provided a model (Fig. 6) by which a majority of the products observed may be derived from phenanthrenes.

In the case of diesel and gasoline engine emissions, the characterization and analysis of the chemical pollutants far exceeds the quantitative estimation of the health hazard which exposure to these agents imposes on humans. In a recent review of epidemiologic data, Schenker (1980) concluded that the information on carcinogenicity of diesel exhaust is limited and inconsistent. Studies that examined railroad workers, transport workers, and miners, all of whom were occupationally exposed to diesel emissions, indicated little or no increased risk for lung cancer. In general, however, the investigations cited by Schenker (1980) were inadequately prepared and had no associated in-depth chemical or biological analyses of the working environment.

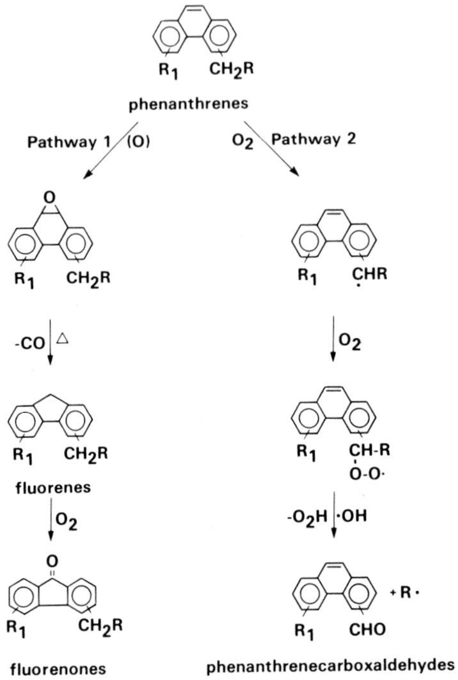

Fig. 6. A mechanism for the formation of fluorenes, fluorenones, and phenanthenecarboxyaldehydes from phenanthrenes. (From Hites *et al.*, 1981, with permission.)

Wei *et al.* (1980) have provided a thoughtful analysis of results obtained with diesel emission condensates and extracts used in the Ames reversion test. These authors note that the mutagenicity detected may be more a product of the collection process than of combustion. They found that crude or HPLC fractionated extracts of filter-bound or electrostatically precipitated material was mutagenic and nontoxic to TA98 and required no S9 activation. Although unsubstituted, PAH have been reported frequently in diesel exhaust (Huisingh *et al.*, 1978; Ohnishi *et al.*, 1980; Hites *et al.*, 1981) and have been used to characterize the levels of diesel emissions in the workplace (Schenker, 1980). Wei and associates (1980) point out that these compounds are a minor component of the mutagenic activity detected in the *Salmonella* assay. In order to cause a significant number of revertants from the B[*a*]P in particulates of diesel emissions, Wei and others (1980) showed that engines would have to produce at least ten times the amount of unsubstituted PAH found in the particulates from the worst case examined by the United States Environmental Protection

Agency. Thus, the direct-acting material probably represents the most serious genetic hazard in these emissions, and the identification and genotoxic characterization of these compounds merit much further investigation.

The evaluation of the hazards to human health posed by air pollutants in occupational and urban environments is obviously a complex problem. It is now relatively easy to determine chemically the quantities of many different compounds found in environmental samples, even when these substances are present at extremely low levels. It has become even less difficult to detect mutagenic activity of these substances using sensitive bioassays. What has become a problem of increasing dimension is the assessment of the probable impact of such materials on human health. Perhaps the information needed to interpret the results of bioassays will come from the detailed study of industrial and other environments in which the human risk is known from epidemiological studies. Studies such as our own, which are designed to combine quantitative chemical analysis with a biological evaluation of the potency of environmental pollutants, will provide essential data that ultimately should allow reasonable predictions of risk.

ACKNOWLEDGMENTS

The foundry study referred to in this chapter is being conducted in collaboration with DOFASCO Inc., Hamilton, Ontario, with financial support from a Wintario grant from the Ontario Ministry of Labour. We are especially indebted to Dr. E. S. Gibson, Mr. N. Lockington, and Mr. A. Kerr of DOFASCO for sampling and for invaluable discussions. Experimental work at McMaster was carried out by Ms C. Kaiser and Mrs. C. Lu and Dr. Clyde Herzman has helped with the statistical analysis.

REFERENCES

Ames, B. (1979). Identifying environmental chemicals causing mutation and cancer. *Science* **204**, 587–593.

Ames, B. N., McCann, J., and Yamasaki, E. (1975). Methods for detecting carcinogens and mutagens with the Salmonella mammalian microsome mutagenicity test. *Mutat. Res.* **31**, 247–364.

Commoner, B., Madyastha, P., Bronsdon, A., and Vithayathil, A. J. (1978). Environmental mutagens in urban air particulates. *J. Toxicol. Environ. Health* **4**, 59–77.

Dehnen, W., Pitz, N., and Tomingas, R. (1977). The mutagenicity of airborne particulate pollutants. *Cancer Lett.* (*Shannon, Irel.*) **4**, 5–12.

Falck, K., Sorsa, M., and Vainio, H. (1980). Mutagenicity in urine of workers in rubber industry. *Mutat. Res.* **79**, 45–52.

Gibson, E. S., Martin, R. H., and Lockington, J. N. (1977). Lung cancer mortality in a steel foundry. *J. Occup. Med.* **19**, 807–812.

Hites, R. A., Yu, M. L., and Thilly, W. G. (1981). Compounds associated with diesel exhaust particulates. *Polynucl. Aromat. Hydrocarbons Int. Symp. Chem. Biol. Effects, 5th Ohio* 455–466.

Hopkin, J. M., and Evans, H. J. (1980). Cigarette smoke-induced DNA damage and lung cancer risks. *Nature (London)* **283**, 388–390.

Hughes, T. J., Pellizzari, E., Little, L., Sparacino, C., and Kolber, A. (1980). Ambient air pollutants: Collection, chemical characterization and mutagenicity testing. *Mutat. Res.* **76**, 51–83.

Huisingh, J. L., Bradow, R., Jungers, R., Claxton, L., Zweidinger, R., Tejada, S., Bumgarner, J., Duffield, F., Waters, M., Simmon, V. F., Hare, C., Rodriguez, C., and Snow, L. (1978). Application of bioassay to the characterization of Diesel particle emissions. *In* "Application of Short-Term Bioassays in the Fractionation and Analysis of Complex Environmental Mixtures, EPA-600/9-78-027" (M. D. Waters, S. Nesnow, J. L. Huisingh, S. S. Sandhu, and L. Claxton, eds.), pp. 1–32.

Huisingh, J. L., and Claxton, L. (1980). Comparative mutagenic activity of particle-bound organics from combustion sources. *Env. Mutagen* **2**, 310.

Kaden, D. A., Hites, R. A., and Thilly, W. G. (1979). Mutagenicity of soot and associated polycyclic hydrocarbons to *Salmonella typhimurium*. *Cancer Res.* **39**, 4152–4159.

Kaiser, C., Kerr, A., McCalla, D. R., Lockington, J. N., and Gibson, E. S. (1980). Mutagenic material in air particles in a steel foundry. *Polynucl. Aromat. Hydrocarbons Int. Symp. Chem. Biol. Effects, 4th, Ohio* pp. 579–588.

Kaiser, C., Kerr, A., McCalla, D. R., Lockington, J. N., and Gibson, E. S. (1981). Use of bacterial mutagenicity assays to probe steel foundry lung cancer hazard. *Polynucl. Aromat. Hydrocarbons Int. Symp. Chem. Biol. Effects, 5th, Ohio* 583–592.

Lee, I. P., Suzuki, K., Lee, S. D., and Dixon, R. L. (1980). Aryl hydrocarbon hydroxylase induction in rat lung, liver, and male reproductive organs following inhalation exposure to diesel emission. *Toxicol. Appl. Pharmacol.* **52**, 181–184.

Löfroth, G. (1978). Mutagenicity assay of combustion emissions. *Chemosphere* **10**, 791–798.

McCoy, E. C., and Rosenkranz, H. S. (1980). Activation of polycyclic aromatic hydrocarbons to mutagens by singlet oxygen: An enhancing effect of atmospheric pollutants? *Cancer Lett. (Shannon, Irel.)* **9**, 35–42.

Møller, M., and Alfheim, I. (1980). Mutagenicity and PAH-analysis of airborne particulate matter. *Atmos. Environ.* **14**, 83–88.

Møller, M., and Dybing, E. (1980). Mutagenicity studies with urine concentrates from coke plant workers. *Scand. J. Environ. Health* **6**, 216–220.

Natusch, D., and Wallace, J. (1974). Urban aerosol toxicity: The influence of particle size. *Science* **186**, 695–699.

Ohnishi, Y., Kachi, K., Sato, K., Tahara, I., Takeyoshi, H., and Tokiwa, H. (1980). Detection of mutagenic activity in automobile exhaust. *Mutat. Res.* **77**, 229–240.

Peters, J., and Seifert, B. (1980). Losses of benzo[a]pyrene under the conditions of high-volume sampling. *Atmos. Environ.* **14**, 117–119.

Pitts, J. N. (1979). Singlet molecular oxygen and photochemistry of urban atmospheres. *Ann. N. Y. Acad. Sci.* **171**, 239–272.

Pitts, J. N., Grosjean, D., and Mischke, T. M. (1977). Mutagenic activity of airborne particulate organic pollutants. *Toxicol. Lett.* **1**, 65–70.

Pitts, J. N., Jr., Van Cauwenberghe, K. A., Grosjean, D., Schmid, J. P., Fitz, D. R., Belser, W. L., Kundson, G. B., and Hynds, P. M. (1978). Atmospheric reactions of polycyclic aromatic hydrocarbons: Facile formation of mutagenic nitro-derivatives. *Science* **202**, 515–519.

Pitts, J. N., Van Cauwenberghe, K. E., Grosjean, D., Schmid, J. P., Fitz, D. R., Belser, W. L., Kundson, G. B., and Hynds, P. M. (1979). Chemical and microbiological studies of mutagenic pollutants in real and simulated atmospheres. *Environ. Sci. Res.* **15**, 353–379.

Rinkus, S. J., and Legator, M. S. (1979). Chemical characterization of 465 known or suspected carcinogens and their correlation with mutagenic activity in the *Salmonella typhimurium* system. *Cancer Res.* **39**, 3289–3381.

Schenker, M. B. (1980). Diesel exhaust: An occupational carcinogen? *J. Occup. Med.* **22**, 41–46.

Schimberg, R. W. (1981). Industrial hygienic measurements of polycyclic aromatic hydrocarbons in foundries. *Polynucl. Aromat. Hydrocarbons Int. Symp. Chem. Biol. Effects, 5th, Ohio* 755–762.

Skytta, E., Schimberg, R., and Vainio, H. (1980). Mutagenic activity in foundry air. *Arch. Toxicol., Suppl.* **4**, 68–72.

Talcott, R., and Harger, W. (1980). Airborne mutagens extracted from particles of respirable size. *Mutat. Res.* **79**, 177–180.

Tola, S., Koskela, R., Hernberg, S., and Järvinen, E. (1979). Lung cancer mortality among iron foundry workers. *J. Occup. Med.* **21**, 753–760.

Wang, C. Y., Lee, M., King, C. M., and Warner, P. (1980). Evidence for nitroaromatics as direct-acting mutagens of airborne particulates. *Chemosphere* **9**, 83–87.

Wei, E. T., Wang, Y. Y., and Rappaport, S. M. (1980). Diesel emissions and the Ames test: A commentary. *J. Air Pollut. Control Assoc.* **30**, 267–271.

Yamasaki, E., and Ames, B. N. (1977). Concentration of mutagens from urine by adsorption with nonpolar resis XAD-2: Cigarette smokers have mutagenic urine. *Proc. Natl. Acad. Sci. U. S. A.* **74**, 3555–3559.

Chapter 5

The Use of Mutagenicity Testing to Evaluate Food Products

H. F. Stich, M. P. Rosin, C. H. Wu, and W. D. Powrie

I. INTRODUCTION

Considering that approximately 4,000,000 chemicals are known and that at least 60,000 are in common use, the available mutagenicity data on a few thousand of them appear very few and carcinogenicity results on a couple of hundred are shockingly inadequate to formulate any ideas about the genotoxic hazards of man's environment. Even this tiny percentage of tested chemicals required the effort of many research institutes and mutagen screening laboratories worldwide. It is becoming only too evident that reasonable progress in this area can

117

MUTAGENICITY:
NEW HORIZONS IN GENETIC TOXICOLOGY

only be achieved by setting strict priorities for examination of the mutagenicity and carcinogenicity of suspected compounds. To assess the percentage contribution of food mutagens to the total mutagenic load of the environment, we have concentrated on their formation during food processing, storage, and cooking. Particular emphasis was placed on those food items in which nonenzymatic or enzymatic browning reactions occurred. This choice was based on the observation that products of browning reactions are very common among a large variety of heated and dried food products and beverages which are consumed in relatively large quantities by virtually all members of a population in a Western nation. Even such a restriction to one aspect of man's environment can only reduce the momentous task of analyses. Today's markets may offer over 10,000 food items which are boiled, fried, roasted, and simmered in uncountable ways, all of which could enhance or reduce their mutagenic capacity.

II. METHODOLOGY

Undoubtedly the introduction of the *Salmonella typhimurium* mutagenicity test by Ames *et al.* (1975) had a major impact in uncovering mutagens in the environment. This simple, reliable, and reproducible method permitted for the first time the screening of a larger number of compounds to which human population groups are exposed daily. In the meantime many other short-term tests for mutagens, and by implication carcinogens, were proposed and partly validated (Stich and San, 1981). Some of these *in vitro* bioassays only duplicate the principle of Ames's *S. typhimurium* test, whereas others are based on different end points, including mitotic recombination, gene conversion, chromosome aberrations, and DNA alterations. This latter group of tests may complement the *S. typhimurium* test which is based on the induction of frameshift and/or base-pair substitution mutation by carcinogens. The use of these assays as a test battery is strongly recommended since at present the genetic lesion(s) involved in neoplastic transformation is unknown. Whether bioassays based on mutations due to frameshift or base-pair substitution or those based on DNA strand breakage are better related to carcinogenesis in man is an unresolved issue (Kinsella and Radman, 1980; Stich *et al.*, 1981c). Thus, all claims that one or the other testing procedure is more relevant in predicting carcinogenesis can be seriously questioned.

Food systems are generally complex colloidal mixtures composed of low-molecular-weight compounds, polymers such as proteins and polysaccharides and particulate matter consisting of aggregated proteins, protein–lipid complexes, and cell fragments. Mutagens in the food systems may be physically absorbed to polymers or may be trapped in the particulate matter. The importance of applying various extraction procedures cannot be underestimated, as seen in Fig. 1. Furthermore, the original foodstuff or each separated fraction can be "spiked" with known amounts of mutagens and their recovery by the various extraction and separation procedures measured. A proper selection of the mutagens/carcinogens used for "spiking" is of vital importance. If these mutagens do not accurately simulate their normal state in a food product, misleading results can easily be obtained. Microbial test organisms can readily become coated with linoleic acid or oleic acid if they are present in the lipophilic extract. Such coated bacteria can become impregnable to many mutagens, including pyrolysis products of amino acids (Hayatsu et al., 1981). However, each test system will produce a few negative results since none of the bioassays can respond to all the various genotoxic effects that a compound can induce (e.g., Stich et al., 1978).

The responses of various test systems to mutagens of various molecular structure can differ because of the peculiarities of the test organism. There are many reasons why an in vitro test can produce a negative result. The mutagen may not penetrate into the test organism. The test organism can carry inactivating enzymes. For example, hydrogen peroxide and H_2O_2-producing mutagens will remain undetected if the test cells are rich in catalase. Thus, any prudent approach in testing unknown chemicals or mixtures of unknown compounds

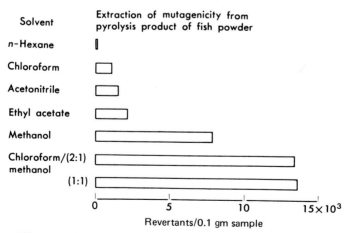

Fig. 1. Efficiency of extraction of mutagens from fish powder pyrolysis products by various solvents. Number of histidine revertants obtained with *Salmonella* strain TA98 in the presence of an S9 liver microsomal preparation (Uyeta et al., 1979).

should include a battery of tests that differ in end point and test organism. To cope with this situation, we chose three short-term tests: (1) the *S. typhimurium* mutagenicity assay as proposed by Ames *et al.* (1975) and modified by Nagao *et al.* (1977a,b, 1979) (Fig. 2 indicates the value of employing a preincubation treatment in studies of the mutagenicity of food products); (2) the gene conversion test on strain D7 of *S. cerevisiae*; and (3) the chromosome aberration test with CHO cells. Any reliance on only one test system will sooner or later lead to erroneous conclusions. The great difference in response to food mutagens can readily be seen from the genotoxic ratios shown in Figs. 3 and 4. Other examples, including the response to reducing agents

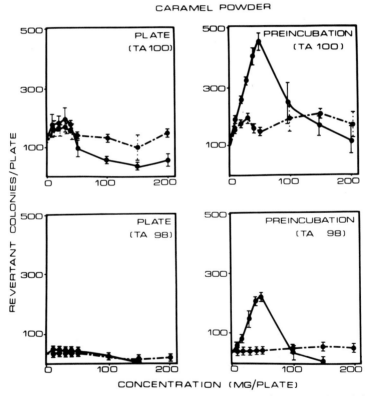

Fig. 2. The influence of a preincubation procedure on the mutagenic activity of a commercial caramel powder in *Salmonella* strains TA100 and TA98. Cultures were treated with either the standard pour plate technique (Ames *et al.*, 1975) or the preincubation modification of this technique (see Fig. 3). Frequencies of reversion to histidine prototrophy ± SD (n = 3) are shown. Solid lines, no S9 mix; dashed lines, S9 mix present.

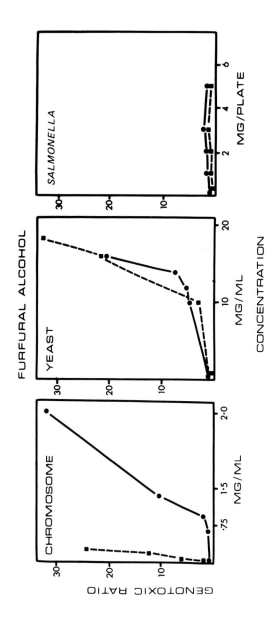

Fig. 3. Genotoxic activity of furfural alcohol as measured by the induction of chromosome aberrations in CHO cells, gene conversion in *Saccharomyces cerevisiae* strain D7 cultures, and reversion to histidine prototrophy in *Salmonella* strain TA100. The *Salmonella* cultures were given a 20-minute preincubation treatment at 37°C to the furfural alcohol in the presence or absence of S9 mixture on minimal agar plates. Data are expressed as a genotoxic ratio (genotoxic frequency of treated cultures/spontaneous genotoxic activity). Genotoxic activity is shown in the presence (dashed lines) and absence (solid lines) of an S9 mix. Maximum genotoxic ratio observed for the *Salmonella* assay was 2.2 (2.5 mg/plate, strain TA100). No increase in the genotoxic frequency was seen with strain TA98 after treatment with furfural alcohol.

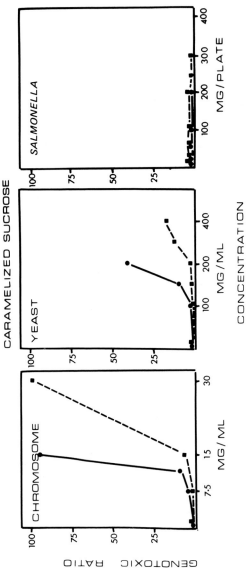

Fig. 4. Genotoxic activity of caramelized sucrose (180°C, 90 minutes) in three test systems. For description see Fig. 3. Maximum genotoxic ratio for *Salmonella* assay was 2.7 (100 mg/plate, strain TA100; strain TA98, no increase in genotoxic ratio).

found in food (Stich *et al.*, 1976, 1978, 1979) and volatile food components such as pyrazines, were previously reported (Stich *et al.*, 1980).

III. MUTAGENS IN NONPROCESSED FOOD

Human populations face an onslaught of mutagenic chemicals which will enter the body through food and beverages. These mutagens can be unintentional contaminants such as mycotoxins, industrial products (e.g., polycyclic aromatic hydrocarbons) or agricultural chemicals (e.g., pesticides), intentionally added food additives, natural components of plants, or products of food processing, cooking, and storage. At present we lack any knowledge about the percentage contribution of each category to the total mutagenic or carcinogenic load of a diet. In the past, screening programs for mutagenicity and carcinogenicity focused on food additives and various food contaminants. The approach was deemed expedient because of the relative ease of regulating hazardous food additives and controlling accidental food contaminations. This attitude contributed to a certain neglect of naturally occurring mutagens and carcinogens. It was generally assumed that only relatively small population segments in exotic countries consume hazardous quantities of carcinogenic food such as cycasin in Polynesia, Kenya, and Kyushu (Laqueur *et al.*, 1963; Laqueur and Matsumoto, 1966), bracken ferns in Japan (Evans and Mason, 1965; Hirono *et al.*, 1973; Pamukcu *et al.*, 1978), pyrrolizidine alkaloids in the Caribbean islands (Schoental, 1976), and betel nuts in India (Hirayama, 1979). This notion must be discarded as false. Each person, regardless of whether they are vegetarian or meat eaters, will daily consume many types of mutagens. A few representatives are shown in Table I. Some of them, such as flavonoids (Hardigree and Epler, 1978; Brown, 1980) and safroles (Miller *et al.*, 1979), are widely distributed throughout the plant kingdom and are consumed by eating fruits, vegetables, and spices. Others, including the furans (Maga, 1979) and pyrazines (Maga and Sizer, 1973), are mainly but by no means exclusively found in heated or dried foods following nonenzymatic or enzymatic browning reactions. The wide distribution of these compounds some of which are mutagenic is exemplified in the case of furans (Table II).

Mutagenicity data of food products should be supplemented by information on human exposure levels and the fate of mutagens following ingestion. Unfortunately, reliable data on these issues are hard to

TABLE I

Examples of Genotoxic Activity of Individual Chemicals Found in Food

Pyrazine[a]	Galangin[e]
2-Methylpyrazine	Kaempferol
2-Ethylpyrazine	Morine
2,5-Dimethylpyrazine	Fisetin
2,6-Dimethylpyrazine	Quercetin
1,5-Dimethyl-2,3,6,7-tetrahydro-1H,5H-	Rhamnetin
biscyclopentapyrazine	Robinetin
	Myricetin
Furan[b]	Tamarixetin
Furfural	
Furfural alcohol	Lucidin[e]
2-Furyl methyl ketone	Purpurin
5-Hydroxymethylfurfural	Ermodin
2,5-Dimethylfuran	Catenarin
5-Methylfurfural	Dichloroanthrarufin
2-Methylfuran	Dinitrochrysazin
Safrole[c]	Thiazolidine[f]
Estragole	2-Methylthiazolidine
β-Asarone	2-Ethylthiazolidine
Anethole	2-n-Propylthiazolidine
Evgenol	2-Isopropylthiazolidine
Myristicin	2-n-Butylthiazolidine
Elemicin	2-Isobutylthiazolidine
3-Amino-1,4-dimethyl-5H-pyrido[4, 3-b]indole (trp-P-1)[d]	Benzo[a]pyrene[g]
3-Amino-1-methyl-5H-pyrido[4, 3-b]indole (Trp-P-2)	Anthracene
2-Amino-6-methyldipyrido[1, 2-a:3′,2′-d]imidazole (Glu-	Benz[a]anthracene
P-1)	Phenanthrene
	Benzo[b]fluoranthene
2-Amino-3-methylimidazo[4,5-f]quinoline (IQ)	Dibenzo[a,h]pyrene
2-Amino-3,4-dimethylimidazo[4, 5-f]quinoline (MeIQ)	Chrysene
	Glyoxal[h]
	Diacetyl
	1,2-Cyclohexanedione
	Maltol
	Ethyl maltol
	Kojic acid

[a] Stich *et al.* (1980).
[b] Stich *et al.* (1981d).
[c] Miller *et al.* (1979).
[d] Sugimura (1979) and Tsuda *et al.* (1980).
[e] Brown (1980).
[f] Mihara and Shibamoto (1980).
[g] Lo and Sandi (1978).
[h] Bjeldanes and Chew (1979).

TABLE II

Food Systems from Which Furan Derivatives Have Been Isolated[a]

Beverages, nonalcoholic	Oil seed products
Cocoa	Rapeseed protein
Coffee	Sesame seed
Tea	Soybeans, deep fat-fried
Beverages, alcoholic	Soybean oil
Beer	Soy sauce
Rum	Soy protein
Wine	Vegetables
Fruits	Asparagus
Arctic bramble	Broccoli
Cloudberry	Cabbage
Cranberry	Cauliflower
Grapes	Celery
Mango	Leek
Orange	Onion
Pineapple	Peppers
Tamarind	Potato products
Meat and poultry products	Baked
Beef broth	Boiled
Beef fat	Chips
Mutton fat	Dehydrated
Boiled beef	Raw
Canned beef	Shallot
Canned beef stew	Tomato
Roast beef	Miscellaneous
Cooked pork liver	Ascorbic acid
Chicken broth	Barley
Cooked chicken	Bread
Eggs	Caramel
Roast turkey	Cod
Milk products	Corn oil
Butter culture	Fish protein concentrate
Dry whole milk	Grape leaf
Nonfat dry milk	Maple syrup
Dry whey	*Mentha* species
Casein	Mushroom
Sodium caseinate	Popcorn
Nut products	Rice
Almond	Rye crisp bread
Filbert	Smoke
Macadamia	Tuna oil
Peanuts	Licorice
Pecans	

[a] From Maga (1979).

find. The following approximation of quantities of mutagenic or carcinogenic food compounds may provide at least some insight into this complex problem. The various methylformylhydrazones occur in quantities varying between 0.2 and 49.9 mg/kg of the false morel, *Gyromitra esculata* (Toth, 1979); 100 ppm of safrole is in black pepper (Concon *et al.*, 1979); several milligrams of ipomeamarone are present in sweet potatoes following damage and infection with the mold *Fusarium solani* (Boyd and Wilson, 1971); and about 1 gm of flavonoids per day is being consumed on the North American continent (Brown, 1980).

IV. MUTAGENS FORMED DURING FOOD PREPARATION AND COOKING

Considerable effort has been exerted to find correlations between cancer frequencies and particular diets. A high fat and protein diet seems to be linked to a high frequency of colon carcinomas in several, but not all, areas of the world; a high consumption of green-yellow vegetables is associated with a low incidence of stomach cancers among Japanese (Hirayama, 1979); and a diet low in fibers is claimed to be related to a higher frequency of colon cancers. In these and many other epidemiological studies, the ingested food is expressed in total weight of beef, pork, vegetables, rice, etc., which is consumed by a population group. Such food consumption data must be considered to be quite incomplete, since they seldom include information on the type of food preparation, the origin of food products, and the various eating patterns. However, these factors can strongly influence the mutagenic and carcinogenic load of a food item. For example, the amount of benzo[a]pyrene in cereals depends strongly on the drying process employed (Table III); the amount of dimethylnitrosamine in malt is influenced by the method of kilning (Table III); the mutagenic activity of a beef hamburger depends on the grilling time (Dolara *et al.*, 1979; Pariza *et al.*, 1979; Weisburger and Spingarn, 1979; Felton *et al.*, 1981); and the clastogenic effect of caramelized sucrose depends on temperature and heating time (Table IV). Moreover, various cooking procedures can lead to different types of mutagens. The pyrolysis products which are formed at relatively high temperatures in the charred parts of protein-containing food (Sugimura *et al.*, 1977; Nagao *et al.*, 1977a; Sugimura, 1979; Yamaizumi *et al.*, 1980; Tsuda *et al.*, 1980) differ from those produced at the lower grilling temperatures (Dolara *et al.*, 1979).

TABLE III

Possible Preventive Measures during Food Processing

Food	Process	Mutagen[c]	
Cereals[a]	Flue gas drying lignite	B[a]P	10.2 μg/kg
	Flue gas drying oil		1.9 μg/kg
	Flue gas drying gas		1.8 μg/kg
	Before drying		0.8–1.4 μg/kg
Dark malt[b]	Kilning, indirect heat	NDMA	± 5 ppb
(beer brewing)	Kilning, direct heat oil		5 ppb
	Kilning, direct heat gas		100–300 ppb

[a] Fritz (1971).
[b] Preussmann et al. (1979).
[c] B[a]P, benzo[a]pyrene; NDMA, dimethylnitrosamine.

At present there is no general agreement about strategies for testing food products. In the past one was inclined to search for a link between one specific carcinogen and a particular cancer of a specific organ. This approach may have detracted from a systematic screening of common food products available in food stores across the nation. Moreover, the lack of interest in the mutagenicity of general food may

TABLE IV

Effect of Temperature and Heating Time on the Clastogenic Activity of Caramelized Sucrose[a]

Temperature (°C)	Time (min)	Percentage metaphase with chromosome aberrations Concentrations (mg/ml)	
		62	31
150	40	2.7 (0.01)[b]	0.8 (0.00)
	90	62.3 (1.18)	11.8 (0.06)
	150	18.8 (0.14)	15.5 (0.09)
	240	3.6 (0.00)	1.2 (0.01)
180	40	M.I.[c]	60.0 (0.03)
	60	M.I.	57.1 (0.69)
	90	M.I.	M.I.
	120	M.I.	50.9 (0.12)

[a] 500 mg/ml solution.
[b] Numbers in parentheses are average number of exchanges per metaphase plate.
[c] M.I., mitotic inhibition. A lower concentration of this sample (16 mg/ml) induced chromosome aberrations in 30.9% of the metaphase plates with 0.04 exchanges per metaphase plate.

have been caused by the so-called GRAS ("generally regarded as safe") list which includes many food items used on a daily basis. Thus, on testing some of the readily available food products for mutagenicity, clastogenicity, recombinogenicity, or convertogenicity, one is somewhat surprised to see the large number of positive items (Table V). These food products not only include solid food but also beverages such as many popular caffeine-containing and caffeine-free soft drinks which have virtually replaced water as a source of fluid (Fig. 5).

Similarly, the number of mutagenic or clastogenic chemicals which have been isolated from food sources is relatively large (Table I). Considering that this is only the beginning of a thorough search for mutagens in food, it is possible that sooner or later mutagens will be found to be integral parts of most if not all food products. Thus, the intake of mutagens and by implication carcinogens through regular diet may exceed by far the amount derived from man-made sources such as industrial pollution, food additives, and pesticides.

TABLE V

Examples of Food Products Found to Give a Positive Genotoxic Activity in at Least Two Tests[a]

Caramelized sugars	Dates
Sucrose	Bananas
Glucose	Caramel product-containing food items
Fructose	Caramel toppings (2 different)
Mannose	Caramel puddings (4 different)
Maltose	Caramel candies (4 different)
Arabinose	Caramel slab
Raffinose	Beverages
Caramel Powders	Caffeine-containing soft drinks
(commercially available)	Caffeine-free soft drinks
Caramel flavors	Root beer
(commercially available)	Coffees I, II, and III
Molasses (commercial product)	Prune juice
Maple syrup	Grape juice
Maple sugars	Apple juice
Dried fruits	Grapes
Prunes	Skin
Figs	Pulp
Raisins	Roasted soya beans
Apricots	

[a] *Salmonella typhimurium* mutagenicity, CHO chromosome aberrations or gene conversion test of yeast D7 strain. For details, see Stich *et al.* (1980, 1981a,b,c).

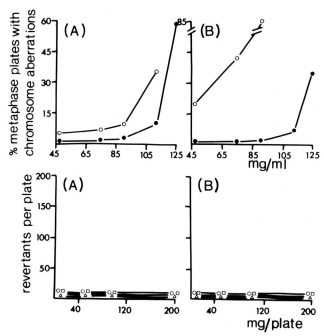

Fig. 5. Clastogenic and mutagenic activities of two caramel-containing carbonated beverages (A and B). Mutagenic activities are expressed as the average number of histidine revertant colonies (minimum of three replicates) in strain TA1537 (open squares), TA100 (open circles), and TA98 (open triangles). Clastogenic activity is presented as the percentage of metaphase plates in treated CHO cells which have at least one chromatid break or exchange figure in them. The presence of an S9 activation mixture is indicated by closed figures, with open figures indicating activities in the absence of the S9 mixture. The mutagenic activity of the two beverages in the presence of S9 is not shown, because this activation mixture had no effect on the observed reversion frequencies. Values are corrected for spontaneous mutation frequencies.

V. MODEL BROWNING REACTIONS

One efficient method of obtaining an insight into the pattern of mutagen formation in foods is the use of model systems (Weisburger and Spingarn, 1979; Mihara and Shibamoto, 1980; Shinohara et al., 1980; Stich et al., 1981a,b; Powrie et al., 1981; W. D. Powrie, C. H. Wu, M. P. Rosin, and H. F. Stich, manuscript submitted to J. Food Sci.). One or two food components rather than the entire complex food item are treated under controlled conditions corresponding to those commonly used in commercial food processing or domestic cooking.

In this way many variable factors such as contamination by environmental mutagens, age, and origin of food products can be eliminated. Heat penetration through the small samples of reaction mixtures can be kept uniform whereas this factor cannot be properly controlled during heating of steaks or hamburgers. The quantity and conceivably also the type of mutagens formed on the surface of steaks and hamburgers will differ from those formed inside, the first fried side will differ from the second one, and the dry parts will differ from the moist regions. To cope with this complexity model systems were introduced which are highly suitable to identify conditions (temperature, pH, reactant concentrations, the use of antioxidants, etc.) that enhance or inhibit the development of the genotoxic component(s).

Currently we are examining the mutagenicity and clastogenicity of products formed by (1) the Maillard reaction and (2) the caramelization of sugars. The model system employed in our laboratory for the Maillard reaction involves the heating of solutions of a sugar and an amino acid at a temperature commonly used in the preparation of foods (Powrie et al., 1981). Figure 6 shows the data obtained from a study in which a 0.8 M solution of glucose and lysine was autoclaved for 1 hour at 121°C and the reaction products assayed for genotoxic activity. Two pH values were used for the initial glucose–lysine solutions, since the pH of the reactants is known to affect the overall rate of this browning reaction. The heated glucose–lysine mixture induced chromatid breaks and exchanges in CHO cells, gene conversion in S. cerevisiae cultures, and reverse point mutation in Salmonella cultures. The genotoxic activity of the reaction products from studies begun at the more alkaline pH was higher than activity at the neutral pH. It has been suggested that the most reactive species of amino acids is the anion which would increase in concentration with an increase in pH above the isoelectric point. This model study has been extended to include various combinations of reducing sugars (glucose, fructose) and three basic amino acids (L-arginine, L-lysine, and L-histidine), one hydroxy amino acid (L-serine), an amide (L-glutamine), and an acidic amino acid (L-glutamic acid) (Powrie et al., 1981). All the reaction products from these combinations contained clastogenic, convertogenic, and/or mutagenic activities. In this connection it is of interest to note the mutagenic activity of cooked meats (e.g., Commoner et al., 1978; Dolara et al., 1979; Pariza et al., 1979; Rappaport et al., 1979; Spingarn and Weisburger, 1979; Spingarn et al., 1980a,b; Felton et al., 1981; Plumlee et al., 1981), dried fruits (Stich et al., 1981b), commercial caramel powder (Stich et al., 1981a), and caramel-containing foods and beverages (Stich et al., 1981a,c).

GLUCOSE–LYSINE PH 7 AND 9

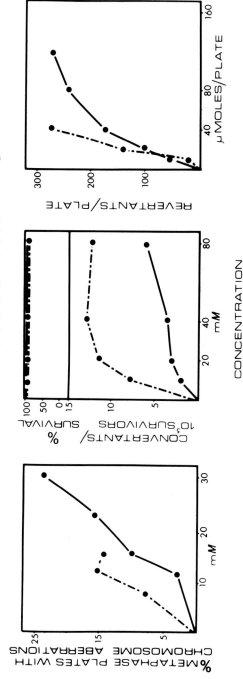

Fig. 6. Genotoxic activity of a heated glucose–lysine mixture in the CHO chromosome aberration test, the *Saccharomyces* gene conversion test, and the modified *Salmonella* assay which employs a preincubation step (see Fig. 3). Gene conversion frequencies are expressed as convertants/plate $\times$ 10^{-1} (filled circles) and as convertants/10^5 survivors (filled squares). *Salmonella* results are shown for two strains, TA100 (filled circles) and TA98 (filled squares) with the data corrected for a spontaneous frequency of histidine reversion of 109 revertant colonies/plate for TA100 and 34 revertant colonies/plate for TA98. The glucose–lysine sample was prepared by heating a 0.8 M solution of glucose and lysine at 121°C for 1 hour. Two pH values were used for the reactants: pH 7, solid line; pH 9, dashed line.

The next step was to fractionate the Maillard reaction products into different chemical groups (e.g., volatiles, Amadori compounds, reductones, melanoidins) and to identify by mass spectroscopy the components of these mixtures. Figure 7 shows the gas chromatogram of the basic fraction of the volatiles obtained from a glucose–lysine model system which was heated at 121°C for 1 hour with an initial pH of 7. The number of volatiles produced by this simple model system consisting of a reducing sugar and one amino acid exemplifies the complexity of the chemical changes that occur during processing of food (Table VI). Volatiles produced in a heated food item will be more numerous than in our simple model system because food products contain several sugars and amino acids, each of these mixtures producing different reaction products upon heating. Once the compounds

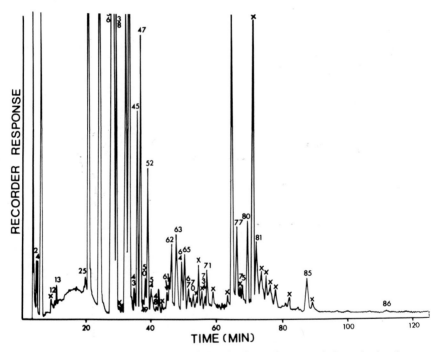

Fig. 7. Gas chromatogram obtained for the basic fraction of the volatiles from a glucose–lysine model browning system, initial pH 7.0, reaction time 1 hour at 121°C. Volatiles were extracted with pH gradient. Column used: 20 ft × 2 mm i.d. 5% Carbowax 20 M on 80/100 Chromosorb WHP. Temperature program 60°–200°C at 2°C/minute. Initial temperature hold for 4 minutes, final temperature hold until end of analysis. Injector separator and line temperature 250°C. Helium 20 ml/minute. All peaks labeled × were unidentified.

TABLE VI

Compounds Identified from Carbonyl, Acidic, Basic, and Neutral Volatile Fractions of a Glucose–Lysine Model Browning System[a]

Aldehydes
- 2 Propanol
- 20 n-Hexanol
- 21 2-Methylbut-2-en-1-ol
- 46 2-Furfural
- 58 5-Methyl-2-furfural
- 82 5-Methyl-2-formylpyrrole

Ketones
- 3 Acetone
- 7 Butan-2-one
- 11 Pentan-3-one
- 15 3-Methyl-2-pentanone
- 16. 2,3-Pentadione
- 19 3-Hexanone
- 26 Cyclopentanone
- 31 3-Hydroxy-2-butanone
- 32 2-Methyltetrahydrofuran-3-one
- 33 Hydroxy-2-propanone
- 40 2,5-Dimethylcyclopent-2-en-1-one
- 48 2,3-Dimethylcyclopent-2-en-1-one
- 52 2-Furyl methyl ketone
- 60 2-Ethyl furyl ketone
- 71 2-Hydroxy-3-methyl-2-cyclopenten-1-one
- 76 2-Acetylpyrrole

Furans
- 5 2-Methylfuran
- 9 2,5-Dimethylfuran
- 10 2,4-Dimethylfuran
- 14 2-Isopropylfuran
- 23 2-Methyl-5-vinyl furan
- 24 2-Vinyl furan
- 28 2-Norpropenylfuran
- 34 Ethyl-2-furfuryl ether
- 35 2-Methyl-5-isopropenyl-2-isobutenylfuran
- 51 2-Methyl-5-acetylfuran
- 56 Furfuryl acetate
- 59 Butylfuran
- 66 5-Methyl-2,2'-methylene difuran

Acids and esters
- 4 Ethyl formate
- 6 Ethyl acetate
- 44 Acetic acid
- 53 Propionic acid
- 85 Diethyl phthalate

Alcohols
- 12 Hexan-3-ol
- 13 Butan-2-ol
- 18 Propanol
- 22 Butan-1-ol
- 39 2-Methylpropan-2-ol
- 63 2-Furfuryl alcohol
- 68 5-Methyl-2-furfuryl alcohol

Pyrazines
- 27 Pyrazine
- 29 2-Methylpyrazine
- 36 2,5-Dimethylpyrazine
- 37 2,6-Dimethylpyrazine
- 38 2,3-Dimethylpyrazine
- 41 2-Ethyl-5-methylpyrazine
- 42 Trimethylpyrazine
- 43 Vinyl pyrazine
- 45 2,5-Dimethyl-3-ethylpyrazine
- 47 2,6-Dimethyl-3-ethylpyrazine
- 49 Tetramethylpyrazine
- 50 2-Methyl-5(6)-vinyl pyrazine
- 54 2,5-Dimethyl-3-vinyl pyrazine
- 62 5-Methyl-6,7-dihydro-(5H)-cyclopentapyrazine
- 64 2,5-Dimethyl-6,7-dihydro-(5H)-cyclopentapyrazine
- 65 3,5-Dimethyl-6,7-dihydro-(5H)-cyclopentapyrazine
- 67 2-Methyl-6,7-dihydro-(5H)-cyclopentapyrazine
- 70 5,6,7,8-Tetrahydroquinoxaline
- 73 2-Methyl-5,6,7,8-tetrahydroquinoxaline
- 77 2-(2'-Furyl)pyrazine
- 79 6-Methylquinoxaline
- 80 2-(2'-Furyl)-5-methylpyrazine
- 84 Diethylmethylpyrazine-2-ethyltrimethylpyrazine

Nitrogen compounds
- 25 2,4,5-Trimethyloxazole
- 55 Pyrrole
- 57 2-Methylpyrrole
- 61 2-Methyl-6-ethylpyridine
- 69 1,2,3,4-Tetrahydroquinoline
- 74 5-Methylfurfuryl-1-pyrrole
- 75 6-Methyl-1,2,3,4-tetrahydroquinoline

(Continued)

Table VI (continued)

Phenols	Miscellaneous
72 Acetoxyphenol	81 Dimethylbenzimidazole
78 Phenol	86 4-(Diethylamino)benzaldehyde
83 o-Cresol	Miscellaneous
	1 Diethyl ether (used as solvent)
	8 Benzene
	17 Toluene
	30 Dichloromethane

^a Numbers designate specific peaks of gas chromatograms.

in reaction mixtures are identified, they can be individually tested for mutagenic, clastogenic, and recombinogenic activity.

Another use for model systems is the study of conditions that lead to a reduced genotoxicity of cooked and processed food products. Model studies were performed on the caramelization of various sugars at various temperatures for different heating times. The clastogenic activity of the heated sucrose increased with higher temperatures and longer duration of heating (Table IV). However, at both 150° and 180°C, there was a decrease in the chromosome-damaging capacity of the products from the longer duration treatments (150°C, 4 minutes; 180°C, 2 minutes). Chemical analyses of such treatment samples may indicate which caramelization products contain clastogenic activity and indicate the time and temperature necessary for the development of such chemicals.

In another model system various fruits or vegetables are homogenized and exposed to air at room temperature prior to sampling for mutagenic, clastogenic, or convertogenic activity. Table VII shows the clastogenic activity of mushrooms at various times after initial homogenization. The chromosome-damaging action of these extracts increases with time prior to sampling. The involvement of an enzymatic browning process in the development of this genotoxicity is suspected but unproved.

VI. INTEGRATED MUTAGENIC LOAD OF DIETS

Intensive screening programs for mutagens in food are being carried out in several laboratories. They will add many new food chemicals which were previously considered to be safe to the fast-growing list of mutagens and, by implication, potential carcinogens. The testing

TABLE VII

Clastogenic Activity of a Fresh, 1-Hour- and 4-Hour-Old Extract of Mushroom[a]

Sampling Time (hours)	Treatment	Percentage metaphase plates with chromosome aberrations[b]
0	− S9	4.76 (0.06)[c]
	+ S9	0.91 (0.01)
1	− S9	24.2 (0.27)
	+ S9	4.5 (0.03)
4	− S9	M.I.[d]
	+ S9	M.I.

[a] Mushrooms obtained from a local supermarket were homogenized and exposed to air for varying times prior to sampling.

[b] Concentration: 750 μl/ml.

[c] Numbers in parentheses are the average number of exchanges per metaphase plate.

[d] M.I., mitotic inhibition. This sample induced chromosome aberrations in 30.0% of the examined metaphase plates with 0.40 exchanges per metaphase plate at a concentration of 375 μl/ml treatment mix.

of isolated purified compounds for mutagenicity, clastogenicity, recombinogenicity, and convertogenicity is, however, only one approach to the assessment of the mutagenic load of daily consumed food products. We would appear to be remiss if we did not point to some of its limitations. For example, coffee contains more than 150 furans (Maga, 1979). Similar numbers of furans will probably occur in most roasted food products. It would be taxing the research and screening facilities of a nation to test all the furans, pyrazines, thiazolidines, reductones, etc., for their capacity to induce point mutations, mitotic recombinations, gene conversion, chromosome aberrations, sister chromatid exchanges, DNA alterations, or neoplastic transformation. Obviously, the mutagenicity of all food chemicals cannot be individually tested within a reasonable time period. Another restriction of this approach is the complex composition of even the simplest food products. They contain not only mutagens but also comutagens, antimutagens (Kada et al., 1978a,b; Rosin and Stich, 1978a,b; Wattenberg, 1980), and polymers (Crosby, 1979) that could adsorb reactive chemicals including mutagens. Thus, the testing of individual compounds and the integration of all the genotoxicity data of single tests to obtain the total genotoxic activity of a food item could easily produce a misleading picture. The mutagenicity of a food item will not only depend on the level of mutagens but also on the activity of modulating factors.

How then do we obtain an estimate of the actual genotoxicity of a food product? Basically there are two choices. Food items consumed at an "average" breakfast, lunch, supper, and snack during one day by one person can be collected, homogenized, and tested for mutagenicity, clastogenicity, or recombinogenicity and convertogenicity. The genotoxicity of individual food items can be measured and the number of positive or negative products summated. Although the estimation of mutagens and clastogens in homogenates of breakfasts, lunches, and suppers appears to be a relatively crude approach, differences between the diets of a population with high gastric cancer risk and those of one with low cancer risk have actually been observed in the Akita prefecture of northern Japan (Fig. 8). The economic technique of examining the genotoxicity of breakfasts, lunches, and suppers can be readily applied to a large number of samples. This approach can be improved by: (1) applying several extraction methods, (2) using several bioassays which depend on different endpoints (e.g., mutations in *S. typhimurium*, recombination and gene conversion in yeast, and chromosome aberrations in CHO cells), (3) including the genotoxicity of volatiles after a nitrogen chase, and (4) separating by filtration procedures the crude extract into large and small molecular fractions and estimating the genotoxicity of each fraction.

Another approach to the assessment of the mutagenic load of food is to set priorities for items to be tested. But what criteria should be applied in this grading process? Within a group of foodstuffs one can arrive at a reasonable decision. A priority rating has actually been proposed for protein-containing food on the basis of content of protein,

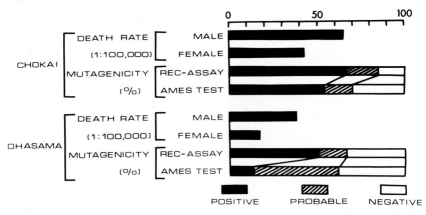

Fig. 8. Age-adjusted stomach cancer mortalities and mutagenicity of diets in Chokai (Akita) and Ohasama (Iwate) villages of Japan (Kamiyama, 1980).

amount consumed, and preparation by heating (Plumlee et al., 1981). The priorities set were based on the assumption that high levels of mutagens will be formed in protein-rich food products following cooking and that a large intake of these food items is more hazardous than a smaller one. Somewhat more ambiguous would be priority settings for all food items and beverages. How, for example, can one compare the genotoxic hazard, if one exists at all, of caramel products in hundreds of foods and soft drinks with that of mutagens in hamburgers?

VII. OUTLOOK

The application of several short-term tests for genotoxicity revealed potent mutagens, clastogens, convertogens, and recombinogens in numerous food items that were previously regarded as safe. Considering that only a very few mutagenicity studies have already produced a long list of genotoxic compounds, the number and amount of daily ingested genotoxic chemicals must be staggering.

At this point we would like to add a nonscientific but pertinent comment (Sugimura, 1978). The detection of the genotoxicity of a compound can lead to its immediate removal if it can be easily replaced. Such an action, if not beneficial, would at least not cause any harm economically or otherwise. The uncovered genotoxicity of common food products, however, does not fall into this category. First, even the most conservative scientific statements about mutagens, clastogens, and potential carcinogens in food and beverages will greatly irritate a public already oversaturated with danger signals released from governmental protection agencies and self-styled health apostles. Second, it may be difficult to substitute in the near future mutagenic food products with nonmutagenic ones. To cope with this dilemma, we strongly recommend placing high priority on studies which aim to assess the genotoxic activity of *entire food mixtures*, to uncover *protection mechanisms* which may be present in mammals but which are absent or inoperative in the *in vitro* bioassays and to examine the role of desmutagens and antimutagens (Kada et al., 1978a; Morita et al., 1978; Lo and Stich, 1978; Rosin and Stich, 1978a,b; Wattenberg, 1980).

One of the most pressing questions concerns the usefulness of *in vitro* genotoxicity data for predicting a genetic hazard to human germ cells or somatic cells. No unequivocal answer can be provided in spite of numerous attempts to prove or disprove a correlation between *in vitro* mutagenicity and *in vivo* carcinogenicity (cf. Chapter 2; Meselson and Russell, 1977).

One of the ongoing discussions concerns the choice of a test system which will provide the best link between carcinogenicity and genotoxicity. There is no a priori reason why frameshift mutations or base-pair substitutions in bacteria should be a better indicator for carcinogenicity than mitotic recombination or gene conversion in yeast or chromosome aberrations in mammalian cells. At present it is unknown whether mutational events are involved in neoplastic transformation and, if this is the case, what type of mutational events play a crucial role.

Another often heated argument deals with the number of "false positives" or "false negatives" which are produced by one or the other *in vitro* bioassay. A careful examination of this problem could easily prove it to be a mere bogus issue. In most cases the results of one *in vitro* bioassay are compared to carcinogenicity data on *several* mammals, including rats, mice, hamsters, and man. The awkward situation is caused by the fact that the various mammals respond differently to different carcinogens. Thus one is virtually at liberty to claim "false positives" or "false negatives" by comparing the *in vitro* results with a responsive or unresponsive rodent species. To illustrate this point, aflatoxin B_1 would be classed as a "false positive" if the positive results of the *S. typhimurium* test are compared to the negative carcinogenicity data on several mouse strains, and ionizing radiation can be said to yield "false negatives" if data from the *S. typhimurium* test are compared to that in rodents and man. This play with claiming "false positives" and "false negatives" can be extended by comparing the results of various *in vitro* test systems with each other. If a chemical fails to induce mutations in a microbial system and elicits chromosome aberrations in cultured human cells, then should one call this a false positive? The important point of this discourse is to keep in mind the relativity of any statement about "false positives and negatives." If the truth is said it all comes down to matters of belief. If one believes that the response of rats, mice, or hamsters to genotoxic agents reflects truthfully the response of man, then a comparison of short-term *in vitro* tests with carcinogenesis data on rodents is acceptable. However, if prediction of the short-lived rodents with latency periods of less than 1 year for most carcinogens should not prove to be as reliable indicators for man as generally assumed, then the importance of obtaining a good relationship between short-term genotoxicity data and rodent carcinogenicity greatly diminishes.

Finally, the results presented on the genotoxicity of many food items clearly reveal the importance of food preparation, cooking, and storage. Thus, epidemiological studies which were restricted to the recording of food products without mentioning their preparation are of questionable value in proving or disproving a role of food mutagens or clastogens in the etiology of human carcinomas. Once we have gained a more profound understanding about the distribution of mutagens in various unprocessed and cooked food products and their fate within man, more precise questions can be formulated which can then be submitted to an epidemiological test.

REFERENCES

Ames, B. N., McCann, J., and Yamasaki, E. (1975). Methods for detecting carcinogens and mutagens with the *Salmonella*/mammalian-microsome mutagenicity test. *Mutat. Res.* **31**, 347–364.

Bjeldanes, L. F., and Chew, H. (1979). Mutagenicity of 1,2-dicarbonyl compounds: Maltol, kojic acid, diacetyl and related substances. *Mutat. Res.* **67**, 367–371.

Boyd, M. R., and Wilson, B. J. (1971). Preparative and analytical gas chromatography of ipomeamarone, a toxic metabolite of sweet potatoes (*Ipomoea batatas*). *J. Agric. Food Chem.* **19**, 547–550.

Brown, J. P. (1980). A review of the genetic effects of naturally occurring flavonoids, anthraquinones and related compounds. *Mutat. Res.* **75**, 243–277.

Commoner, B., Vithayathil, A. J., Dolara, P., Nair, S., Madyastha, P., and Cura, G. C. (1978). Formation of mutagens in beef and beef extract during cooking. *Science* **201**, 913–916.

Concon, J. M., Newburg, D. S., and Swerczek, T. W. (1979). Black pepper (*Piper nigrum*): Evidence of carcinogenicity. *Nutr. Cancer* **1**, 22–26.

Crosby, L. (1979). Fiber-standardized source. *Nutr. Cancer* **1**, 15–26.

Dolara, P., Commoner, B., Vithayathil, A., Cura, G., Tuley, E., Madyastha, P., Nair, S., and Kriebel, D. (1979). The effect of temperature on the formation of mutagens in heated beef stock and cooked ground beef. *Mutat. Res.* **60**, 231–237.

Evans, I. A., and Mason, J. (1965). Carcinogenic activity of bracken. *Nature (London)* **208**, 913–914.

Felton, J. S., Healy, S., Stuermer, D., Berry, C., Timourian, H., Hatch, F. T., Morris, M., and Bjeldanes, L. F. (1981). Mutagens from the cooking of food. I. Improved extraction and characterization of mutagenic fractions from cooked ground beef. *Mutat. Res.* **88**, 33–44.

Fritz, W. (1971). Extent and origin of contamination of foodstuffs by carcinogenic hydrocarbons. *Ernaehrungsforschung* **16**, 547–557.

Hardigree, A. A., and Epler, J. L. (1978). Comparative mutagenesis of plant flavonoids in microbial systems. *Mutat. Res.* **58**, 231–239.

Hayatsu, H., Arimoto, S., Togawa, K., and Makita, M. (1981). Inhibitory effect of the ether extract of human feces on activities of mutagens: Inhibition by oleic and linoleic acids. *Mutat. Res.* **81**, 287–293.

Hirayama, T. (1979). The epidemiology of gastric cancer. In "Gastric Cancer" (C. J. Pfeiffer, ed.), pp. 61–82. G. Witzstrock Publications, New York.

Hirono, I., Fushimi, K. Mori, H., Miwa, T., and Haga, M. (1973). Comparative study of carcinogenic activity in each part of bracken. *J. Natl. Cancer Inst.* **50**, 1367–1371.

Kada, T., Sadaie, Y., and Hara, M. (1978a). Analysis of mutagen-antimutagen reactions in food and food additives by the rec-assay and reversion-assay procedures. *Mutat. Res.* **53**, 206–207.

Kada, T., Morita, K., and Inoue, T. (1978b). Anti-mutagenic action of vegetable factor(s) on the mutagenic principle of tryptophan pyrolysate. *Mutat. Res.* **53**, 351–353.

Kamiyama, S. (1980). Gastric cancer mortality and dietary mutagenicity. *Presented at the U. S.-Jpn. Coop. Workshop Gastrointest. Tract Cancer, Honolulu, March 1980.*

Kinsella, A. R., and Radman, M., (1980). Inhibition of carcinogen-induced chromosomal aberrations by an anticarcinogenic protease inhibitor. *Proc. Natl. Acad. Sci. U. S. A.* **77**, 3544–3547.

Laqueur, G. T., and Matsumoto, H. (1966). Neoplasms in female Fischer rats following intraperitoneal injection of methylazoxymethanol. *J. Natl. Cancer Inst.* **37**, 217–232.

Laqueur, G. T., Mickelsen, O., Whiting, M., and Kurland, L. T. (1963). Carcinogenic properties of nuts from *Cycas circinalis* L. indigenous to Guam. *J. Natl. Cancer Inst.* **31**, 919–951.

Lo, L. W., and Stich, H. F. (1978). The use of short-term tests to measure the preventive action of reducing agents on formation and activation of carcinogenic nitroso compounds. *Mutat. Res.* **57**, 57–67.

Lo, M.-T., and Sandi, E. (1978). Polycyclic aromatic hydrocarbons (polynuclear) in foods. *Residue Rev.* **69**, 35–86.

Maga, J. A. (1979). Furans in food. *CRC Crit. Rev. Food Sci. Nutr.* **11**, 355–400.

Maga, J. A., and Sizer, C. E. (1973). Pyrazines in food. *CRC Crit. Rev. Food Sci. Nutr.* **4**, 39–115.

Meselson, M., and Russell, Y. (1977). Comparison of carcinogenic and mutagenic potency. *In* "Origins of Human Cancer" (H. H. Hiatt, J. D. Watson, and J. A. Winsten, eds.), pp. 1473–1482. Cold Spring Harbor Lab., Cold Spring Harbor, New York.

Mihara, S., and Shibamoto, T. (1980). Mutagenicity of products obtained from cysteamine-glucose browning model systems. *J. Agric. Food Chem.* **28**, 62–66.

Miller, J. A., Swanson, A. B., and Miller, E. C. (1979). The metabolic activation of safrole and related naturally occurring alkenylbenzenes in relation to carcinogenesis by these agents. *In* "Naturally Occurring Carcinogens—Mutagens and Modulators of Carcinogenesis" (E. C. Miller, J. A. Miller, I. Hirono, T. Sugimura, and S. Takayama, eds.), pp. 111–125. Univ. Park Press, Baltimore, Maryland.

Morita, K., Hara, M., and Kaday, T. (1978). Studies on natural desmutagens: Screening for vegetable and fruit factors active in inactivation of mutagenic pyrolysis products from amino acids. *Agric. Biol. Chem.* **42**, 1235–1238.

Nagao, M., Honda, M., Seino, Y., Yahagi, T., and Sugimura, T. (1977a). Mutagenicities of smoke condensates and the charred surface of fish and meat. *Cancer Lett. (Shannon, Irel.)* **2**, 221–226.

Nagao, M., Yahagi, T., Seino, Y., Sugimura, T., and Ito, N. (1977b). Mutagenicities of quinoline and its derivatives. *Mutat. Res.* **42**, 335–342.

Nagao, M., Honda, M., Seino, Y., Yahagi, T., and Sugimura, T. (1979). Mutagens in coffee and tea. *Mutat. Res.* **68**, 101–106.

Pamukcu, A. M., Ertürk, E., Yalciner, S., Milli, V., and Bryan, G. T. (1978). Carcinogenic and mutagenic activities of milk from cows fed bracken fern (*Pteridium aquilinum*). *Cancer Res.* **38**, 1556–1560.

Pariza, M. W., Ashoor, S. H., Chu, F. S., and Lund, D. B. (1979). Effects of temperature and time on mutagen formation in pan-fried hamburger. *Cancer Lett. (Shannon, Irel.)* **7**, 63–69.

Plumlee, C., Bjeldanes, L. F., and Hatch, F. T. (1981). Food item priority assessment for studies of mutagen production during cooking. *J. Am. Diet. Assoc.* **79**, 446–449.

Powrie, W. D., Wu, C. H., Rosin, M. P., and Stich, H. F. (1981). Clastogenic and mutagenic activities of Maillard reaction model systems. *J. Food Sci.* **46**, 1433–1438, 1445.

Preussmann, R., Spiegelhalder, B., Eisenbrand, G., and Janzowski, C. (1979). N-Nitroso compound in food. *In* "Naturally Occurring Carcinogens—Mutagens and Modulators of Carcinogenesis" (E. C. Miller, J. A. Miller, I. Hirono, T. Sugimura, and S. Takayama, eds.), pp. 185–194. Univ. Park Press, Baltimore, Maryland.

Rappaport, S. M., McCartney, M. C., and Wei, E. T. (1979). Volatilization of mutagens from beef during cooking. *Cancer Lett. (Shannon, Irel.)* **8**, 135–145.

Rosin, M. P., and Stich, H. F. (1978a). The inhibitory effect of cysteine on the mutagenic activities of several carcinogens. *Mutat. Res.* **54**, 73–81.

Rosin, M. P., and Stich, H. F. (1978b). The inhibitory effect of reducing agents on N-acetoxy- and N-hydroxy-2-acetylaminofluorene-induced mutagenesis. *Cancer Res.* **38**, 1307–1310.

Schoental, R. (1976). Carcinogens in plants and microorganisms. *In* "Chemical Carcinogens" (C. E. Searle, ed.), pp. 626–689. Amer. Chem. Soc., Washington, D. C.

Shinohara, K., Wu, R.-T., Jahan, N., Tanaka, M., Morinaga, N., Murakami, H., and Omura, H. (1980). Mutagenicity of the browning mixtures of amino-carbonyl reactions on *Salmonella typhimurium* TA100. *Agric. Biol. Chem.* **44**, 671–672.

Spingarn, N. E., and Weisburger, J. H. (1979). Formation of mutagens in cooked foods. I. Beef. *Cancer Lett. (Shannon, Irel.)* **7**, 259–264.

Spingarn, N. E., Slocum, L. A., and Weisburger, J. H. (1980a). Formation of mutagens in cooked foods. II. Foods with high starch content. *Cancer* *(Shannon, Irel.)* **9**, 7–12.

Spingarn, N. E., Kasai, H., Vuolo, L. L., Nishimura, S., Yamaizumi, Z., Sugimura, T., Matsushima, T., and Weisburger, J. H. (1980b). Formation of mutagens in cooked foods. III. Isolation of a potent mutagen from beef. *Cancer Lett. (Shannon, Irel.)* **9**, 177–183.

Stich, H. F., and San, R. H. C., eds. (1981). "Short-Term Tests for Chemical Carcinogens." Springer-Verlag, Berlin and New York.

Stich, H. F., Karim, J., Koropatnick, J., and Lo, L. (1976). Mutagenic action of ascorbic acid. *Nature (London)* **260**, 722–724.

Stich, H. F., Wei, L., and Lam, P. (1978). The need for a mammalian test system for mutagens: action of some reducing agents. *Cancer Lett. (Shannon, Irel.)* **5**, 199–204.

Stich, H. F., Wei, L., and Whiting, R. F. (1979). Enhancement of the chromosome-damaging action of ascorbate by transition metals. *Cancer Res.* **39**, 4145–4151.

Stich, H. F., Stich, W., Rosin, M. P., and Powrie, W. D. (1980). Mutagenic activity of pyrazine derivatives: A comparative study with *Salmonella typhimurium*, *Saccharomyces cerevisiae* and Chinese hamster ovary cells. *Food Cosmet. Toxicol.* **18**, 581–584.

Stich, H. F., Stich, W., Rosin, M. P., and Powrie, W. D. (1981a). Clastogenic activity of caramel and caramelized sugars. *Mutat. Res.* **91**, 129–136.

Stich, H. F., Rosin, M. P., Wu, C. H., and Powrie, W. D. (1981b). Clastogenic activity of dried fruits. *Cancer Lett. (Shannon, Irel)* **12**, 1–8.

Stich, H. F., Rosin, M. P., San, R. H. C., Wu, C. H., and Powrie, W. D. (1981c). Intake, formation and release of mutagens by man. *In* "Gastrointestinal Cancer: Endogenous Factors" (W. R. Bruce, P. Correa, M. Lipkin, S. R. Tannenbaum, and T. D. Wilkins, eds.), pp. 247–266. Cold Spring Harbor Lab, Cold Spring Harbor, New York, Banbury Report 7.

Stich, H. F., Rosin, M. P., Wu, C. H., and Powrie, W. D. (1981d). Clastogenicity of furans found in food. *Cancer Lett. (Shannon, Irel.)* **13**, 89–95.

Sugimura, T. (1978). Let's be scientific about the problem of mutagens in cooked food. *Mutat. Res.* **55**, 149–152.

Sugimura, T. (1979). Naturally occurring genotoxic carcinogens. In "Naturally Occurring Carcinogens—Mutagens and Modulators of Carcinogenesis" (E. C. Miller, J. A. Miller, I. Hirono, T. Sugimura, and S. Takayama, eds.), pp. 241–261. Univ. Park Press, Baltimore, Maryland.

Sugimura, T., Nagao, M., Kawachi, T., Honda, M., Yahagi, T., Seino, Y., Sato, S., Matsukura, N., Matsushima, T., Shirai, A., Sawamura, M., and Matsumoto, H. (1977). Mutagens-carcinogens in food, with special reference to highly mutagenic pyrolytic products in broiled foods. In "Origins of Human Cancer" (H. H. Hiatt, J. D. Watson, and J. A. Winsten, eds.), Book C, pp. 1561–1577. Cold Spring Harbor Lab., Cold Spring Harbor, New York.

Toth, B. (1979). Mushroom hydrazines: Occurrence, metabolism, carcinogenesis, and environmental implications. In "Naturally Occurring Carcinogens—Mutagens and Modulators of Carcinogenesis" (E. C. Miller, J. A. Miller, I. Hirono, T. Sugimura, and S. Takayama, eds.), pp. 57–65. Univ. Park Press, Baltimore, Maryland.

Tsuda, M., Takahashi, Y., Nagao, M., Hirayama, T., and Sugimura, T. (1980). Inactivation of mutagens from pyrolysates of tryptophan and glutamic acid by nitrite in acidic-solution. *Mutat. Res.* **78**, 331–339.

Uyeta, M., Kanada, T., Mazaki, M., Taue, S., and Takahashi, S. (1979). Assaying mutagenicity of food pyrolysis products using the Ames test. In "Naturally Occurring Carcinogens—Mutagens and Modulators of Carcinogenesis" (E. C. Miller, J. A. Miller, I. Hirono, T. Sugimura, and S. Takayama, eds.), pp. 169–176. Univ. Park Press, Baltimore, Maryland.

Wattenberg, L. W. (1980). Inhibition of chemical carcinogenesis by antioxidants. In "Modifiers of Chemical Carcinogenesis: An Approach to the Biochemical Mechanism and Cancer Prevention" (T. J. Slaga, ed.), pp. 85–98. Raven, New York.

Weisburger, J. H., and Spingarn, N. E. (1979). Mutagens as a function of mode of cooking of meats. In "Naturally Occurring Carcinogens—Mutagens and Modulators of Carcinogenesis" (E. C. Miller, J. A. Miller, I. Hirono, T. Sugimura, and S. Takayama, eds.), pp. 177–184. Univ. Park Press, Baltimore, Maryland.

Yamaizumi, Z., Shiomi, T., Kasai, H., Nishimura, S., Takahashi, Y., Nagao, M., and Sugimura, T. (1980). Detection of potent mutagens, Trp-P-1 and Trp-P-2, in broiled fish. *Cancer Lett. (Shannon, Irel.)* **9**, 75–83.

Chapter 6

Transformation of Somatic Cells in Culture

Andrew Sivak and Alice S. Tu

MUTAGENICITY:
NEW HORIZONS IN GENETIC TOXICOLOGY

I. INTRODUCTION

A. General Usefulness and Necessity of "Short-Term" Tests

A wide variety of assay procedures have been developed over the last decade to assess the potential for induction of genetic lesions resulting from exposure to environmental agents, both chemical and physical (Hsie *et al.*, 1979; Sivak, 1979; Mishra *et al.*, 1980). There were several driving forces behind this process. Although the specific importance of each factor may not be quantitatively estimated, some can be identified. Two that seem to be particularly relevant are (1) the increased sensitivity and perception of those in both the scientific and general population that substances capable of causing deleterious effects are in the environment and (2) the technology to perform the necessary genetic assays has come of age. A third factor that also relates strongly to this surge of development in genetic toxicology is the view professed by some that mutagens identified in these genetic assays are putative carcinogens (McCann *et al.* 1975; but see also Sivak, 1976; Rubin, 1980).

Although few would claim that a "short-term" genetic assay would by itself provide a basis for regulatory decisions, the appropriate application of a set (tier, battery) of these assays can provide valuable information at modest expense and time in comparison to long-term experimental animal studies. Indeed, data from these kinds of assays have been incorporated into decision sequences by several Federal regulatory agencies. The key issue in the use of such data is the relevance to the health effect being evaluated. This issue is of special concern with respect to the use of information from one kind of assay (mutagenesis) to derive estimates of health effects in a qualitatively different biological phenomenon (carcinogenesis). The availability of short-term assay procedures for neoplastic transformation in cell culture, which measure in a more direct manner the events associated with carcinogenesis *in vivo*, provides a set of tools to bridge this gap. The discussion of the rationale for these neoplastic transformation assays, a consideration of their relevance to the *in vivo* process of carcinogenesis, and the key problem areas that still require experimental attention are the primary subjects of this chapter.

B. A Brief Historical Perspective of *in Vitro* Neoplastic Transformation

The concept that the process of carcinogenesis induced by exposure to exogenous chemicals could be studied in cell culture systems is not a recent view. From the time of the earliest successful cultivation of mammalian cells *in vitro* over 40 years ago, a number of investigators began to examine the effects of the known chemical carcinogens of that time on cells in culture. Perhaps the largest single body of work came from the laboratories of Earle at the U.S. National Cancer Institute (Earle and Voegtlin, 1938; Earle, 1943; Earle and Nettleship, 1943; Earle *et al.*, 1950). Earle and associates, in fact, did demonstrate that a chemical carcinogen such as 3-methylcholanthrene could induce changes in cellular behavior that could be recognized in culture and that these cellular changes were associated with oncogenicity of the altered cell populations in appropriate experimental animal hosts. However, the occurrence of morphologically altered cells in the populations of mouse C3H fibroblasts also occurred in control cultures, albeit with a lower frequency and at a later time than in carcinogen-treated cultures. The key observation Earle made was that characteristic morphological changes in cell culture were associated with oncogenicity of these cells *in vivo*, and this concept remains one of the important arguments for the use of the neoplastic transformation assays (Earle and Nettleship, 1943; Earle *et al.*, 1950).

The door to the present era of neoplastic transformation studies in cell culture was opened by the observations of Berwald and Sachs (1963) who reported that they could obtain morphological transformation in clonal cultures of early passage Syrian hamster cells with exposure to chemical carcinogens. Since then significant progress has been made toward an understanding of the mechanism of neoplastic transformation using both *in vivo* and *in vitro* model systems. However, much remains to be discovered with respect to the early diagnosis of neoplasia, the interruption, and ultimately the prevention of the carcinogenic process. The identification of chemical carcinogens present in the environment represents a major element in steps taken toward the control and prevention of carcinogenesis; the development of short-term bioassays, including the cell-culture neoplastic transformation assays, is a significant contribution toward this end.

II. *IN VITRO* AND *IN VIVO* STUDIES OF BASIC MECHANISM OF NEOPLASTIC TRANSFORMATION

The demonstration of neoplastic transformation *in vitro* has facilitated the use of cell-culture systems to augment experimental animal models in the studies on the mechanism of action of carcinogenesis on a cellular and molecular level. This process of chemically induced neoplastic transformation can be separated into three phases: (1) metabolic activation of those agents requiring it, (2) macromolecular interaction of these agents in target cells and molecular and/or cellular repair, and (3) expression of the neoplastic phenotype.

A. Metabolic Activation

Metabolic studies have shown that those chemical carcinogens (procarcinogens) that are not direct-acting agents are metabolically converted to highly reactive electrophiles (Miller and Miller, 1977). These are best exemplified by the conversion of polycyclic aromatic hydrocarbons to diol-epoxides (Sims et al., 1974), the aromatic amines to aryl hydroxylamines (Miller and Miller, 1969), and the nitrosamines to α-hydroxynitrosamines (Magee, 1977). These reactions are believed to be catalyzed by the cytochrome P-450 monooxygenases which are a complex of enzymes that are membrane-bound in microsomes, exist in multiple forms, and are generally inducible. Marked qualitative and quantitative differences in monooxygenases in various tissues and species have been reported (Lotlikar et al., 1967; Lower and Bryan, 1973; Weekes and Brusick, 1975; Selkirk et al., 1976). This appears to account in part for differences in susceptibilities to cancer induction. Nevertheless, metabolic activation of a carcinogen should not be equated with its carcinogenicity, since other reactions are possible and often deactivating pathways are in operation simultaneously with the activating pathways. The balance between kinetics of activation and deactivation pathways may ultimately determine the carcinogenic potential of chemicals that require metabolic activation (Oesch et al., 1977).

Many cells in culture and particularly those used as target cells for transformation studies have nondetectable or only a limited capability to metabolize many types of procarcinogens. This represents a major limitation in the development of in vitro cell assay systems to detect chemicals of carcinogenic potential, since metabolic activation is of paramount importance in the conversion of a procarcinogen to a directly active carcinogen. A number of enzyme sources have been used to mimic in vivo metabolism of test compounds. These include utilization of agents to induce endogenous levels of xenobiotic metabolizing enzymes; the supplementation using liver microsomal homogenates isolated from animals pretreated with various inducers; and the cocultivation of target cells with whole-cell preparations that are rich in procarcinogen metabolizing enzymes. In most cases, these metabolic activation systems have been used only with limited success in neo-plastic transformation assays.

B. Macromolecular Interaction and Cellular Repair

The second phase in the process of chemical transformation involves the macromolecular interaction of electrophiles of the direct-acting or activated carcinogen with any electron-rich groups in the cellular environment. This includes RNA, DNA, proteins, and other small molecules. The interaction of electrophiles with DNA has been extensively studied, because it is presumed to be the critical target for an occurrence of a mutagenic/carcinogenic event (Miller and Miller, 1974). Recent evidence shows that a great number of carcinogens form adducts with DNA. The number of sites with which a carcinogen can react in the DNA varies with each carcinogen. Although substances such as aflatoxin B_1 and methylmethane sulfonate produce primarily single major adducts in DNA (Croy et al., 1978; Hemminki, 1980) other compounds such as benzo[a]pyrene diol-epoxide and N-methyl-N'-nitro-N-nitrosoguanidine (MNNG) interact with DNA at multiple sites (King et al., 1975; Singer, 1976; Eastman and Bresnick, 1979). With MNNG, as many as 15 different reaction products with DNA have been identified (Singer, 1976). Although some adducts formed are minor, their biological significance nevertheless could be significant. The extent of carcinogen binding to DNA therefore does not necessarily predict the carcinogenic potential of a chemical.

The DNA repair synthesis as a cellular response to damage induced by carcinogens has been studied with cells in culture. Individuals affected with the rare genetic disorder xeroderma pigmentosum are predisposed to cancer and their cells in culture are found to be deficient in excision-type repair, thus lending support to the potential role of altered DNA in the carcinogenic process (Cleaver and Bootsma, 1975). However, a battery of cellular repair mechanisms that are dependent on the type of lesions induced by the carcinogens have been proposed (Regan and Setlow, 1974; Higgins et al., 1976). Furthermore, DNA repair capacity differs greatly among various cell types. In general, the ability of cells to perform excision repair decreased with the differentiated state of the cells and with the expected life-span of the species of origin (Hart and Setlow, 1974; Holliday and Pugh, 1975). It is not known if cells in vitro that are replicative have different repair capacity than cells that are nonreplicative in vivo. This imposes further limitations on the interpretation of results for in vitro cell models as compared to in vivo situations.

C. Expression of Neoplastic State

Although a large body of information has accumulated concerning the active metabolites of various carcinogens and their interaction with cellular DNA, comparatively little is known concerning the steps that lead to the expression of morphological transformation and ultimately to tumorigenicity. In this respect, the major approach has been to identify phenotypic alterations associated with the carcinogenic process. Although many structural as well as functional changes have been described in virtually all cell systems examined, it has not been possible to identify a property of the transformed cell populations that is an unequivocal marker related to a specific step in neoplastic development. It is evident, however, that the development of neoplastic state is a multistep process both *in vitro* and *in vivo*. This concept is well illustrated in the experimental model systems of liver carcinogenesis *in vivo* (Farber, 1980) and of transformation of Syrian hamster embryo cells *in vitro* (Ts'o, 1979).

In the *in vivo* model of liver carcinogenesis, the preneoplastic phase includes the initial induction of molecular alterations, a relatively long period of growth and selection of putative initiated cells forming foci with constitutive enzyme alterations. This is followed by a premalignant phase in which the "initiated" cells acquired some autonomy represented by appearance of hyperplastic nodules. This process leads to appearance of foci of malignant neoplasia that invade and ultimately metastasize. Throughout the process, architectural, vascular, cytological, enzymatic, and physiological changes all play roles in the determination of behavioral patterns of the progressing lesions (Farber, 1980).

The progressive nature of neoplastic transformations has been extensively studied in the diploid Syrian hamster cells and in the aneuploid cell lines. Earlier studies of Kakunaga (1974) and Borek and Sachs (1967) showed that at least one cell division is necessary to fix transformation and several more cell divisions are required for the expression of transformed state (Borek and Sachs, 1968). Mondal and Heidelberger (1970) further demonstrated that every cell of an established cell line treated with carcinogen had the potential to give rise to transformed clones. In assays that utilize viral-chemical interactions, the virus component plays an active role in the chemical transformation. In contrast, virus apparently does not play an important part in chemical transformation of cell lines such as BALB/c-3T3 and C3H-10T$\frac{1}{2}$ cells, because no detectable production of murine leukemic sarcoma virus, viral specific antigens, or reverse transcriptase have been found (Rapp *et al.*, 1975).

In addition to the morphological criteria initially used to identify the transformed phenotype, a number of changes in membrane characteristics, growth behavior, and cellular properties have been observed in transformed populations. Tumor-specific transplantation antigens were first demonstrated in chemically transformed cells in which syngeneic hosts became immune to further challenge following surgical excision of subcutaneous growth derived from initial challenge of transformed cells (Mondal et al., 1970). The neoantigens were immunologically distinct in that there was no cross-protection between transformed lines derived from separate transformed clones. The unique specificities of neoantigens expressed as a response to carcinogens argue that they arise from specific interaction between carcinogen and "target cell." Other cell-surface changes such as expression of embryonic surface antigens (Coggin and Anderson, 1974) and loss of the cell-surface glycoprotein fibronectin (Kahn and Shin, 1974) also have been reported. Transformed cells often have low serum requirement for growth, and exhibit high cloning efficiency and saturation density when grown on attached surface. Other cellular changes from their nonmalignant counterparts include decreased organization of intracellular structural components (actin), lower level of cyclic nucleotides, and increased proteolytic activity such as the production of plasminogen activator and the property of anchorage-independent growth (Jones et al., 1976; Bertram et al., 1977).

Barrett and Ts'o (1978) examined the sequence of appearance of other markers in serially passaged morphologically transformed Syrian hamster cells and found increased fibrinolytic activity preceded the ability of cells to grow in soft agar. Of the in vitro markers examined, growth in soft agar was found to correlate best with tumorigenic growth potential in vivo in this system (Barrett et al., 1979). Saxholm (1979) also found that passaging of C3H-10T$\frac{1}{2}$ transformed populations in culture is required for the ultimate expression of tumorigenicity. In most cases, however, the relationship between morphological alterations and other markers of neoplastic transformation is uncertain. Furthermore, most of the properties expressed by transformed cells can be dissociated from their in vivo growth potential. New antigens can be induced in methylcholanthrene-treated rat embryo cells without obvious evidence of their having become transformed (Embleton and Baldwin, 1979). Absence of fibronectin and presence of plasminogen activator have been found in both normal and malignant human mammary epithelial cells (Yang et al., 1980). More importantly, there are even exceptions to the rule of correlation between anchorage-independent growth and in vivo tumorigenicity. Discordance between malignancy and anchorage-independent growth has been reported in human carcinoma cells (Marshall *et al.*, 1977; Ossowski and Reich, 1980), adenovirus-transformed cell lines (Gillimore *et al.*, 1977), and a variety of other transformed cell lines (Stiles *et al.*, 1976). In other words, no obligatory step has yet been found in the multistep progressive nature of neoplasia either *in vitro* or *in vivo*. Continued research in this area will undoubtedly contribute to the understanding of the basic mechanism of neoplasia and improve methodologies for cell transformation assays.

III. TYPES OF CELL TRANSFORMATION SYSTEMS

A. Transformation Assays Currently Available

As the methodologies for performing neoplastic transformation assays have evolved, it became clear that no single cell type or population had all of the characteristics that were desired. It has been often argued that, because the preponderance of neoplastic disease in humans involved cell types of epithelial origin, cell culture systems that are derived from this embryonal layer would be the most relevant to predict for potential carcinogenic activity of a chemical agent. Neoplastic transformation has been demonstrated in a few epithelial cell systems, notably, those derived from rat or mouse liver cells and from primary cultures of mouse epidermal cells (Williams, 1976; Mondal and Sala, 1978; Colburn *et al.*, 1978). A common problem in the use of epithelial cells as target cells appeared to be the lack of suitable measurable parameters for the detection of transformed phenotypes shortly after treatment. The demonstration of malignancy *in vivo*, which often requires several months of passage of cells prior to testing for this property, remains the only sure criterion for neoplastic transformation. Therefore, this lack of early phenotypic markers inhibits the use of these sorts of cells as targets in a direct neoplastic transformation assay, especially as a screening tool to identify chemical carcinogens. In the absence of suitable epithelial cell systems for this purpose, the majority of the investigations in this field have been performed with lines of fibroblastlike cells and with strains of mixed cell types usually derived from early passage Syrian hamster embryo cell cultures.

Of the various cell systems that have been demonstrated to be susceptible to transformation by chemical carcinogens, several appear to be less practical than others for development into screening assays. For example, mass culture transformation assays such as those with human fibroblasts (Kakunaga, 1978a,b; Milo and DiPaolo, 1978) and guinea pig embryo cells (Evans and DiPaolo, 1975) require extensive cell passage prior to being recognized as transformed. Other cells such as human osteocarcinoma (Rhim et al., 1975) or BHK-21 (Styles, 1977) are already transformed by several criteria and the meaning of their alteration by carcinogens is unclear. The following represent more defined systems that have been validated to a greater or lesser degree for use as in vitro short-term bioassays.

1. Syrian hamster embryo (SHE) cells (cell strain)
2. BALB/c-3T3; C3H-10T½ and C3H prostate cells (mouse cell lines)
3. RLV/Fischer Rat, AKR/mouse embryo cells and SA7/SHE (viral-chemical interactions)

B. Methodologies and Criteria for Transformation

1. Syrian Hamster Embryo (SHE) Cells

The induction of morphological transformation in early passage Syrian hamster embryo cells (SHE) has been used extensively as a model system to study the mechanism of in vitro neoplastic transformation and as a bioassay to screen for potential carcinogenic chemicals (DiPaolo et al., 1969, 1971a,b, 1972; Pienta et al., 1977). In general, the assay involves exposing a small number (300–500) of cells from a freshly prepared culture or secondary cryopreserved cultures to a graded concentration of chemicals. A feeder layer of lethally irradiated cells (6–8 × 10⁴/50 mm flask) is usually used in the assay to improve the cloning efficiency of the SHE cells and in many cases may contribute to the metabolism of chemicals. Morphologically transformed colonies derived from single cells are scored after the cells are incubated with the chemical for 7–9 days.

More recently, a focus assay was described by Casto et al. (1977) using the SHE cells. In this assay, cells are plated in mass culture at 5×10^4/50 mm dish. The cells were treated with two feedings of test chemicals in a period of 6 days. The cells are then incubated for another 20–25 days prior to staining. Morphologically transformed foci over the background monolayer are scored.

2. BALB/c-3T3, C3H-10T½, and C3H Prostate Cells

At about the same time that in vitro transformation was first demonstrated in SHE cells, Todaro and Green (1963) developed the Swiss-3T3 cell line which possesses the property of density-dependent inhibition of cell growth. These cells were used extensively to study neoplastic transformation with papovaviruses. Subsequently, similar cell lines were developed by Aaronson and Todaro (BALB/c-3T3) and Heidelberger and associates (C3H-10T½ and C3H prostate). The use of these cell lines for neoplastic transformation by chemicals was demonstrated by Kakunaga (1973) in BALB/c-3T3 cells, Reznikoff et al. (1973) in C3H-10T½ cells, and Chen and Heidelberger (1969) in adult C3H ventral prostate cells as well as the later work of Marquardt et al. (1972) in this cell system.

Although each of these systems with cell lines differ in minor ways with respect to response to chemicals and in experimental details of protocol, the cell transformation assays which were developed empirically are essentially the same for the three cell lines. The assay involved plating the target cells in mass culture at 10^3–10^4 cells/plate. One day after cell plating, the test chemical is introduced at various concentrations for a period of 1–3 days depending on the cell line used. Subsequently, the cells are further incubated and fed with fresh medium at regular intervals for a period of 4–6 weeks in the absence of chemicals. The plates are then fixed, stained, and scored for morphological transformed foci.

3. *RLV/1706, AKR/Mouse, SA7/SHE*

The viral–chemical interactions assays utilize either RNA or DNA viruses as agents to induce or enhance chemical transformation. The role of an RNA virus in the induction of leukemia in irradiated mice was first found by Lieberman and Kaplan (1959). Subsequently, viral enhancement of chemical carcinogenesis by C-type virus was developed by various laboratories to screen chemical carcinogens (Freeman et al., 1973; Rhim et al., 1974; Hetrick and Kos, 1975). Among them, a stable Fischer rat cell line 1706 infected with Rauscher leukemia virus (RLV/1706) and mouse embryo cell strains infected with AKR virus (AKR/mouse) are two most studied assays. The general protocol of RLV/1706 assay involves plating target cells at 10^5 cells/flask. The cultures are infected with virus at day 0 or chronically infected. The cells are subsequently subcultured and treated with chemicals for varied periods of time. After treatment the cells are passaged fortnightly until transformed foci can be observed. This may require 6–8 weeks. Sometimes no discrete foci are seen, rather the entire culture appears to be transformed, probably due to the uniform dispersion of normal and transformed cells during cell passage.

Stoker (1963) first reported the enhancement of polyoma DNA virus transformation of BHK 21 cells by X-irradiation with the adenovirus SV40. The exact mechanism for the chemical–viral interaction is not known, although it has been postulated that agents interacting with cellular DNA may create additional sites for the integration of viral DNA which resulted in an increased viral transformation frequency. The enhancement of adenovirus transformation as a screening assay has been developed mainly by Casto *et al.* (1974), in the Syrian hamster cells (SA7/SHE). The following procedure is generally used. Target cells at 3–4 $\times$ 10^6 cells/60 mm dish are treated with various dilutions of chemicals for 2 or 18 hours. After chemical treatment the virus is added and adsorption carried out for 3 hours. The cells are then trypsinized and replated at 2 $\times$ 10^5 cells. After 5–6 days incubation, the cultures are changed into low calcium medium containing 0.5% Bacto-agar. These conditions permit the optimal development of transformed foci in virus-treated but not non-virus-treated cells. Foci can be scored in 2–3 weeks.

C. Evaluation and Comparison of Various Test Systems

1. Assay Performance

a. Duration of Assay. Some of the desirable qualities for short-term tests are that they should be simple, rapid, and inexpensive to perform. Operationally, none of the transformation assays or mammalian cell short-term tests, in general, is as simple, rapid, or inexpensive as tests using microbial mutagenesis systems. Obviously, however, they are considerably more rapid and inexpensive than the long-term *in vivo* carcinogenicity bioassay and are surely more relevant than the microbial tests. The length of time one takes to complete an assay reflects the space requirement and the labor intensity which are translatable to the cost of the assay. Among the transformation assays, the durations vary substantially. For example, the SHE clonal assay is completed in about 10 days without medium renewal or cell manipulation, and the RLV/1706 assay requires as long as 2 months to complete with multiple cell passages.

b. Target Cells. Theoretically, the most desirable type of target cell for neoplastic transformation assay is a diploid cell with normal physiological and structural characteristics and capable of undergoing a recognizable morphological change after exposure to a carcinogen. In many respects, Syrian hamster embryo cells are the only ones that fulfill these requirements to any satisfactory degree. Yet even among these cell populations that are derived from midstage (12–14 day) embryo tissue, there is an heterogeneity of response with respect to transformation that is not well understood. According to Poiley *et al.* (1980), not all litters provide cells that are transformable in the standard assay and there even appears to be considerable variation among the pups in a single litter with respect to the facility of cell populations from these individual pups to be transformed by a chemical carcinogen. It is evident in the process from original derivation of cells in culture to the completion of the assay that there are several very strong selection events that may influence the final result. The first selection pressure is in the original primary culture that is prepared at an extraordinarily high cell density, usually greater than 5×10^6 cells per 100-mm culture dish. In fact, the successful cultures for transformation are those that can be cultured initially at these densities. The second selection pressure is the cloning process accompanying the actual performance of the assay in which, in the absence of feeder cells, fewer than 10% of the cells survive and would be at risk for transformation. Whether the heterogeneity of neoplastic transformation response among populations of Syrian hamster embryo cells is a consequence of these selection pressures is not known and suggests an area of research endeavor to explore the factors that determine the transformability of a specific cell population.

In contrast to this property of nonuniform target cell pools in the Syrian hamster embryo system, the several cell-line systems do provide a considerably higher degree of uniformity. The price paid for this increase in consistent behavior among different cell pools is the necessity of working with cells that are unquestionably not "normal." As Hayflick (1967) had indicated a number of years ago, the alteration of a cell population from a strain to a line is accompanied by acquisition of several properties associated with neoplastic cells such as aneuploidy, "immortality," and high cloning efficiency and often actual oncogenicity. Thus, the relationship between a morphological alteration in a cell-line culture resulting from exposure to a carcinogen and the events leading to a tumor *in vivo* is not clear.

Even though a few populations of rodent cell lines have been obtained that are relatively stable in their ability to be transformed morphologically by carcinogens (BALB/c-3T3, C3H-10T$\frac{1}{2}$), this property of susceptibility to chemical transformation appears to be a rare event, at least for BALB/c-3T3 cells. "Variants" that are not susceptible to transformation have been isolated from a population which is transformable by chemical carcinogens (Kakunaga and Crow, 1980). In contrast, transformation with either DNA (SV40, polyoma) or RNA (mouse sarcoma, Rauscher leukemia) tumor viruses seems to be regularly demonstrable with a variety of populations in both cell strains and cell lines.

In sum then, the selection of either cell strains or lines for the study of the carcinogenic process and/or the identification of putative carcinogens is accompanied by compromises dictated by certain properties of these different cell populations which are, as yet, unexplained. The key unanswered question is what makes a cell transformable by a chemical carcinogen.

c. End Points. The primary end point used for diagnosis of neoplastic transformation in cell culture bioassays remains a morphological one, be it expressed as an altered single colony in a clonal culture or as a focus in a background monolayer in a mass culture. The altered cell populations are recognized as variants that exhibit properties of disorganization as compared to a regular orientation of untransformed cells as well as a tendency to achieve multiple layers of cell growth as compared to a monolayer.

In the mass culture assay such as in the RLV/1706 assay and SHE/ focus assay, transformed cells are often dispersed uniformly with nontransformed cells during cell passage, and the recognition of transformed phenotype can be quite difficult and subjective. In the SHE/ clonal assay, the clonal appearance of control cells is not uniform. Single- and multiple-layered cell growth are often not clearly distinguishable among different clones, even in control cultures, and the diagnosis of transformation is most often made on the morphological configuration of cells at the periphery of a clone. In the focus assays using BALB/c-3T3 or C3H-10T½ cells, the transformed foci are easily distinguishable from the untransformed background monolayer. However, this advantage is accompanied by the fact that this assay procedure imposes certain restrictions in the interpretation of data as will be considered later.

The validation for the use of morphological alteration in cell culture as a diagnostic for neoplastic transformation comes from the observation in most systems that transformed cell populations often give rise to tumors when implanted into appropriate hosts. This tumorigenic response, however, is not one that is absolute or immediate (Barrett and Ts'o, 1978; Saxholm, 1979). A number of investigators have shown that passage of transformed cell populations in vitro results in an increasing likelihood of these cells exhibiting markers of transformation (discussed in Section II,C). The difficulty in interpreting these validation experiments is the inability to demonstrate whether these progressions to increased oncogenicity is the result of some selection process that enriches the population for the transformed phenotype or whether the population actually acquired these properties with increasing passage in culture. Unfortunately, experiments specifically designed to test this hypothesis have not been performed and a clear answer to this question is not available.

d. Metabolic Activation. One of the common problems that is shared by all short-term tests *in vitro* is the limited capability of many of the target cells to metabolize procarcinogens. This necessitates the incorporation into the assays of enzyme complexes isolated from mammalian tissues as the metabolic activation system. The most widely used metabolic supplement is an Aroclor-induced rat liver microsomal homogenate often referred to as S9 fraction. Although the S9 fraction has been used successfully to activate procarcinogens such as aflatoxins and aromatic amines to mutagenic compounds in the microbial Salmonella assay (McCann et al., 1975) and in certain mammalian cell gene mutation assays, it has not been as successful for use in the transformation assays, mainly due to the high inherent toxicity of S9 fraction to the target cells.

A primary limitation to the use of exogenous metabolic activation systems has been attributed to an artificial ratio between activation and deactivation enzymes as compared to the ratio *in vivo* in the intact organ or even intact cell in the case of hepatocytes (Schmeltz et al., 1978; Bigger et al., 1980). This is further complicated by the findings that the ability of S9 fractions to metabolize many procarcinogens is dependent on the species and strain of animals as well as the tissues and inducers used (Hutton et al., 1979; Müller et al., 1980).

Recently, hamster hepatocytes have been used as a metabolic activation system in the SHE clonal transformation assay with some success (Poiley et al., 1979), and rat liver hepatocytes may be of use with BALB/c-3T3 cell transformation assays (A. S. Tu and A. Sivak, unpublished results). Overall, however, the application of appropriate metabolic systems for the various transformation assays remains a problem that requires considerable experimental effort to approach a widely utilizable and generic system.

2. Interpretation of Data

a. Assay Quantitation. The experimental protocols by which each of the transformation assays are conducted determine to a considerable degree their amenability to a quantitative analysis of data. For example, the clonal assays provide for both cytotoxicity and transformation evaluation in the same experimental vessel, thus allowing a direct calculation of transformation frequency on a cell survivor basis. In contrast, the nature of a focus-type assay raises uncertainties with respect to the origin of transformed foci as well as the impact of the background monolayer of untransformed cells on the transformation response.

A key factor is the measurement of cytotoxicity resulting from carcinogen exposure that directly affects the estimation of the size of target cell population at risk. Because both the method of determining cytotoxicity (cell counts, direct cloning, delayed cloning) and the chemical nature of the carcinogen influence the outcome of this measurement, apparently independently of each other, no simple relationship between cytotoxicity and transformation response is evident. Moreover, since the carcinogen is often present for several cell cycles, an estimation of the actual size of the target cell population at risk is further confounded (Sivak and Tu, 1980b).

Another factor that requires consideration in focus assays is the role of target-cell density. In addition to the influence that this parameter has on determination of cytotoxicity, it also affects the actual transformation response. In experiments with both chemical and physical carcinogenic agents, the calculated transformation frequency decreases with an increase in the target cell population, whether calculated either as a function of initial plating density or as surviving cells by some measure of cytotoxicity (Haber *et al.*, 1977; Kakunaga, 1978a,b; Sivak and Tu, 1980b). This variable further complicates a meaningful quantitation of transformation frequency on a cellular basis.

Related to this factor of cell density in focus assays is the individual behavior of the few transformed cells in a population containing untransformed cells at a level of several orders of magnitude greater. With BALB/c-3T3 cells, reconstruction experiments have demonstrated that the expression of the neoplastic phenotype of transformed cells is inhibited by an excess of untransformed cells and that cells of transformed foci do not exhibit a tendency for migration and establishment of secondary foci (A. S. Tu and A. Sivak, unpublished results). In contrast, results with C3H-10T$\frac{1}{2}$ cells suggest the likelihood for migration and seeding of secondary foci of transformed cells in confluent plates, thus raising another complicating factor in the attempt to express transformation frequency in a quantitative manner in focus assays (C. Heidelberger, personal communication).

In the chemical–viral interaction assays, the procedures require either serial passaging of treated cells and/or chronic infection with virus so that the target cell size is almost impossible to quantitate. In the DNA virus assay, the virus itself induced transformation, and the number of foci induced by the combination of virus and chemical is often expressed as an enhancement ratio. Although this enhancement has been associated with DNA damage induced by the carcinogen, the relationship of the response to the carcinogenic process is not clear. Furthermore, cytotoxicity of a chemical in these assays also introduced uncertainties, since chemical-induced cytotoxicity can result in fewer transformants in treated (virus plus chemical) than in control cultures (virus alone).

b. Dose–Response Effect. In the event that a chemical induces a positive response in the transformation assay, it is desirable that the chemical should do so in a dose-dependent manner. In validation studies of the different transformation assays, a dose–response effect was not observed in many instances. This could be due in part to an inappropriate concentration range of chemical used in the test, insolubility, or inactivation of the chemical in the aqueous medium. However, other factors inherent in the transformation assays could also account for this observation. For example, in the SHE assay, less than 25% of the chemicals that induced a positive response have been reported to show a dose-dependent effect (Pienta *et al.,* 1977). In this case, the heterogeneity of the target cell populations in their responsiveness to carcinogens could be an explanation. In the BALB/c-3T3 assay, transformation frequency is expressed as foci/plate or plates with foci and not on a per cell basis due to the various factors that affect the quantitation of transformation frequency in a focus assay (Section III,C,2,a). Therefore, the cytotoxic effect of a chemical may result in no apparent increase in transformation with increasing concentration of chemical because of the decrease in actual target cell size.

c. Spontaneous Transformation Frequency. The magnitude of the background transformation frequency as determined in the untreated or solvent controls varies in each of the transformation assays and in different laboratories. Pienta *et al.* (1977) did not report any spontaneous transformed colonies in SHE cells in over 29,000 control colonies. Ts'o (1979), however, reported a spontaneous transformation frequency of $1/10^4$ in SHE cells. Since the assay used is a clonal assay, with perhaps as many as 2000 colonies screened at one time, the occurrence of spontaneous transformation appears to be a relatively rare event.

In the focus transformation assays with established cell lines, C3H-10T$\frac{1}{2}$ cells have a lower spontaneous transformation frequency than BALB/c-3T3 cells. The reason for this is not known, since both cell lines are quite similar in several respects in that they both are derived from mouse embryo fibroblasts and are aneuploid. However, even within a single cell line, populations have been isolated which varied in their spontaneous- and induced-transformation frequencies (Kakunaga and Crow, 1980). In general, it appears that the higher frequencies of carcinogen-induced transformation are observed in populations that exhibit elevated frequencies of spontaneous transformation (Sivak and Tu, 1980a,b). The occurrence of spontaneous transformation does necessitate more stringent quality control on the performance of the assay. These include pretesting of target cell pools and fetal calf serum lots to determine an acceptable range of spontaneous-transformation frequency.

3. Data Base

The usefulness of any of the short-term assays as a means to identify chemicals of carcinogenic potential will ultimately be determined from the accumulated data base. Validation of the transformation assays with *in vivo* carcinogenicity data will resolve the sensitivity and specificity of the assays. Unfortunately, in many instances *in vivo* data are not available or are deficient, so that a comprehensive evaluation may not yet be possible. A major effort to explore this issue is the Gene-Tox program of the United States Environmental Protection Agency.

Of the assortment of transformation assays described which have been validated under double-blind study, the largest number of chemicals have been tested in four of the assays which are representative of the different types of assay that are available: BALB/c-3T3, SHE/clonal, RLV/1706, and SA7/SHE. Of the chemicals that have been tested, the overall correlation with the available *in vivo* carcinogenicity data appears good in the four transformation assays. The majority of false negatives are found in assays that have not yet utilized reliable metabolic activating systems.

D. Future Directions

Some of the limitations to the use of *in vitro* cell culture systems to study mechanisms of *in vivo* neoplasia have been identified above. Mainly, the *in vitro* model of chemical transformation does not take into account the pharmacokinetic distribution, tissue specific metabolism, differential repair mechanisms, humoral factors, and immunologic processes that occur *in vivo*. Nevertheless, one cannot fail to recognize some of the parallel phenomena that occur between the *in vitro* model of transformation and the *in vivo* situation.

Both the *in vitro* and *in vivo* neoplastic transformations are found to be a multistep progressive process. Some of the cellular markers of transformation are recognized both *in vitro* and *in vivo*. Among them are the behavior of uncontrolled growth, the ability to grow in semisolid media, and the exhibition of transformation related cell surface antigens. Agents such as phorbol esters which have been found to promote tumors *in vivo* also enhance transformation *in vitro* (Mondal *et al.*, 1976; Sivak and Tu, 1980a). Some known inhibitors of *in vivo* carcinogenesis such as retinoids and protease inhibitors have been found to inhibit *in vitro* transformation (Kuroki and Drevon, 1979; Merriman and Bertram, 1979). Furthermore, it has been demonstrated that transformed cells in culture induced by carcinogens can produce tumors in appropriate host animals. Thus, the use of transformation assays as screening tools to identify carcinogenic chemicals appears to have relevant end points with respect to the process of carcinogenicity *in vivo*. However, additional research and developmental activities are necessary to elucidate the biological mechanisms and factors that influence the expression of the carcinogenic phenotype. Several questions are pertinent for the further understanding and development of the *in vitro* transformation assays.

The cellular properties associated with a capability to undergo carcinogen-induced neoplastic transformation are not known. The heterogeneity of neoplastic transformation response among populations of Syrian hamster embryo cells and the clonal variability among various established cell lines suggest an area of research endeavor to explore the factors that determine the transformability of a specific cell population. Thus far, properties such as differential cytotoxicity, transformability by Kristen murine sarcoma virus, induction of mutants, chromosome constitution, or differences in capability to metabolize polycyclic aromatic hydrocarbons do not appear to be critical ones contributing to the cellular susceptibility to undergo transformation in the BALB/c-3T3 cell line (Sivak *et al.*, 1980).

The applicability of metabolic activation systems is paramount in order to improve the predictive value of the transformation assays. The use of rat liver S9 as a universal metabolic activation system has been seriously questioned in view of the evidence accumulated on the differential metabolism between intact cells and enzyme homogenates. Another approach that may be necessary is the development of a number of metabolic activation systems which are optimal for certain particular classes of chemicals and/or utilization of specific types of transformation assays because of unique endogenous sensitivity to certain chemical classes.

Most of the short-term bioassays such as the microbial and mammalian mutagenicity assays that have been developed for screening carcinogenic chemicals implicitly assume that mutagens are carcinogens. The use of transformation assays, however, does not presuppose the occurrence of a genetic event since the end point chosen is morphological. In many respects morphological criteria alone are also not satisfactory, especially with cells in culture whose morphology can easily be manipulated by experimental conditions. While *in vivo* tumorigenicity remains the final validation of transformed phenotype, the identification of other dependable early transformation markers will undoubtedly strengthen the morphological criteria.

IV. REGULATORY AND COMMERCIAL USES OF NEOPLASTIC TRANSFORMATION ASSAYS

A. Regulatory

The combination of an increased awareness concerning possible human health hazards resulting from exposure to chemicals in the environment and the legislative regulatory actions stemming from this awareness has provided the environment for the proliferation of short-term bioassays as tools to assess these hazards. Because these tests are relatively inexpensive and of short duration in comparison to the chronic exposure studies that have long been the mainstay of experimental toxicology, they could provide a means to assist both regulatory and commercial organizations in making decisions concerning health risks associated with exposure to chemical agents.

The initial experimental thrust was focused on a variety of direct (gene mutation, chromosome aberration) and indirect (DNA repair, sister chromatid exchange) measures of toxic effects on the genome. Characteristic of these sorts of "genotoxic" events is that they are single-step events that can be measured as a direct consequence of an interaction between a chemical and the DNA of a target cell. The substantial coincidence between mutagenic and carcinogenic activity, at least in some laboratory experimental systems, has led to proposals that regulatory actions with respect to prevention of carcinogenic risk could be taken on the basis of positive responses in one or a few genetic assays.

This view has led to the construction of correlations (McCann *et al.*, 1975; Melselson and Russell, 1977) that fail to appreciate the innate biological differences between mutagenic and carcinogenic processes. Mutagenesis is a single-step process whereas carcinogenesis is clearly a multistep, multifactorial process, even though it, like mutagenesis, can be initiated under some conditions by a single exposure to a chemical or physical agent. It is in this context of providing an assay that displays a series of biological events leading to a phenotype associated with neoplastic behavior that the cell-culture transformation assays find their niche in the short-term test armamentarium.

The question of their proper use in regulatory decisions depends in part on the type of regulatory action contemplated and in part on the amount and quality of associated information. For identification of chemical entities that might be expected to induce tumors in experimental animals, and possibly in humans, a single or set of cell-culture transformation assays would suffice. A more ambitious use of the results of cell transformation assays would be directly in a risk assessment evaluation. Beyond the issues of limitations of transformation assays mentioned above, there are at least two additional key areas that will contribute to the development of a risk assessment profile for a chemical. Both are related to the question of exposure that appears to be the determining factor in such calculations.

The first area deals with the external factors that control the level of an environmental chemical that is available for interaction with a target organism, humans in the case of a risk assessment for a health hazard. This complex and extremely important area is undergoing rapid development and sophistication in modeling that is often not appreciated by the practicing experimental toxicologist, even though it provides the basic information on which interpretation of biological effects and pharmacokinetic data are dependent in the risk assessment process (Anderson et al., 1980). Although this review is not the proper vehicle to consider this component of exposure in detail, those who would embark on the performance of risk assessment must incorporate this type of information in order for the assessments to have practical meaning.

The second area deals with estimations of internal or target tissue/cell/molecule exposure, i.e., pharmacokinetics. Ideally, one could envision the acquisition of such data at sufficient sensitivity and discrimination to approach an estimation of cellular exposure associated with some level of environmental interaction taking into account the routes of exposure. The combination of this type of pharmacological information with biological effects data from one or more cellular transformation assays might provide a means to estimate human risk with uncertainties no greater than those obtained with the use of experimental animal carcinogenesis data. Even though the pioneering experimental work of investigators such as Gehring and associates (1979) and the conceptual contributions of Cornfield (1977) and Whittemore (1980), for example, have improved our ability to utilize experimental animal data in risk assessment, much work remains to be done to apply pharmacokinetic principles to *in vivo* systems, with considerable additional effort required before these principles could be applied in association with cell culture assays.

B. Commercial

The primary commercial use of transformation assays stems from their application as a means to identify potential chemical carcinogens. For internal uses, information obtained from these and other short-term tests could have value in the selection of substances of lower potential toxicity from a group of congeners of approximately equivalent efficacy. Another possible use would be as part of a decision network to either continue or stop development of a material because of its genotoxic potential.

Externally, the transformation assays are useful in addressing a variety of regulatory needs for various Federal agencies. In particular, data from such assays could be useful in support of pesticide registrations or premanufacture notifications for the Environmental Protection Agency or in support of drug or food additive applications to the Food and Drug Administration.

V. SUMMARY

One of the major attractions of the cell-culture neoplastic transformation systems is that biological events relating to the carcinogenic process can be followed at the cellular or cell population level in the relatively well-defined and controlled atmosphere of a culture vessel. Although such a situation does provide a high degree of specificity with respect to the types of measurements that can be performed in contrast to those that are possible in whole animals, considerable caution should be exercised in the interpretation of cell-culture data as models for intact animal phenomena related to carcinogenesis. It is very likely that complex immunological, hormonal, and nutritional influences that occur in intact animals are the major controlling factors in the carcinogenic process. Thus, the absence of these factors in cell neoplastic transformation systems raises questions about how far *in vitro* mechanism studies can be extrapolated to events occurring in intact animals. Nevertheless, validation studies have shown that the *in vitro* neoplastic transformation assays measure a relevant end point of carcinogenic process and that they are useful in the identification of chemicals that are putative carcinogens. With further development in metabolic activation systems and the standardization of the assays among different laboratories, these transformation assays will undoubtedly have significant value as a short-term screening test for environmental carcinogens.

REFERENCES

Anderson, M. W., Hoel, D. G., and Kaplan, N. L. (1980). A general scheme for the incorporation of pharmacokinetics in low-dose risk estimation for chemical carcinogenesis: Example—vinyl chloride. *Toxicol. Appl. Pharmacol.* **55**, 154–161.

Barrett, J. C., and Ts'o, P. O. P. (1978). Evidence for the progressive nature of neoplastic transformation *in vitro*. *Proc. Natl. Acad. Sci. U.S.A.* **75**, 3761–3765.

Barrett, J. C., Crawford, B. D., Mixter, L. O., Schechtman, L. M., Ts'o, P. O. P., and Pollack, R. (1979). Correlation of *in vitro* growth properties and tumorigenicity of Syrian hamster cell lines. *Cancer Res.* **39**, 1504–1510.

Bertram, J. S., Libby, P. R., and Stourgeon, W. M. (1977). Changes in nuclear action levels with change in growth state of C3H-10T ½ cells and the lack of response in malignantly transformed cells. *Cancer Res.* **37**, 4104–4111.

Berwald, Y., and Sachs, L. (1963). *In vitro* transformation with chemical carcinogens. *Nature (London)* **200**, 1182–1184.

Bigger, C. A. H., Tomaszewski, J. E., Dipple, A., and Lake, R. S. (1980). Limitations of metabolic activation systems used with *in vitro* tests for carcinogens. *Science* **209**, 503–505.

Borek, C., and Sachs, L. (1967). Cell susceptibility to transformation by X-irradiation and fixation of the transformed state. *Proc. Natl. Acad. Sci. U.S.A.* **57**, 1522–1527.

Borek, C., and Sachs, L. (1968). The number of cell generations required to fix the transformed state in X-ray induced transformation. *Proc. Natl. Acad. Sci. U.S.A.* **59**, 83–85.

Casto, B. C., Pieczysnki, W. J., and DiPaolo, J. A. (1974). Enhancement of adenovirus transformation of hamster cells *in vitro* by chemical carcinogens. *Cancer Res.* **34**, 72–78.

Casto, B. C., Janosko, N., and DiPaolo, J. A. (1977). Development of a focus-assay model for transformation of hamster cells *in vitro* by chemical carcinogens. *Cancer Res.* **37**, 3508–3515.

Chen, T. T., and Heidelberger, C. (1969). Quantitative studies on the malignant transformation of mouse prostate cells by carcinogenic hydrocarbons *in vitro*. *Int. J. Cancer* **4**, 166–178.

Cleaver, J., and Bootsma, D. (1975). Xeroderma pigmentosum: Biochemical and genetic characteristics. *Annu. Rev. Genet.* **9**, 19–38.

Coggin, J. H., and Anderson, N. G. (1974). Cancer, differentiation and embryonic antigens: Some central problems. *Adv. Cancer Res.* **19**, 105–165.

Colburn, N. H., Vorder Brugge, W. F., Bates, J., and Yuspa, S. H. (1978). Epidermal cell transformation *in vitro*. In "Carcinogenesis: Mechanism of Tumor Promotion and

Cocarcinogenesis" (T. J. Slaga, A. Sivak, and R. K. Boutwell, eds.), Vol. 2. pp. 257–271. Raven, New York.

Cornfield, J. (1977). Carcinogenic risk assessment. *Science* **198,** 693.

Croy, R. G., Essigmann, J. M., Reinhold, V. N., and Wogan, G. N. (1978). Identification of the principal aflatoxin B_1-DNA adduct formed *in vivo* in rat liver. *Proc. Natl. Acad. Sci. U.S.A.* **75,** 1745–1749.

DiPaolo, J. A., Donovan, P. J., and Nelson, R. L. (1969). Quantitative studies of *in vitro* transformation by chemical carcinogens. *J. Natl. Cancer Inst.* **42,** 867–876.

DiPaolo, J. A., Donovan, P. J., and Nelson, R. L. (1971a). *In vitro* transformation of hamster cells by polycyclic hydrocarbons. Factors influencing the number of cells transformed. *Nature (London) New Biol.* **230,** 240–242.

DiPaolo, J. A., Nelson, R. L., and Donovan, P. J. (1971b). Morphological, oncogenic, and karyological characteristics of Syrian hamster embryo cells transformed *in vitro* by carcinogenic polycyclic hydrocarbons. *Cancer Res.* **31,** 1118–1127.

DiPaolo, J. A., Nelson, R. L., and Donovan, P. J. (1972). *In vitro* transformation of Syrian hamster embryo cells by diverse chemical carcinogens. *Nature (London)* **235,** 270–280.

Earle, W. R. (1943). Production of malignancy *in vitro*. IV. The mouse fibroblast cultures and changes seen in the living cells. *J. Natl. Cancer Inst.* **4,** 165–212.

Earle, W. R., and Nettleship, A. (1943). "Production of malignancy *in vitro*. V. Results of injections of cultures into mice. *J. Natl. Cancer Inst.* **4,** 213–227.

Earle, W. R., and Voegtlin, C. (1938). The mode of action of methylcholanthrene on cultures of normal tissues. *Am. J. Cancer* **34,** 373–390.

Earle, W. R., Shelton, E., and Schilling, E. L. (1950). Production of malignancy *in vitro*. XI. Further results from reinjection of *in vitro* cell strains into strain C3H mice. *J. Natl. Cancer Inst.* **10,** 1105–1113.

Eastman, A., and Bresnick, E. (1979). Metabolism and DNA binding of 3-methylcholanthrene. *Cancer Res.* **39,** 4316–4321.

Embleton, M. J., and Baldwin, R. W. (1979). Tumor-related antigen specifications associated with 3-methylcholanthrene-treated rat embryo cells. *Int. J. Cancer* **23,** 840–845.

Evans, C. H., and DiPaolo, J. A. (1975). Neoplastic transformation of guinea pig cells in culture induced by chemical carcinogen. *Cancer Res.* **35,** 1035–1044.

Farber, E. (1980). The sequential analysis of liver cancer induction. *Biochem. Biophys. Acta. Rev. Cancer* **605,** 149–166.

Freeman, A. E., Weisburger, E. K., Weisburger, J. H., Wolford, R. G., Maryak, J. M., and Huebner, R. J. (1973). Transformation of cell cultures as an indication of the carcinogenic potential of chemicals. *J. Natl. Cancer Inst.* **51,** 799–808.

Gehring, P. J., Watanabe, P. G., and Blau, G. E. (1979). Risk assessment of environmental carcinogens utilizing pharmacokinetic parameters. *Ann. N.Y. Acad. Sci.* **329,** 137–152.

Gillimore, P. H., McDougall, J. K., and Chen, L. B. (1977). *In vitro* traits of adenovirus-transformed cell lines and their relevance to tumorigenicity in nude mice. *Cell* **10,** 669–678.

Haber, D. A., Fox, D. A., Dynan, W. S., and Thilly, W. G. (1977). Cell density dependence of focus formation in the C3H/10T$\frac{1}{2}$ transformation assay. *Cancer Res.* **37,** 1644–1648.

Hart, R. W., and Setlow, R. B. (1974). Correlation between deoxyribonucleic acid excision-repair and life-span in a number of mammalian species. *Proc. Natl. Acad. Sci.* **71,** 2169–2173.

Hayflick, L. (1967). Oncogenesis *in vitro*. *Natl. Cancer Inst. Monogr. No.* **26**, 355–377.

Hemminki, K. (1980). Identification of guanine-adducts of carcinogens by their fluorescence. *Carcinogenesis* **1**, 311–316.

Hetrick, F. M., and Kos, W. L. (1975). Transformation of cell culture as a parameter for detecting the potential carcinogenicity of antischistosomal drugs. *J. Toxicol. Environ. Health* **1**, 323–327.

Higgins, N. P., Kato, K., and Strauss, B. (1976). A model for replication repair in mammalian cells. *J. Mol. Biol.* **101**, 417–425.

Holliday, R., and Pugh, J. (1975). DNA modification mechanisms and gene activity during development. *Science* **187**, 226–232.

Hsie, A. H., O'Neill, J. P., and McElheny, V. K., eds. (1979). *Banbury Rep.* **2**, 1–504.

Hutton, J. J., Meier, J., and Hackney, C. (1979). Comparison of the *in vitro* mutagenicity and metabolism of dimethyl-nitrosamine and benzo[a]pyrene in tissues from inbred mice treated with phenobarbital, 3-methylcholanthrene or polychlorinated biphenyls. *Mutat. Res.* **66**, 75–94.

Jones, P. A., Laug, W. E., Gardner, A., Nye, C. A., Fink, L. M., and Benedict, W. F. (1976). *In vitro* correlate of transformation in C3H-10T ½ clone 8 mouse cells. *Cancer Res.* **36**, 2863–2867.

Kahn, P., and Shin, S. I. (1979). Cellular tumorigenicity in nude mice. Test of associations among loss of cell-surface fibronectin, anchorage independence and tumor-forming ability. *J. Cell Biol.* **82**, 1–16.

Kakunaga, T. (1973). Quantitative system for assay of malignant transformation by chemical carcinogens using a clone derived from BALB/3T3. *Int. J. Cancer* **12**, 463–473.

Kakunaga, T. (1974). Requirement for cell replication in the fixation and expression of the transformed state in mouse cells treated with 4-nitroquinoline-1-oxide. *Int. J. Cancer* **14**, 736–742.

Kakunaga, T. (1978a). Factors affecting polycyclic hydrocarbon-induced cell transformation. *In* "Polycyclic Hydrocarbons and Cancer" (H. Gelboin and P. O. P. Ts'o, eds.), Vol. 2, pp. 293–304. Academic Press, New York.

Kakunaga, T. (1978b). Neoplastic transformation of human diploid fibroblast cells by chemical carcinogens. *Proc. Natl. Acad. Sci. U.S.A.* **75**, 1334–1338.

Kakunaga, T., and Crow, J. D. (1980). Cell variants showing differential susceptibility to ultraviolet light-induced transformation. *Science* **209**, 505–507.

King, H. W. S., Thompson, M. H., and Brookes, P. (1975). The benzo[a]pyrene deoxyribonucleoside products isolated from DNA after metabolism of benzo[a]pyrene by rat liver microsomes in the presence of DNA. *Cancer Res.* **34**, 1263–1269.

Kuroki, T., and Drevon, C. (1979). Inhibition of chemical transformation in C3H/10T½ cells by protease inhibitors. *Cancer Res.* **39**, 2755–2761.

Lotlikar, P. D., Enomoto, M., Miller, J. A., and Miller, E. C. (1967). Species variation in the N- and ring-hydroxylation of 2-acetylaminofluorene and effects of 3-methylcholanthrene pretreatment. *Proc. Soc. Exp. Biol. Med.* **125**, 341–346.

Lower, G. M., Jr., and Bryan, G. T. (1973). Enzymic N-acetylation of carcinogenic aromatic amines by liver cytosol of species displaying different organ susceptibilities. *Biochem. Pharmacol.* **22**, 1581–1588.

Lieberman, M., and Kaplan, H. S. (1959). Leukemogenic activity of filtrates from radiation-induced lymphoid tumors of mice. *Science* **130**, 387–388.

McCann, J., Choi, E., Yamasaki, E., and Ames, B. N. (1975). Detection of carcinogens as mutagens in the *Salmonella*/microsome test: Assay of 300 chemicals. *Proc. Natl. Acad. Sci. U.S.A.* **72**, 5135–5139.

Magee, P. N. (1977). Evidence for the formation of electrophilic metabolites from N-nitroso compounds. *In* "Origins of Human Cancer" (H. H. Hiatt, J. D. Watson, and J. A. Winston, eds.), Book B, pp. 629–638. Cold Spring Harbor Lab., Cold Spring Harbor, New York.

Marquardt, H., Kuroki, T., Huberman, E., Selkirk, J. K., Heidelberger, C., Grover, P. L., and Sims, P. (1972). Malignant transformation of cells derived from mouse prostate by epoxides and other derivatives of polycyclic hydrocarbons *Cancer Res.* **32**, 716–720.

Marshall, C. J., Franks, L. M., and Carbonell, A. W. (1977). Markers of neoplastic transformation in epithelial cell lines derived from human carcinomas. *J. Natl. Cancer Inst.* **58**, 1743–1747.

Melselson, M., and Russell, K. (1977). Comparisons of carcinogenic and mutagenic potency. *In* "Origins of Human Cancer" (H. H. Hiatt, J. D. Watson, and J. A. Winsten, eds.), Book C, pp. 1473–1481. Cold Spring Harbor Lab., Cold Spring Harbor, New York.

Merriman, R. L., and Bertram, J. S. (1979). Reversible inhibition by retinoids of 3-methylcholanthrene-induced neoplastic transformation in C3H/10T $\frac{1}{2}$ clone 8 cells. *Cancer Res.* **39**, 1661–1666.

Miller, E. C., and Miller, J. A. (1974). Biochemical mechanisms of chemical carcinogenesis. *In* "Molecular Biology of Cancer" (H. Busch, ed.), pp. 377–402. Academic Press, New York.

Miller, J. A., and Miller, E. C. (1969). The metabolic activation of carcinogenic aromatic amines and amides. *Prog. Exp. Tumor Res.* **11**, 273–301.

Miller, J. A., and Miller, E. C. (1977). Ultimate chemical carcinogens as reactive mutagenic electrophiles. *In* "Origins of Human Cancer" (H. H. Hiatt, J. D. Watson, and J. A. Winston, eds.), Book B, pp. 605–627. Cold Spring Harbor Lab., Cold Spring Harbor, New York.

Milo, G. E., Jr., and DiPaolo, J. A. (1978). Neoplastic transformation of human diploid cells *in vitro* after chemical carcinogen treatment. *Nature (London)* **275**, 130–132.

Mishra, N., Dunkel, V., and Mehlman, M., eds. (1980). "Advances in Modern Environmental Toxicology: Mammalian Cell Transformation by Chemical Carcinogens," Vol. 1. Senate Press, Inc., Princeton Junction, New Jersey.

Mondal, S., and Heidelberger, C. (1970). *In vitro* malignant transformation by methylcholanthrene of the progeny of single cells derived from C3H mouse prostate. *Proc. Natl. Acad. Sci. U.S.A.* **65**, 219–225.

Mondal, S., and Sala, M. (1978). Malignant transformation of a mouse epithelial cell line with polycyclic aromatic hydrocarbons. *Nature (London)* **274**, 370–372.

Mondal, S., Iype, P. T., Griesbach, L. M., and Heidelberger, C. (1970). Antigenicity of cells derived from mouse prostate cells after malignant transformation *in vitro* by carcinogenic hydrocarbons. *Cancer Res.* **30**, 1593–1597.

Mondal, S., Brankow, D. W., and Heidelberger, C. (1976). Two-stage chemical oncogenesis in cultures of C3H/10T $\frac{1}{2}$ cells. *Cancer Res.* **36**, 2254–2260.

Müller, D., Nelles, J., Deparade, E., and Arni, P. (1980). The activity of S9-liver fractions from seven species in the *Salmonella*/mammalian-microsome mutagenicity test. *Mutat. Res.* **70**, 279–300.

Oesch, F. M., Raphael, D., Schwind, H., and Glatt, H. R. (1977). Species differences in activating and inactivating enzymes related to the control of mutagenic metabolites. *Arch. Toxicol.* **39**, 97–108.

Ossowski, L., and Reich, E. (1980). Loss of malignancy during serial passage of human carcinoma in culture and discordance between malignancy and transformation parameters. *Cancer Res.* **40,** 2310–2315.

Pienta, R. J., Poiley, J. A., and Lebherz, W. B., III. (1977). Morphological transformation of early passage golden Syrian hamster embryo cells derived from cryopreserved primary cultures as a reliable *in vitro* bioassay for identifying carcinogens. *Int. J. Cancer* **19,** 642–655.

Poiley, J. A., Raineri, R., and Pienta, R. J. (1979). Use of hamster hepatocytes to metabolize carcinogens in an *in vitro* bioassay. *J. Natl. Cancer Inst.* **63,** 519–524.

Poiley, J. A., Raineri, R., Cavanaugh, D. M., Ernst, M. K., and Pienta, R. J. (1980). Correlation between transformation potential and inducible enzyme levels of hamster embryo cells. *Carcinogenesis* **1,** 323–328.

Rapp, U. R., Nowinski, R. C., Reznikoff, C. A., and Heidelberger, C. (1975). The role of endogenous oncornaviruses in chemically induced transformation. 1. Transformation independent of virus production. *Virology* **65,** 392–409.

Regan, J., and Setlow, R. (1974). Two forms of repair in the DNA of human cells damaged by chemical carcinogens and mutagens. *Cancer Res.* **34,** 3318–3325.

Reznikoff, C. A., Bertram, J. S., Brankow, D. W., and Heidelberger, C. (1973). Establishment and characterization of a cloned line of C3H mouse embryo cells sensitive to post-confluence inhibition of division. *Cancer Res.* **33,** 3231–3238.

Rhim, J. S., Park, D. K., Weisburger, E. K., and Weisburger, J. H. (1974). Evaluation of an *in vitro* assay system for carcinogens based on prior infection of rodent cells with non-transforming RNA tumor virus. *J. Natl. Cancer Inst.* **52,** 1167–1173.

Rhim, J. S., Kim, C. M., Arnstein, P., Huebner, R. J., Weisburger, E. K., and Nelson-Rees, W. A. (1975). Transformation of human osteosarcoma cells by a chemical carcinogen. *J. Natl. Cancer Inst.* **55,** 1291–1294.

Rubin, H. (1980). Is somatic mutation the major mechanism of malignant transformation? *J. Natl. Cancer Inst.* **64,** 995–1000.

Saxholm, H. J. K. (1979). The oncogenic potential of three different 7,12-dimethylbenz[a]anthracine transformed C3H/10T $\frac{1}{2}$ cell clones at various passages and the importance of the mode of immunosuppression. *Eur. J. Cancer* **15,** 515–526.

Schmeltz, I., Tosk, J., and Williams, G. (1978). Comparison of the metabolic profiles of benzo[a]pyrene obtained from primary cell cultures and subcellular fractions derived from normal and methycholanthrene-induced rat liver. *Cancer Lett. (Shannon, Irel.)* **5,** 81–89.

Selkirk, J. K., Croy, R. G., Weibel, F. J., and Gelboin, H. V. (1976). Differences in benzo[a]pyrene metabolism between rodent liver microsomes and embryonic cells. *Cancer Res.* **36,** 4476–4479.

Sims, P., Grover, P. L., Swaisland, A., Pal, K., and Hewer, A. (1974). Metabolic activation of benzo[a]pyrene proceeds by a diol-epoxide. *Nature (London)* **252,** 326–328.

Singer, B. (1976). All oxygens in nucleic acids react with carcinogenic ethylating agents. *Nature (London)* **264,** 333–339.

Sivak, A. (1976). "The Ames Assay." *Science* **193,** 272–274.

Sivak, A. (1979). Overview and status of *in vitro* transformation. *J. Assoc. Off. Anal. Chem.* **62,** 889–899.

Sivak, A., and Tu, A. S. (1980a). Cell culture tumor promotion experiments with saccharin, phorbol myristate acetate and several common food materials. *Cancer Lett. (Shannon, Irel.)* **10,** 27–32.

Sivak, A., and Tu, A. S. (1980b). Factors influencing neoplastic transformation by chemical carcinogens in BALB/c-3T3 cells. In "The Predictive Value of Short-Term Screening Tests in Carcinogenicity Evaluation" (G. M. Williams, R. Kroes, H. W.

Waaijers, and K. W. van de Poll, eds.) Elsevier/North-Holland Biomedical Press, Amsterdam.

Sivak, A., Charest, M. C., Rudenko, L., Silveira, D. M., Simons, I., and Wood, A. M. (1980). BALB/c-3T3 cells as target cells for chemically induced neoplastic transformation. In "Advances in Modern Environmental Toxicology: Mammalian Cell Transformation by Chemical Carcinogens" (N. Mishra, V. Dunkel and M. Mehlman, eds.), Vol. I, pp. 133–180. Senate Press, Inc., Princeton Junction, New Jersey.

Stiles, C. D., Desmond, W., Chuman, L. M., Sato, G., and Saier, M. M. (1976). Relationship of cell behavior in vitro to tumorigenicity in athymic nude mice. Cancer Res. **36**, 3300–3305.

Stoker, M. (1963). Effect of X-irradiation on susceptibility of cells to transformation by polyoma virus. Nature (London) **200**, 756–758.

Styles, J. A. (1977). A method for detecting carcinogenic organic chemicals using mammalian cells in culture. Br. J. Cancer **36**, 558–563.

Todaro, G. J., and Green, H. (1963). Quantitative studies of the growth of mouse embryo cells in culture and their development into established lines. J. Cell Biol. **17**, 299–313.

Ts'o, P. O. P. (1979). Current progress in the study of the basic mechanisms of neoplastic transformation. Banbury Rep. **2**, 385–406.

Weekes, U., and Brusick, D. (1975). In vitro metabolic activation of chemical mutagens. II. The relationship among mutagen formation, metabolism and carcinogenicity of dimethyl- and diethyl-nitrosamine in the livers, kidneys, and lungs of BALB/cJ, C57BL/6J and RF/J mice. Mutat. Res. **31**, 175–183.

Whittemore, A. S. (1980). Mathematical models of cancer and their use in risk assessment. J. Environ. Path. Toxicol. **3**, 353–362.

Williams, G. M. (1976). The use of liver epithelial cultures for the study of chemical carcinogens. Am. J. Pathol. **85**, 739–754.

Yang, N. S., Kirkland, W., Jorgensen, T., and Furmanski, P. (1980). Absence of fibronectin and presence of plasminogen activator in both normal and malignant human mammary epithelial cells in culture. J. Cell Biol. **84**, 120–130.

Chapter 7

Mutagenicity Testing with Cultured Mammalian Cells: Cytogenetic Assays

A. T. Natarajan and G. Obe

MUTAGENICITY:
NEW HORIZONS IN GENETIC TOXICOLOGY

I. INTRODUCTION

Simultaneously with the discovery of the ability of mustard gas to induce mutations in *Drosophila* by Auerbach (1943), the chromosome breaking ability of urethane in *Oenothera* was also discovered by Oehlkers (1943). Because most of the mutagens induce chromosomal aberrations, a test for chromosome breaking ability can unravel the possible mutagenic effect of a test chemical. Tests for induction of point mutations in higher eukaryotic systems are time consuming when compared to an *in vitro* cytogenetic test. Hence, this test is being recommended by many legislative bodies as one of the short-term tests for detection of mutagenicity for newly introduced chemicals.

Mammalian cells in culture offer a good system to evaluate cytogenetic effects of different mutagens, in view of the ease with which treatments can be made, the wide range of concentrations that can be employed (including near-lethal concentrations) for treatment, and the possibility of quantifying the effects with greater accuracy in contrast to mammalian *in vivo* systems.

Mutagenic agents induce lesions in DNA of target cells and these lesions are subjected to cellular repair. Unrepaired lesions or misrepaired lesions will lead to point mutations and/or chromosomal aberrations. The primary effects of mutagenic agents on cellular DNA can be measured directly by biochemical methods, such as estimation of base damage, strand breaks, pyrimidine dimers, cross-links, etc. DNA repair can be estimated by either biochemical methods, such as repair replication, reduction in the frequency of primary lesions, unscheduled DNA synthesis (scintillation counting), or by cytological methods such as unscheduled DNA synthesis using tritium microautoradiography. Sister chromatid exchanges probably represent cellular repair associated with normal DNA replication. Chromosomal aberrations that may reflect unrepaired or misrepaired lesions can be quantified by estimating different classes of chromosomal aberrations scored either in metaphase, anaphase, or interphase (micronuclei). In this chapter, we shall discuss the following cytological end points measured in mammalian cells in culture:

1. Chromosomal aberrations in metaphase,
2. Chromosomal aberrations in anaphase,
3. Micronuclei in interphase,
4. Sister chromatid exchanges (SECs) in metaphase,
5. Unscheduled DNA synthesis as measured by microautoradiography.

Utility of these techniques for purposes other than mutagenicity testing will also be discussed. In view of the importance of human blood lymphocytes in these studies, they are considered in greater detail in the second part of this chapter.

II. CYTOLOGICAL END POINTS USED

A. Chromosomal Aberrations in Metaphase

Two main classes of aberrations are recognized depending on whether both or one of the chromatids in a metaphase chromosome is involved in an aberration. These are chromosome-type and chromatid-type aberrations. The type of aberration observed in metaphase depends on the stage of the cell cycle in which the treatment was made and the type of mutagenic agent used. Agents such as X-rays induce chromosome-type aberrations if the cell is treated in G_1 stage [i.e., prior to DNA synthetic stage (S)] and chromatid-type aberrations, if the cell is treated in G_2 (post S stage), and a mixture of both types, if the treatment is made in S stage (Evans, 1962). Most of the chemical agents and UV light induce chromatid-type aberrations irrespective of the cell stage in which the treatment is made. In addition, if G_2 cells are treated, no aberrations can be observed during the oncoming mitosis, but only in the next mitosis, indicating that the cells must go through an S stage before the aberrations become fixed (Evans and Scott, 1969).

Based on these observations, two classes of chromosome-breaking agents are recognized, namely, the S-dependent and S-independent ones. The aberrations induced by S-independent agents are fixed immediately irrespective of the cell stage at which the treatment was made, and these aberrations are assumed to arise due to nonrepair or misrepair of the lesions. In contrast, the aberrations induced by S-dependent agents are assumed to be formed due to misreplication (Evans and Scott, 1969; Kihlman, 1977). These definitions are only operational ones and exceptions can be found for these generalities.

Three types of aberrations are generally scored in metaphase: gaps, exchanges, and breaks. These can be either chromatid type or chromosome type. In some instances, isochromatid breaks, i.e., involving both chromatids, are encountered and these cannot, in practice, be distinguished from chromosome breaks.

Gaps can also be found to be either isochromatid or chromatid ones. Gaps are difficult types of aberrations to score in the sense that they are more variable in the frequencies from experiment to experiment and the true nature of gaps is not understood.

Electron microscopic studies have shown that in so-called gaps there are still connections that are not visible under the light microscope between the stained regions (Scheid and Traut, 1970; Mace et al., 1978). Some chemicals, especially those that inhibit synthesis of DNA, usually induce a very high frequency of gaps when cells are treated in S or G_2 stage. Many investigators do not take gaps into consideration for quantitative comparisons. In isochromatid breaks, sometimes sister unions occur, i.e., rejoining of the two sister chromatids at the point of break. These sister unions can involve both the ends of the broken chromatids or only one of them, proximal or distal in relation to the centromere. Exchanges between chromatids of two different chromosomes are called chromatid interchanges, which can be either symmetrical or asymmetrical, and complete or incomplete ones depending on whether all the broken ends are rejoined or not.

In the chromosome-type aberration, asymmetrical exchange between two chromosomes will give rise to a dicentric and a fragment, whereas symmetrical exchanges will lead to reciprocal translocations. The latter are difficult to score accurately, without employing banding techniques to distinguish individual chromosomes. A chromosomal intrachange will give rise to a ring and a fragment. Inversions within the chromosome are difficult to score, without banding of the chromosomes.

In view of these difficulties, while scoring for chromosome type of aberrations it is usual to score for gaps, breaks, and asymmetrical exchanges for comparing the effects of different mutagens or different concentrations of one chemical. Some typical chromosomal aberrations are illustrated in Fig. 1. A description of different types of aberration is given by Savage (1975).

B. Chromosomal Aberrations in Anaphase

The type of aberrations which can be recognized in anaphase are breaks and bridges. Breaks can be single or paired (i.e., those that were chromatid breaks or chromosome breaks in metaphase), and bridges are single or double bridges stretched between anaphase groups. Dicentrics will give rise to double bridges and asymmetrical chromatid intra- or interchanges will give rise to single bridges. Centric rings with one SCE will also give rise to a double bridge. Multipolar spindles (i.e., more than two) and lagging chromosomes can also be scored in anaphase. Scoring for chromosomal aberrations in anaphase is easy, but to get large numbers of anaphases is difficult as one cannot accumulate anaphase cells as it can be done with metaphases by treatment with spindle poisons.

C. Micronuclei in Interphase Cells

A proportion of fragments or whole chromosomes which lag in anaphase will not be included in the main nucleus and, in the next interphase, they can condense to form a small nucleus, which is called the micronucleus. Micronuclei are easy to score. Sometimes micronuclei can become prematurely condensed, giving the appearance of extended chromosome or pulverised nucleus (Obe et al., 1975b). Several factors, such as the number of cell cycles passed following treatment and the proportion of cells which have divided, must be taken into account before quantitative comparisons based on the frequencies of micronuclei can be made.

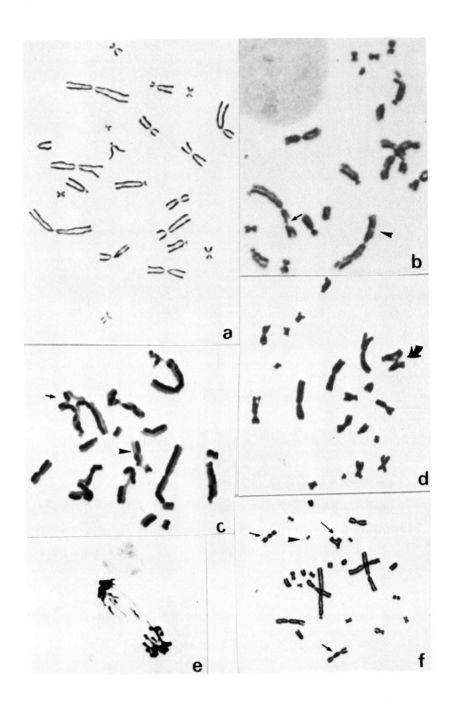

D. Sister Chromatid Exchanges

Sister chromatid exchanges (SCEs) represent exchange of DNA between replication products at homologous points. At metaphase, these represent symmetrical exchanges between sister chromatids at identical loci, which can only be visualized if the sister chromatids can be distinguished either by radioactive labeling or differential staining following incorporation of 5-bromodeoxyuridine. Although several models for the origin of SCEs have been proposed (Evans, 1977b; Shafer, 1977; Stetka, 1979; Painter, 1980), the molecular basis of their formation has not been elucidated. These may represent some repair processes associated with DNA replication. Agents which induce chromosomal aberrations in an S-dependent manner are also very efficient in inducing SCEs in contrast to the S-independent agents which are poor inducers of SCEs. Sister chromatid exchanges are easy to score and this method appears to be a sensitive technique to detect certain classes of chemicals.

E. Unscheduled DNA Synthesis

When nonsynchronized cells are treated with a mutagen, DNA repair occurs in all cells. By providing a DNA precursor, such as tritiated thymidine, one can evaluate the extent of repair in the cells by measuring the radioactivity in the DNA. Although the cells in S stage are heavily labeled, the other cells are lightly labeled and these can be identified by autoradiography of the microscopic preparation of fixed cells (Rasmussen and Painter, 1964).

The "S"-phase cells will appear dark due to heavy labeling and the few grains on "non-S"-phase cells can be counted and quantified.

III. METHODOLOGY

A. Cell Lines Used

For cytological detection of chromosomal aberrations and sister chromatid exchanges, cells with large sized and small numbers of chromosomes are preferable. Thus, most of the routine studies have been

Fig. 1. Chinese hamster ovary cell. (a) A normal cell with 22 chromosomes; (b) treated cell showing a deletion (arrow) and a gap (arrow head); (c) a cell stained to reveal sister chromatid differentiation. Arrow indicates a chromatid exchange and arrow head indicates a sister chromatid exchange; (d) a cell showing a subchromatid exchange (arrow); (e) an anaphase with a single bridge and a fragment; (f) a cell with multiple aberrations; short arrows indicate dicentrics, long arrow indicates an exchange (triradial) and arrow head indicates isochromatid breaks (minutes).

carried out on Chinese hamster cells. Most of the established cell lines of Chinese hamster are near diploid and contain one or two marker chromosomes. Chinese hamster ovary (CHO) cells have 21 to 22 chromosomes, with one X chromosome intact, and a large acrocentric marker chromosome (Fig. 1). These cells are easy to synchronize, because the metaphase cells round off and detach when shaken (mitotic shake-off; Terasima and Tolmach, 1961). The mitotic cells once plated are in G_1 stage after about 2 hours. The synchrony obtained is about 90–95%. However, by growing CHO cells in a roller bottle and controlling the shake-off procedure by increasing the speed of rotation 99.5% pure mitotic population can be obtained (T. S. B. Zwanenberg, unpublished). The V79 cells ($2n = 22$–23), which are derived from lungs of Chinese hamster originally and several of its derivatives, are employed in cytological studies. These cells are more difficult to synchronize than CHO cells by the mitotic shake-off method. There is a difference for posttreatment caffeine sensitivity between these two cell lines, V79 being more sensitive to caffeine. Some laboratories use their own established Chinese hamster cell lines or other cell lines such as DON (Ishidate and Odashima, 1977). Syrian or Golden hamster cells with a chromosome number of 44 ($2n$) are used occasionally (Perry, 1980). The mouse cell lines are usually not stable in their chromosome number, and the chromosomes are small and uniform in size when compared to Chinese hamster cells. However, in special circumstances, such as studying the role of heterochromatin in aberration induction or in vivo and in vitro comparisons, the cells of ascites tumors are especially useful (Schöneich et al., 1970; Natarajan and Raposa, 1975).

Human cells derived from fetal tissues or from the foreskin are also employed by several investigators. The number of chromosomes is high ($2n = 46$). Many transformed cells, which are fast growing, are available, but their chromosome number is variable. Investigations on human syndromes that are repair deficient or supersensitive to mutagens with these cell lines (Wolff et al., 1977; Natarajan and Meijers, 1979; Natarajan et al., 1981) are often conducted.

Such cells are mostly of skin fibroblast origin. Human lymphoid cell lines are also employed (Huang and Furukawa, 1978). Human lymphocytes are often used and these cells are specially treated in the latter sections of this chapter.

Primary cell cultures and transformed cell lines of the same origin may exhibit different sensitivities to mutagens. For example, human fibroblasts derived from one XP patient react differently to chemical mutagens such as EMS when compared to SV40 transformed cells from the same origin, the latter being more sensitive (Wolff et al., 1977; Heddle and Arlett, 1980).

Different cell types have different cell cycle times and also different durations spent at various stages of the cell cycle. Human cells have a cell cycle of about 24 hours and Chinese hamster cells about 11–12 hours. The protocols to detect the chromosome-breaking ability of chemicals should take into account the cell cycle durations of different cell lines.

B. Experiments to Detect Chromosome-Breaking Ability

Since most of the chemical mutagens break the chromosomes in an S-dependent way, it is essential that the cells are treated for a short duration and allowed to recover for a long enough time to pass through an S phase before they are fixed for cytological analysis. Many chemicals are unstable in aqueous solution and therefore care should be taken to treat the cells immediately after dilution or to use solvents such as DMSO and ethanol. A range of concentrations of the mutagen under test should be employed, and two to three fixations should be made in order to compensate for the delay induced at higher concentrations. For CHO cells, a fixation between 6 and 16 hours, following treatment, will cover the cells which were at S phase or G_1 phase during treatment. For each experimental point 100 cells are usually scored. Concentration-dependent increase in the frequencies of aberrations is taken as an indicator of chromosome-breaking ability of a chemical. If one desires to know which stage in the cell cycle the chemical in question is most active, a pulse labeling at the time of treatment is performed and the cytological preparations, after scoring for chromosomal aberrations (aceto-orcein stained), are processed for microautoradiography to detect the labeling pattern in the mitotic cells carrying chromosomal aberrations. If all aberration-carrying cells are labeled, this would indicate that S cells are preferentially affected.

C. Experiments to Detect Induction of Sister Chromatid Exchanges

The current procedure to obtain sister chromatid differentiation is to incorporate 5-bromodeoxyuridine (BUdR) in cells for two cycles (Latt, 1974; Perry and Wolff, 1974). This will generate cells with chromosomes in metaphase, in which DNA in one chromatid is bifiliarly substituted with BUdR (BB) and the other chromatid contains DNA unifiliarly substituted with BUdR (TB). One can also employ a single round of BUdR incorporation followed by another cycle in normal medium, in which case the chromosomes will contain DNA in one chromatid unsubstituted with BUdR (TT) and the DNA in the sister chromatid unifiliarly substituted with BUdR (TB) (Kihlman *et al.*, 1978). Differential BUdR content between the sister chromatids can be visualized microscopically by staining with the fluorochrome Hoechst 33258 and viewing through a fluorescent microscope or by exposing the Hoechst stained slides to visible light or near-UV light, followed by incubation in $2\times$ saline sodium citrate (SSC) at 65°C for about 2 hours and staining with Giemsa solution.

The chromatid containing less BUdR (in terms of strands) will fluoresce brighter or stain stronger with Giemsa in comparison to its sister. Any switch between segments of sister chromatids can easily be detected and evaluated. In these experiments, the cells are usually treated for a short duration followed by incubation in a BUdR-containing medium for a period long enough to cover about $1\frac{1}{2}$ to 2 cell cycles. In the one-cycle method (Natarajan and van Kesteren-van Leeuwen, 1981), the cells are grown first in BUdR for one cycle and then treated with the chemical in question, followed by incubation in normal medium for another cycle. The latter method is advantageous in two ways, namely, the cells recover better in medium containing thymidine, and, second, in such a protocol, the occurrence of chromosomal breaks and SCEs can be evaluated in the same cells, or in the same populations of cells which makes comparisons between these two end points more reliable. To estimate the frequencies of SCEs 25–50 cells are scored.

D. Method for Study of Aberrations at Anaphase

The cells are grown on cover glasses and fixed *in situ*, without any hypotonic treatment, or only with mild hypotonic treatment (Nichols *et al.*, 1977). Since there is no easy way to accumulate anaphases, and the time spent at this phase by cells is short, one finds low frequencies of this stage and therefore screening of several preparations is necessary, especially in cases where the mitotic index is low. The spontaneous as well as induced frequencies of aberrations detected are usually higher in anaphase than in metaphase (A. T. Natarajan, unpublished).

E. Detection of Micronuclei

The easiest way to make preparations to detect micronuclei in fibroblast cultures is to grow the cells on cover glasses and fix them *in situ* with methanol. For detecting induced micronuclei, the cells should be allowed to go through at least one mitosis following treatment. Thus, the fixations have to be carried out taking into account the induced mitotic delay. No hypotonic treatment is necessary. For quantitative evaluation it is important to know how many cells have gone through mitosis. Pulse labeling with tritiated thymidine will give an idea of whether the micronucleus-carrying cells have gone through a mitosis. Staining with a dye specific for DNA (e.g., Giemsa, Feulgen, Fluorochrome) can be used to stain the slides. Usually 2000 cells are evaluated per experimental point. The frequencies of micronuclei are around 10% of the frequencies of aberrations observed in metaphase (Iskandar, 1981).

F. Unscheduled DNA Synthesis

The basic method is to detect incorporation of tritiated precursors of DNA in the nucleus of the treated cell which is not in the S phase. Such an incorporation can be estimated by incubating these cells in tritiated thymidine in the presence of hydroxyurea, which inhibits semiconservative DNA replication, and isolating DNA and employing scintillation counting of the radioactivity. This system is not very clean. Cytologically the extent of UDS can be evaluated by counting the number of grains on non-S stage cells. For this test it is easier to use nonproliferating cells, such as human lymphocytes or hepatocytes, in which case the problem of the presence of S cells can be avoided.

G. Metabolic Activation

One of the problems of using cells in culture is their inability to activate mutagenic carcinogens, which need such an activation. Primary embryonic cells have some capacity to activate such carcinogens. The same is true with hepatocytes and to a lesser extent with human lymphocytes (see Table I and Section XV). However, this can be overcome by introducing a drug-activating system, such as S9 fraction from induced rat liver, during treatment (Natarajan et al., 1976). Such a schedule, when employed, can detect most of the indirect-acting mutagens. This technique is similar to the Ames test with minor modifications such as using HEPES buffer instead of phosphate buffer in the S9 mixture. Use of S9 mixture can be eliminated if cells which can activate mutagens are used as target cells. A systematic study using rat hepatocytes for treatment and detection of unscheduled DNA synthesis has been made by Williams (1977). There are some permanent cell lines of liver origin which have retained some capacity to activate carcinogens. These are RL1 developed by Dean and Hodson-

TABLE I

Activating and Deactivation Capacities of Different Cell Lines[a]

Cell line	Activating enzyme [benzo[a]pyrene monooxygenase]	Deactivating enzyme [3-OH-benzo[a]pyrene glucuronyltransferase]
A 549 (human)	+	+ + +
NC 37 Ba EV (human)	−	±
RAG (rat)	−	+ + + ,
H-4-II-E (rat)	+ + +	+ + +
3T3/BALB (mouse)	+	+ + +
V79 (Chinese hamster)	−	±
BHK-TK⁻ (Syrian hamster)	+ +	+ +

[a] Based on Wiebel *et al.* (1980).

Walker (1979) and H39, developed by G. Veldhuisen (personal communication).

Baby hamster kidney cells have been reported to have some capacity to activate mutagens. In experiments involving *in vitro* metabolic activation, some deactivating enzymes may also be present in the mixture, which may reduce the effect of indirectly and directly acting agents, e.g., 4NQO (see Table IV). Some problems can also be encountered when the reactive species is short lived or easily deactivated.

IV. COMPARISON BETWEEN CHROMOSOME ABERRATIONS AND SISTER CHROMATID EXCHANGES

Since the introduction of fluorescent plus Giemsa technique (FPG), a large amount of work has been done to evaluate chemicals for their ability to induce SCEs. This technique has been a preferred one over the scoring of classical chromosomal aberrations. This is because of the fact that it is easy to score SCEs in contrast to scoring chromosomal aberrations, which need expertise to identify and quantify different classes of aberrations, and only a few cells, usually 25, need to be scored, in contrast to at least 100 cells per point for the detection of chromosomal aberrations.

These arguments are valid; however, it is clear that scoring for SCEs cannot replace the scoring of chromosomal aberrations in order to evaluate the mutagenicity of a chemical. Although all agents that induce SCEs also induce chromosome aberrations, the reverse is not true. It is well known that agents such as ionizing radiations, bleomycin, and streptonygrin, which directly induce DNA strand breaks, do not increase the frequencies of SCEs to the same extent as they induce chromosomal aberrations. A strategy for testing chemicals, in our opinion, seems to be to test for induction of SCEs, in the first instance, and, if negative, to test for induction of chromosomal aberrations.

Two arguments in favor of using techniques to detect SCEs over chromosomal aberrations are (1) it is more sensitive than the testing for chromosomal aberrations and (2) SCEs may reflect mutagenic events when compared to chromosomal aberrations. Both these statements are not entirely true. Perry and Evans (1975) made a comparison of chromosome-breaking ability and SCE-inducing ability of several compounds, which showed that SCEs occurred at far higher frequencies when compared to chromosomal aberrations. However, these comparisons were made with the frequencies of aberrations induced in the second cell cycle following treatment, a stage in which it is known that the majority of aberrations induced in the first cell division would have disappeared. Second, the spontaneous frequencies of chromosomal aberrations are very low, about 0.03–0.05 per cell, in comparison to 6–12 per cell encountered for SCEs. A doubling of the frequencies of SCEs can be demonstrated by studying 20 cells or even less, whereas a doubling of chromosomal aberrations can be demonstrated only after scoring hundreds of cells. Thus, the sensitivity of SCE technique is due to the ease with which the cells can be scored. In a critical comparison of concentrations required to increase the frequencies of SCEs and chromosomal aberrations by five different monofunctional alkylating agents, in CHO cells, it has been shown that there is no difference in the required concentrations to double these end points (Vogel and Natarajan, 1981).

V. TYPES OF LESIONS LEADING TO SISTER CHROMATID EXCHANGES AND/OR CHROMOSOMAL ABERRATIONS

DNA strand breaks lead to chromosomal aberrations induced by ionizing radiations and visible light on BUdR-substituted chromosomes. Agents that directly break chromosomes, such as X-rays and bleomycin, are poor inducers of SCEs. That lesions leading to SCEs may be different, or partly different, at least in the repair of lesions comes from the observation of two human diseases. Fanconi's anemia and Bloom's syndrome are characterized by increased frequencies of spontaneous chromosomal aberrations, but only Bloom's syndrome is associated with increased frequencies of SCEs. Posttreatment with caffeine increases efficiently the frequencies of aberrations induced by alkylating agents, but not the frequencies of SCEs (Kihlman, 1977). The latter observation may indicate differences in repair pathways. With agents that induce both chromosomal aberrations and SCEs effectively it is difficult to discriminate between the lesions responsible for the induction of these two end points. The observation that with monofunctional alkylating agents the concentration required to double the frequencies of chromosomal aberrations or SCEs is similar may indicate that the lesions for these two end points are the same or similar (Vogel and Natarajan, 1981). Although it was suggested that chromosomal aberrations induced by shortwave UV arise due to the induced pyrimidine dimers, but not the SCEs, and based on the observation that photoreactivation of UV-irradiated cells lead to a decrease in the frequencies of induced dimers but not SCEs in chicken embryonic fibroblasts (cell line possessing PR enzyme), later experiments have proved this to be wrong (Wolff, 1978; Natarajan *et al.*, 1980b). Experiments with fibroblasts of *Xenopus laevis* have shown that PR reduces the frequencies of UV-induced pyrimidine dimers, SCEs, chromosomal aberrations, and point mutations (van Zeeland *et al.*, 1980).

At present there is no clear evidence to demonstrate specific differences between lesions induced by agents that produce chromosomal aberrations and SCEs. Since the chromatid-type aberrations are predominantly induced by DNA cross-linking agents, it is proposed that intrastrand cross-links give rise to chromosome aberrations and interstrand cross-links give rise to SCEs (Scott, 1980). There is no direct evidence for this because UV-induced pyrimidine dimers (lesions analogous to intrastrand cross-links) lead very efficiently to SCEs (Natarajan *et al.*, 1980b).

VI. INFLUENCE OF TUMOR PROMOTERS OR INHIBITORS ON INDUCTION OF CHROMOSOMAL ALTERATIONS

It has been reported that a tumor-promoting agent (TPA–12-*O*-tetradecanoyl phorbol 13-acetate) induced very high frequencies of SCEs at least in a proportion of cells by Kinsella and Radman (1978). This led them to the theory that tumor promoters act by initiating a recombinationlike process in the cell and such recombinant events lead to the expression of heterozygotes of recessive genes by becoming homozygous. This concept is not a new one. Work carried out in several laboratories including ours has not been able to reproduce these results, namely, very high induction of SCEs by TPA, though a small increase has been observed in some instances (Loveday and Latt, 1979; Fujiwara et al., 1980; Gentil et al., 1980).

In addition, we have observed a small, but significant increase in the frequencies of aberrations in cells treated with TPA. Subsequently, an inhibitor of tumor promotion, namely, antipain, has been reported not to affect the frequencies of SCEs, but decrease the frequencies of exchange-type aberrations induced by MNNG, indicating that antipain may suppress some recombinationlike processes in the cell (Kinsella and Radman, 1980). This observation was not confirmed in another laboratory (DiPaolo et al., 1980).

In combination treatments in in vitro cytogenetic experiments, extreme care has to be taken to avoid screening different populations of cells and drawing conclusions. A short treatment, followed by multiple fixation, or treatment of synchronized cells followed by two to three fixations to cover the oncoming mitosis, or a pulse labeling during treatment and processing the slides after scoring through microautoradiography to identify the stage of the cell cycle in which the cells were during treatment, should be followed.

VII. CHROMOSOMAL ALTERATIONS AND POINT MUTATIONS

It has been postulated by some authors that chromosomal aberrations are related to cytotoxicity, whereas SCEs are related mutagenesis in in vitro cells. Two types of observations are used for this postulation.

1. Among the different sites of alkylation in DNA, O-6 alkylation of guanine has been implicated to be the lesion responsible for point mutations. In XP variant cells, it has been demonstrated that methyl nitroso urea induced higher frequencies of SCEs than in normal cells and parallelly this cell line was not efficient in removing O-6 alkyl group of guanine. This may indicate that O-6 alkylation may lead to mutations as well as sister chromatid exchanges (Goth-Goldstein, 1977; Wolff, 1978).

2. When dose–response curves for induction of $HGPRT^-$ mutations and SCEs were compared, both gave similar linear regression lines indicating that they may be related processes (Carrano et al., 1978).

Among the monofunctional alkylating agents, one can select compounds which induce relatively high frequencies of O-6 alkylations compared to N-7 alkylations and the reverse types as well. Agents which react by S_N1 type of action induce higher proportion of O-6 alkylations when compared to agents which react with S_N2 type. If chromosomal aberrations and SCEs are induced by a different primary lesion following treatment with monofunctional alkylating agents, one would expect the S_N1-type agents to be potent inducers of SCEs, whereas agents acting in an S_N2-type reaction would be potent inducers of chromosomal aberrations. When such a comparison was made, it was found that S_N2-type agents were efficient in inducing both chromosome aberrations and SCEs. With all compounds tested, the concentration required to double the spontaneous frequencies of aberrations or SCEs were similar (Vogel and Natarajan, 1981). However, the induction of point mutations was high following treatment with S_N1-type compounds, when compared to S_N2-type compounds (J. W. I. M. Simons et al., in preparation), results comparable to those observed in Drosophila and other organisms (Vogel and Natarajan, 1979a,b).

Thus, as far as monofunctional alkylating agents are concerned, there is no strong evidence for different lesions being responsible for chromosomal aberrations and SCEs. These results corroborate well with those obtained with UV and UV–photoreactivation experiments, in which it could be demonstrated that pyrimidine dimers are involved in the production of chromosomal aberrations and SCEs (Natarajan et al., 1980b; van Zeeland et al., 1980).

VIII. MUTAGENIC COMPOUNDS ASSAYED BY CYTOGENETIC METHODS

A comprehensive evaluation of chemicals tested by cytogenetic methods is being prepared by the Gene-Tox Program (see Waters, 1979) and is not yet available. Several lists have been published compiling different classes of chemicals that have been tested (see Table II). In general all chemicals which interact with DNA (genotoxic chemicals) can be detected by the three main cytogenetic methods, namely, by induction of chromosomal aberrations, SCEs, and UDS.

IX. DETECTION OF CARCINOGENS

Most of the mutagens are carcinogenic. Both direct- and indirect-acting carcinogens that interact with DNA can be detected by in vitro cytogenetic methods. Among 20 compounds comprised of both carcinogens and noncarcinogens which were blind tested in our labo-

TABLE II

Published Reviews Listing Chemicals Tested in Cytogenetic Test Systems *in Vitro*

Test system	Number of compounds	References
Chromosomal aberrations	134	Ishidate and Odashima (1977)
	16	Matsuoka *et al.* (1979)
	40	Hollstein *et al.* (1979)
	33	Abe and Sasaki (1977)
SCEs	93	Perry (1980)
	35	Hollstein *et al.* (1979)
	33	Abe and Sasaki (1977)
	26	Wolff (1977)
UDS	44	IARC (1980)
	14	Williams (1977)
	87	Stich *et al.* (1975)
	33	Williams (1980)
	227	Swenberg and Petzold (1979)

ratory, about 75–80% reliability in predicting carcinogenicity was obtained (see Table III). Of special interest are compounds such as diethylstilbestrol which are supposed to be nongenotoxic and induce chromosomal aberrations, which may indicate that the cytogenetic tests may prove valuable in picking up potential carcinogens when compared to other short-term assays.

X. APPLICATION OF CYTOGENETIC TESTS TO DETECT HUMAN RECESSIVE DISORDERS

Several human recessive diseases are associated with increased spontaneous frequencies of cancer. Some of these are also associated with increased spontaneous frequencies of chromosomal aberrations and/or SCEs. In addition, they are sensitive to one or more classes of mutagens. This sensitivity to specific mutagens can be exploited to detect these syndromes cytologically, especially in suspected cases. Some of the typical syndromes and the mutagens to which they are sensitive are presented in Table IV.

There is evidence suggesting that AT heterozygotes have a higher risk of getting malignant tumors compared to normal individuals (Swift, 1977). It is not yet clear whether the heterozygotes are also

TABLE III

Results from a Blind Study to Evaluate Mutagenicity by Chromosome Aberrations (CA) and Sister Chromatid Exchanges (SCE) Test in CHO Cells of 20 Compounds with Known Carcinogenic Activity[a]

Chemical	Carcinogenicity	CA	SCE
4-Nitroquinoline-N-oxide	+	+	+
3-Methyl-4-nitroquinoline-N-oxide	−	+	+
Benzidine	+	+	+
3,3′:5,5′-Tetramethylbenzidine	−	−	−
4-Dimethylaminoazobenzene (butter yellow)	+	±	−
4-Dimethylaminoazobenzene-4-sulfonic acid sodium salt	−	−	−
Benzo[a]pyrene	+	+	+
Pyrene	−	−	−
Dimethylcarbamoyl chloride	+	+	+
Dimethylformamide	−	−	−
2-Naphthylamine	+	+	+
1-Naphthylamine	−	+	+
DL-Ethionine	+	−	−
Methionine	−	−	−
Hydrazine sulfate	+	−	+
Hexamethylphosphoramide (HMPA)	+	−	−
Ethylenethiourea	+	−	−
Diethylstilbestrol	+	−	+
Safrole	+	−	−
Epichlorhydrin	+	+	+

[a] The tests were run with and without metabolic activation.

supersensitive to specific mutagens. Cytogenetic methods can be employed to detect heterozygotes (Auerbach *et al.*, 1979). Further work is necessary to confirm these findings.

XI. DETECTION OF COMPLEMENTATION GROUPS OF RECESSIVE DISORDERS

When recessive disorders such as XP are detected in different geographic locations it is of interest to find out whether these genetic changes are identical. The XP cells when irradiated with UV cannot perform excision repair because of the lack of the action of the endonuclease which makes a nick near the dimers. If this function is governed by more than one allele it should be possible to restore this

TABLE IV

Human Recessive Disorders Associated with Increased Frequencies of Spontaneous Chromosomal Alterations and/or Increased Sensitivities to Mutations[a]

Recessive disorder	Elevated frequencies of		Increased sensitivities to
	Spontaneous chromosomal alterations	SCEs	
Xeroderma pigmentosum	−	−	uv[1]; 4NQO[2]; DEB[3]; Furo-coumarins + UVA[4]
Fanconi's anemia	+ +[5]	−	MMC[5,13]; X-rays[3]; DEB[3]; acetaldehyde[6]
Bloom's syndrome	+ +[12]	+	EMS (?)[7]
Ataxia telangiectasia	+	−	X-rays[8,9,14]; bleomycin[10]; furocoumarins + UVA[4]
Blackfan–Diamond	+[11]	−	X-rays[11]

[a] Key to references (superscript numbers):
[1] Cleaver (1968); [2] Zelle and Bootsma (1980); [3] A. T. Natarajan (unpublished); [4] Natarajan et al. (1981); [5] Sasaki and Tonomura (1973); [6] Obe et al. (1979); [7] Krepinsky et al. (1980); [8] Taylor et al. (1976); [9] Natarajan and Meijers (1979); [10] Taylor et al. (1979); [11] Iskandar et al. (1980); [12] Chaganti et al. (1974); [13] Latt et al. (1975); [14] Natarajan et al. (1980a).

function by making somatic hybrids between different types of XP. This repair can be measured by evaluating UDS by tritium auto-radiography. In a typical experiment two different cell lines of XP are fused together and UV irradiated. After fusion the cells are post-incubated in a medium containing tritiated thymidine. Two types of hybrids are produced by fusion, namely, the homokaryons and the heterokaryons. If there is no UDS in all the hybrid cells it would indicate that both the cell lines belong to the same complementation group. If some of the hybrids, namely, heterokaryons, exhibit UDS it would mean that the cell lines belong to different complementation groups. Using this method seven complementation groups have been found in XP (Bootsma, 1979). A similar approach can also be employed successfully to detect complementation groups of other human disorders listed in Table IV.

XII. EFFECT OF COCULTIVATION OF TWO DIFFERENT TYPES OF CELLS

Cells deficient in one type of metabolic function can be rectified by cocultivating them with another type of cells proficient in that special function, provided that the metabolic activity is mediated by one or more diffusible enzymes. Cells not capable of activating indirect-acting carcinogens can be grown together with cells proficient in this activity (cell-mediated mutagenesis).

Cells from Bloom's syndrome which exhibit a high spontaneous frequency of SCEs when cocultivated with normal cells show a lower frequency of SCEs, indicating that the defect in Bloom's cells can be modified at least partly by a diffusible substance from the normal cells. Similar experiments have also been done with cells with Fanconi's anemia and AT cells (van Buul et al., 1978; Shaham et al., 1980).

XIII. QUANTIFICATION OF THE BIOLOGICAL EFFECTS IN RELATION TO PRIMARY EFFECTS IN THE DNA

The conventional method of expressing dose for chemical mutagens is to calculate concentration applied to the test system multiplied by the time of the treatment. However, when comparing different cell types or different chemicals this way of calculating dose is not accurate. This is because of the different reaction mechanisms, stability of compounds, and different biological parameters. To circumvent this it is better to measure the reaction product between a test chemical and DNA. Using tritium-labeled EMS this type of experiment has been conducted in which alkylation per unit DNA has been determined. In parallel experiments conducted on identical conditions, induced frequencies of SCEs and point mutations ($HGPRT^-$) in V79 cells have been determined. These results allow the possibility of comparing the genetic effects obtained in bacteria with mammals, etc., treated with EMS. Such comparisons can lead to predictions of effects in vivo based on experiments with mammalian cells in vitro.

These types of data are valuable for estimating risk at least in somatic cells due to exposure to chemicals (Aaron et al., 1978; Mohn et al., 1980).

XIV. IN VITRO CYTOGENETIC ASSAYS TO DETECT MUTAGENS IN COMPARISON TO OTHER SHORT-TERM TESTS

Among the short-term tests to detect mutagenicity of chemicals, the Ames test, E. coli, yeast, and mammalian cells in vitro and in vivo (cytogenetics, point mutations, and *Drosophila*) are recommended by legislative bodies. All these tests have some advantages and some disadvantages which has led to the usage of a battery of tests. Experience has shown that *in vitro* cytogenetic tests can pick up some of the mutagens which are difficult to be identified by other tests. For example, diethylnitrosamine, urethane, and procarbazine are not mutagenic as evaluated with the Ames test but are effective in *in vitro* cytogenetic tests (Hollstein *et al.*, 1979). Base analogues and intercalating agents are difficult to detect in *Drosophila* but are positive in *in vitro* cytogenetic tests. Similarly, compounds such as diethylnitrosamine, MNNG, and caffeine, which are negative in the mouse *in vivo* micronucleus test, are positive in *in vitro* cytogenetic tests (Hollstein *et al.*, 1979). These observations indicate that *in vitro* cytogenetic tests should take a central role in a test battery.

XV. THE HUMAN LYMPHOCYTE

A. Introduction

Human peripheral lymphocytes represent a cell population, which is almost exclusively in a presynthetic stage of the cell cycle, i.e., G_0 phase, in which the biochemical and physiological activities are highly suppressed.

Only 0.2% or less of the peripheral lymphocytes have been shown to replicate their DNA, i.e., are in the autosynthetic cell cycle (Bond *et al.*, 1958; Trepel, 1976). These cycling cells probably come from the pool of large lymphoid cells, representing stimulated T lymphocytes or immature plasma cells, which represent activated B lymphocytes (up to 15% of the peripheral lymphocytes) (Ling and Kay, 1975; Trepel, 1976). Cells from this group may give rise to the rarely found mitoses in the peripheral blood (Bessis, 1973).

Nowell (1960) was the first to show that peripheral "human leukocytes" can be stimulated by phytohemagglutinin (PHA) to undergo mitoses *in vitro*, and Carstairs (1962) could show that the "small lymphocytes" are the target cells for the mitogenic activity of PHA. The small lymphocytes are not cells in an end stage of cellular differentiation but they are rather "pluripotential" (Carstairs, 1962).

The small lymphocytes are just packages of DNA wrapped up in a minimal set of cytoplasmic constituents. This package is unpacked upon stimulation with antigens *in vivo* or with mitogens *in vitro*, leading to a "normal" cycling cell. Some characteristics of these cells which are important for their use in mutagenicity research are discussed below. The typical appearance of these cells in the resting stage

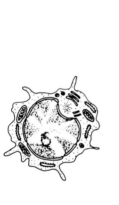

SMALL LYMPHOCYTE

ANTIGENS
MITOGENS

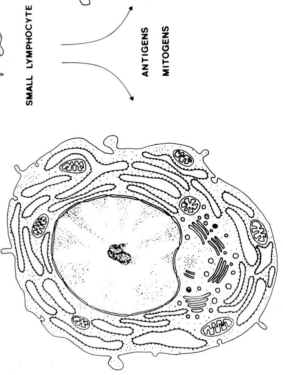

LYMPHOBLAST (activated T Lymphocyte)

PLASMA CELL (activated B Lymphocyte)

and in the stimulated stage can be seen in Fig. 2, which has been drawn after data and figures from Lentz (1971), Bessis (1973), Rhodin (1974), David (1977), and Krstić (1978).

B. The Lymphocytes

Small lymphocytes have a dense nucleus with little cytoplasm surrounding it. The majority have a small, often ring-shaped nucleolus, indicating low RNA synthesis (Busch and Smetana, 1970; Bessis, 1973; Beek and Obe, 1977). Especially in newborns but also in infants more multinucleolar lymphocytes can be seen than in adults (Busch and Smetana, 1970). Electron microscopic studies reveal a nucleus with highly condensed regions especially at the nuclear membrane. There are only a few mitochondria and a small Golgi region. Ribosomes are located at the nuclear membrane and free in the cytoplasm. The endoplasmatic reticulum (ER) is rudimentary. A pair of centrioles is

---

Fig. 2. Schematic drawings of the ultrastructure of a small lymphocyte, a plasma cell, and a lymphoblast; the latter two cell types originate from a small lymphocyte by antigenic or mitogenic stimulation.

The *small lymphocytes* are in a presynthetic resting stage of the cell cycle (G$_0$); the stimulated (transformed) cells are in an autosynthetic cell cycle. Even the small lymphocytes possess a pair of centrioles indicating that they are partially prepared to undergo mitosis. The small lymphocyte has a nucleus with most of the chromatin highly condensed (heterochromatic) and a small ring-shaped nucleolus. The small lymphocyte has only little cytoplasm containing some mitochondria and some profiles of rough endoplasmatic reticulum and granula. There are free ribosomes in the cytoplasm. In a nuclear indentation a pair of centrioles is situated; nearby is a small Golgi region. The cell surface can be smooth or can be covered with microvilli.

The *plasma cell* represents an activated B lymphocyte (representing one cell type of the small lymphocytes) and is, together with lymphoblasts, occurring after stimulation with pokeweed mitogen (PWM). The egg-shaped cell is actively producing immunoglobulins. The granular ER is highly developed with the cisternae being more or less extended. In the nuclear indentation a pair of centrioles can be seen, which is surrounded by a large Golgi field. There are few mitochondria and the cytoplasm is filled with free ribosomes. The cell surface is covered with few microvilli and bloblike extrusions. The arrangement of the heterochromatin and euchromatin looks like a cartwheel, a characteristic that can be used for diagnosing such cells in the light microscope. The nucleus contains a typical nucleolus.

The *lymphoblast* represents an activated T lymphocyte (representing one cell type of the small lymphocytes) and is predominantly occurring after stimulation with phytohemagglutinin (PHA). The roundish cell has a nucleus with dispersed chromatin. The heterochromatin is mainly confined to the nuclear membrane. There are large hypertrophied nucleoli. In an indentation of the nucleus there is a pair of centrioles and in its vicinity a small Golgi field. The cytoplasm contains mitochondria and a few profiles of ER; it is filled with many polyribosomes indicating that the cell is actively synthesizing proteins. The cell surface is irregular with some microvilli.

situated in an indentation of the nucleus (Bessis, 1973; Trepel, 1976). Peripheral lymphocytes have a diameter of around 6 μm and the volume can be estimated to be around 170 μm^3 (Trepel, 1976). Two main types of lymphocytes can be distinguished, i.e., T and B lymphocytes. Both types of lymphocytes originate ontogenetically from immunologically incompetent stem cells in the yolk sac which eventually settle in the bone marrow. Such undifferentiated stem cells migrate into the thymus, multiply there, and most probably, by somatic mutations, become immunocompetent T lymphocytes responsible for the cellular immunity. In birds, stem cells migrate to the bursa of Fabricius and again by multiplying and accumulating somatic mutations they become B lymphocytes which are responsible for the humoral immunity. In mammals, a bursa of Fabricius is missing but there exists a bursa equivalent in which B lymphocytes are formed from undeterminated stem cells. This bursa equivalent is considered to be the bone marrow itself. The contact of the body with an antigen provokes an immunological answer in form of the proliferation and differentiation of single effector cells which recognize the antigen.

This clonal proliferation leads in the B lymphocytes to the formation of plasmoblasts and plasma cells which synthesize and secrete antibodies. The T lymphocytes upon antigenic contact form lymphoblasts which give rise to clones of sensitized T lymphocytes. Both the clones of plasmoblasts and of T lymphoblasts give rise to memory cells (T and B memory cells), which in case of a second contact with the antigen, lead to a much quicker and stronger immune response by the formation of immunocompetent effector cells (sensitized T lymphocytes and plasma cells).

This scheme of the immune system implies that there are two different types of stem cells in the lymphatic system: (1) undetermined immunologically incompetent stem cells and (2) immunologically determined stem cells of the T (originating in the thymus) and of the B (originating in the bone marrow, or better the bursa equivalent) cell type. The B lymphocytes can be recognized by the presence of immunoglobulin receptors on their surface and T lymphocytes by their ability to bind sheep erythrocytes (T cell rosettes) (Trepel, 1975, 1976; Wara and Ammann, 1977; Jungermann and Möhler, 1980).

The lymphocyte concentrations in the peripheral blood are variable. In a 21-year-old adult the mean count is estimated to be 2500/mm^3 (range 1000–4800/mm^3), in newborns it is more than doubled, i.e., 5500/mm^3 (range 2000–11,000/mm^3), and in 6-month-old children it is even higher (7300/mm^3, range 4000–13,500/mm^3) (see Wara and Ammann, 1977). Normally up to 90% of the peripheral lymphocytes are small lymphocytes; some 5% are medium-sized lymphocytes and up to 15% are large lymphoid cells (Ling and Kay, 1975; Trepel, 1976).

In the peripheral blood of adults 70% of the lymphocytes are of the T and 30% are of the B type (Trepel, 1976, Wara and Ammann, 1977). The content of T and B lymphocytes in the peripheral blood is dependent on age (Augener et al., 1974; Gajl-Peczalska et al., 1974; Smith et al., 1974; Diaz-Jouanen et al., 1975; Steel et al., 1975).

In old age there seems to be a deficiency of T cells (Augener et al., 1974; Smith et al., 1974; Diaz-Jouanen et al., 1975), and this absolute deficiency of T lymphocytes may be responsible for the reported decreased stimulation capacities of PHA in cultures of lymphocytes from old people (Pisciotta et al., 1967).

The contents of B but not of T lymphocytes seem to be higher in cord blood and in the blood of children as compared to adults and old aged individuals (Smith et al., 1974). Gajl-Peczalska et al. (1974) report on 32% B lymphocytes in infants and 22% in middle-aged people; the B lymphocyte count in 4- to 12-month-old children was $2704/mm^3$ and $382/mm^3$ in 91- to 100-year-old people. There has also been found a lymphocytosis after vigorous exercise, with the rapidly mobilized lymphocytes being non-rosette-forming cells, i.e., B lymphocytes (Steel et al., 1974). There are even diurnal and seasonal variations in the B to T lymphocyte ratios (Steel et al., 1974).

Variations in the contents of T and B lymphocytes in the peripheral blood may well influence the rate of spontaneous and induced mutations in peripheral lymphocytes (see Section XV,E).

The total number of lymphocytes of a healthy young adult has been estimated to be some 500×10^9. Only about 2% of the lymphocytes (2×10^9) are present in the peripheral blood, and the others are located in the thymus, lymph nodes, tonsils, lymphatic tissues of the intestines, spleen, bone marrow, and other tissues (see Trepel, 1975, 1976). The lifetime of the lymphocytes is not determined as it is for erythrocytes. Most lymphocytes (90%) are long-lived ones, with a half-life of about 3 years, the other 10% have a half-life of 1–10 days (Buckton et al., 1967; Trepel, 1975, 1976). The B lymphocytes seem to have a somewhat shorter lifetime than the T lymphocytes (Trepel, 1976). The mean rate of renewal of the lymphocytes in the body can be estimated to be 2–5% per day (Trepel, 1976).

With respect to the interpretation of analyses of in vivo-induced chromosomal aberrations and other types of mutations in man it is of great importance that the bulk of the peripheral lymphocytes (at least 80%) belongs to the so-called redistributional pool (Trepel, 1975, 1976). They leave the blood and pass through the spleen, lymph nodes, and other tissues, and reenter the blood. These recirculating cells mainly represent small lymphocytes of the long-living T-cell type. The mean time that a given lymphocyte of the redistributional pool is present in the peripheral blood is about 30 minutes. Not only the blood lymphocytes but also lymphocytes in the other lymphatic organs are recirculating. It has been estimated that about 50%, i.e., 250×10^9 lymphocytes, belong to the redistributional pool, and that the overall recirculation time is about 12 hours (see Trepel, 1975, 1976).

This means that lymphocytes with mutations that have been induced somewhere in the body can eventually show up in the peripheral blood. With the human lymphocyte test system not only mutations that have been induced in the lymphocytes in the peripheral blood proper can be picked up, but also mutations that have been induced in lymphocytes distributed in different organs throughout the body.

C. Stimulation

1. Phytohemagglutinin

Most of the peripheral lymphocytes are in a resting stage of the cell cycle (G_0) and have the diploid DNA content of about 5.06 pg (see Strauss, 1974). These cells can be stimulated in vitro by phytohem-agglutin (PHA), a protein derived from Phaseolus vulgaris, to undergo mitotic divisions (see Ling and Kay, 1975). Under the influence of PHA the lymphocytes are transformed to blastoid cells. As compared to unstimulated lymphocytes, in these blastoid cells the volume of the nucleus and of the whole cell is about 3 times larger. The number of the nuclear pores per μm^2 of the nucleus remains constant but the absolute number of pores is nearly doubled. The cytoplasmic volume is 4 times higher than before the stimulation and the total volume of mitochondria rises by 5 times. The volume of the granular ER increases about 4 times and that of the agranular ER with the Golgi apparatus about 20 times, and the total volume of the lysosomes increases by about 10-fold (David, 1977). There are many polyribosomes in the cytoplasm; the ER is in most cases not very extended (Trepel, 1976, Fig. 1).

A further morphological correlate of the activation of lymphocytes by PHA is the formation of several large and typically structured nucleoli, a phenomenon that can be seen in light and electron microscopic preparations (Busch and Smetana, 1970; Bessis, 1973; Gani, 1976; Trepel, 1976; Beek and Obe, 1977; Obe and Beek, 1981). Biochemically and electron microscopically it can be demonstrated that under the influence of PHA the cells synthesize RNA and proteins

(Pogo, 1972; Sören and Biberfeld, 1973; Ahern and Kay, 1975). A few minutes after addition of PHA, the nuclei of the lymphocytes show a higher binding to acridine orange, probably as a consequence of an unmasking of the dye-binding groups in the nuclear DNA, which may be a necessary condition for the intensive RNA synthesis induced by PHA (Killander and Rigler, 1965, 1969; Rigler and Killander, 1969). The DNA-melting experiments also indicate a weakening of protein DNA complexes (Rigler and Killander, 1969). These findings correlate well with the euchromatinization of the chromatin under the influence of PHA, which can be seen in the light and electron microscope (Busch and Smetana, 1970; Bessis, 1973; Trepel, 1976).

The nonhistone proteins also change in stimulated as compared to nonstimulated lymphocytes (Yeoman et al., 1976). The changes in the nuclear structure have been taken as one explanation for the differential radiosensitivities of PHA-stimulated lymphocytes in their first in vitro G_1 phase (Beek and Obe, 1977), and for the higher sensitivity against chromosome-breaking activities of chemicals in stimulated as compared to unstimulated lymphocytes (Obara, 1974; Obe et al., 1976b). In an autoradiographic analysis, Beek and Obe (1975) found two RNA synthesis peaks during the first G_1 phase of lymphocytes cultured in Ham's F-10 medium and PHA, one around 14 and the other at 19 hours following stimulation. Loeb et al., (1970) analyzed enzyme activities connected with the DNA replication in PHA-stimulated lymphocytes and found the following: there is little DNA polymerase activity in unstimulated lymphocytes. With the beginning of the DNA synthesis as measured by [^{3}H]TdR incorporation, the DNA polymerase activity rises in parallel with the DNA synthetic activity and reaches up to a 150-fold increase as compared to the unstimulated stage. Likewise parallel to the DNA synthesis there was a rise in DNase activity as well. Associated with DNA synthesis there was also an increase of the activities of thymidine kinase and TMP kinase. Deoxyguanosine kinase and GMP kinase activities are not stimulated by PHA. The authors showed (Loeb et al., 1970) that one of the earliest biochemical activities in PHA stimulated lymphocytes is RNA synthesis ([^{3}H]uridine incorporation) followed by an increase in protein synthesis (L-[^{14}C]leucine incorporation), and by the DNA synthesis ([^{3}H]thymidine incorporation). Using different culture media (Ham's F-10; Medium TC 199, MEM) we analyzed the frequencies of [^{3}H]TdR labeled interphase nuclei by autoradiography and mitotic indexes in the course of PHA-induced lymphocyte stimulation and found different characteristics for the curves of the labeling and mitotic indexes (Dudin et al., 1974; Obe et al., 1975a, 1976a).

In Ham's F-10 medium the DNA synthesis starts about 26 hours after culture initiation, and the first mitoses are found about 34 hours after culture initiation. There are two peaks of DNA-synthetic activity, one at 34 and a second at 40 hours, and two peaks of mitotic activity, one at around 44 and a second at around 49 hours. Together with our analysis concerning the RNA synthesis (see above) we postulated the existence of two subpopulations of cells which show a different stimulation pattern in a culture set up with Ham's F-10 and PHA (see also Obe and Beek, 1981).

We found a higher frequency of chromatid translocations (RB') in the first as compared to the second subpopulation after treating the cells with an alkylating agent (Beek and Obe, 1974) and higher frequencies of dicentric chromosomes in the first as compared to the second population after irradiation in G_0 (Beek and Obe, 1976). For further discussion of different mutagen sensitivities in lymphocytes see Obe and Beek (1981). It is difficult to determine the frequency of stimulated cells in a lymphocyte culture, because there is a high variability depending on intrinsic and extrinsic factors (Ling and Kay, 1975). Wilson and Thomson (1968) analyzed the PHA-induced transformation of lymphocytes from eight donors after 3 days culture of isolated cells in TC 199 medium and found values from 45 to 73% with an average of 60%. The PHA predominantly stimulates T lymphocytes (Janossy and Greaves, 1972; Ling and Kay, 1975). Probably under the influence of the stimulated T cells, B cells seem to be stimulated at later culture times (Vischer, 1972; Phillips and Roitt, 1973; Ling and Kay, 1975; Steffen et al., 1978).

Lymphocytes with prolonged G_1 phases in PHA cultures may represent stimulated B lymphocytes (Steffen and Stolzmann, 1969; Jasińska et al., 1970; Sören, 1973; Steffen et al., 1978; Riedel and Obe, 1980).

Polgar et al. (1968) found that in lymphocyte cultures, in which the PHA had been washed off after 24 hours, the cells reverted to small lymphocytes at day 9 of culture, a phenomenon associated with the cessation of DNA synthesis, and these small lymphocytes could be restimulated by addition of PHA.

2. Concanavalin A

Concanavalin A (Con A) is a protein from the jack bean (Canavalia ensiformis) and is the best analyzed plant lectin (Edelman et al., 1972). It stimulates preferably T lymphocytes (Powell and Leon, 1970; Edelman et al., 1972; Pauli and Strauss, 1973; Ling and Kay, 1975; Cunningham et al., 1976). Morphology, physiology, and biochemistry of the stimulation process are similar to those with PHA. Concanavalin

A-stimulated lymphocytes were found to bind 15 times more [9-^{14}C]N-acetoxy-2-acetylaminofluorene than unstimulated lymphocytes (Scudiero *et al.*, 1976).

The stimulation by Con A is dependent on a permanent binding of the protein with cell membrane constituents. Removal of the bound Con A by treating the cells with medium containing α-D-mannoside (MAM) leads to a suppression of DNA synthesis without killing the cells (Powell and Leon, 1970). When transferred to a MAM-containing medium after stimulation in Con A, the cells return to a resting stage. The resting cells can be restimulated when transferred again to Con A-containing medium (Pauli and Strauss, 1973). This Con A–MAM system has been successfully used to perform liquid holding experiments with peripheral lymphocytes (Obe *et al.*, 1980a).

3. *Pokeweed Mitogen*

Pokeweed mitogen (PWM) is a lymphocyte mitogen from pokeweed (*Phytolacca americana*). Pokeweed mitogen-stimulated cells not only resemble the ones originating preferably in PHA-stimulated cultures (PHA-blast cells, Douglas and Fudenberg, 1969), but also to cells which resemble plasma cells originating from B lymphocytes. Cultures incubated 7–10 days showed 10–20% of cells with a pronounced development of the rough ER with clearly formed cisternae (see Fig. 2). The other cells in such cultures resemble the cells originating in PHA cultures in that they have an only slightly developed rough ER without cisternal dilatations (see Fig. 2) (Douglas and Fudenberg, 1969; see also Trepel, 1976). Pokeweed mitogen stimulates both T and B lymphocytes (Ling and Kay, 1975) and may be of importance in analyzing differences in mutagen sensitivities between T and B lymphocytes (see Section XV,E). As compared to PHA and Con A, PWM stimulates less effectively and more slowly (Trepel, 1976; Dudin *et al.*, 1974, 1976).

D. Repair

With cytological and biochemical methods it has been shown that unstimulated lymphocytes are able to perform excision repair (unscheduled DNA synthesis) after treatment with UV light, X-, and γ-rays, and a variety of chemical mutagens (Evans and Norman, 1968; Spiegler and Norman, 1969; Darzynkiewicz, 1971; Clarkson and Evans, 1972; Lieberman and Dipple, 1972; Norman, 1972; Slor, 1973; Cleaver and Painter, 1975; Scudiero *et al.*, 1976; Lavin and Kidson, 1977; Evans *et al.*, 1978; Lavin *et al.*, 1978; Lambert *et al.*, 1979; Lewensohn *et al.*, 1979; Tuschl *et al.*, 1980). This repair capacity is clearly higher in stimulated as compared to unstimulated lymphocytes (Spiegler and Norman, 1969; Darzynkiewicz, 1971; Norman, 1972; Scudiero et al., 1976; Lavin and Kidson, 1977; Lavin et al., 1978; Lewensohn et al., 1979).

Lewensohn et al., (1979) irradiated different cell cycle stages of lymphocytes in vitro with UV and found in autoradiographs 5.3 grains/cell in G_0, 17.7 grains/cell in G_1, and 32.7 grains/cell in G_2 cells. The repair capacity in unstimulated lymphocytes irradiated with UV seems to be negatively correlated with the age of the lymphocyte donor (Lambert et al., 1979). Tuschl et al. (1980) analyzed the repair capacity of UV-irradiated unstimulated lymphocytes from personnel working in a region of Gastein, Austria, with a high natural background irradiation from ^{222}Rn, estimated to be 400–800 mrad in 6 months. As compared to controls they found a clearly higher rate of repair. This finding indicated that low doses of ionizing radiation induce an increase of DNA repair capacities in unstimulated lymphocytes in vivo (Tuschl et al., 1980). This capacity of peripheral lymphocyte to react to chronic exposures with an elevation of their repair activities should be taken into consideration when lymphocytes from chronically exposed humans are analyzed with respect to genetic damage.

A further possibility for analyzing repair capacities of lymphocytes is to perform liquid-holding (LH) experiments with unstimulated (Evans and Vijayalaxmi, 1980) or with restimulated (Obe et al., 1980a) lymphocytes in vitro. Since the peripheral lymphocytes are nearly all in the same stage of the cell cycle (G_0) they are an ideal system for performing LH experiments. One problem arising in these types of analyses are probable differential rates of cell death, which may result in the predominance of special cell populations which may simulate a LH effect.

E. Different Sensitivities of T and B Lymphocytes against Mutagens

Survival studies of unstimulated lymphocytes after irradiation with γ-rays and X-rays have revealed a higher sensitivity of B as compared to T lymphocytes (Santos Mello et al., 1974; Prosser, 1976; Kwan and Norman, 1977). Yew and Johnson (1978) measured the survival of T and B lymphocytes after UV irradiation and found a much higher sensitivity of T as compared to B lymphocytes, with the B lymphocytes having a D_0 of 14 J/m² and the T lymphocytes a D_0 of 4 J/m². The survival curve of unseparated peripheral lymphocytes was biphasic; 75% of the cells had a D_0 of 5 J/m² and 25% a D_0 of 20 J/m², probably reflecting the T and B lymphocyte fractions of the peripheral blood. Using sedimentation characteristics of nuclear bodies (nucleoids), Yew and Johnson (1978) were able to show that T lymphocytes have a slower repair mechanism against UV induced lesions as compared to B lymphocytes. Unstimulated lymphocytes from untreated chronic lymphatic leukemia (CLL) have been found to have a higher repair activity after UV- and X-irradiation as compared to controls and are able to remove thymine dimers more effectively when UV irradiated in the PHA stimulated state (Huang et al., 1972). The CLL lymphocytes are mostly of the B lymphocyte type (Wilson and Nossal, 1971; Trepel, 1976) and the finding of Huang et al., (1972) may be interpreted in the sense of B lymphocytes having higher repair capacities than T lymphocytes. Santesson et al. (1979) found a higher "spontaneous" SCE rate in T (11.9 ± 0.28) as compared to B lymphocytes (6.3 ± 0.36) and a stronger inhibition of the proliferation of T as compared to B lymphocytes using 15 μg/ml bromodeoxyuridine (BUdR) and a culture time of 72 hours. The T lymphocytes but not B lymphocytes showed SCE differentiation after incubation with only 1.5 μg/ml BUdR (Santesson et al., 1979).

We treated whole-blood cultures set up with Ham's F-10 medium and PHA from 15 to 20 hours incubation time with Trenimon in a final concentration of 0.5×10^{-6} M (Riedel and Obe, 1980). At a culture time of 51 hours, two cultures each received Colcemid every 4 hours and were then fixed. This experimental schedule resulted in 17 fixation times starting with 55 hours and ending with 119 hours. At all fixation times we found first, second, third, and further post-treatment metaphases, even at 119 hours; 0.5% first- and 5.5% second-posttreatment metaphases were found.

The SCE frequencies dropped from values between 36.7 and 58.4 up to the fixation time of 87 hours to values between 21.5 and 29.6 up to the fixation time of 115 hours (119 hours gave no results). There were clearly higher frequencies of structural chromosomal aberrations at earlier (55 and 67 hours) as compared to later (87, 95, and 111 hours) fixation times in first posttreatment metaphases; a similar drop in the aberration frequencies was found in second posttreatment metaphases. These results were interpreted as due to differences in repair capacities between T and B lymphocytes, the former representing the fast- and the latter the slow-growing fraction of cells (Riedel and Obe, 1980). In experiments with slowly growing lymphocytes we did not find such differences in the SCE frequencies after treating the cells with Trenimon (Beek and Obe, 1979) (see also Section XV,C). Lewensohn et al., (1981) analyzed the DNA repair synthesis (UDS) in unstimulated T and B lymphocyte enriched cell populations. The cells were treated for 30 minutes with nitrogen mustard $(10^{-4} M)$, methylmethanesulfonate $(10^{-3} M)$, or UV $(19.2 J/m^2)$. The authors found no statistically significant differences between the mean grain counts in T and B lymphocytes, neither in the controls nor in the treated cells. The interindividual variations in the grain counts of the treated cells from the six individuals tested was always greater in the B as compared to the T lymphocytes (Lewensohn et al., 1981).

F. Drug Metabolism

The drug-metabolizing enzyme system that has been studied in lymphocytes to a considerable extent is the aryl hydrocarbon hydroxylase (AHH), which is part of the microsomal cytochrome P-448 or P-450 containing mixed-function oxygenase. The AHH system is inducible by polycyclic aromatic hydrocarbons (PAH) such as 3-methylcholanthrene (MC), benz[a]anthrazene (BA), or benzo[a]pyrene (BA); AHH is activating and detoxifying PAH. Unstimulated lymphocytes have a very low AHH activity, which is not inducible by MC (Busbee et al., 1972; Gurtoo et al., 1975). The enzyme activity increases when the cells have been stimulated with PHA or PWM and is highest when induced in addition with MC, BA, or dibenz[a,h]anthracene (DBA) for 24 to 30 hours in vitro after a previous culture time of 48 or 72 hours (Busbee et al., 1972; Whitlock et al., 1972; Gurtoo et al., 1975; Okano et al., 1979).

The AHH enzyme of stimulated lymphocytes is a typical mixed-function oxygenase in that it needs NADPH and is inhibited by CO and by 7,8-benzoflavone (Whitlock et al., 1972). Whitlock et al. (1972) reported the higher inducibility of AHH in lymphocytes stimulated with PWM as compared to lymphocytes stimulated with PHA and discussed their results as being probably a biochemical correlate of plasma cells (i.e., B lymphocytes; see Section XV,C) which are stimulated by PWM and have a highly developed ER (see Fig. 2). Kellermann et al., (1973a) found no effect of PWM as compared to PHA stimulation.

Kellermann and co-workers consider their findings an indication that the inducibility of the AHH system is under genetic control which can be formally described to be due to a single locus, resulting in low, intermediate, and high AHH inducibility (Kellermann, 1977; Kellermann et al., 1973a,b).

This lymphocyte-bound AHH system offers an excellent possibility to study some aspects of drug metabolism in human cells in vitro.

G. Lymphocytes in Mutagenicity Research

In this section we will not discuss the wealth of literature existing on the use of the peripheral lymphocytes for the analysis of the mutagenic activities of various chemical, biological, and physical mutagens *in vitro*, because different methods for doing this, including the mammalian metabolism in the test system, have been extensively discussed recently (Madle and Obe, 1980; Obe and Natarajan, 1980; Natarajan and Obe, 1980; Obe and Madle, 1981; Obe and Beek, 1981). We want, however, to point out briefly the advantages of the peripheral lymphocytes for analyzing the chromosome-breaking activities of such chemicals *in vivo* in man which act in a S-dependent manner (Evans, 1977a, Natarajan and Obe, 1978).

The peripheral lymphocytes are an ideal system for analyzing the effects of chronic exposures to low doses of mutagens, because (1) most of the cells are long living and this can lead to an accumulation of lesions in their DNA over a number of years, (2) most of the peripheral lymphocytes are in a defined stage of the cell cycle (G_0) and can be transformed *in vitro* under controllable conditions to undergo mitoses, and (3) the peripheral lymphocytes have a lower repair activity than cycling cells which allows DNA lesions to accumulate. Recurrent induction of chemical lesions in the DNA of the small lymphocytes *in vivo* may have different consequences:

1. The lesions are repaired but more slowly than in normal cells.
2. During the repair, DNA single- and double-strand breaks may occur. In rare cases it may happen that two double-strand breaks in one cell lead to the formation of dicentric or ring chromosomes by misrepair (see Evans, 1977a). This means that a chronic exposure to chemicals leads to a very slow accumulation of chromosome-type aberrations which are stable and can only disappear with the death of the cell carrying them.
3. The lesions are still in the DNA when the cells are cultured and stimulated. This will eventually lead to the formation of chromatid type aberrations by misreplication (see Evans, 1977a).
4. After cessation of the chronic exposure the rate of chromatid-type aberrations will decrease because the lesions leading to their formation are slowly repaired.

In a recent study of the frequencies of exchange-type aberrations in alcoholics, we found evidence for the events discussed above most probably occur (Obe *et al.*, 1980b). Alcohol is an active mutagen via its first metabolite acetaldehyde which is a cross-linking agent (see

Obe and Ristow, 1979). We found chromatid-type exchanges, which was expected, but we found also chromosome-type aberrations (dicentric and ring chromosomes) in chronic alcoholics. In dry alcoholics (in our case alcoholics on therapy with Antabuse for more than 2 years), the rate of chromatid-type exchanges but not of chromosome-type exchanges dropped to the control level. This may indicate that some of the chromosome-type aberrations induced persisted for a long time.

ACKNOWLEDGMENTS

The authors' own research cited in this review has been partly financed by the Deutsche Forschungsgemeinschaft (GO, SFB 29, Berlin), Koningin Wilhelmina Funds (SG 77.44), the Association between Euratom and the University of Leiden Contract No. 052/64/1 BIAN, and IAEA Contract No. 1745/R4/RB.

REFERENCES

Aaron, C. S., van Zeeland, A. A., Mohn, G. R., and Natarajan, A. T. (1978). Molecular dosimetry of the chemical mutagen ethyl methanesulfonate in Escherichia coli and in V-79 Chinese hamster cells. Mutat. Res. **50**, 419–426.

Abe, S., and Sasaki, M. (1977). Chromosome aberrations and sister chromatid exchanges in Chinese hamster cells exposed to various chemicals. J. Natl. Cancer Inst. **58**, 1635–1640.

Ahern, T., and Kay, J. E. (1975). Protein synthesis and ribosome activation during the early stages of phytohemagglutinin lymphocyte stimulation. Exp. Cell Res. **92**, 513–515.

Auerbach, A. D., Warburton, D., Bloom, A. D., Chaganti, R. S. K. (1979). Prenatal detection of Fanconi's anaemia gene by cytogenetic methods. Am. J. Hum. Genet. **31**, 77–81.

Auerbach, C., Robson, J. M., and Carr, J. G. (1943). The chemical production of mutations. Science **105**, 243–247.

Augener, W., Cohnen, G., Reuter, A., and Brittinger, G. (1974). Decrease of T lymphocytes during ageing. Lancet **I**, 1164.

Beek, B., and Obe, G. (1974). The human leukocyte test system. II. Different sensitivities of subpopulations to a chemical mutagen. Mutat. Res. **24**, 395–398.

Beek, B., and Obe, G. (1975). The human leukocyte test system IV. The RNS-synthesis pattern in the first G_1-phase of the cell cycle after stimulation with PHA. Mutat. Res. **29**, 165–168.

Beek, B., and Obe, G. (1976). The human leukocyte test system X. Higher sensitivity to X-irradiation in the G_0 stage of the cell cycle of early as compared to late replicating cells. Hum. Genet. **35**, 57–70.

Beek, B., and Obe, G. (1977). Differential chromosomal radiosensitivity within the first G_1-phase of the cell cycle of early-dividing human leukocytes in vitro after stimulation with PHA. Hum. Genet. **35**, 209–218.

Beek, B., and Obe, G. (1979). Sister chromatid exchanges in human leukocyte chromosomes: Spontaneous and induced frequencies in early- and late-proliferating cells *in vitro. Hum. Genet.* **49,** 51–61.

Bessis, M. (1973). "Living Blood Cells and their Ultrastructure." Springer–Verlag, Berlin and New York.

Bond, V. P., Cronkite, E. P., Fliedner, T. M., and Schork, P. (1958). Deoxyribonucleic acid synthesizing cells in peripheral blood of normal human beings. *Science* **128,** 202–203.

Bootsma, D. (1979). DNA repair deficiencies in man. *Radiat. Res. Proc. 6th Int. Cong., Tokyo,* 472–475.

Buckton, K. E., Smith, P. G., and Court Brown, W. M. (1967). The estimation of lymphocyte lifespan from studies on males treated with X-rays for ankylosing spondylitis. *In* "Human Radiation Cytogenetics" (H. J. Evans, W. M. Court Brown, and A. S. McLean, eds.), pp. 106–114. North-Holland, Amsterdam.

Busbee, D. L., Shaw, C. R., and Cantrell, E. T. (1972). Aryl hydrocarbon hydroxylase induction in human leukocytes. *Science* **178,** 315–316.

Busch, H., and Smetana, K. (1970). "The Nucleolus." Academic Press, New York.

Carrano, A. V., Thompson, L. H., Lindl, P. A., and Minkler, J. L. (1978). Sister chromatid exchange as an indicator of mutagenesis. *Nature (London)* **271,** 551–553.

Carstairs, K. (1962). The human small lymphocyte: Its possible pluripotential quality. *Lancet* **I,** 829–832.

Chaganti, R. S. K., Schönberg, S., and German, J. (1974). A many fold increase in sister chromatid exchanges in Bloom's syndrome lymphocytes. *Proc. Natl. Acad. Sci. U.S.A.* **71,** 4508–4512.

Clarkson, J. M., and Evans, H. J. (1972). Unscheduled DNA synthesis in human leucocytes after exposure to UV light, γ-rays, and chemical mutagens. *Mutat. Res.* **14,** 413–430.

Cleaver, J. E. (1968). Defective repair replication in xeroderma pigmentosum. *Nature (London)* **218,** 652–656.

Cleaver, J., and Painter, R. (1975). Absence of specificity in inhibition of DNA repair replication by DNA-binding agents, cocarcinogens, and steroids in human cells. *Cancer Res.* **35,** 1773–1778.

Cunningham, B. A., Sela, B.-A., Yahara, I., and Edelman, G. M. (1976). Structure and activities of lymphocyte mitogens. *In* "Mitogens in Immunobiology" (J. J. Oppenheim and D. L. Rosenstreich, eds.), pp. 13–30. Academic Press, New York.

Darzynkiewicz, Z. (1971). Radiation-induced DNA synthesis in normal and stimulated human lymphocytes. *Exp. Cell Res.* **69,** 356–360.

David, H. (1977). "Quantitative Ultrastructural Data of Animal and Human Cells." Fischer, Stuttgart.

Dean, B. J., and Hodson-Walker, G. (1979). An *in vitro* chromosome assay using cultured rat-liver cells. *Mutat. Res.* **64,** 329–337.

Diaz-Jouanen, E., Williams, R. C., Jr., and Strickland, R. G. (1975). Age-related changes in T and B cells. *Lancet* **I,** 688–689.

DiPaolo, J. A., Amsbaugh, S. C., and Popescu, N. C. (1980). Antipain inhibits N-methyl-N-nitro-N-nitrosoguanidine-induced transformation and increases chromosomal aberrations. *Proc. Natl. Acad. Sci. U.S.A.* **77,** 6649–6653.

Douglas, S. D., and Fudenberg, H. H. (1969). *In vitro* development of plasma cells from lymphocytes following pokeweed mitogen stimulation: A fine structural study. *Exp. Cell Res.* **54,** 277–279.

Dudin, G., Beek, B., and Obe, G. (1974). The human leukocyte test system I. DNA synthesis and mitosis in PHA-stimulated 2-day cultures. *Mutat. Res.* **23,** 279–281.

Dudin, G., Beek, B., and Obe, G. (1976). The human leukocyte test system VIII. DNA synthesis and mitoses in 3-day cultures stimulated with pokeweed mitogen. *Hum. Genet.* **32**, 323–327.

Edelman, G. M., Cunningham, B. A., Reeke, G. N., Jr., Becker, J. W., Waxdal, M. J., and Wang, J. L. (1972). The covalent and three-dimensional structure of concanavalin A. *Proc. Natl. Acad. Sci. U.S.A.* **69**, 2580–2584.

Evans, H. J. (1962). Chromosome aberrations induced by ionizing radiations. *Int. Rev. Cytol.* **13**, 221–231.

Evans, H. J. (1977a). Molecular mechanisms in the induction of chromosome aberrations. *In* "Progress in Genetic Toxicology" (D. Scott, B. A. Bridges, and F. H. Sobels, eds.), pp. 57–74. Elsevier North-Holland Biochemical Press, Amsterdam.

Evans, H. J. (1977b). What are sister chromatid exchanges? *In* "Chromosomes Today" (A. de la Chapelle and M. Sorsa, eds.), Vol. 6, pp. 315–326. Elsevier/North-Holland, Amsterdam.

Evans, H. J., and Scott, D. (1969). The induction of chromosome aberrations by nitrogen mustard and its dependence on DNA synthesis. *Proc. R. Soc. London Ser. B* **173**, 491–512.

Evans, H. J., and Vijayalaxmi (1980). Storage enhances chromosome damage after exposure of human leukocytes to mitomycin C. *Nature (London)* **284**, 370–372.

Evans, H. J., Adams, A. C., Clarkson, J. M., and German, J. (1978). Chromosome aberrations and unscheduled DNA synthesis in X- and UV-irradiated lymphocytes from a boy with Bloom's syndrome and a man with xeroderma pigmentosum, *Cytogenet. Cell Genet.* **20**, 124–140.

Evans, R. G., and Norman, A. (1968). Radiation stimulated incorporation of thymidine into the DNA of human lymphocytes. *Nature (London)* **217**, 455–456.

Fujiwara, Y., Kano, Y., Tatsumi, M., and Paul, P. (1980). Effects of a tumor promoter and an anti-promoter on spontaneous and UV-induced 6-thioguanine-resistant mutations and sister chromatid exchanges in V-79 Chinese hamster cells. *Mutat. Res.* **71**, 243–251.

Gajl-Peczalska, K. J., Hallgren, H., Kersey, J. H., Zusman, J., and Yunis, E. J. (1974). B lymphocytes during ageing. *Lancet* **II**, 163.

Gani, R. (1976). The nucleoli of cultured human lymphocytes. I. Nucleolar morphology in relation to transformation and the DNA cycle. *Exp. Cell Res.* **97**, 249–258.

Gentil, A., Renault, G., and Margot, A. (1980). The effect of the tumor promotor 12-O-Tetradecanoyl-Phorbol-13 acetate (TPA) on UV- and MNNG-induced sister chromatid exchanges in mammalian cells. *Int. J. Cancer* **26**, 517–521.

Goth-Goldstein, R. (1977). Repair of DNA damaged by alkylating carcinogens is defective in xeroderma pigmentosum-derived fibroblasts. *Nature (London)* **267**, 81–82.

Gurtoo, H. L., Bejba, N., Minowada, J. (1975). Properties, inducibility, and an improved method of analysis of aryl hydrocarbon hydroxylase in cultured human lymphocytes. *Cancer Res.* **35**, 1235–1243.

Heddle, J. A., and Arlett, C. F. (1980). Untransformed xeroderma pigmentosum cells are not hypersensitive to sister chromatid exchange production by ethyl methanesulphonate—Implications for the use of transformed cell lines and for the mechanism by which SCE arise. *Mutat. Res.* **72**, 119–125.

Hollstein, M., McCann, J., Angelosanto, F. A., and Nichols, W. W. (1979). Short-term tests for carcinogens and mutagens. *Mutat. Res.* **65**, 133–226.

Huang, C. C., and Furukawa, M. (1978). Sister chromatid exchanges in human lymphoid cell lines cultured in diffusion chambers in mice. *Exp. Cell Res.* **111**, 458–461.

Huang, A. T., Kremer, W. B., Laszlo, J., and Setlow, R. B. (1972). DNA repair in human leukaemic lymphocytes. *Nature (London) New Biol.* **240**, 114–116.

International Agency for Research of Cancer (*IARC*) (1980). Long-term and short-term screening assays for carcinogens: A critical appraisal. *IARC Sci. Publ. Suppl.* **2**, 201–221.

Ishidate, M., Jr., and Odashima, S. (1977). Chromosome tests with 134 compounds on Chinese hamster cells *in vitro*—A screening for chemical carcinogens. *Mutat. Res.* **48**, 337–354.

Iskandar, O. (1981). Induction of micronuclei in human lymphocytes. Ph.D. Thesis, University of Indonesia, Jakarta, Indonesia.

Iskandar, O., Jager, M. J., Willemze, R., and Natarajan, A. T. (1980). A case of pure red cell aplasia with a high incidence of spontaneous chromosome breakage: A possible X-ray sensitive syndrome. *Hum. Genet.* **55**, 337–340.

Janossy, G., and Greaves, M. F. (1972). Lymphocyte activation. II. Discriminating stimulation of lymphocyte subpopulations by phytomitogens and heterologous anti-lymphocyte sera. *Clin. Exp. Immunol.* **10**, 525–536.

Jasińska, J., Steffen, J. A., and Michalowski, A. (1970). Studies on *in vitro* lymphocyte proliferation in cultures synchronized by the inhibition of DNA synthesis. II. Kinetics of the initiation of the proliferative response. *Exp. Cell Res.* **61**, 333–341.

Jungermann, K., and Möhler, H. (1980). "Biochemie," Chapter 10.2, pp. 603–626. Immunantwort, Springer–Verlag, Berlin and New York.

Kellermann, G. (1977). Hereditary factors in human cancer. *In* "Origins of Human Cancer" (H. H. Hiatt, J. D. Watson, and J. A. Winsten, eds.), Book B, pp. 837–845. Cold Spring Harbor Lab., Cold Spring Harbor, New York.

Kellermann, G., Cantrell, E., and Shaw, C. R. (1973a). Variations in extent of aryl hydrocarbon hydroxylase induction in cultured human lymphocytes. *Cancer Res.* **33**, 1654–1656.

Kellermann, G., Luyten-Kellermann, M., and Shaw, C. R. (1973b). Genetic variation of aryl hydrocarbon hydroxylase in human lymphocytes. *Am. J. Hum. Genet.* **25**, 327–331.

Kihlman, B. A. (1977). "Caffeine and Chromosomes," pp. 504. Elsevier, Amsterdam.

Kihlman, B. A., Natarajan, A. T., and Andersson, H. C. (1978). Use of the 5-bromodeoxy-uridine-labelling technique for exploring mechanisms involved in the formation of chromosomal aberrations. I. G_2 Experiments with root tips of *Vicia faba*. *Mutat. Res.* **52**, 181–198.

Killander, D., and Rigler, R. (1965). Initial changes of deoxyribonucleoprotein and synthesis of nucleic acid in phytohemagglutinin-stimulated human leukocytes *in vitro*. *Exp. Cell Res.* **39**, 701–704.

Killander, D., and Rigler, R. (1969). Activation of deoxyribonucleoprotein in human leukocytes stimulated by phytohemagglutinin. I. Kinetics of the binding of acridine orange to deoxyribonucleoprotein. *Exp. Cell Res.* **54**, 163–170.

Kinsella, A., and Radman, M. (1978). Tumor promoter induces sister chromatid exchanges: Relevance to mechanisms of carcinogenesis. *Proc. Natl. Acad. Sci. U.S.A.* **75**, 6149–6153.

Kinsella, A. R., and Radman, M. (1980). Inhibition of carcinogen-induced chromosomal aberrations by an anticarcinogenic protease inhibitor. *Proc. Natl. Acad. Sci. U.S.A.* **77**, 3544–3547.

Krepinsky, A. B., Rainbow, A. J., and Heddle, J. A. (1980). Studies on the ultraviolet light sensitivity of Bloom's syndrome fibroblasts. *Mutat. Res.* **69**, 357–368.

Krstić, R. V. (1978). "Die Gewebe des Menschen und der Säugetiere. Ein Atlas zum Studium für Mediziner und Biologen." Springer–Verlag, Berlin and New York.

Kwan, D. K., and Norman, A. (1977). Radiosensitivity of human lymphocytes and thymocytes. *Radiat. Res.* **69**, 143–151.

Lambert, B., Ringborg, U., and Skoog, L. (1979). Age-related decrease of ultraviolet light-induced DNA repair synthesis in human peripheral leukocytes. *Cancer Res.* **39**, 2792–2795.

Latt, S. A. (1974). Sister chromatid exchanges, indices of human chromosome damage and repair: Detection by fluorescence and induction by mitomycin C. *Proc. Natl. Acad. Sci. U.S.A.* **71**, 3162–3166.

Latt, S. A., Stetten, G., Juergens, L. A., Buchanan, G. R., and Gerald, P. S. (1975). Induction by alkylating agents of sister chromatid exchanges and chromatid breaks in Fanconi's anemia. *Proc. Natl. Acad. Sci. U.S.A.* **72**, 4066–4070.

Lavin, M. F., and Kidson, C. (1977). Repair of ionizing radiation induced DNA damage in human lymphocytes. *Nucleic Acids Res.* **4**, 4015–4021.

Lavin, M. F., Chen, P. C., and Kidson, C. (1978). Ataxia telangiectasia: Characterization of heterozygotes. *In* "DNA Repair Mechanisms" (P. C. Hanawalt, E. C. Friedberg, and C. F. Fox, eds.), pp. 651–654. Academic Press, New York.

Lentz, T. L. (1971). "Cell Fine Structure: An Atlas of Drawings of Whole Cell Structure." Saunders, Philadelphia, Pennsylvania.

Lewensohn, R., Killander, D., Ringborg, U., and Lambert, B. (1979). Increase of UV-induced DNA repair synthesis during blast transformation of human lymphocytes. *Exp. Cell Res.* **123**, 107–110.

Lewensohn, R., Ringborg, U., Baral, E., and Lambert, B. (1981). DNA repair synthesis in subpopulations of human lymphocytes. (personal communication)

Lieberman, M. W., and Dipple, A. (1972). Removal of bound carcinogen during DNA repair in nondividing human lymphocytes. *Cancer Res.* **32**, 1855–1860.

Ling, N. R., and Kay, J. E. (1975). "Lymphocyte Stimulation." North-Holland, Amsterdam.

Loeb, L. A., Ewald, J. L., and Agarwal, S. S. (1970). DNA polymerase and DNA replication during lymphocyte transformation. *Cancer Res.* **30**, 2514–2520.

Loveday, K. S., and Latt, S. (1979). The effect of tumor promoter 12-O-tetra-decanoyl-phorbol-13-acetate (TPA) on sister chromatid exchange formation in cultured Chinese hamster cells. *Mutat. Res.* **67**, 343–348.

Mace, M. L., Jr., Daskal, Y., and Wray, W. (1978). Scanning-electron microscopy of chromosome aberrations. *Mutat. Res.* **52**, 199–206.

Madle, S., and Obe, G. (1980). Methods for analysis of the mutagenicity of indirect mutagens/carcinogens in eukaryotic cells. *Hum. Genet.* **56**, 7–20.

Matsuoka, A., Mayash, M., and Ishidate, M., Jr. (1979). Chromosomal aberration tests on 29 chemicals combined with S-9 mix *in vitro*. *Mutat. Res.* **66**, 277–290.

Mohn, G. R., van Zeeland, A. A., Knaap, A. G. A. C., Glickman, B. W., Natarajan, A. T., Brendel, M., de Serres, F. J., and Aaron, C. S. (1980). Quantitative molecular dosimetry of ethyl methanesulfonate (EMS) in several genetic test systems. *In* "Short Term Test Systems for Detecting Carcinogens" (K. H. Norpoth and R. C. Garner, eds.), pp. 160–169. Springer–Verlag, Berlin and New York.

Natarajan, A. T., and Meijers, M. (1979). Chromosomal radiosensitivity of ataxia telangiectasia at different cell cycle stages. *Hum. Genet.* **52**, 127–132.

Natarajan, A. T., and Obe, G. (1978). Molecular mechanisms involved in the production of chromosomal aberrations. I. Utilization of *Neurospora* endonuclease for the study of aberration production in G_2 stage of the cell cycle. *Mutat. Res.* **52**, 137–149.

Natarajan, A. T., and Obe, G. (1980). Screening of human populations for mutations induced by environmental pollutants: Use of human lymphocyte system. *Exotoxicol. Env. Safety* **4**, 468–481.

Natarajan, A. T., and Raposa, T. (1975). Heterochromatin and chromosome aberrations: A comparative study of three mouse cell lines with different karyotype and heterochromatin distribution. *Hereditas* **80**, 83–90.

Natarajan, A. T., and van Kesteren-van Leeuwen, A. C. (1981). The mutagenic activity of 20 coded compounds in chromosome aberrations/sister chromatid exchanges assay using Chinese hamster ovary (CHO) cells. Proc. International Program on the Evaluations of Short Term Tests for Predicting Carcinogenicity, St. Simon Island. *Progress in Mutation Res.* **1**, 551–559.

Natarajan, A. T., Tates, A. D., van Buul, P. P. W., Meijers, M., and de Vogel, N. (1976). Cytogenetic effects of mutagens/carcinogens after activation in a microsomal system *in vitro*. I. Induction of chromosome aberrations and sister chromatid exchanges by diethylnitrosamine (DEN) and dimethylnitrosamine (DMN). *Mutat. Res.* **37**, 83–90.

Natarajan, A. T., Obe, G., and Dulout, F. N. (1980a). The effect of caffeine post-treatment on X-ray induced chromosomal aberrations in human blood lymphocytes *in vitro*. *Hum. Genet.* **54**, 183–184.

Natarajan, A. T., van Zeeland, A. A., Verdegaal-Immerzeel, P. A. M., and Filon, A. R. (1980b). Studies on the influence of photoreactivation on the frequencies of UV-induced chromosomal aberrations, sister-chromatid exchanges and pyrimidine dimers in chicken embryonic fibroblasts. *Mutat. Res.* **69**, 307–317.

Natarajan, A. T., Verdegaal-Immerzeel, E. A. M., Meijers, M., Zoetelief, J., and Ashwood-Smith, M. J. (1981). Cytogenetic response of ataxia telaniectasia cells to physical and chemical mutagens. In "Ataxia-Telengiectasia" (B. A. Bridges and D. Harnden, eds.) pp. 219–226. Wiley, New York.

Nichols, W. W., Miller, R. C., and Bradt, C. (1977). In vitro anaphase and metaphase preparations in mutation testing. In "Handbook of Mutagenicity Test Procedures" (B. J. Kilbey, M. Legator, E. Nichols, and C. Ramel, eds.), pp. 225–233. Elsevier, Amsterdam.

Norman, A. (1972). DNA repair in lymphocytes and some other human cells. In "DNA-Repair Mechanisms" pp. 9–16. Schattauer, Stuttgart.

Nowell, P. C. (1960). Phytohemagglutinin: An initiator of mitosis in cultures of normal human leukocytes. *Cancer Res.* **20**, 462–466.

Obara, Y. (1974). The effect of bleomycin on PHA-stimulated human peripheral lymphocytes: Cytogenetic and autoradiographic studies. *Sci. Rep. Hirosaki Univ.* **21**, 57–66.

Obe, G., and Beek, B. (1981). The human leukocyte test system. *Chem. Mutagens* **7**, 337–400.

Obe, G., and S. Madle. (1981). Prüfung der Mutagenität von Medikamenten beim Menschen. *Prax. Pneumonol.* (in press).

Obe, G., and Natarajan, A. T. (1980). Umweltbedingte Mutationen beim Menschen, *Klinkarzt* **9**, 271–289.

Obe, G., and Ristow, H. (1979). Mutagenic, cancerogenic, and teratogenic effects of alcohol. *Mutat. Res.* **65**, 229–259.

Obe, G., Beek, B., and Dudin, G. (1975a). The human leukocyte test system. V. DNA synthesis and mitoses in PHA-stimulated 3-day cultures. *Humangenetik* **28**, 295–302.

Obe, G., Beek, B., and Vaidya, V. G. (1975b). The human leukocyte test system. III. Premature chromosome condensation from chemically and X-ray induced micronuclei. *Mutat. Res.* **27**, 89–101.

Obe, G., Brandt, K., and Beek, B. (1976a). The human leukocyte test system. IX. DNA synthesis and mitoses in PHA stimulated 2-day cultures set up with Eagle's minimal essential medium (MEM). *Hum. Genet.* **33**, 263–268.

Obe, G., Beek, B., and Slacik-Erben, R. (1976b). The use of the human leukocyte test system for the evaluation of potential mutagens. *Excerpta Medica Int. Congr. Ser.* **376**, 118–126.

Obe, G., Natarajan, A. T., Meijers, M., and den Hertog, A. (1979). Induction of chromosomal aberrations in peripheral blood lymphocytes of human blood *in vitro* and SCEs in bone marrow cells of mice *in vivo* by ethanol and its metabolite acetaldehyde. *Mutat. Res.* **68**, 291–294.

Obe, G., Natarajan, A. T., and den Hertog, A. (1980a). Studies on the influence of liquid holding in Con-A stimulated human peripheral blood lymphocytes on mitosis and X-ray induced chromosome aberrations. *Hum. Genet.* **54**, 385–390.

Obe, G., Göbel, D., Engeln, H., Herha, J., and Natarajan, A. T. (1980b). Chromosomal aberrations in peripheral lymphocytes of alcoholics. *Mutat. Res.* **73**, 377–386.

Oehlkers, F. (1943). Die Auslösung von Chromosomenmutationen in der Meiosis durch Einwirkung von Chemikalien. *Zid. Absb. Vererbuugsl.* **81**, 313–341.

Okano, P., Miller, H. N., Robinson, R. C., and Gelboin, H. V. (1979). Comparison of benzo[a]pyrene and (−)-trans-7,8-dihydroxy-7,8-dihydrobenzo[a]pyrene metabolism in human blood monocytes and lymphocytes. *Cancer Res.* **39**, 3184–3193.

Painter, R. B. (1980). A replication model for sister-chromatid exchange. *Mutat. Res.* **70**, 337–341.

Pauli, R. M., and Strauss, B. S. (1973). Proliferation of human peripheral lymphocytes. Characteristics of cells once stimulated or restimulated by concanavalin A. *Exp. Cell Res.* **82**, 357–366.

Perry, P. E. (1980). Chemical mutagens and sister chromatid exchange. *Chem. Mutagens* **6**, 1–39.

Perry, P., and Evans, H. J. (1975). Cytological detection of mutagen carcinogen exposure by sister chromatid exchange. *Nature (London)* **258**, 121–125.

Perry, P., and Wolff, S. (1974). New Giemsa method for differential staining of sister chromatid exchanges. *Nature (London)* **251**, 156–158.

Phillips, B., and Roitt, I. M. (1973). Evidence for transformation of human B lymphocytes by PHA. *Nature (London) New Biol.* **241**, 254–256.

Pisciotta, A. V., Estring, D. W., De Prey, C., and Walsh, B. (1967). Mitogenic effect of phytohemagglutinin at different ages. *Nature (London)* **215**, 193–194.

Pogo, B. G. T. (1972). Early events in lymphocyte transformation by phytohemagglutinin. I. DNA-dependent RNA polymerase activities in isolated lymphocyte nuclei. *J. Cell Biol.* **53**, 635–641.

Polgar, P. R., Kibrick, S., and Foster, J. M. (1968). Reversal of PHA-induced blastogenesis in human lymphocyte cultures. *Nature (London)* **218**, 596–597.

Powell, E. A., and Leon, M. A. (1970). Reversible interaction of human lymphocytes with the mitogen concanavalin A. *Exp. Cell Res.* **62**, 315–325.

Prosser, J. (1976). Survival of human T and B lymphocytes after X-irradiation. *Int. J. Radiat. Biol.* **30**, 459–465.

Rasmussen, R. E., and Painter, R. B. (1964). Evidence for repair of ultraviolet damaged deoxyribonucleic acid in cultured mammalian cells. *Nature (London)* **203**, 1360–1362.

Rhodin, J. A. G. (1974). "Histology." Oxford Univ. Press, London and New York.

Riedel, L., and Obe, G. (1980). Trenimon-induced SCEs and structural chromosomal aberrations in early- and late-dividing lymphocytes. *Mutat. Res.* **73**, 125–131.

Rigler, R., and Killander, D. (1969). Activation of deoxyribonucleoprotein in human leukocytes stimulated by phytohemagglutinin. II. Structural changes of deoxyribonucleoprotein and synthesis of RNA. *Exp. Cell Res.* **54**, 171–180.

Santesson, B., Lindahl-Kiessling, K., and Mattson, A. (1979). SCE in B and T lymphocytes. Possible implications for Bloom's syndrome. *Clin. Genet.* **16**, 133–135.

Santos Mello, R., Kwan, D. K., and Norman, A. (1974). Chromosome aberrations and T-cell survival in human lymphocytes. *Radiat. Res.* **60**, 482–488.

Sasaki, M. S., and Tonomura, A. (1973). A high susceptibility of Fanconi's anemia to chromosome breakage by DNA crosslinking agents. *Cancer Res.* **33**, 1829–1836.

Savage, J. R. K. (1975). Classification and relationship of induced chromosomal structural changes. *J. Med. Genet.* **12**, 103–122.

Scheid, W., and Traut, H. (1970). Ultraviolet-microscopical studies on achromatic lesions ("gaps") induced by X-rays in the chromosomes of *Vicia faba*. *Mutat. Res.* **10**, 159–161.

Schöneich, J., Michaelis, A., and Rieger, R. (1970) Coffein und die chemische Induktion von Chromatidenaberrationen bei *Vicia faba* und Ascitestumoren der Maus. *Biol. Zentralb.* **89**, 49–63.

Scott, D. (1980). Molecular mechanisms of chromosome structural changes. *Toxicol. Environ. Sci.* **7**, 101–113.

Scudiero, D., Norin, A., Karran, P., and Strauss, B. (1976). DNA excision repair deficiency of human peripheral blood lymphocytes treated with chemical carcinogens. *Cancer Res.* **36**, 1397–1403.

Shafer, D. A. (1977). Replication bypass model of sister chromatid exchanges and implications for Bloom's syndrome and Fanconi's anemia. *Hum. Genet.* **39**, 177–190.

Shaham, M., Becker, Y., and Cohen, N. M. (1980). A diffusable clastogenic factor in ataxia telangiectasia, *Cytogenet. Cell Genet.* **27**, 155–161.

Slor, H. (1973). Induction of unscheduled DNA synthesis by the carcinogen 7-bromethyl-benz[a] anthracene and its removal from the DNA of normal and xeroderma pigmentosum lymphocytes. *Mutat. Res.* **19**, 231–235.

Smith, M. A., Evans, J., and Steel, C. M. (1974). Age related variation in proportion of circulating T cells. *Lancett* **II**, 922–924.

Sören, L. (1973). Variability of the time at which PHA-stimulated lymphocytes initiate DNA synthesis. *Exp. Cell Res.* **78**, 201–208.

Sören, L., and Biberfeld, P. (1973). Quantitative studies on RNA accumulation in human PHA-stimulated lymphocytes during blast-transformation. *Exp. Cell Res.* **79**, 1–9.

Spiegler, P., and Norman, A. (1969). Kinetics of unscheduled DNA synthesis induced by ionizing radiation in human lymphocytes. *Radiat. Res.* **39**, 400–410.

Steel, C. M., Evans, J., and Smith, M. A. (1974). Physiological variation in circulating B cell: T cell ratio in man. *Nature (London)* **247**, 387–389.

Steel, C. M., Evans, J., and Smith, M. A. (1975). Age-related changes in T and B cells. *Lancet* **I**, 914–915.

Steffen, J. A., and Stolzmann, W. M. (1969). Studies on *in vitro* lymphocyte proliferation in cultures synchronized by the inhibition of DNA synthesis. I. Variability of S plus G_2 periods of first generation cells. *Exp. Cell Res.* **56**, 453–460.

Steffen, J. A., Swierkowska, K., Michalowski, A., Kling, E., and Nowakowska, A. (1978). *In vitro* kinetics of human lymphocytes activated by mitogens. In "Mutagen-Induced Chromosome Damage in Man" (H. J. Evans and D. C. Lloyd, eds.), pp. 89–107. Edinburgh Univ. Press, Edinburgh, Scotland.

Stetka, D. G., Jr. (1979). Further analysis of the replication bypass model for sister chromatid exchange. *Hum. Genet.* **49**, 63–69.

Stich, H. F., Lam, P., Lo, L. W., Koropatnick, D. J., and San, R. H. C. (1975). The search for relevant short term bioassays for chemical carcinogens: The tribulation of a modern Sisyphus. *Can. J. Genet. Cytol.* **17**, 471–492.

Strauss, B. S. (1974). Nuclear DNA. In "The Cell Nucleus" (H. Busch, ed.), pp. 3–33. Academic Press, New York.

Swenberg, J. A., and Petzold, G. L. (1979). "Strategies for Short-Term Testing for Mutagens/Carcinogens," p. 77. CRC Press, West Palm Beach, Florida.

Swift, M. (1977). Malignant neoplasam in heterozygous carriers of genes for certain autosomal recessive syndromes. In "Genetics of Human Cancer" (J. J. Mulvihill, R. W. Miller, and J. F. Fraumeni, Jr., eds.), pp. 209–215. Raven, New York.

Taylor, M. R., Metcalfe, J. A., Oxford, J. M., and Harnden, D. G. (1976). Is chromatid-type damage in ataxia telangiectasia after irradiation at G_0 a consequence of defective repair? Nature (London) 260, 441–443.

Taylor, A. M. R., Rosney, C. M., and Campbell, J. B. (1979). Unusual sensitivity of ataxia telangiectasia cells to bleomycin. Cancer Res. 39, 1046–1050.

Terasima, T., and Tolmach, L. J. (1961). Changes in X-ray sensitivity of HeLa cells during the division cycle. Nature (London) 190, 1210–1211.

Trepel, F. (1975). Kinetik lymphatischer Zellen. In "Lymphocyt und klinische Immunologie" (H. Theml and H. Begemann, eds.), pp. 16–26. Springer-Verlag, Berlin and New York.

Trepel, F. (1976). Das lymphatische Zellsystem: Struktur, allgemeine Physiologie und allgemeine Pathophysiologie. In "Blut und Blutkrankheiten, Teil 3, Leukocytäres und retikuläres System I" (H. Begemann ed.), pp. 1–191. Springer-Verlag, Berlin and New York.

Tuschl, H., Altman, H., Kovac, R., Topaloglou, A., Egg, D., and Günther, R. (1980). Effects of low-dose radiation on repair processes in human lymphocytes. Radiat. Res. 81, 1–9.

van Buul, P. P. W., Natarajan, A. T., and Verdegaal-Immerzeel, E. A. M. (1978). Suppression of the frequencies of sister chromatid exchanges in Bloom's syndrome fibroblasts by co-cultivation with Chinese hamster cells. Hum. Genet. 44, 187–189.

van Zeeland, A. A., Natarajan, A. T., Verdegaal-Immerzeel, E. A. M., and Filon, A. R. (1980). Photoreactivation of UV induced cell killing, chromosome aberrations, sister-chromatid exchanges, mutations and pyrimidine dimers in Xenopus laevis fibroblasts. Mol. Gen. Genet. 180, 495–500.

Vischer, T. (1972). Mitogenic factors by lymphocyte activation effect on T- and B-cells. J. Immunol. 109, 401–402.

Vogel, E., and Natarajan, A. T. (1979a). The relation between reaction kinetics and mutagenic action of monofunctional alkylating agents in higher eukaryotic systems. I. Recessive lethal mutations and translocations in Drosophila. Mutat. Res. 62, 51–100.

Vogel, E., and Natarajan, A. T. (1979b). The relation between reaction kinetics and mutagenic action of monofunctional alkylating agents in higher enkaryotic systems. II. Total and partial sex-chromosome loss in Drosophila. Mutat. Res. 62, 101–123.

Vogel, E., and Natarajan, A. T. (1981). The relation between reaction kinetics and mutagenic action of monofunctional alkylating agents in higher eukaryotic systems. III. Interspecies Comparisons. Chem. Mutagens 7, 295–336.

Wara, D. W., and Ammann, A. J. (1977). Immunological disorders of childhood. In "Pediatrics" (A. M. Rudolph, H. L. Barnett, and A. H. Einhorn, eds.), 16th ed., pp. 299–327. Appleton, New York.

Waters, M. D. (1979). The gene-tox program. Banbury Rep. 2, 449–467.

Whitlock, J. O., Jr., Cooper, H. L., and Gelboin, H. V. (1972). Aryl hydrocarbon (benzopyrene) hydroxylase is stimulated in human lymphocytes by mitogens and benz[a]anthracene. Science 177, 618–619.

Wiebel, F. J., Schwartz, L. R., and Goto, T. (1980). Mutagen metabolizing enzymes in mammalian cell cultures: Possibilities and limitations for mutagenicity screening.

In "Short Term Test Systems for Detecting Carcinogens" (K. H. Norpoth and R. C. Garner, eds.), 209–225. Springer-Verlag, Berlin and New York.

Williams, G. M. (1977). Detection of chemical carcinogens by unscheduled DNA synthesis in rat liver primary cell cultures. *Cancer Res.* **37**, 1845–1851.

Williams, G. M. (1980). The predictive value of DNA damage and repair assays of carcinogenicity. *In* "The Predictive Value of Short Term Screening Tests in Carcinogenicity Evaluation" (G. M. Williams, R. Kroes, H. W. Waaijers, and K. W. van de Poll, eds.), Elsevier/North Holland, Amsterdam, 213–230.

Wilson, J. D., and Nossal, G. J. V. (1971). Identification of human T and B lymphocytes in normal peripheral blood and in chronic lymphocytic leukemia. *Lancett* **II**, 788–791.

Wilson, J. D., and Thomson, A. E. R. (1968). Death and division of lymphocytes. Neglected factors in assessment of P.H.A.-induced transformation. *Lancet* **II**, 1120–1123.

Wolff, S. (1977). Sister chromatid exchange. *Ann. Rev. Genet.* **11**, 183–201.

Wolff, S. (1978). Relationship between DNA repair, chromosome aberrations, and sister chromatid exchanges. *ICN-UCLA Symp. Mol. Cell. Biol.* **9**, 751–760.

Wolff, S., Rodin, B., and Cleaver, J. E. (1977). Sister chromatid exchanges induced by mutagenic carcinogens in normal and xeroderma pigmentosum cells. *Nature (London)* **265**, 347–349.

Yeoman, L. C., Seeber, S., Taylor, C. W., Fernbach, D. J., Falletta, J. M., Jordan, J. J., and Busch, H. (1976). Differences in chromatin proteins of resting and growing human lymphocytes. *Exp. Cell Res.* **100**, 47–55.

Yew, F.-H., and Johnson, R. T. (1978). Human B and T lymphocytes differ in UV-induced repair capacity. *Exp. Cell Res.* **113**, 227–231.

Zelle, B., and Bootsma, D. (1980). Repair of DNA damage after exposure to 4-nitroquinoline-1-oxide in heterokaryons derived from xeroderma pigmentosum cells. *Mutat. Res.* **70**, 373–381.

Chapter 8

Measurement of Mutations in Somatic Cells in Culture

Veronica M. Maher and
J. Justin McCormick

I. IMPORTANCE OF USING MAMMALIAN CELLS IN CULTURE FOR MUTAGENESIS STUDIES

There is now a convincing body of evidence pointing to the involvement of environmental agents in the origin of human cancer and a growing conviction that the process of carcinogenesis results directly or indirectly from heritable alterations in the somatic cells of the body (e.g., *"Origins of Human Cancer,"* 1977, edited by Hiatt *et al.*). The

215

MUTAGENICITY:
NEW HORIZONS IN GENETIC TOXICOLOGY

apparent lack of correlation between strong mutagens and carcinogens, which prevailed for so many years (Burdette, 1955; Kaplan, 1959), reflected a lack of understanding of the need for metabolic activation of carcinogens into reactive forms capable of interacting with cellular macromolecules. However, it also reflected the lack of suitable test systems for demonstrating the mutagenic potential of such strong carcinogens as 7,12-dimethylbenz[a]anthracene or 2-acetylaminofluorene. (The majority of the early assays made use of microorganisms which lacked the ability to metabolize the chemicals.)

Since that time, progress has been made on both fronts. For example, the fact that most chemical carcinogens require cellular activation into electrophilic forms is now well established (e.g., Miller and Miller, 1977; "Polycyclic Hydrocarbons and Cancer," 1978, edited by Gelboin and Ts'o; "Carcinogenesis: Fundamental Mechanisms and Environmental Effects," 1980, edited by Pullman et al., "Chemical Carcinogens and DNA," 1979, edited by Grover). Second, a large number of new test systems have been developed and these have been coupled with various systems providing metabolic activation (see "In Vitro Metabolic Activation in Mutagenesis Testing," 1976, edited by deSerres et al., and the "Chemical Mutagens" series 1971–1981, edited by Hollaender). Many of these tests utilize microorganisms because of the advantages offered, e.g., well characterized genetics, speed of assay, and minimum cost. Yet, there are now a substantial number of assays that employ mammalian cells in culture, including diploid human cells. These assays in mammalian cells are more expensive and time consuming, but offer the advantage that the target cells are more closely related to the cells at risk in the human body. Therefore, the results obtained are likely to be more instructive for weighing the dangers of environmental agents than are results derived from microbial assays. It is recognized that studies of mutagenicity and risk assessment (e.g., Generoso et al., 1980a) using intact animals allows for pharmacodynamic factors such as absorption, transport, metabolic activation, and deactivation, storage, elimination, etc. (Gillette, 1976), which cannot be easily mimicked in cell culture. Nevertheless, mutation studies conducted with mammalian cells in culture offer unique advantages. First of all, they can be used as part of a battery of short-term test systems to identify potential mutagens (carcinogens) and to compare the potency of one agent with that of another (Waters, 1979). Second, studies in mammalian cells in culture can provide important insights into the mechanisms of mutagenesis and carcinogenesis (cf. McCormick et al., 1980; Ts'o, 1980; Kakunaga et al., 1980). Similar mechchanistically oriented studies usually cannot be carried out in whole animals because of their complexity.

II. OVERVIEW OF SELECTED ASPECTS OF MAMMALIAN CELL MUTAGENESIS STUDIES

A. Examples of Cell Types Employed

The number of cell types being utilized for such studies is increasing continually. It is inappropriate to attempt to characterize each of them here. However, the majority of mutagenesis studies have been carried out using only a few cell types or lines, viz., Chinese hamster cell lines (CHO and V79), a mouse lymphoma cell line (L5178Y), and diploid or quasidiploid human cells. These particular target-cell lines will be discussed briefly. However, for specific information on these cell types as well as others in current use for mutagenesis studies, the reader is referred to such sources as the "Chemical Mutagens" series, edited by Hollaender, to the "Banbury Report, 1" (1979), edited by McElheny and Abrahamson and "Banbury Report, 2" (1979), edited by Hsie *et al.*; to the forthcoming summary of the EPA Gene-Tox Program (Evaluation of Current Status of Bioassays in Genetic Toxicology) (Waters, 1979); and to published reviews of individual systems.

1. Chinese Hamster Cell Lines

Chinese hamster cell lines with unlimited life-spans, derived from ovary (CHO cells) (Puck and Kao, 1967) or from male lung tissue (V79 cells) (Chu and Malling, 1968), were among the earliest established cell lines employed for mutagenesis studies and have proved extremely useful for measuring the effects of a very wide range of mutagens. Chinese hamster ovary cells have the advantage that they are able to grow in suspension as well as in monolayers attached to plastic, have a short doubling time, and exhibit very high cloning efficiency. These characteristics and the fact that they can be assayed for resistance to 6-thioguanine (TG) at relatively high cell densities without loss of the mutant cells by metabolic cooperation (see Subak-Sharp *et al.*, 1969; Corsaro and Midgeon, 1977a,b) have made them especially suited to yielding quantitative data on the mutagenic potency of a large number of agents. In addition, CHO cells have proved uniquely suited to selection of recessive mutants, a characteristic that at least in part has been attributed to their being functionally hemizygous for many markers (Siminovitch, 1976). For examples of the use of CHO cells see Carver *et al.* (1976a,b), Siminovitch (1976), O'Neill *et al.* (1977a,b), Hsie *et al.* (1977, 1978a,b,c), Gupta and Siminovitch (1978a,b,c, 1980a,b,c,d), O'Neill and Hsie (1979), Puck (1979), Clarkson and Mitchell (1979), Thompson *et al.* (1979), Waldren *et al.* (1979), and Burki *et al.* (1980).

The V79 cell lines also have unlimited life-spans in culture, short doubling times, and high cloning efficiency. They have been well characterized and widely used for mutagenesis studies. Examples of studies carried out using V79 cells include Chu (1971), Arlett *et al.* (1975), Huberman (1975), van Zeeland and Simons (1975, 1976), Chu and Powell (1976), Huberman *et al.* (1976a,b), Huberman and Sachs (1976), Newbold and Brookes (1976), Chang *et al.* (1978a,b), McMillan and Fox (1979), and Suter *et al.* (1980).

2. *Mouse Lymphoma Cell Lines*

One particular cell line, L5178Y, has been extensively used for screening of mutagenic agents for induction of forward mutations to loss of thymidine kinase (Clive *et al.*, 1972, 1973; Clive and Spector, 1975; Knapp and Simons, 1975; Clive and Voytek, 1977; Cole and Arlett, 1976, 1978). This transformed, pseudodiploid cell line grows only in suspension, has a very short generation time, and forms colonies in soft agar with high efficiency. Other mouse lymphoma cell lines have also been used (e.g., Anderson and Fox, 1974; Friedrick and Coffino, 1977).

3. *Diploid Human Cells*

a. Skin Fibroblasts. Extensive studies are now being conducted using diploid human cells. Albertini and DeMars (1970, 1973) were the first to describe the conditions needed to isolate 8-azaguanine (AG) resistant diploid human skin fibroblasts. These investigators used cells derived from foreskin material of newborns and used X-irradiation as the source of mutagen. The system was extended to a chemical carcinogen by Maher and Wessel (1975) who demonstrated a dose-dependent increase in the frequency of AG-resistant human skin fibroblasts following exposure to N-acetoxy-2-acetylaminofluorene. Cox and Masson (1976) reported a dose-dependent increase in 6-thioguanine (TG) resistance induced in these cells by X-rays.

Use of human diploid fibroblasts for mutation studies not only offers the advantage that the cells are more closely related to cells in the human body than are cells from other mammals, but also that they can be used in comparative studies with skin fibroblasts derived from xeroderma pigmentosum (XP) patients with an inherited predisposition to skin cancer induced by sunlight (Robbins *et al.*, 1974), which are deficient in DNA excision repair (Cleaver, 1969) or "postreplication repair" (Lehmann *et al.*, 1975). Maher and McCormick and co-workers developed methods for comparing normal and XP fibroblasts for sensitivity to the cytotoxic and mutagenic effects of ultraviolet radiation or chemical carcinogens (cf. the cited references by Maher and McCormick; Maher *et al.*; McCormick and Maher, 1978; Heflich *et al.*, 1980; Yang *et al.*, 1980). Comparative studies of the UV-induced mutagenesis have also been conducted by Glover *et al.* (1979) and Myhr *et al.* (1979). Such comparative studies can give information on mechanism (see below). Diploid human fibroblasts can be cloned on plastic surfaces with efficiences of 20–70%, have a doubling time in culture of 24 hours, and are very responsive to contact inhibition which facilitates studies on DNA repair (see Section III,B).

b. Lymphoblastoid Cell Lines. Thilly and collaborators developed techniques necessary to carry out quantitative mutagenesis studies on diploid or quasidiploid human lymphoblastoid lines (e.g., Thilly *et al.*, 1976, 1980). In theory, such lymphoblastoid cell lines can be prepared from peripheral blood lymphocytes of any human donor by transformation with Epstein–Barr virus, and have unlimited life-spans. The particular cell lines used by Thilly and co-workers offer many advantages for quantitative studies, including the ability to grow in suspension and to clone in soft agar, the lack of density interference with recovery of TG resistant cells caused by cell–cell interaction, potential to be adapted to semiautomated techniques, and the availability of lymphocytes from XP patients. The techniques used have been reviewed by Thilly (1977, 1979).

c. Peripheral Blood Lymphocytes. Albertini (1979) has developed techniques for quantifying the frequency of AG-resistant human peripheral blood lymphocytes (Strauss and Albertini, 1979). This system which is the subject of Chapter 11 will not be described here, except to point out that it serves as a unique approach toward assessing mutations *in vivo* in the human species. The importance of reliable, rapid, *in vivo* mutagenesis assays that can determine if human beings have been exposed to significant doses of mutagenic agents need not be underscored here.

4. Other Cell Types

Mutagenesis assays using many other cell types are being developed, for example, diploid Syrian hamster embryo (SHE) cells (Barrett and Ts'o, 1978a; Barrett *et al.*, 1978a; Ts'o, 1979), a rat epithelial cell line (San and Williams, 1977), a mouse embryo cell line 10T$\frac{1}{2}$ (Taylor and Jones, 1979), human diploid endothelial cells (Reznikoff and DeMars, 1981), as well as a hybrid Chinese hamster CHO cell line containing a single human chromosome (Waldren *et al.*, 1979).

B. Examples of Genetic Markers Employed

The term mutation, as used in this discussion, refers to any event involving a heritable change in the primary structure of DNA, i.e., base-pair change, frame shift, deletion, insertion, or rearrangement of primary structure (cf. DeMars, 1974; Siminovitch, 1976). Until recently, it has not been feasible in mammalian cells to determine specific physical alterations in the primary structure of DNA induced by various mutagens, although this has been and remains a clear goal (cf. Chasin and Urlaub, 1979). However, a large number of selectable markers and nonselectable markers have been developed for detecting heritable changes. Representative examples are listed in Table I with one or more selected references.

TABLE I

Selected Genetic Markers for Mammalian Cell Mutagenesis

Phenotype	Recessive	Dominant or codominant	References
Resistant to			
Guanine analogues (8-azaguanine, 6-thioguanine)	+		Carver et al. (1976a, 1976b) Caskey and Kruh (1979) Szybalski (1958), DeMars (1974), Huberman et al. (1976b), Hsie et al. (1978b, 1979), Krahn (1979), Maher et al. (1979), Jennsen and Ramel (1980), Thilly et al. (1980)
Adenine analogues (8-azaadenine, 2,6-diaminopurine)	+		Chasin (1974), DeMars (1974) Jones and Sargent (1974) Taylor et al. (1977, 1979)
Adenosine analogues (toyocamycin, tubericidin)	+		Gupta and Siminovitch (1978b)
Thymidine analogues (5-bromodeoxyuridine, trifluorothymidine, iododeoxyuridine)	+		Adair and Carver (1979) Clive et al. (1972, 1973, 1979) Kaufmann and Davidson (1979) Skopek et al. (1978)

TABLE I. (*continued*)

Phenotype	Inheritance pattern		References
	Recessive	Dominant or codominant	
Methotrexate		+	Chasin and Urlaub (1979)
Hydroxyurea		+	Lewis and Wright (1979)
Ouabain		+	Baker *et al.* (1974), Mankowitz *et al.* (1974), Arlett *et al.* (1975), Buchwald (1977), Chang *et al.* (1978a)
α-Amanitin		+	Chan *et al.* (1972), Buchwald and Ingles (1976), Ingles *et al.* (1976)
Emetine	+		Siminovitch (1976), Gupta *et al.* (1978), Gupta and Siminovitch (1978c, 1980d), Campbell and Worton (1979)
Diphtheria toxin		+	Moehring and Moehring (1977), Gupta and Siminovitch (1978a, 1980a,c), Glover *et al.* (1979), Burki *et al.* (1980)
Podophyllotoxin		+	Gupta (1980, 1981)
5,6-dichloro-1β-D-ribofuranosyl benzimidazale		+	Gupta and Siminovitch (1980b)
Auxotrophic Requiring glycine	+		Kao and Puck (1968), Taylor and Hanna (1977)
Conditional lethal Temperature-sensitive aminoacyl-tRNA synthetase	+		Fenwick and Caskey (1975)
Cell surface antigens	+		Waldren *et al.* (1979), Kavathas *et al.* (1980)
Loss of anchorage dependence		?	Bellett and Younghusband (1979), Silinskas *et al.* (1981), Maher *et al.* (1982)

There is indirect evidence that these phenotypes result from a genetic change. It has been shown for many traits that they exhibit
concentration-dependent induction by mutagens, that this increase
does not result from selective pressure exerted by the mutagen, that
the variant cells are stable in the absence of selection, and that the
frequencies of induction exhibit the expected ploidy effect (e.g.,
Chasin, 1973, 1974). For some traits, there is evidence that a change
in the structure of the affected protein gene product has occurred, e.g.,
temperature sensitivity or the presence of cross-reacting material or
altered protein structure (cf. Beaudet *et al.*, 1973; Fenwick and Caskey,
1975; Caskey and Kruh, 1979). Each of the markers listed in Table I
has its own set of proper procedures for making certain that the cells
selected or isolated represent genetic variants rather than cells that
have merely survived the selection or isolation technique. Details of
these can be found in the specific references or review articles.

It is not surprising that one can select for dominant or codominant
traits since in theory this requires mutating only one of two alleles
and should occur at a reasonable frequency. Similarly, TG or AG
resistance should occur with reasonable frequency because the gene
which is responsible for this phenotype, viz., the *hpt* gene—coding
for hypoxanthine (guanine) phosphoribosyltransferase (HPRT), is located on the X chromosome and, thus, appears as a single active allele.
However, in CHO cell lines, it has been possible to isolate many
variants with traits which proved to be recessive. One possible reason
for this, viz., functional hemizygosity, is discussed by Siminovitch
(1976), Gupta *et al.* (1978), and Gupta and Siminovitch (1980d).

The two most widely studied genetic markers in mammalian cell
mutagenesis are resistance to TG or AG, which can result from loss
of a functional HPRT enzyme, and resistance to thymidine analogues
such as 5-bromodeoxyuridine or trifluorothymidine, which can result
from loss of a functional thymine kinase. The gene for TK is not sex-
linked and therefore full resistance to these thymidine analogues re-
quires two steps, viz., $TK^{+/+} \rightarrow TK^{+/-} \rightarrow TK^{-/-}$. However, if one
works with cell lines with nonlimited life-spans, such as CHO, V79,
L5178Y, or human lymphoblastoid lines, it is possible to isolate the
heterozygous intermediate strains and use them for the target cell.
Both of these sets of variants, $HPRT^-$ and $TK^{-/-}$, offer the advantage
of having a selectable reversion marker viz., ability to grow in medium
containing hypoxanthine, aminopterin, and thymidine. Thus, these
two markers provide very flexible systems which, when used with
proper precautions, can yield quantitative data on the frequency of
forward and reverse mutations.

C. Examples of Methods Used to Provide Necessary Metabolic Activation of Promutagens

1. Microsomal Fraction-Mediated Methods

Because the majority of chemical carcinogens and mutagens require metabolic activation before they can interact with cellular macromolecules, it has proved necessary to use chemically synthesized reactive metabolites or derivatives or to provide activation when the cells used for mutagenesis assays lack this ability. Rodent microsomes or the 9000 g supernatant from rodent liver homogenates have been adapted for use with bacteria (Malling, 1971; Garner et al., 1972; Ames et al., 1973; McCann et al., 1975) and are now widely used for screening potential mutagens. They have more recently been adapted for use with mammalian cells (e.g., King et al., 1975; Krahn and Heidelberger, 1977; Kuroki et al., 1977; Dent, 1979; Couch et al., 1979; Krahn, 1979).

There is no question that such fractions have made it possible to detect mutagenic action by hundreds of chemicals that would otherwise prove negative. However, if the purpose of using these activation systems is to assist in determining whether or not unknown mixtures of compounds or previously untested new compounds are mutagenic, and more importantly, are likely to be metabolized into a mutagenic form in human beings or other intact mammals, there are a number of different questions that must be answered. For example, are the metabolites formed by the disrupted subcellular components comparable to those produced by intact cells (deSerres et al., 1976; Gillette, 1976; Selkirk, 1977)? From what animal species or species should the tissue homogenate be prepared? From which organs of the body (Weekes and Brusick, 1975; Bartsch et al., 1977)? What enzyme inducers are most appropriate? At what concentration should the homogenate be applied? What cofactors should be supplied? Is activation being emphasized at the expense of deactivation? (See Dent, 1979, and Couch et al., 1979, for examples of the controversy over proper choice of the activation portion of mammalian cell assays which are to be used to predict the carcinogen potency of various agents and how the choice of conditions for metabolic activation can affect the results obtained.)

Another question should be considered. Are the DNA adducts formed after S9 activation the same as those formed by the cells from which the S9 fraction was made? For example, it has been demonstrated that isolated microsomes for activation of benzo[a]pyrene (B[a]P) can produce B[a]P–DNA adducts via the 4,5-oxide or the 7,8-

diol-9,10-epoxide of B[a]P (Grover, et al., 1972; Selkirk, 1977; Shinohara and Cerutti, 1977). However, only the latter are the precursors
of the DNA adducts formed by B[a]P in intact hamster embryo cells
(Baird et al., 1975), mouse skin (Daudel et al., 1975; Grover et al.,
1976), human bronchus (Hsu et al., 1978), or human colon tissue
explants (Autrup et al., 1978). Yet the B[a]P–DNA adducts formed by
each of these metabolites correlate with mutation induction (Newbold
and Brookes, 1976).

2. Cell-Mediated Methods

Several groups of investigators have succeeded in using intact mammalian cells that retain metabolic activation functions as "metabolizing
layers." These cells are incubated with the target cells which are to
be used to quantitate mutations (e.g., Huberman and Sachs, 1974;
Newbold et al., 1977; San and Williams, 1977; Langenbach, 1978a,b;
Aust et al., 1980, 1981).

Others have successfully used human tissue explants (Hsu et al.,
1978) or have attempted to adapt the host-mediated assay (Gabridge
and Legator, 1969) for use with mammalian target cells (Capizzi et al.,
1974; Hsie et al., 1978a). Each of these is an attempt to validate and
extend the applicability of mammalian cell mutagenesis studies. Those
making use of intact human cells (Aust et al., 1980, 1981) or human
tissue explants (Hsu et al., 1978) have contributed information that
can be compared with results obtained by the several other systems.
Much basic research remains to be done, however, if these kinds of
activation schemes are to be used in valid risk assessment studies.

III. UTILIZATION OF MUTAGENESIS STUDIES WITH CULTURED MAMMALIAN CELLS

A. Identification of Potential Mutagens in the Environment

As noted the past decade has seen the rapid development of test
systems designed not only to detect the mutagenic potential of chemicals in the environment but to begin to explore the problems connected with assessment of the potential risk to humans from environmental mutagens and carcinogens. The 1978 Banbury conference

discussed the means of evaluating mutagenic risks to humans from exposure to chemicals; considered which existing test systems are of use for extrapolating risks; and attempted to "develop a scientific summary of the state-of-the-art in mutagenic risk assessment" (Mc-Elheny and Abrahamson, 1979). The consensus was that "existing methodology can be utilized to derive risk estimates for human exposure to chemical mutagens." However, such estimates "consist of certain unavoidable assumptions and undefined levels of uncertainty that can be reduced with better understanding of mutational phenomena in eucaryotes and more information regarding the genetic disease mechanisms of humans." As part of their summary, the conferees emphasized, among other points, the need for greater understanding of the differences in human susceptibility resulting from such factors as metabolism, DNA repair capabilities, etc.; for insights into the role of DNA repair in eukaryotic mutagenesis; and for a concerted approach to determining the doses of chemicals which actually reach the target cells in intact animals. Although epidemiological and whole-animal experimental studies will be crucial to the solution of these kinds of questions, mutagenicity studies conducted in mammalian cells, including diploid human cells, can be expected to contribute significantly.

A Banbury conference held the following year (Hsie *et al.*, 1979) critically examined the predictive value of existing mammalian cell mutagenesis test systems; the biochemical, molecular, and genetic evidence for a genetic basis of the mutations being measured; the methods used to quantitate mutation induction and the feasibility of applying these systems to screening tests to determine the mutagenic activity of industrial and environmental chemicals. The consensus reached was that great progress has been made in defining systems and that although no one system was completely adequate, application of these assays to the critical problems of mutagen testing had to be pursued at the present time.

B. Probing the Mechanisms of Somatic Cell Mutagenesis

1. Quantitation of the Mutagenic Effect

In an effort to assess the mutagenicity of various agents, one encounters the problem of how to express the mutagenic activity of one agent compared with another, or again, of how to compare the results obtained in one test system or cell strain with those obtained in another. This kind of evaluation is useful, not only for assessing the relative danger of exposure to various mutagens/carcinogens, but also for determining mechanisms of action of compounds or for evaluating various assays. A traditional approach has been to compare the relative mutagenic activity of various agents as a function of the concentration applied to the test organism, but this does not take into consideration the effective concentration which reaches the cellular target being measured. It gives information on relative effect, but does not indicate mutagenicity as a function of extent of reaction.

In an attempt to compare the mutagenicity of one agent with another, many investigators, ourselves included, have considered that chemicals closely related in structure will cause equal cytotoxicity, if they have reacted equally with the cell targets. Given this as a working hypothesis, one can compare the relative mutagenicity of various agents as a function of the number of mean lethal events sustained by the cells from the initial exposure to the agent (see e.g., Maher and McCormick, 1976; Maher et al., 1976a,b, 1977; Newbold and Brookes, 1976; Newbold et al., 1977; Simon et al., 1981). This approach is useful for certain purposes, but for investigating mechanisms of mutagenesis (McCormick and Maher, 1978) or comparing the mode of action of various agents, it is far more informative to determine the extent of reaction of the agents with DNA as the critical target and be able to compare mutagenicity as a function of specific DNA lesions (e.g., Bodell et al., 1979; Yang et al., 1980; Newbold et al., 1980; Suter et al., 1980; Simon et al., 1981).

2. Role of Excision Repair

Many important insights into the mechanism of action of mutagenic agents in mammalian cells can be obtained by comparing their effect in cells which differ in their capacity to excise or repair the lesions induced in DNA. We have conducted such studies in diploid human fibroblasts and demonstrated that the excision repair process for lesions induced by UV radiation and for a number of chemical carcinogens is virtually "error-free" (Maher and McCormick, 1976, 1977, 1980; Maher et al., 1976a,b, 1977, 1979, 1982; Yang et al., 1980). The availability of excision repair defective cells derived from XP patients has made these studies possible. Recently, several investigators have isolated DNA-repair-deficient mutant cell lines derived from other species and these should prove very useful not only in such studies of mechanism but also for screening of environmental mutagen/carcinogens (Sato and Hieda, 1979; Schultz et al., 1981; Thompson et al., 1979).

3. Role of Postreplication Repair

Studies are in progress in many laboratories to determine the role of DNA replication in the mutagenesis process. Our approach has been to synchronize human cells and mutagenize them at various times prior to the onset of S phase (Konze-Thomas et al., 1982; Yang et al., 1982) or to treat them in a nonreplicating G_0 state and allow various lengths of time before putting the cells into the replicating conditions of culture (Maher et al., 1979; Heflich et al., 1980; Yang et al., 1980). The results of these studies indicate clearly that the time available for excision repair before the cells enter DNA synthesis is critical for mutagenesis. Examples of studies on the role of the cell cycle in mammalian cell mutagenesis, other than human fibroblasts, include Carver et al. (1976b), Riddle and Hsie (1978), Clarkson and Mitchell (1979), and Burki et al. (1980).

C. Investigating the Mechanisms of Carcinogenesis

1. Evidence That Human Carcinogenesis Is a Multistepped Process

Studies of the induction of cancer in animals by chemical carcinogens suggest that cancer is induced by a multistepped process (Berenblum, 1954). Similarly, epidemiologists studying the induction of human cancer from radiation, chemicals, or unknown causes find that only a process with a number of steps fits the available evidence. Epidemiologists use the word multistage to describe this phenomenon since "stages" are defined mathematically according to their methods. The epidemiologist, Peto (1977), defines the "essential" multistage hypothesis:

> a few distinct changes (each heritable when cells carrying them divide) are necessary to alter a normal cell into a malignant cell . . . human cancer usually arises from the proliferation of a clone derived from a single cell that suffered all the necessary changes and then started to proliferate malignantly.

The long latent period, for example, between exposure to asbestos and the appearance of pleural mesotheliomas is supportive of this hypothesis (Selikoff et al., 1965). Similar data for the induction of tumors by other environmental carcinogens as well as by cigarette smoking support this thesis. Analysis of epidemiological data suggests that between three and seven stages exist in human cancer, but such research cannot define a "stage" except in terms of its rate (see Peto,

1977). Armitage and Doll (1957) have shown, however, that even a seven-stage model is compatible with a two mutational model for the origin of cancer, if a singly mutated cell had a somewhat faster growth rate. Furthermore, studies by the epidemiologist Knudsen (1977) and Knudsen *et al.* (1975) on the genetic predisposition of humans to retinoblastoma and some other childhood tumors suggest that such tumors involve two mutations, the first being prezygotic. Thus, although we are not yet able to identify with certainty the biological events that underlie malignancy, there is good evidence to suggest that there may be more than one step, but only a limited number of steps. It may be that one or more of these steps involves mechanisms for escaping the immunological surveillance of the body.

2. Investigation of the Number and Nature of Steps Involved in Transformation of Diploid Human Cells Using "in Vitro" Transformation Systems

The complexity of the problem of carcinogenesis in *vivo* has prompted the development of *in vitro* transformation systems in which the process can be more readily examined. Although the malignant (tumorigenic) transformation of cells in culture may not fully represent the in *vivo* process of carcinogenesis, such an approach permits a more direct experimental manipulation and quantitation of the individual steps involved and can yield information on the nature of these steps. A further advantage of using cell-culture systems is that they allow experimental studies with human cells, which would otherwise be impossible for obvious ethical reasons. In fact, dissection of the steps involved in the neoplastic transformation of human cells in culture represents, perhaps the only experimental approach to obtaining such information.

In Chapter 6 of this volume, Sivak and Tu describe the application of assays of transformation of somatic cells in culture as short-term tests for carcinogen screening and their application in studies of risk assessment. Our purpose here is to point out that there is a close connection between studies of somatic cell mutagenesis and studies of the mechanisms of transformation of diploid cells derived from primary cell culture into cells capable of producing tumors in appropriate host animals. As outlined in Section II,A, most mammalian cell mutagenesis studies have been carried out using quasidiploid cell lines with unlimited life-spans in culture because of their various specific characteristics adapted for mutagenesis. In contrast, the majority of cell-transformation studies have employed diploid cells of early passage derived from primary cultures of Syrian hamster embryos (e.g.,

Huberman, 1975; Pienta et al., 1977; Barrett and Ts'o, 1978a,b; DiPaolo, 1980), although permanent cell lines have also been used. More recently, early passage diploid cells from mouse embryos (Bellett and Younghusband, 1979) and from human skin fibroblasts (Kakunaga, 1978; Milo and DiPaolo, 1978; McCormick et al., 1980; Silinskas et al., 1981; Maher et al., 1982) have been used to investigate the mechanisms of transformation.

What has been established from these studies and from studies of tumor-forming cells by Freedman and Shin (1974) and Shin et al. (1975) is that there is a very high correlation between the ability of cells to form colonies in semisolid medium (loss of anchorage dependence) and their ability to form tumors in appropriate hosts. This correlation holds true for Syrian hamster embryo-derived fibroblasts, but in contrast to mouse or human cells, hamster cells do not lose anchorage dependence until more than 30 cell doublings subsequent to exposure to a mutagen/carcinogen (Barrett and Ts'o, 1978b). Significantly, Bellett and Younghusband (1979) and more recently, McCormick and co-workers (McCormick et al., 1980; Silinskas et al., 1981; Maher et al., 1982) have demonstrated that, at least in diploid mouse and human cells, loss of anchorage dependence acts as a typical genetic event induced *early* in direct response to exposure to a mutagen/carcinogen. They have carried out fluctuation tests (Bellett and Younghusband, 1979) which indicate that this trait is induced by the mutagen. There is a short expression period, i.e., approximately four cell doublings in the mouse cells and eight in the human cells. The phenotype is stable in the absence of selection. These anchorage-independent cells isolated from soft agar invariably produce tumors upon injection into appropriate host animals. (The human cells, assayed in sublethally irradiated athymic mice, give rise to fibrosarcomas composed of cells with a human karyotype and these give tumors upon reinjection into mice.) Further evidence that loss of anchorage dependence with its subsequent tumorigenicity is induced as a genetic event in human diploid cells derives from a recent study in our laboratory (Maher et al., 1982) comparing the frequency of induction in normal human cells and in cells derived from a xeroderma pigmentosum patient from complementation group D. The latter are very deficient in rate of excision repair of UV-induced DNA lesions (Robbins et al., 1974). The XP cells irradiated with 8-fold lower doses of UV (254 nm) radiation than the normal cells exhibited the same frequency of UV-induced anchorage-independent cells in the population as did the normal cells. Both sets of anchorage-independent cells gave rise to fibrosarcomas in athymic mice. These data together with those of

Bellett and Younghusband (1979) indicate that diploid mouse and human somatic cells in culture can be used to investigate and quantitate the role of mutagenesis in the process of carcinogenesis.

IV. CONCLUSION

Progress in the development and application of mutagenesis assays based on cultured mammalian cells has been rapid during the past decade. Nevertheless, the need for valid methods of determining the risk to human beings from the ever-increasing number environmental chemicals remains as formidable a problem as ever. Not only is there the question of somatic cell mutations leading to human cancer, but also the concern regarding induction of mutations in the germinal cells of the gonads which could then be genetically transmitted to future generations. One approach to this complex situation would be to discern the nature of the mechanisms of mutagenesis in cultured mammalian cells from nonhumans and compare these with the mechanisms involved in cultured diploid human cell mutagenesis. If the mechanisms are the same and if such mechanisms can be shown to be operative within the somatic and germinal cells of intact mammals and humans, cell culture studies can offer a solid base for extrapolation. Validation studies in intact animals are required, of course, to take into consideration the many factors mitigating against the hazardous effects of exposure of whole animals to toxic mutagens.

Another approach is directed toward discerning the mechanism(s) involved in the carcinogenesis process in human beings. If a causal relationship between the two processes, mutagenesis and carcinogenesis, could be determined this would strengthen the basis for risk estimates. This is without doubt a very large undertaking and, even were it completed, there would remain the problems connected with the extreme variation in exposure among human beings because of differences in occupational and social habits, as well as inherited differences in sensitivity based on such factors as variations in carcinogen metabolism and DNA repair capability, etc. Nevertheless, such approaches are being developed. The coming decade of work will no doubt bring even greater progress than did the previous one.

ACKNOWLEDGMENT

We thank Carol Howland for excellent editorial and clerical assistance in the preparation of this manuscript. This work was supported in part by DHHS Grants CA 21247, CA 21253, and CA 21289 and by D.O.E. Contract EV-78-S-02-4659.

REFERENCES

Adair, G. M., and Carver, J. H. (1979). Unstable, non-mutational expression of resistance to the thymidine analogue, trifluorothymidine in CHO cells. *Mutat. Res.* **60,** 207–213.

Albertini, R. M. (1979). Direct mutagenicity testing with peripheral blood lymphocytes. *Banbury Rep.* **2,** 359–376.

Albertini, R. J., and DeMars, R. (1970). Diploid azaguanine-resistant mutant of cultured human fibroblasts. *Science* **169,** 482–485.

Albertini, R. J., and Demars, R. (1973). Somatic cell mutation detection and quantification of x-ray-induced mutation in cultured, diploid human fibroblasts. *Mutat. Res.* **18,** 199–224.

Ames, B. N., Durston, W. E., Yamasaki, E., and Lee, F. D. (1973). Carcinogens are mutagens: A simple test system combining liver homogenates for activation and bacteria for detection. *Proc. Natl. Acad. Sci. U.S.A.* **70,** 2281–2285.

Anderson, D., and Fox, M. (1974). The induction of thymidine- and IUdR-resistant variants in $P_3$88 mouse lymphoma cells by x-rays. UV and mono- and bi-functional alkylating agents. *Mutat. Res.* **25,** 107–122.

Arlett, C. F., Turnbull, D., Harcourt, S. A., Lehmann, A. R., and Colella, C. M. (1975). A comparison of the 8-azaguanine and ouabain-resistance systems for the selection of induced mutant Chinese hamster cells. *Mutat. Res.* **33,** 261–278.

Armitage, P., and Doll, R. (1957). A two-stage theory of carcinogenesis in relation to the age distribution of human cancer. *Br. J. Cancer* **11,** 161–166.

Aust, A. E., Falahee, K. J., Maher, V. M., and McCormick, J. J. (1980). Human cell-mediated benzo[a]pyrene cytotoxicity and mutagenicity on human diploid fibroblasts. *Cancer Res.* **40,** 4070–4075.

Aust, A. E., Antczak, M. R., Maher, V. M., and McCormick, J. J. (1981). Identifying human cells capable of metabolizing various classes of carcinogens. *J. Supramol. Struct. Cell. Biochem.* **16,** 269–279.

Autrup, H., Harris, C. C., Trump, B. F., and Jeffrey, A. M. (1978). Metabolism of benzo[a]pyrene and identification of the major benzo[a]pyrene-DNA adducts in cultured human colon. *Cancer Res.* **38,** 3689–3696.

Baird, W. M., Harvey, R. G., and Brookes, P. (1975). Comparison of the cellular DNA-bound products of benzo[a]pyrene with the products formed in the reaction of benzo[a]pyrene 4,5-oxide with DNA. *Cancer Res.* **35,** 54–57.

Baker, R. M., Brunette, D. M., Mankovitz, R., Thompson, L. H., Whitmore, G. F., Siminovitch, L., and Till, J. E. (1974). Ouabain-resistant mutants of mouse and hamster cells in culture. *Cell* **1,** 9–21.

Barrett, J. C., and Ts'o, P. O. P. (1978a). The relationship between somatic mutation and neoplastic transformation. *Proc. Natl. Acad. Sci. U.S.A.* **75,** 3297–3301.

Barrett, J. C., and Ts'o, P. O. P. (1978b). Evidence for the progressive nature of neoplastic transformation *in vitro. Proc. Natl. Acad. Sci. U.S.A.* **75,** 3761–3765.

Barrett, J. C., Bias, N. E., and Ts'o, P.O.P. (1978a). A mammalian cellular system for the concomitant study of neoplastic transformation and somatic mutation. *Mutat. Res.* **50,** 121–136.

Barrett, J. C., Tsutsui, T., and Ts'o, P. O. P. (1978b). Neoplastic transformation induced by a direct perturbation of DNA. *Nature (London)* **274,** 229–232.

Bartsch, H., Margison, G. P., Malaveille, C., Camus, A. M., Brun, G., Margison, J. M., Kolar, G. F., and Wiessler, M. (1977). Some aspects of metabolic activation of chemical carcinogens in relation to their organ specificity. *Arch. Toxicol.* **39,** 51–59.

Beaudet, A. L., Roufa, D. J., and Caskey, C. T. (1973). Mutations affecting the structure of hypoxanthine: Guanine phosphoribosyl transferase in cultured Chinese hamster cells. *Proc. Natl. Acad. Sci. U.S.A.* **70**, 320–324.

Bellett, A. J. D., and Younghusband, H. B. (1979). Spontaneous, mutagen-induced and adenovirus-induced anchorage independent tumorigenic variants of mouse cells. *J. Cell Physiol.* **101**, 33–48.

Berenblum, I. (1954). A speculative review: The probable nature of promoting action and its significance in the understanding of the mechanism of carcinogenesis. *Cancer Res.* **14**, 475–447.

Bodell, W. J., Singer, B., Thomas, G. H., and Cleaver, J. E. (1979). Evidence for removal at different rates of O-ethyl pyrimidines and ethylphotriesters in two human fibroblast cell lines. *Nucleic Acids Res.* **6**, 2819–2829.

Buchwald, M. (1977). Mutagenesis at the ouabain-resistance locus in human diploid fibroblasts. *Mutat. Res.* **44**, 401–411.

Buchwald, M., and Ingles, C. J. (1976). Human diploid fibroblast mutants with altered RNA polymerase II. *Somatic Cell Genet.* **2**, 225–233.

Burdette, W. J. (1955). The significance of mutation in relation to the origin of tumours: A review. *Cancer Res.* **15**, 201–226.

Burki, H. J., Lam, C., and Wood, R. D. (1980). UV light-induced mutations in synchronous CHO cells. *Mutat. Res.* **69**, 347–356.

Campbell, C. E., and Worton, R. G. (1979). Evidence obtained by induced mutation frequency analysis for functional hemizygosity at the *emt* locus in CHO cells. *Somatic Cell Genet.* **5**, 51–65.

Capizzi, R. L., Papirmeister, B., Mullins, J. M., and Cheng, E. (1974). The detection of chemical mutagens using the L5178Y/Asn-murine leukemia *in vitro* and in a host-mediated assay. *Cancer Res.* **34**, 3073–3082.

Carver, J. H., Dewey, W. C., and Hopwood, L. E. (1976a). X-Ray induced mutants resistant to 8-azaguanine. I. Effects of cell density and expression time. *Mutat. Res.* **34**, 447–464.

Carver, J. H., Dewey, W. C., and Hopwood, L. E. (1976b). X-Ray induced mutants resistant to 8-azaguanine. II. Cell cycle dose response. *Mutat. Res.* **34**, 465–480.

Caskey, C. T., and Kruh, G. D. (1979). The HPRT locus. *Cell* **16**, 1–9.

Chan, V. L., Whitmore, G. F., and Siminovitch, L. (1972). Mammalian cells with altered forms of RNA polymerase II. *Proc. Natl. Acad. Sci. U.S.A.* **69**, 3119–3123.

Chang, C. C., Trosko, J. E., and Akera, T. (1978a). Characterization of ultraviolet light-induced ouabain-resistant mutations in Chinese hamster cells. *Mutat. Res.* **51**, 85–98.

Chang, C. C., D'Ambrosio, S. M., Schultz, R., Trosko, J. E., and Setlow, R. B. (1978b). Modification on UV-induced mutation frequencies in Chinese hamster cells by dose fractionation, cycloheximide and caffeine treatments. *Mutat. Res.* **52**, 231–245.

Chasin, L. A. (1973). The effect of ploidy on chemical mutagenesis in cultured Chinese hamster cells. *J. Cell Physiol.* **82**, 299–308.

Chasin, L. A. (1974). Mutations affecting adenine phosphoribosyl transferase activity in Chinese hamster cells. *Cell* **2**, 37–41.

Chasin, L. A., and Urlaub, G. (1979). Selection of recessive mutants at diploid loci by titration of the gene product: CHO cell mutants deficient in dihydrafolate reductase activity. *Banbury Rep.* **2**, 201–209.

Chu, E. H. Y. (1971). Induction and analysis of gene mutations in mammalian cells in culture. *Chem. Mutagens* **2**, 411–444.

Chu, E. H. Y., and Malling, H. V. (1968). Mammalian cell genetics. II. Chemical induction of specific locus mutations in Chinese hamster cells *in vitro*. *Proc. Natl. Acad. Sci. U.S.A.* **61**, 1306–1312.

Chu, E. H. Y., and Powell, S. S. (1976). Selective systems in somatic cell genetics. *Adv. Hum. Genet.* **7**, 189–258.

Clarkson, J. M., and Mitchell, D. L. (1979). The recovery of mammalian cells treated with methyl methanesulfonate nitrogen mustard or UV light. *Mutat. Res.* **61**, 333–342.

Cleaver, J. E. (1969). Xeroderma pigmentosum: A human disease in which an initial stage of DNA repair is defective. *Proc. Natl. Acad. Sci. U.S.A.* **63**, 428–435.

Clive, D., and Spector, J. F. S. (1975). Laboratory procedure for assessing specific locus mutations at the TK locus in cultured L5178Y mouse lymphoma cells. *Mutat. Res.* **31**, 17–29.

Clive, D., and Voytek, P. (1977). Evidence for chemically induced structural gene mutations at the thymidine kinase locus in cultured L5178Y mouse lymphoma cells. *Mutat. Res.* **44**, 269–278.

Clive, D., Flamm, W. G., Machesko, M. R., and Bernheim, N. J. (1972). A mutational assay system using the thymidine kinase locus in mouse lymphoma cells. *Mutat. Res.* **16**, 77–87.

Clive, D., Flamm, W. G., and Patterson, J. B. (1973). Specific locus mutational assay systems for mouse lymphoma cells. *Chem. Mutagens* **3**, 79–102.

Clive, D., Johnson, K. O., Spector, J. F. S., Batson, A. G., and Brown, M. M. M. (1979). Validation and characterization of the L5178Y/TK$^{+/-}$ mouse lymphoma mutagen assay system. *Mutat. Res.* **59**, 61–108.

Cole, J., and Arlett, C. F. (1976). Ethyl methanesulfonate mutagenesis with L5178Y mouse lymphoma cells: A comparison of ouabain, thioguanine, and excess thymidine resistance. *Mutat. Res.* **34**, 507–526.

Cole, J., and Arlett, C. F. (1978). Methyl methanesulfonate mutagenesis in L5178Y mouse lymphoma cells. *Mutat. Res.* **50**, 111–120.

Corsaro, C. M., and Migeon, B. R. (1977a). Contact-mediated communication of ouabain-resistance in mammalian cells in culture. *Nature (London)* **268**, 737–739.

Corsaro, C. M., and Migeon, B. R. (1977b). Comparison of contact-mediated communication in normal and transformed human cells in culture. *Proc. Natl. Acad. Sci. U.S.A.* **74**, 4476–4480.

Couch, D. B., Bermudez, E., Decad, G. M., and Dent, J. G. (1979). The influence of activation systems on the metabolism of 2,4-dinitrotoluene. *Banbury Rep.* **2**, 303–309.

Cox, R., and Masson, W. K. (1976). X-Ray induced mutation to 6-thioguanine resistance in cultured human diploid fibroblasts. *Mutat. Res.* **37**, 125–136.

Daudel, P., Duquesne, M., Vigny, P., Grover, P. L., and Sims, P. (1975). Fluorescence spectral evidence that benzo[a]pyrene-DNA products in mouse skin arise from diol-epoxides. *FEBS Lett.* **57**, 250–253.

DeMars, R. (1974). Resistance of cultured human fibroblasts and other cells to purine and pyrimidine analogues in relation to mutagenesis detection. *Mutat. Res.* **24**, 335–364.

Dent, J. G. (1979). Choice of activating systems for *in vitro* mutagenesis assays. *Banbury Rep.* **2**, 295–302.

deSerres, F. J., Fouts, J. R., Bend, J. R., Philpot, R. M., eds. (1976). "*In Vitro* Metabolic Activation in Mutagenesis Testing." Elsevier/North-Holland Biomedical Press, Amsterdam.

DiPaolo, J. A. (1980). Quantitative *in vitro* transformation of Syrian golden hamster embryo cells with the use of frozen stored cells. *J. Natl. Cancer Inst.* **64**, 1485–1489.

Fenwick, R. G., Jr., and Caskey, C. T. (1975). Mutant Chinese hamster cells with a thermosensitive hypoxanthine-guanine phosphoribosyl transferase. *Cell* **5**, 115–122.

Freedman, V. H., and Shin, S.-I. (1974). Cellular tumorigenicity in nude mice: Correlation with cell growth in semi-solid medium. *Cell* **3**, 355–359.

Friedrick, U., and Coffino, P. (1977). Mutagenesis in S49 mouse lymphoma cells: Induction of resistance to ouabain, 6-thioguanine and dibutyryl cyclic AMP. *Proc. Natl. Acad. Sci. U.S.A.* **74**, 679–683.

Gabridge, M. G., and Legator, M. S. (1969). A host-mediated assay for the detection of mutagenic compounds. *Proc. Soc. Exp. Biol. Med.* **130**, 831–841.

Garner. R. C., Miller, E. C., and Miller, J. A. (1972). Liver microsomal metabolism of aflatoxin B_1 to a reactive derivative toxic to *Salmonella typhimurium* TA 1530. **Cancer Res. 32**, 2058–2066.

Gelboin, H. V., and Ts'o, P. O. P., eds. (1978). "Polycyclic Hydrocarbons and Cancer: Environment, Chemistry, Molecular and Cell Biology," Vols. 1, 2. Academic Press, New York.

Generoso, W. M., Cain, K. T., Krishna, M., and Gryder, R. M. (1980a). Heritable translocation and dominant-lethal mutation induction with ethylene oxide in mice. *Mutat. Res.* **73**, 133–142.

Generoso, W. M., Shelby, M. D., and deSerres, F. J., eds. (1980b). "DNA Repair and Mutagenesis in Eukaryates." Plenum, New York.

Gillette, J. R. (1976). Activating systems—Characteristics and drawbacks—Comparison of different organ, tissues, problems with toxification-detoxification balance in various tissues, extrapolation from *in vitro* to *in vivo*. Pharmacokinetics, absorption, distribution, excretion, metabolism. *In* "In Vitro Metabolic Activation in Mutagenesis Testing" (F. J. deSerres, J. R. Fouts, J. R. Bend, and R. M. Philpot, eds.), pp. 13–53. Elsevier/North-Holland, Amsterdam.

Glover, T. W., Chang, C. C., Trosko, J. E., and Li, S. L. (1979). Ultraviolet light induction of diphtheria toxin-resistant mutants in normal and xeroderma pigmentosum human fibroblasts. *Proc. Natl. Acad. Sci. U.S.A.* **76**, 3982–3986.

Grover, P. L., ed. (1979). "Chemical Carcinogens and DNA," Vol. 1, 2. C.R.C. Press, Boca Raton, Florida

Grover, P. L., Hewer, A., and Sims, P. (1972). Formation of K-region epoxides as microsomal metabolites of pyrene and benzo[a]pyrene. *Biochem. Pharmacol.* **21**, 2713–2726.

Grover, P. L., Hewer, A., Pal, K., and Sims, P. (1976). The involvement of a diol epoxide in the metabolic activation of benzo[a]pyrene in human bronchial mucosa and in mouse skin. *Int. J. Cancer* **18**, 1–6.

Gupta, R. S. (1980). Podophyllotoxin resistance: A codominant selection system for quantitative mutagenesis studies in mammalian cells. *Mutat. Res.* **83**, 261–270.

Gupta, R. S. (1981). Resistance to the microtubule inhibitor podophyllotoxin: Selection and partial characterization of mutants in CHO cells. *Somatic Cell Genet.* **7**, 59–71.

Gupta, R. S., and Siminovitch, L. (1978a). Isolation and characterization of mutants of human diploid fibroblasts resistant to diphtheria toxin. *Proc. Natl. Acad. Sci. U.S.A.* **75**, 3337–3341.

Gupta, R. S., and Siminovitch, L. (1978b). Genetic and biochemical studies with the adenosine analogs toyocamycin and tubercidin: Mutation at the adenosine kinase locus in Chinese hamster cells. *Somatic Cell Genet.* **4**, 715–735.

Gupta, R. S., and Siminovitch, L. (1978c). An *in vitro* analysis of the dominance of enetine sensitivity in Chinese hamster ovary cell hybrids. *J. Biol. Chem.* **253**, 3978–3982.

Gupta, R. S., and Siminovitch, L. (1980a). Diphtheria toxin resistance in human fibroblast cell strains from normal and cancer-prone individuals. *Mutat. Res.* **73**, 331–338.

Gupta, R. S., and Siminovitch, L. (1980b). DRB resistance in Chinese hamster and human cells: Genetic and biochemical characteristics of the selection system. *Somatic Cell Genet.* **6**, 151–169.

Gupta, R. S., and Siminovitch, L. (1980c). Diphtheria toxin resistance in Chinese hamster cells: Genetic and biochemical characteristics of the mutants affected in protein synthesis. *Somatic Cell Genet.* **6**, 361–369.

Gupta, R. S., and Siminovitch, L. (1980d). Genetic markers for quantitative mutagenesis studies in Chinese hamster ovary cells: Characteristics of some recently developed selective systems. *Mutat. Res.* **69**, 113–126.

Gupta, R. S., Chan, D. Y. H., and Siminovitch, L. (1978). Evidence for functional hemizygosity at the *emt* locus in CHO cells through segregation analysis. *Cell* **14**, 1007–1013.

Heflich, R. H., Hazard, R. M., Lommel, L., Scribner, J. D., Maher, V. M., and McCormick, J. J. (1980). A comparison of the DNA binding, cytotoxicity, and repair synthesis induced in human fibroblasts by reactive derivatives of aromatic amide carcinogens. *Chem. Biol. Interact.* **29**, 43–56.

Hiatt, H. H., Watson, J. D., and Winsten, J. A., eds. (1977). "Origins of Human Cancer," Vols. A, B, C. Cold Spring Harbor Lab., Cold Spring Harbor, New York.

Hollaender, A., ed. (1971–1981). *Chem. Mutagens* **1–8.**

Hsie, A. W., Brimer, P. A., Machanoff, R., and Hsie, M. H. (1977). Further evidence for the genetic origin of mutations in mammalian somatic cells: The effects of ploidy level and selection stringency on dose-dependent chemical mutagenesis to purine analogue resistance in Chinese hamster ovary cells. *Mutat. Res.* **45**, 271–282.

Hsie, A. W., Machanoff, R., Couch, D. B., and Holland, J. M. (1978a). Mutagenicity of dimethyl nitrosamine and ethyl methanesulfonate as determined by a quantitative host-mediated CHO/HGPRT assay. *Mutat. Res.* **51**, 77–84.

Hsie, A. W., Couch, D. B., O'Neill, J. P., San Sebastian, J. R., Brimer, P. A., Machanoff, R., Riddle, J. C., Li, A. P., Fuscoe, J. C., Forbes, N. L., and Hsie, M. H. (1978b). Utilization of a quantitative mammalian cell mutation system, CHO/HGPRT, in experimental mutagenesis and genetic toxicology. In "Strategies for Short-Term Testing for Mutagens/Carcinogens" (B. E. Butterworth, ed.), pp. 39–54. CRC Press, Cleveland, Ohio.

Hsie, A. W., O'Neill, J. P., Couch, D. B., San Sebastian, J. R., Brimer, P. A., Machanoff, R., Fuscoe, J. C., Riddle, J. C., Li, A. P., Forbes, N. L., and Hsie, M. H. (1978c). Quantitative analysis of radiation and chemical-induced lethality and mutagenesis in Chinese hamster ovary cells. *Radiat. Res.* **76**, 471.

Hsie, A. W., O'Neill, P. J., and McElheny, V. K., eds. (1979). "Mammalian Cell Mutagenesis: The Maturation of Test Systems," pp. 1–477. Cold Spring Harbor Lab., Cold Spring Harbor, New York, Banbury Report 2.

Hsu, I. C., Stoner, G. D., Autrup, H., Trump, B. F., Selkirk, J. K., and Harris, C. C. (1978). Human bronchus-mediated mutagenesis of mammalian cells by carcinogenic polynuclear aromatic hydrocarbons. *Proc. Natl. Acad. Sci. U.S.A.* **75**, 2003–2007.

Huberman, E. (1975). Mammalian cell transformation and cell-mediated mutagenesis by carcinogenic polycyclic hydrocarbons. *Mutat. Res.* **29**, 285–291.

Huberman, E., and Sachs, L. (1974). Cell-mediated mutagenesis of mammalian cells with chemical carcinogens. *Int. J. Cancer* **13**, 326–333.

Huberman, E., And Sachs, L. (1976). Mutability of different genetic loci in mammalian cells by metabolically activated carcinogenic polycyclic hydrocarbons. *Proc. Natl. Acad. Sci. U.S.A.* **73**, 188–192.

Huberman, E., Mager, R., and Sachs. L. (1976a). Mutagenesis and transformation of normal cells by chemical carcinogens. *Nature (London)* **264**, 360–361.

Huberman, E., Sachs, L., Yang, S. K., and Gelboin, H. V. (1976b). Indentification of mutagenic metabolites of benzo[a]pyrene in mammalian cells. *Proc. Natl. Acad. Sci. U.S.A.* **73**, 607–611.

Ingles, C. J., Guialis, A., Lam, J., and Siminovitch, L. (1976). α-Amanitin resistance of RNA polymerase II in mutant Chinese hamster ovary cell lines. *J. Biol. Chem.* **251**, 2729–2734.

Jenssen, D., and Ramel, C. (1980). Relationship between chemical damage of DNA and mutations in mammalian cells I. Dose-response curves for the induction of 6-thioguanine-resistant mutants by low doses of monofunctional alkylating agents, X-rays and UV radiation in V79 Chinese hamster cells. *Mutat. Res.* **73**, 339–347.

Jones, G. E., and Sargent, P. A. (1974). Mutants of cultured Chinese hamster cells deficient in adenine phosphoribosyltransferase. *Cell* **2**, 43–54.

Kakunaga, T. (1978). Neoplastic transformation of human diploid fibroblast cells by chemical carcinogens. *Proc. Natl. Acad. Sci. U.S.A.* **75**, 1334–1338.

Kakunaga, T., Lo, K.-Y., Leavitt, J., and Ikenaga, M. (1980). Relationship between transformation and mutation in mammalian cells. *In* "Carcinogenesis: Fundamental Mechanisms and Environmental Effects" (B. Pullman, P. O. P. Ts'o, and H. V. Gelboin, ed.), pp. 527–541. D. Reidel Publishing Co., Dordrecht.

Kao, F.-T., and Puck, T. T. (1968). Genetics of somatic mammalian cells. VII. Induction and isolation of nutritional mutants in Chinese hamster cells. *Proc. Natl. Acad. Sci. U.S.A.* **60**, 1275–1281.

Kaplan, S. H. (1959). Some implications of indirect induction mechanisms in carcinogenesis: A review. *Cancer Res.* **19**, 791–803.

Kaufmann, E. F., and Davidson, R. L. (1979). Mechanisms of resistance to thymidine analogs in mammalian cells. *Banbury Rep.* **2**, 225–235.

Kavathas, P., Bach, F. H., and DeMars, R. (1980). Gamma ray-induced loss of expression of HLA and glyoxalase I alleles in lymphoblastoid cells. *Proc. Natl. Acad. Sci. U.S.A.* **77**, 4251–4255.

King, H. W. S., Thompson, M. H., and Brookes, P. (1975). The benzo[a]pyrene deoxyribonucleoside products isolated from DNA after metabolism of benzo[a]pyrene by rat liver microsomes. *Cancer Res.* **35**, 1263–1269.

Konze-Thomas, B., Hazard, R. M., Maher, V. M., and McCormick, J. J. (1981). Extent of excision repair before DNA synthesis determines the mutagenic but not the lethal effect of UV radiation. *Mutat. Res.* (submitted).

Knapp, A. G. A. C., and Simons, J. W. I. M. (1975). A mutational assay system for L5178Y mouse lymphoma cells, using hypoxanthine guanine phosphoribosyl transferase deficiency as a marker. The occurrence of a long expression time for mutations induced by X-rays and EMS. *Mutat. Res.* **30**, 97–110.

Knudsen, A. G. (1977). Genetic predisposition to cancer. *In* "Origins of Human Cancer" (H. H. Hiatt, J. D. Watson, J. A. Winsten, ed.), Vol. A, pp. 45–54. Cold Spring Harbor Lab., Cold Spring Harbor, New York.

Knudsen, A. G., Hethcote, H. W., and Brown, B. W. (1975). Mutation and childhood cancer: A probabilistic model for the incidence of retinoblastoma. *Proc. Natl. Acad. Sci. U.S.A.* **72**, 5116–5120.

Krahn, D. F. (1979). Utilization of the CHO/HGPRT system: Metabolic activation and method for testing gases. *In* "Mammalian Cell Mutagenesis: The Maturation of Test Systems" (A. W. Hsie, J. P. O'Neill, and V. K. McElheny, eds.), pp. 251–261. Cold Spring Harbor Lab., Cold Spring Harbor, New York, Banbury Report 2.

Krahn, D. F., and Heidelberger, C. (1977). Liver homogenate-mediated mutagenesis in

Chinese hamster V79 cells by polycyclic aromatic hydrocarbons and aflatoxins. *Mutat. Res.* **49**, 27–44.

Kuroki, T., Drevon, C., and Montesano, R. (1977). Microsome-mediated mutagenesis in V79 Chinese hamster cells by various nitrosamines. *Cancer Res.* **37**, 1044–1050.

Langenbach, R., Freed, H. J., and Huberman, E. (1978a). Liver cell-mediated mutagenesis of mammalian cells by liver carcinogens. *Proc. Natl. Acad. Sci. U.S.A.* **75**, 2864–2867.

Langenbach, R., Freed, H. J., Ranek, D., and Huberman, E. (1978b). Cell specificity in metabolic activation of aflatoxin B_1 and benzo[a]pyrene to mutagens for mammalian cells. *Nature (London)* **276**, 277–280.

Lehmann, A. R., Kirk-Bell, S., Arlett, C. F., Paterson, M. C., Lohman, P. H. M., de Weerd-Kastelein, E. A., and Bootsma, D. (1975). Xeroderma pigmentosum cells with normal levels of excision repair have a defect in DNA synthesis after UV irradiation. *Proc. Natl. Acad. Sci. U.S.A.* **72**, 219–223.

Lewis, W. H., and Wright, J. A. (1979). Isolation of hydroxyurea-resistant CHO cells with altered levels of ribonucleotide reductase. *Somatic Cell Genet.* **5**, 83–96.

McCann, J., Choi, E., Yamasaki, E., and Ames, B. N. (1975). Detection of carcinogens as mutagens in the *Salmonella*/microsome test: Assay of 300 chemicals. *Proc. Natl. Acad. Sci. U.S.A.* **72**, 5135–5139.

McCormick, J. J., and Maher, V. M. (1978). Mammalian cell mutagenesis as a biological consequence of DNA damage. *ICN-UCLA Symp. Mol. Cell. Biol.* **9**, 739–749.

McCormick, J. J., Silinskas, K. C., and Maher, V. M. (1980). Transformation of diploid human fibroblasts by chemical carcinogens. *13th Jerusalem Symp. Quantum Chem. Biochem.* pp. 491–498.

McElheny, V. K., and Abrahamson, S., eds. (1979). "Assessing Chemical Mutagens: The Risk to Humans," pp. 1–367. Cold Spring Harbor Lab., Cold Spring Harbor, New York, Banbury Report 1.

McMillan, S., and Fox, M. (1979). Failure of caffeine to influence mutation frequencies and the independence of cell killing and mutation induction in V79 Chinese hamster cells. *Mutat. Res.* **60**, 91–107.

Maher, V. M., and McCormick, J. J. (1976). Effect of DNA repair on the cytotoxicity and mutagenicity of UV irradiation and of chemical carcinogens in normal and xeroderma pigmentosum cells. In "Biology of Radiation Carcinogenesis" (J. M. Yuhas, R. W. Tennant, and J. B. Regan, eds.), pp. 129–145. Raven, New York.

Maher, V. M., and McCormick, J. J. (1977). DNA repair and carcinogenesis. In "Chemical Carcinogens and DNA" (P. L. Grover, ed.), pp. 133–158. CRC Press, Cleveland, Ohio.

Maher, V. M., and McCormick, J. J. (1980). Comparison of the mutagenic effect of ultraviolet radiation and chemicals in normal and DNA-repair-deficient human cells in culture. *Chem. Mutagens* **6**, 309–329.

Maher, V. M., and Wessel, J. E. (1975). Mutations to azaguanine resistance induced in cultured diploid human fibroblasts by the carcinogen, N-acetoxy-2-acetylaminofluorene. *Mutat. Res.* **28**, 277–284.

Maher, V. M., Curren, R. D., Ouellette, L. M., and McCormick, J. J. (1976a). Role of DNA repair in the cytotoxic and mutagenic action of physical and chemical carcinogens. In "In Vitro Metabolic Activation in Mutagenesis Testing" (F. J. deSerres, J. R. Fouts, J. R. Bend, R. M. Philpot, eds.), pp. 313–336. Elsevier/North-Holland Biomedical Press, Amsterdam.

Maher, V. M., Ouellette, L. M., Curren, R. D., and McCormick, J. J. (1976b). Frequency of ultraviolet light-induced mutations is higher in xeroderma pigmentosum variant cells than in normal human cells. *Nature (London)* **261**, 593–595.

Maher, V. M., McCormick, J. J., Grover, P. L., and Sims, P. (1977). Effect of DNA repair on the cytotoxicity and mutagenicity of polycyclic hydrocarbon derivatives in normal and xeroderma pigmentosum human fibroblasts. *Mutat. Res.* **43**, 117–138.

Maher, V. M., Dorney, D. J., Mendrala, A. L., Konze-Thomas, B., and McCormick, J. J. (1979). DNA excision repair processes in human cells can eliminate the cytotoxic and mutagenic consequences of ultraviolet irradiation. *Mutat. Res.* **62**, 311–323.

Maher, V. M., Rowan, L. A., Silinskas, K. C., Kately, S. A., and McCormick, J. J. (1982) *Proc. Natl. Acad. Sci. U.S.A.* (in press).

Malling, H. V. (1971). Dimethylnitrosamines: Formation of mutagenic compounds by interaction with mouse liver microsomes. *Mutat. Res.* **13**, 425–429.

Mankovitz, R., Buchwald, M., and Baker, R. M. (1974). Isolation of ouabain-resistant human diploid fibroblasts. *Cell* **3**, 221–226.

Miller, J. A., and Miller, E. C. (1977). Ultimate chemical carcinogens as reactive mutagenic electrophiles. *Cold Spring Harbor Conf. Cell Proliferation* **4**, 605.

Milo, G. E., and DiPaolo, J. A. (1978). Neoplastic transformation of human diploid cells *in vitro* after chemical carcinogen treatment. *Nature (London)* **275**, 130–132.

Moehring, T. J., and Moehring, J. M. (1977). Selection and characterization of cells resistant to diphtheria toxin and pseudomonas exotoxin A: Presumptive translational mutants. *Cell* **11**, 447–454.

Myhr, B. C., Turnbull, D., and DiPaolo, J. A. (1979). Ultraviolet mutagenesis of normal and xeroderma pigmentosum variant human fibroblasts. *Mutat. Res.* **62**, 341–353.

Newbold, R. F., and Brookes, P. (1976). Exceptional mutagenicity of benzo[a]pyrene diol epoxide in cultured mammalian cells. *Nature (London)* **261**, 52–54.

Newbold, R. F., Wigley, C. B., Thompson, M. H., and Brookes, P. (1977). Cell-mediated mutagenesis in cultured Chinese hamster cells by carcinogenic polycyclic hydrocarbons: Nature and extent of the associated hydrocarbon-DNA reaction. *Mutat. Res.* **43**, 101–116.

Newbold, R. F., Warren, W., Medcalf, A. C. S., and Amos, J. (1980). Mutagenicity of carcinogenic methylating agents is associated with a specific DNA modification. *Nature (London)* **283**, 596–599.

O'Neill, J. P., and Hsie, A. W. (1979). CHO/HGPRT mutation assay: Adaptation for mutagen screening. *Banbury Rep.* **2**, 311–317.

O'Neill, J. P., Brimer, P. A., Machanoff, R., Hirsch, G. P., and Hsie, A. W. (1977a). A quantitative assay of mutation induction at the hypoxanthine-guanine phosphoribosyl transferase locus in Chinese hamster ovary cells (CHO/HGPRT system): Development and definition of the system. *Mutat. Res.* **45**, 91–101.

O'Neill, J. P., Couch, D. B., Machanoff, R., San Sebastian, J. R., Brimer, P. A., and Hsie, A. W. (1977b). A quantitative assay of mutation induction at the hypoxanthine-guanine phosphoribosyl transferase locus in Chinese hamster ovary cells (CHO/HGPRT system): Utilization with a variety of mutagenic agents. *Mutat. Res.* **45**, 103–109.

Peto, R. (1977). Epidemiology, multi-stage models, and short-term mutagenicity tests. In "The Origins of Human Cancer" (H. H. Hiatt, J. D. Watson, and J. A. Winsten, eds.), pp. 1403–1407. Cold Spring Harbor Lab., Cold Spring Harbor, New York.

Pienta, R. J., Poiley, J. A., and Lebherz, W. B. (1977). Morphological transformation by early passage golden Syrian hamster embryo cells derived from cryopreserved primary cultures as a reliable in vitro bio-assay for identifying diverse carcinogens. *Int. J. Cancer* **19**, 642–655.

Puck, T. T. (1979). Historical perspectives on mutation studies with somatic mammalian cells. *Banbury Rep.* **2**, 3–14.

Puck, T. T., and Kao, F. T. (1967). Genetics of mammalian cells. V. Treatment with 5-

bromodeoxyuridine and visible light for isolation of nutritionally deficient mutants. *Proc. Natl. Acad. Sci. U.S.A.* **58**, 1227.

Pullman, G., Ts'o, P. O. P., and Gelboin, H. V., eds. (1980). "Carcinogenesis: Fundamental Mechanisms and Environmental Effects." D. Reidel Publishing Co., Dordrecht.

Reznikoff, C. A., and DeMars, R. (1981). *In vitro* chemical mutagenesis and viral transformation of a human endothelial cell strain. *Cancer Res.* **41**, 1114–1126.

Riddle, J. C., and Hsie, A. W. (1978). An effect of cell-cycle position on ultraviolet-light-induced mutagenesis in Chinese hamster ovary cells. *Mutat. Res.* **52**, 409–420.

Robbins, J. H., Kraemer, K. H., Lutzner, M. A., Festoff, B. W., and Coon, H. G. (1974). Xeroderma pigmentosum—An inherited disease with sun sensitivity, multiple cutaneous neoplasms, and abnormal DNA repair. *Ann. Intern. Med.* **80**, 221–248.

San, R. H. D., and Williams, G. (1977). Rat hepatocyte primary cell culture-mediated mutagenesis of adult rat liver epithelial cells by procarcinogens. *Proc. Soc. Exp. Biol. Med.* **156**, 534.

Sato, K., and Hieda, N. (1979). Isolation and characterization of a mutant mouse lymphoma cell sensitive to methyl methanesulfonate and X-rays. *Radiat. Res.* **78**, 167–171.

Schultz, R. A., Trosko, J. E., and Chang, C.-C. (1981). Isolation and partial characterization of mutagen-sensitive and DNA repair mutants of Chinese hamster fibroblasts. *Environ. Mutagen.* **3**, 53–64.

Selikoff, I. J., Chung, J., and Hammond, E. C. (1965). Relation between exposure to asbestos and mesothelioma. *N. Engl. J. Med.* **272**, 560–565.

Selkirk, J. K. (1977). Divergence of metabolic activation systems for short-term mutagenesis assays. *Nature (London)* **270**, 604–607.

Shin, S.-I., Freedman, V. H., Risser, R., and Pollack, R. (1975). Tumorigenicity of virus-transformed cells in nude mice is correlated specifically with anchorage independent growth *in vitro*. *Proc. Natl. Acad. Sci. U.S.A.* **72**, 4435–4439.

Shinohara, K., and Cerutti, P. A. (1977). Formation of benzo[a]pyrene-DNA adducts in peripheral human lung tissue. *Cancer Lett. (Shannon, Irel.)* **3**, 303–309.

Silinskas, K. C., Kateley, S. A., Tower, J. E., Maher, V. M., and McCormick, J. J. (1981). Induction of anchorage independent growth in human fibroblasts by propane sultone. *Cancer Res.* **41**, 1620–1627.

Simon, L., Hazard, R. M., Maher, V. M., and McCormick, J. J. (1981). Enhanced cell killing and mutagenesis by ethylnitrosourea in xeroderma pigmentosum cells. *Carcinogenesis* **2**, 567–570.

Siminovitch, L. (1976). On the nature of hereditable variation in cultured somatic cells. *Cell* **7**, 1–7.

Skopek, T. R., Liber, H. L., Penman, B. W., and Thilly, W. G. (1978). Isolation of a human lymphoblastoid line heterozygous at the thymidine kinase locus: Possibility for a rapid human cell mutation assay. *Biochem. Biophys. Res. Commun.* **84**, 411–416.

Strauss, G. H., and Albertini, R. J. (1979). Enumeration of 6-thioguanine resistant peripheral blood lymphocytes in man as a potential test for somatic cell mutations arising *in vivo*. *Mutat. Res.* **61**, 353–379.

Subak-Sharpe, H., Burk, R. R., and Pitts, J. D. (1969). Metabolic cooperation between biochemically marked mammalian cells in tissue culture. *J. Cell Sci.* **4**, 353–367.

Suter, W., Brennand, J., McMillan, S., and Fox, M. (1980). Relative mutagenicity of antineoplastic drugs and other alkylating agents in V79 Chinese hamster cells, independence of cytotoxic and mutagenic responses. *Mutat. Res.* **73**, 171–181.

Szybalski, W. (1958). Resistance to 8-azaguanine, a selective marker for a human cell line. *Microb. Genet. Bull.* **16**, 30.

Taylor, S. M., and Jones, P. A. (1979). Multiple new phenotypes induced in 10T½ and 3T3 cells treated with 5-azacytidine. *Cell* **17**, 771–779.

Taylor, R. T., and Hanna, M. L. (1977). Folate-dependent enzymes in cultured Chinese hamster cells: Folyl polyglutamate synthetase and its absence in mutants auxotropic for glycine + adenosine + thymidine. *Arch. Biochem. Biophys.* **181**, 331–344.

Taylor, M. W., Pipkorn, J. H., Tokito, M. K., and Pozzatti, R. O. (1977). Purine mutants of mammalian cell lines. III. Control of purine biosynthesis in adenine phosphoribosyl transferase mutants of CHO cells. *Somatic Cell Genet.* **3**, 195–206.

Taylor, M., Hershey, H. V., and Simon, A. E. (1979). An analysis of mutation at the adenine phosphoribosyl transferase locus. *Banbury Rep.* **2**, 211–223.

Thilly, W. G. (1977). Chemical mutation in human lymphoblasts. *J. Toxicol. Environ. Health* **2**, 1343–1352.

Thilly, W. G. (1979). Study of mutagenesis in diploid human lymphoblasts. *Banbury Rep.* **2**, 341–349.

Thilly, W. G., DeLuca, J. G., Hoppe, H., IV, and Penman, B. W. (1976). Mutation of human lymphoblasts by methylnitrosourea. *Chem. Biol. Interact.* **15**, 33–50.

Thilly, W. G., DeLuca, J. G., Furth, E. E., Hoppe, H., IV, Kaden, D. A., Krowlewski, J. J., Liber, H. L., Skopek, T. R., Slapikoff, S. A., Tizard, R. J., and Penman, B. W. (1980). Gene-locus mutation assays in diploid human lymphoblast lines. *Chem. Mutagens* **6**, 331–364.

Thompson, L. H., Rubin, J., Cleaver, J. E., and Brookman, K. (1979). A screening method for isolating DNA repair-deficient mutants of CHO cells. *Somatic Cell Genetics* **6**, 391–405.

Ts'o, P. O. P. (1979). Current progress in the study of the basic mechanisms of neoplastic transformation. *Banbury Rep.* **2**, 385–406.

Ts'o, P. O. P. (1980). Neoplastic transformation, somatic mutation, and differentiation. In "Carcinogenesis: Fundamental Mechanisms and Environmental Effects" (B. Pullman, P. O. P. Ts'o, and H. V. Gelboin, eds.), pp. 297–310. D. Reidel Publishing Co., Dordrecht.

Waldren, C., Jones, C., and Puck, T. T. (1979). Measurement of mutagenesis in mammalian cells. *Proc. Natl. Acad. Sci. U.S.A.* **76**, 1358–1362.

Waters, M. D. (1979). The gene-tox program. *Banbury Rep.* **2**, 449–467.

Weekes, U., and Brusick, D. J. (1975). In vitro metabolic activation of chemical mutagens. II. Relations among mutagen formation, metabolism and carcinogenicity for dimethylnitrosamine and diethylnitrosamine in liver, kidney, and lung of BALB/cJ, C57BL/6J and RF/J mice. *Mutat. Res.* **31**, 175–183.

Yang, L. L., Maher, V. M., and McCormick, J. J. (1980). Error-free excision of the cytotoxic and mutagenic N^2-deoxyguanosine DNA adduct formed in human fibroblasts by ($\pm$)-7β,8α-dihydroxy-9α,10α-epoxy-7,8,9,10-tetrahydrobenzo[a]pyrene. *Proc. Natl. Acad. Sci. U.S.A.* **77**, 5933–5937.

Yang, L. L., Maher, V. M., and McCormick, J. J. (1982). Relationship between excision repair and the cytotoxic and mutagenic effect of the "anti" 7,8-diol-9,10-epoxide of benzo(a)pyrene in human cells. *Mutat. Res.* (in press).

van Zeeland, A. A., and Simons, J. W. I. M. (1975). The effect of calf serum on the toxicity of 8-azaguanine. *Mutat. Res.* **27**, 135–138.

van Zeeland, A. A., and Simons, J. W. I. M. (1976). Linear dose-response relationships after prolonged expression times in V79 Chinese hamster cells. *Mutat. Res.* **35**, 129–138.

Chapter 9

Chromosomal Aberrations Induced in Occupationally Exposed Persons

Maria Kučerová

241

MUTAGENICITY:
NEW HORIZONS IN GENETIC TOXICOLOGY

I. INTRODUCTION: POSSIBILITIES OFFERED BY CYTOGENETIC METHODS FOR TESTING THE EFFECTS OF SMALL DOSES OF CHEMICALS APPLIED CHRONICALLY TO MAN

A. History

In the early 1960s the cytogenetic methods became widely used in analyzing human chromosomes and thus opened the way to screening various mutagens for their effects directly on human somatic cells *in vivo*. Chromosomal analysis was first used for this purpose in 1960 by Tough *et al.* in their radiation studies. Radiation has since become a "classical" mutagen. A great deal of research effort was applied to the study of its cytogenetic effects on human chromosomes during the 1960s and the knowledge and experience acquired during this period have provided the basis for testing the effects of chemical mutagens in the early 1970s.

Studies and testing of the chemically induced human chromosome changes occurring *in vivo* are more difficult to perform than the radiocytogenetic studies (for details, see Chapter 13 and Evans, 1980), but this area of research is of particular interest and importance for human welfare.

The major differences between the radiation and chemical cytogenetic effects may be summarized as follows:

1. The cytogenetic effects of X-rays are fairly uniform; no marked differences between the sensitivity of cells from different persons were found *in vitro* (Kučerová *et al.*, 1972; Kučerová and Polívková, 1977; Dolphin, 1978; Scott and Lyons, 1979) or *in vivo* (Kučerová *et al.*, 1976b; Buckton *et al.*, 1978; Nordenson *et al.*, 1980b).

2. Comparison of the cytogenetic effects detected after the same dose exposure *in vitro* and *in vivo* (whole-body exposure) also gives very similar results (Buckton *et al.*, 1971; Bauchinger, 1978).

In contrast to the radiation effects, the cytogenetic effects of chemicals differ in both features mentioned above. There are significant differences between the cytogenetic effects of the same chemical in the cells from different human individuals not only *in vivo* (Gebhart *et al.*, 1980; Kučerová *et al.*, 1980), but also *in vitro* (Kučerová and Polívková, 1977, 1978; Nevstad, 1977). There are also marked differences between the mutagenic effects of various chemicals according to their different chemical structure (Evans, 1977).

The wide variety of chemicals and the specificity of their effects, combined with differences in the sensitivity of individuals dictates a special research approach to each chemical in order to provide detailed data on their properties and to permit a careful evaluation of the results obtained. It is very difficult to make general conclusions that would be valid for more than one chemical. Individual testing of each chemical and for each person, if possible, is recommended.

Despite all these difficulties, the cytogenetic analysis of human lymphocytes is used for chemical mutagenicity testing in an ever increasing number of studies. The reason is that the speed with which the human environment becomes contaminated by chemicals urgently requires a control of the extent to which the DNA is damaged in humans exposed to chemicals and an estimation of what the consequences of this damage will be. Cytogenetics is one of the available methods that can be used for this purpose.

B. Cytogenetic Techniques Used for Mutagenicity Testing and Types of Chromosomal Damage Detected by Them

1. The Conventional or Classical Cytogenetic Technique

Homogenous staining of chromosomes was the first classical cytogenetic technique for testing the mutagenic effects on human chromosomes. A modification of Hungerford's method (Hungerford, 1965) has been most frequently used. It includes short-term cultivation (2 or 3 days) of peripheral whole blood samples in standard media, calf serum, or AB human serum, antibiotics and phytohemagglutinin, blockade of mitosis in metaphase by colcemid or its derivatives or other spindle-destroying factors (vincristine, vinblastine), hypotonization by KCl and repeated fixation in Carnoy's solution, and staining with Giemsa or orcein. The conventional technique without differential staining permits rapid overall analysis of tested cells, i.e., checking of the chromosomal number and registration of most of the chromosomal damage. All cells with so-called unstable aberrations (Cu) (Buckton et al., 1962) could be detected by this method: chromosomal and chromatid breaks, deletions, fragments, rings and dicentrics with or without fragments, and chromatid exchanges. These types of aberrations have a lethal effect on the cell. Some of the aberrant cells with so-called stable aberrations (Cs) (Buckton et al., 1962), which do not interfere with the division of chromosomes and permit the cell to survive, escape detection by this conventional technique, e.g., peri- and paracentric inversions, insertions, small deletions, and many reciprocal translocations.

2. Banding (Differential) Techniques

These techniques were introduced by Caspersson et al., in 1970, and were developed thereafter in numerous variants by many investigators. The method in routine use is the G-banding technique. It was originally described by Sumner et al. (1971) and modified by Seabright (1971). Seabright's modification uses trypsin as a banding factor. It is technically simple and ensures a good standard quality of the bands. This technique is not normally used in monitoring studies and has been applied only in radiation studies; the mutagenic effects of chemicals in vivo have not yet been tested by this method, most probably for two main reasons: (a) complete analysis of banded chromosomes in mitotic plates is time consuming and requires experienced scorers so that the number of individuals that can be tested by this method is usually limited, and (b) the spectrum of chromosomal aberrations induced by chemicals shows that the majority of them can be detected by the conventional technique, without using G banding.

3. Types of Chromosomal Aberrations Induced by Different Mutagens

The typical postradiation chromosomal aberrations are deletions, terminal and interstitial; dicentrics and rings, with or without fragments; translocations; and inversions.

The scale of aberrations induced by chemicals, detected thus far in human cells, is quite different from postradiation changes. The most frequent of these aberrations are the chromatid breaks, which remain open, and then the chromosome breaks which are usually also open. Chromatid and chromosomal exchanges and rearrangements are very rare (Fig. 1–5).

Some chemicals have a specific additional effect. For instance, mitomycin C causes subchromatid breaks of the centromeric region of chromosomal heterochromatin. They can be seen as special tips near the centromere (Fig. 5). Some chemicals (epichlorhydrin and vinyl chloride) sometimes despiralize selected chromosomes in some cells (Fig. 5).

The in vitro study of chromosomal damage after exposure of human lymphocytes to X-rays and to alkylating agents (Kučerová and Polívková, 1976) using conventional and G-banding techniques for scoring in parallel showed that 19–38% of the aberrations are missed by the conventional technique in radiation experiments, but only 3–11% in experiments with alkylating agents.

The G-banding technique was used for in vivo monitoring of two large groups of people irradiated in the past (Buckton et al., 1978;

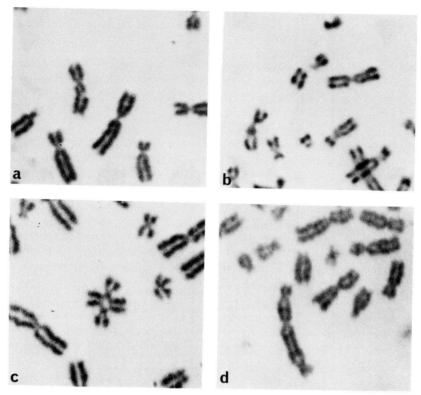

Fig. 1. Chromosomal aberrations induced by chemicals *in vivo* (usual types): (a) chromatid break; (b) chromosome break; (c) chromatid exchange; (d) chromosome exchange.

Sofuni *et al.*, 1978). In these two studies the G-banding and the conventional technique were used in parallel. A comparison of the aberration yield detected by these two methods revealed that 19–30% of the aberrations were lost when the conventional technique was only used for scoring. These results are very close to the *in vitro* data.

4. Harlequin (or FPG or SCE) Technique

This was first described as a sensitive, convenient method for routine chemical mutagenicity testing by Latt (1974), Wolff and Perry (1974), and Perry and Evans (1975). This technique, upon addition of thymidine analogues for two rounds of replication, produces chromosomes with chemically different chromatids which can be distinguished by Giemsa or other dyes.

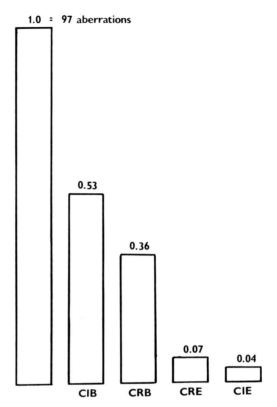

Fig. 2. Chromosomal aberration spectrum induced *in vivo* by vinyl chloride (Kučerová *et al.*, 1979). CIB, chromatid break; CRB, chromosome break; CIE, chromatid exchange; CRE, chromosome exchange.

An increased number of sister chromatid exchanges (SCE) (Fig. 6) was noticed even when the dose of a chemical was below the concentration required for the detection of the classical chromosomal aberrations. This means that this method is more sensitive than the classical one for some chemicals. However, this does not hold for other chemicals. The sensitivity of the SCE method is more pronounced *in vitro* than *in vivo*, as was demonstrated by Nevstad in studies on adriamycin (1977), by Hansteen (1977) on vinyl chloride, by Meretoja *et al.* (1978) and Uggla *et al.* (1980) on styrene, by Kučerová and Polívková (1977) on imuran, and by Kram *et al.* (1979) on mitomycin C. The increase in SCE incidence after radiation *in vitro* and *in vivo* is very low.

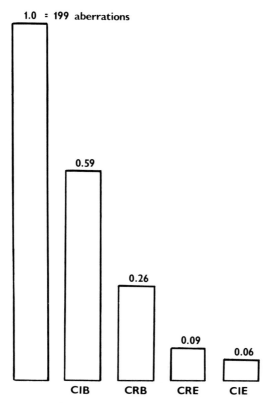

Fig. 3. Chromosomal aberration spectrum induced *in vivo* by epichlorhydrin (Kučerová *et al.*, 1977). CIB, chromatid break; CRB, chromosome break; CIE, chromatid exchange, CRE, chromosome exchange.

In summary, we can say that the harlequin technique provides another possible approach to human population monitoring of the chemical mutagenic effects. The detection of SCE in cells requires a lower level of technical skills and is more rapid than scoring of the classical aberrations. We also know that the SCEs occur by a different mechanism than the classical aberrations; therefore, we are testing two different types of DNA damage. For these reasons, the harlequin technique is convenient for large screening studies.

5. Automatic Scoring of Chromosomal Aberrations

During the past 15 years the semiautomatic system for analysis of human chromosomes has been developed (Castleman *et al.* 1976;

MARIA KUČEROVÁ

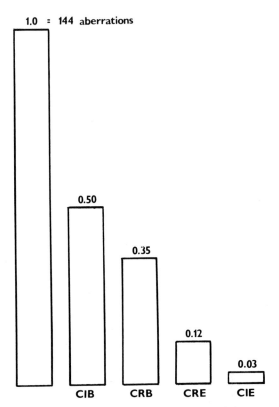

Fig. 4. Chromosomal aberration spectrum induced *in vivo* by Imuran and prednisone (Kučerová *et al.*, 1980). CIB, chromatid break; CRB, chromosome break; CIE, chromatid exchange; CRE, chromosome exchange.

Mason and Rutovitz, 1978). The automatic chromosomal analysis offers a possibility to control systematically large groups of occupationally exposed people. However, this method is still in the experimental stage and is not yet ready for widespread use.

II. STUDIES OF WORKERS OCCUPATIONALLY EXPOSED TO CHEMICALS

A. Possible Approaches to Human Monitoring Studies

Population cytogenetic monitoring is one of the ways in which the effects of environmental mutagens may be detected in man. The chromosomal aberrations and SCEs are used as markers in such studies.

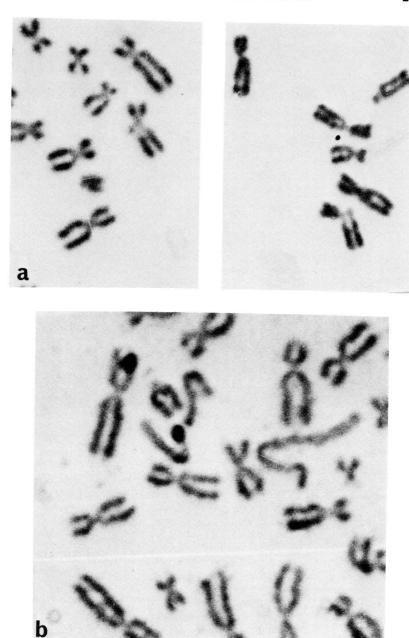

Fig. 5. Chromosomal aberrations induced by chemicals *in vivo* (unusual types): (a) "tips" induced by mitomycin-C in centromeric heterochromatin; (b) despiralization of individual chromosomes induced by alkylating agents.

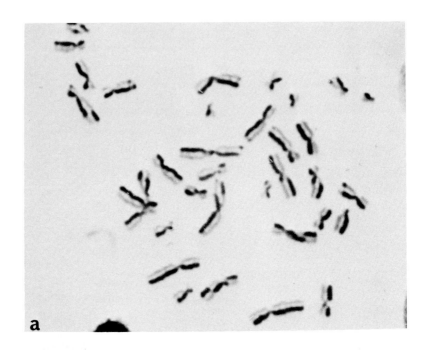

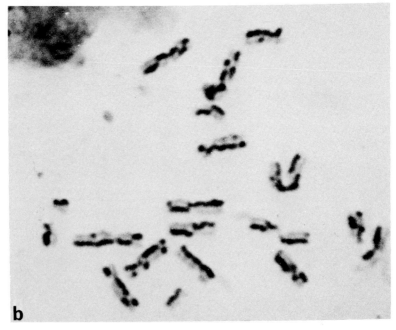

Workers who have undergone long-term occupational exposure to high levels of a test chemical are supposed to be at high risk. Controls for such persons are needed, and a cytogenetic analysis is suitable for this purpose. A small peripheral blood sample provides enough cells to be scored for chromosomal aberrations. Blood collections are technically easy and it is therefore possible to carry out repeated periodical sampling of exposed workers.

An evaluation of chromosomal changes detected in cells of exposed workers requires reliable control data. It is most convenient to use the cells from the same person before exposure, as a control, and then, if possible, at various time intervals during and after exposure to the chemical tested (Cohen and Bloom, 1971; Sorsa, 1981). Such controls are optimal but they are seldom available when occupationally exposed persons are tested. Therefore, blood samples from other persons must be used as controls. If possible the control should be matched in terms of sex, age, and race. Control blood samples should be cultured under exactly the same conditions as samples from exposed people (e.g., medium, serum, chemicals) and coded slides should be scored.

A number of factors, such as the use of drugs, drinking and smoking habits, irradiation, and viral or other infections in the last 3 months before sampling, need to be taken into account since they can have a profound influence on the results. These individuals who may not recall exposure to any of these factors may have an elevated level of chromosomal aberrations. In addition, exposure to elevated levels of radiation and chemicals in the remote past is not convenient for use as a control since some types of aberrations may survive for years in the human body.

The use of chromosomal aberrations for monitoring means that we are studying the response of single, randomly chosen cells; each aberrant cell is a member of the lymphocyte population. Therefore, it is possible to obtain significant data from only a few persons if enough randomly chosen cells are scored. Cytogenetic monitoring is, in fact, the monitoring of a population (cells) in a population (group of people).

The average percentage of aberrant cells in peripheral blood lymphocytes of healthy adult people, who are not exposed to unusual doses of mutagens, is 1 or 2% in most laboratories, if gaps are not included (K. E. Buckton, N. P. Bochkov, and M. Kučerová, unpublished data). Young children and infants usually have less than 1% of aberrant cells.

Fig. 6.   Sister chromatid exchanges (SCE). (a) The cell with usual number of SCEs; (b) the cell with highly increased number of SCEs.

B. Examples of Monitoring Studies

1. Vinyl Chloride Monomer

Vinyl chloride monomer (VCM) is a chemical that has been most properly studied by cytogenetic monitoring. This chemical has been manufactured in many factories and in many countries for more than 40 years. It attracted the cytogeneticists' attention after its carcinogenic effects had been described.

Originally it was supposed that VCM had no effect in humans. However, in 1971, Viola et al. reported on the carcinogenic effects of VCM in experimental animals and, 3 years later, in 1974, Creech and Johnson found endotheliomas of the livers in workers who had long-term exposure to high doses of VCM. The carcinogenic effect of VCM has since been systematically investigated, and over 50 other endotheliomas of the liver were described in workers with long-term exposure by different investigators in different countries. The average time of exposure of these workers with tumors was about 15 years and the dose of exposure was extremely high (500 ppm or more) compared to present-day exposures. Nevertheless, the manufacture of VCM is still increasing (Bartsch and Montesano, 1975) because VCM is being widely used for the production of polyvinyl chloride and as a propellant for pesticides, deodorants, and other sprays. Vinyl chloride monomer is most readily absorbed by inhalation, but it can also be absorbed by skin or by ingestion.

The mutagenic effects of VCM were first detected by chromosomal aberrations in peripheral lymphocytes of exposed workers in 1975 (Ducatman et al., 1975; Funes-Cravioto et al., 1975) and in 1976 (Kučerová, 1976; Purchase et al., 1976). All these studies comprised smaller groups of the most exposed workers and were followed by other more detailed studies (Hansteen, 1977; Hansteen et al., 1978; Purchase et al., 1978; Natarajan et al., 1978; Kučerová et al., 1979; Anderson et al., 1980). Most of the cytogenetic studies have shown a significant increase of chromosomal aberrations in the group of workers exposed for a long time to high doses as compared to the matched controls.

Only the recent investigations of Hansteen et al. (1978) on a group of subjects working in industrial plants with low-dose exposure to VCM gave negative results. They also found that workers who had previously had significant elevations in aberration frequency had returned to control levels within 2 years of the exposure being reduced to a minimum (0.1 ppm).

This finding was also supported by the results of Anderson et al. (1980), who noticed that the aberrations in peripheral lymphocytes of VCM-exposed workers had decreased $3\frac{1}{2}$ years after the reduction of exposure below 5 ppm. Similarly, Natarajan et al. (1978) did not find any increase in aberrations in lymphocytes of workers exposed to low doses of VCM (1 ppm). The effect of VCM on human chromosomes in vivo is summarized in Table I.

The harlequin technique was used in the most recent studies of VCM effects (Hansteen, 1977; Hansteen et al., 1978; Kučerová et al., 1979). Hansteen et al. (1978) analyzed the SCE numbers in cells of workers exposed to reduced doses of VCM (below 0.1 ppm) and did not find any difference in SCE numbers between exposed workers and controls. Kučerová et al. (1979) investigated the SCE numbers in a group of workers still exposed to relatively high doses of VCM (between 15 and 20 ppm). Five out of the seven highly exposed workers had significantly elevated numbers of SCEs per cell.

One can conclude from the results of all cytogenetic VCM studies in vivo that VCM increases the number of somatic mutations in humans if the exposure is long-term and if the doses are high enough. Nevertheless, all cytogenetic studies have shown that there is considerable individual variability in sensitivity to VCM, because even two persons exposed to the same dose of VCM over the same period of time may have a different degree of chromosomal damage to their lymphocytes, whether measured by the classical aberrations or the SCEs. As already mentioned, this individually different response in vivo is typical for chemical mutagenicity.

2. Epichlorhydrin

An alkylating agent, epichorhydrin (ECHH), has become the object of cytogenetic monitoring because its production is increasing in modern industry. Mutagenicity of ECHH was detected in 1948 using Drosophila (Rapoport, 1948) and then in other studies on lower organisms and also on human cells in vitro (Kučerová et al., 1976a).

In vivo, the 2-year prospective study (Kučerová et al., 1977) in which each worker's own control (collected before the beginning of ECHH production) was used has shown that the classical chromosomal aberrations gradually increased in the lymphocytes of exposed workers. These results were confirmed by Picciano (1979), who tested a larger group of workers occupationally exposed to ECHH and also found a significant increase in chromosomal aberrations in their peripheral lymphocytes. The spectrum of chromosomal aberrations was similar

TABLE I

Chromosomal Aberrations in Lymphocytes of Persons Occupationally Exposed to Vinyl Chloride

Reference	No. of workers	Length of exposure (average years)	Dose of exposure (average ppm)	No. of breaks/ 100 cells (workers)	No. of controls	No. of breaks/ 100 cells (controls)	No. of cells scored/ person
Funes-Cravioto et al. (1975)	7	16.5	20–30	12	3	2.8	150–200
Ducatman et al. (1975)	11	15	<500	5.9[a]	10	2.1[a]	50
Purchase et al. (1978)	17	10.7	$\doteq 350$	3.8	24	1.3	100
Hansteen et al. (1978)	39	13	$\doteq 500$	3.4	16	1.7	100
Kučerová et al. (1979)	8	16.5	100–500	5.8	7	2.1	100

[a] Including gaps.

in both studies. Picciano (1979) used a matched control because the subject's own control was not available. Results of these two studies (Kučerová et al., 1977; Picciano, 1979) are summarized in Table II.

Šrám et al. (1980) repeated the cytogenetic control of some ECHH workers, who had been previously analyzed by Kučerová et al. (1977). They investigated 21 of the 31 workers included in the original group. Kučerová et al. (1977) noted that the average percentage of cells with chromosomal aberrations increased from 1.37, which was the control level before exposure, to 1.90 after 1 year of exposure and to 2.69 after 2 years of exposure. Šrám et al. (1980) found 3.02% of aberrant cells in their group of workers after 4-year exposure.

Even though the group tested by Šrám et al. (1980) comprised only two-thirds of the workers from the original group and even though it is rather difficult to compare the results of two different laboratories [different culture conditions, different experiences in cytogenetic scoring, and different personal errors of individual scorers—as reported by Zavala et al. (1979)], an increase in aberrant cells was observed although it was not as steep as in the first 2 years of exposure. All three cytogenetic studies in vivo have demonstrated that ECHH has a mutagenic effect on somatic cells of persons occupationally exposed to it.

3. Styrene and Other Chemicals

Styrene is widely used in the plastic industry and relatively large numbers of workers are thus exposed to it. The mutagenic effect of styrene was studied in detail by the Scandinavian cytogeneticists (Meretoja et al., 1977, 1978; Sorsa et al., 1979; Uggla et al., 1980; Norppa et al., 1980). Cytogenetic analysis of peripheral lymphocytes of exposed workers showed that they had highly elevated frequencies of the classical aberrations but the number of SCEs in their cells did not differ from the controls. These results have again supported the hypothesis that a different induction mechanism is involved in the classical aberrations and in SCEs. Results of all Scandinavian research groups obtained in different factories are in good agreement. The mean percentage of aberrant cells in those studies was very high (15–26%) when compared to that resulting from the effects of other chemicals. With the finding of the carcinogenic effect of a styrene oxide in experiments on animals the idea has emerged that workers with the high percentage of aberrant cells may run the risk of malignant disease.

Some other chemicals have also been tested by cytogenetic methods in occupationally exposed workers, but such studies are rare and only isolated groups at risk have been investigated. Tough and Court-Brown

TABLE II

Chromosomal Aberrations in Lymphocytes of Persons Occupationally Exposed to Epichlorhydrin

Reference	No. of workers	Dose of exposure (average ppm)	No. of breaks/ 100 cells (workers)	Percentage of aberrant cells (workers)	No. of controls	No. of breaks/ 100 cells (controls)	Percentage of aberrant cells (controls)	No. of cells scored/ person
Kučerová et al. (1977)	31	0.13–1.3	3.26	2.69	35	1.6	1.3	170
Picciano (1979)	93	0.5–5.0	5.5	4.25	75	2.8	2.38	200

(1965) and Forni et al. (1971) described an increase of chromosomal aberrations after exposure to benzene. Fungicides were the object of monitoring in the studies of Pilinskaja (1970, 1974), and the mutagenic effect on chromosomes was confirmed. The same was found after exposure to pesticides (Yoder et al., 1973), lead (Deknudt et al., 1977), and sulfur dioxide (Nordenson et al., 1980a). Funes-Cravioto et al. (1977) have demonstrated an increased frequency of chromosomal aberrations in workers in chemical laboratories and in rotoprinting factories. The results of all these studies have proved the usefulness of cytogenetic monitoring.

4. Complex Studies

Cytogenetic monitoring substitutes for the experimental approach which is not feasible in humans. Cytogenetic methods offer the possibility of detecting recent mutations in somatic cells of exposed people. Although the real consequences of accumulation of somatic mutations in the human body are not yet known, the possibility must be considered that the malignant process may be initiated. Moreover, knowledge of what the genetic consequences of mutagenic exposure might be for succeeding generations, for the progeny of exposed persons, is of immense importance.

One can suppose that persons with increased levels of chromosomal mutations in somatic cells will also have an increased frequency of mutations in the gonads. Nevertheless, the correlation between the frequency of these two types of mutation is not yet known. In addition, one should also expect that point mutations occur together with chromosomal mutations in both somatic and gonadal cells. All this might be proved only indirectly by analyzing the progeny of exposed people.

Studies that combine cytogenetic monitoring of occupationally exposed workers with testing of their progeny (including spontaneous abortions, perinatal deaths, and congenital malformations) are very useful and promising. Such complex studies are still very rare, because they are difficult to perform and are time consuming, but the first results of such studies are encouraging.

A complex study of this type has been published in six parts by Nordström, Nordenson, and co-workers in 1978–1979 (Nordström et al., 1978a,b, 1979a,b; Nordenson et al., 1978a,b). They investigated, from many points of view, the workers exposed to different metals in a smelter in northern Sweden. Cytogenetic studies have shown that the frequency of chromosomal aberrations was increased in peripheral lymphocytes of workers exposed to lead and arsenic. Population studies of the workers' families revealed an increase in spontaneous abortions, malformed children, and decreased birth weight of children in female employees. Most probably combined mutagenic and teratogenic effects could have been responsible for the observed changes, because a pure mutagenic effect might be discovered only if the progeny of exposed male workers were controlled. In practice, however, the mutagenic and teratogenic effects are often combined. Another interesting study was done by Funes-Cravioto *et al.* (1977), who found an increased frequency of chromosomal aberrations and SCEs in children of female workers in the chemical laboratories.

Complex studies cannot provide clear-cut results. The human environment is too complex and it cannot be simplified for experimental purposes. Therefore it is difficult to detect the mutagenic effect of a chemical which is suspected to act as mutagenic in man. Nevertheless, such studies are the only way in which the whole complex of possible detrimental effects of mutagen may be discovered and studied.

III. OPEN QUESTIONS IN THE FIELD OF CYTOGENETIC MONITORING OF OCCUPATIONALLY EXPOSED PEOPLE

A. Different Individual Sensitivity

In chemical mutagenicity studies considerable individual variability in chromosomal changes within the groups of tested persons has been described and discussed (Funes-Cravioto *et al.*, 1977; Natarajan *et al.*, 1978; Purchase, 1978; Berg, 1979; Kilian and Picciano, 1979; Gebhart *et al.*, 1980; Kučerová *et al.*, 1979, 1980; Sorsa, 1979, 1981; Drake, 1980; Evans, 1980). This finding is not surprising because the mutagen undergoes a long and complicated process between the first contact with the human body and the appearance of the ultimate chromosomal change, which we are able to detect as a chromosomal aberration in the lymphocyte.

According to Vogel (1978) this process proceeds at three levels: (1) entrance into the human body, (2) metabolization of the mutagen, and (3) direct contact with DNA. These three levels are considered as follows:

1. Entrance of the mutagen into the human body and also its excretion have a strong influence on the actual dose of the mutagen, i.e., its concentration in different tissues. The inherited or acquired changes in the function of the lungs, digestive tract, kidneys, and other organs can change the qualitative and quantitative effects of the mutagen.

2. For metabolization of the mutagen, the detoxicating or activating ability of the liver and other organs in the body is important and can be influenced genetically or by environmental factors (for instance, a

common disease, such as infectious hepatitis, in the past may change the activity of the liver).

3. Each individual differs most probably in the ability to repair the DNA damage. We do not know how this process occurs in human cells in detail, but from the inherited partial defects in repair, which were described in some genetic diseases (e.g., Bloom's syndrome, Fanconi's anemia, xeroderma pigmentosum, ataxia telangiectasia), it is obvious that a complicated enzymatic system is involved (Cleaver, 1977).

B. The Problem of Valid Controls

We have already mentioned that the subject's own control is very probably the most appropriate in chemical monitoring studies. Control of this kind avoids the difficulties encountered with differing sensitivities of individuals tested, and the comparison of observed aberrations with the controls is quite exact. It is technically easy to obtain control from a patient when the mutagenicity of drugs is tested. However, control is not available in workers occupationally exposed to chemicals for many years although their cytogenetic investigation is important. Controls matched for sex and age are the only possibility in such studies. Sometimes it is difficult to decide what type of person should be chosen for this purpose. An ideal subject, who has not been exposed at all to any kind of mutagen, does not exist in technically developed countries.

Some investigators are collecting control samples from subjects working in neighboring plants, or from persons belonging to the administrative staff and working in the same factories as the tested workers. Perhaps the people working in the same factory, even as administrative workers, are not the best controls, because one cannot be certain that chemical contamination is not present in the same building or in its neighborhood. Most probably the people working or living in other parts of the same city are the best possible matched controls. However, even such controls are influenced by individual sensitivity to the mutagen and different individual recall, and these factors may distort the results. For these reasons, matched controls are only a second choice.

C. Evaluation of the Real Consequences of Chromosomal Damage in Individuals

It has been already mentioned that the consequences of elevated levels of chromosomal aberrations observed in some persons exposed to chemicals are still unknown. Theoretically, an acceleration of some unfavorable processes, such as aging, onset of atherosclerosis and vascular diseases, and increased risk of cancer may be assumed. The danger of a neoplastic process is supposed to be greater in individuals with clones of aberrant cells.

Even if the tested persons with high levels of aberrant cells are healthy, we must accept this finding as an indication that these persons had been exposed to some kind of mutagen. On the basis of experience in this field, we can determine whether the mutagen to which a person had been exposed was radiation or a chemical because their effects differ at the chromosomal level. However, we cannot distinguish between the cytogenetic effects of chemicals and viruses, because their chromosomal effects are similar. The considerable interindividual variability in sensitivity to chemicals makes the proper dose estimate from the chromosomal changes impossible. Therefore, chromosomal aberrations are at present only a qualitative indicator of mutagenic effects.

D. Possible Utilization of the Results from Cytogenetic Studies in Occupational Health Service

Monitoring of individuals occupationally exposed to chemicals opens the question of what to do with individuals with highly elevated chromosomal aberrations. Keeping in mind that there may be a heritable increase of individual sensitivity to mutagens, we should ascertain whether one or more members of the group tested have increased levels of chromosomal aberrations. If it appears that chromosomal aberrations have increased, exposure must be reduced, because these individuals may also have other types of genetic damage (e.g., point mutations in somatic and germinal cells and chromosomal aberrations in germinal cells). Workers should be advised not to plan to conceive a child within 1 year for women or within the next 3 months for men.

Results of Hansteen et al. (1978) and Anderson et al. (1980) have demonstrated that chromosomal changes are reversible, and that chromosomal aberrations may revert to control levels in 2–3 years after reduction of the exposure to a mutagen. One should keep in mind that the process of replacement of aberrant cells by normal cells is slow (usually 1–2% per year according to our own experience at the genetic counseling center). Therefore, if very high levels of aberrant cells are observed, the return to normal levels will be slower. Lymphocytes, which are the most common object of our testing, divide very rarely in the human body so that the replacement process is slow; most probably the replacement of cells in other tissues proceeds more rapidly.

IV. CONCLUSIONS: SUMMARY OF CURRENT KNOWLEDGE AND FUTURE TRENDS

During the last 10 years cytogenetic monitoring of workers occupationally exposed to chemicals has proved to be a useful tool for detecting the chemical mutagenic effects. Cytogenetic analysis of human chromosomes in peripheral lymphocytes allows direct detection of fresh mutations in somatic cells. Any individual with higher chromosomal aberration levels has been either exposed to unusually high doses of a mutagen, or is more susceptible to the effect of that mutagen. In either case, steps should be taken to reduce exposure or to eliminate the mutagen from the environment of the tested person.

Conventional and harlequin techniques have proved to be the most convenient methods for cytogenetic monitoring of people at risk. Considerable interindividual variability in sensitivity to chemical mutagens is the reason why the subjects' own controls are preferable in these studies. If the own controls are not available, controls matched for age and sex should be used.

The harlequin technique is more sensitive for detecting the mutagenic effects of some chemicals, but this does not hold for all chemical mutagens. The classical conventional technique has shown that chromatid breaks are the most common aberration after chemical exposure, while the second most common are chromosome breaks; chromatid and chromosome exchanges are very rare.

Theoretically, chromosomal aberrations in lymphocytes may serve as a screening indicator that carriers of these aberrations are in danger also of possible premature aging, carcinogenicity, and increased genetic pathology in the progeny. Therefore, a systematic cytogenetic control of persons occupationally exposed to chemicals is recommended.

What is the possible future development of cytogenetic monitoring in this field? We may expect that other, even more sensitive cytogenetic methods will be developed and that further progress in automatic analyzing of chromosomal aberrations and/or SCEs will be made. There is still so much to be done along these two lines that I do not think we shall be able in the foreseeable future to put such new techniques to practical use. In the meantime, a systematic analysis of large groups of occupationally exposed individuals by methods routinely used, conventional and harlequin techniques, in combination with the analysis of the progeny of exposed persons are most desirable. The more we can control the people and the chemicals in this way, the fewer will be the questions to be answered in this field of mutagenic research. The present state of knowledge is still not sufficient to allow us to propose wide, reasonable preventive measures for the regulation of human exposure.

REFERENCES

Anderson, D., Richardson, C. R., Wight, T. M., Purchase, I. F. H., and Adams, W. G. F. (1980). Chromosomal analyses in vinyl chloride exposed workers. Results from analysis 18 and 42 months after an initial sampling. *Mutat. Res.* **79**, 151–162.

Bartsch, H., and Montesano, R. (1975). Mutagenic and carcinogenic effects of vinyl chloride. *Mutat. Res.* **32**, 93–114.

Bauchinger, M. (1978). Chromosome aberrations in human lymphocytes as a quantitative indication of radiation exposure. In "Mutagen-Induced Chromosome Damage in Man" (H. J. Evans and D. C. Lloyd, eds.), pp. 9–14. Edinburgh University Press, Edinburgh, Scotland.

Berg, K. (1979). Inherited variation in susceptibility and resistance to environmental agents. In "Genetic Damage in Man Caused by Environmental Agents" (K. Berg, ed.), pp. 1–25. Academic Press, New York.

Buckton, K. E., Jacobs, P. A., Court-Brown, W. M., and Doll, R. (1962). A study of the chromosome damage persisting after X-ray therapy for ankylosing spondylitis. *Lancet* **I**, 676–680.

Buckton, K. E., Langlands, A. O., Smith, P. G., Woodcock, G. A., Looby, P. C., and McLelland, J. (1971). Further studies on chromosome aberrations production after whole-body irradiation in man. *Int. J. Radiat. Biol.* **19**, 369–378.

Buckton, K. E., Hamilton, G. E., Paton, L., and Langlands, A. O. (1978). Chromosome aberrations in irradiated ankylosing spondylitis patients. In "Mutagen-Induced Chromosome Damage in Man" (H. J. Evans and D. C. Lloyd, eds.), pp. 142–151. Edinburgh University Press, Edinburgh, Scotland.

Caspersson, T., Zach, Z., and Johanson, C. (1970). Analysis of human metaphase chromosomes set by aid of DNA-binding fluorescent agents. *Exp. Cell Res.* **62**, 690–692.

Castleman, K. R., Melnyk, J., Frieden, H. J., Persinger, G. W., and Wall, R. J. (1976). Karyotype analysis by computer and its application to mutagenicity testing of chemicals. *Mutat. Res.* **41**, 153–162.

Cleaver, J. E. (1977). DNA repair processes and their impairment in some human diseases. In "Progress in Genetic Toxicology" (D. Scott, B. A. Bridges and F. H. Sobels, eds.), pp. 29–43. Elsevier, Amsterdam.

Cohen, M. M., and Bloom, A. D. (1971). Monitoring for chromosomal abnormality in man. In "Monitoring, Birth Defects and Environment: The Problem of Surveillance" (E. B. Hook, D. T. Janerich, and I. H. Porter, eds.), pp. 249–274. Academic Press, New York.

Creech, J. L., and Johnson, M. N. (1974). Angiosarkoma of the liver in the manufacture of polyvinyl chloride. *J. Occup. Med.* **16**, 150–151.

Deknudt, Gh., Manuel, Y., and Cerber, G. B. (1977). Chromosomal aberrations in workers professionally exposed to lead. *J. Toxicol. Environ. Health* **3**, 885–891.

Dolphin, G. W. (1978). A review of *in vitro* dose-effect relationship. *In* "Mutagen-Induced Chromosome Damage in Man" (H. J. Evans and D. C. Lloyd, eds.), pp. 1–9. Edinburgh University Press, Edinburgh, Scotland.

Drake, J. W. (1980). Future directions in environmental mutagenesis: Pitfalls and opportunities. *In* "Progress in Environmental Mutagenesis" (M. Alačević, ed.), pp. 333–339. Elsevier, Amsterdam.

Ducatman, A., Hirschhorn, K., and Selikoff, J. J. (1975). Vinyl chloride exposure and human chromosome aberrations. *Mutat. Res.* **31**, 163–168.

Evans, H. J. (1977). Molecular mechanism in the induction of chromosome aberrations. *In* "Progress in Genetic Toxicology" (D. Scott, B. A. Bridges, and F. H. Sobels, eds.), pp. 57–77. Elsevier, Amsterdam.

Evans, H. J. (1980). How effects of chemicals might differ from those of radiation in giving rise to genetic ill-health in man. *In* "Progress in Environmental Mutagenesis" (M. Alačević, ed.), pp. 3–23. Elsevier, Amsterdam.

Forni, A. M., Cappellini, A., Pacifico, E., and Vigliani, E. C. (1971). Chromosome changes and their evolution in subjects with past exposure to benzene. *Arch. Environ. Health* **23**, 385–391.

Funes-Cravioto, F., Lambert, B., Lindsten, J., Ehrenberg, L., Natarajan, A. T., and Osterman-Golkar, S. (1975). Chromosome aberrations in workers exposed to vinyl chloride. *Lancet* **I**, 459.

Funes-Cravioto, F., Zapata-Gayon, C., Kolmodin-Hedman, I. B., Lambert, B., Lindsten, J., Norberg, E., Nordenskjöld, M., Olin, R., and Swensson, A. (1977). Chromosome aberrations and sister chromatid exchanges in workers in chemical laboratories and roto-printing factory and in children of women laboratory workers. *Lancet* **II**, 322–325.

Gebhart, E., Lösing, J., Windolph, B., and Wopfner, F. (1980). Chromosome- and SCE-studies in patients with cytostatic interval therapy. Abstract, EEMS 9th Annual Meeting, Tučepi-Makarska, 1979. *Mutat. Res.* **74**, 193.

Hansteen, I. (1977). Chromosome breakage frequency and sister chromatid exchanges in PVC workers two years after exposure. A follow-up study. Abstract, Helsinki Chromosome Conference, 1977.

Hansteen, I., Hillestad, L., Thiis-Evensen, E., and Heldaas, S. S. (1978). Effects of vinyl chloride in man—A cytogenetic follow-up study. *Mutat. Res.* **51**, 271–278.

Hungerford, D. A. (1965). Leucocytes cultured from small inocula of whole blood and preparation of metaphase chromosomes by treatment with hypotonic KCl. *Stain Technol.* **40**, 333–338.

Kilian, J., and Picciano, D. J. (1979). Monitoring chromosomal damage in exposed industrial populations. *In* "Genetic Damage in Man Caused by Environmental Agents" (K. Berg, ed.), pp. 101–115. Academic Press, New York.

Kram, D., Schneider, E. L., Senula, G. C., and Nakamishi, Y. (1979). Spontaneous and mitomycin C induced sister-chromatid-exchanges. *Mutat. Res.* **60**, 339–347.

Kučerová, M. (1976). Cytogenetic analysis of human chromosomes and its value for the estimation of genetic risk. *Mutat. Res.* **41**, 123–130.

Kučerová, M., and Polívková, Z. (1976). Banding techniques and the detection of chromosomal aberrations induced by radiation and alkylating agents TEPA and epichlorhydrin. *Mutat. Res.* **34**, 279–290.

Kučerová, M., and Polívková, Z. (1977). Mutagenic effects of different mutagens on human chromosomes *in vitro* as detected by conventional and "Harlequin" methods. *In* "Progress in Genetic Toxicology" (D. Scott, B. A. Bridges, and F. H. Sobels, eds.), pp. 319–325. Elsevier, Amsterdam.

Kučerová, M., and Polívková, Z. (1978). Cell sensitivity to physical and chemical mutagens. *In* "Mutagen-induced Chromosome Damage in Man" (H. J. Evans and D. C. Lloyd, eds.), pp. 185–191. Edinburgh University Press.

Kučerová, M., Anderson, A. J. B., Buckton, K. E., and Evans, H. J. (1972). X-ray-induced chromosome aberrations in human peripheral blood leucocytes: The response to low levels of exposure *in vitro*. *Int. J. Radiat. Biol.* **21,** 389–396.

Kučerová, M., Polívková, Z., Šrám, R., and Matoušek, V. (1976a). Mutagenic effect of epichlorhydrin. I. Testing on human lymphocytes *in vitro* in comparison with TEPA. *Mutat. Res.* **34,** 271–278.

Kučerová, M., Polívková, Z., and Hradcová, L. (1976b). Influence of diagnostic roentgen doses on human chromosomes and influence of age on the aberration yield. *Acta Radiol. Ther. Phys. Biol.* **15,** 91–96.

Kučerová, M., Zhurkov, V. S., Polívková, Z., and Ivanova, J. E. (1977). Mutagenic effect of epichlorhydrin. II. Analysis of chromosomal aberrations in lymphocytes of persons occupationally exposed to epichlorhydrin. *Mutat. Res.* **48,** 355–360.

Kučerová, M., Polívková, Z., and Bátora, J. (1979). Comparative evaluation of the frequency of chromosomal aberrations and the SCE numbers in peripheral lymphocytes of workers occupationally exposed to vinyl chloride monomer. *Mutat. Res.* **67,** 97–100.

Kučerová, M., Polívková, Z., Reneltová, I., Kočandrle, V., and Smetanová, J. (1980). Dynamics of chromosomal aberrations and SCE in patients treated with Imuran and Cyclophosphamide. Abstract, EEMS 9th Annual Meeting, Tučepi-Makarska, 1979. *Mutat. Res.* **74,** 173.

Latt, S. A. (1974). Sister chromatid exchange indices of human chromosome damage and repair. Detection by fluorescence and induction by mitomycin C. *Proc. Natl. Acad. Sci. U.S.A.* **71,** 3162–3165.

Mason, D., and Rutovitz, D. (1978). The economics of automatic aberration scoring. *In* "Mutagen-Induced Chromosome Damage in Man" (H. J. Evans and D. C. Lloyd, eds.), pp. 339–345. Edinburgh University Press, Edinburgh, Scotland.

Meretoja, T., Vainio, H., Sorsa, M., and Härkönen, H. (1977). Occupational styrene exposure and chromosomal aberrations. *Mutat. Res.* **56,** 193–197.

Meretoja, T., Järventaus, H., Sorsa, M., and Vainio, H. (1978). Chromosome aberrations in lymphocytes of workers exposed to styrene. *Scand. J. Work. Environ. Health Suppl. 2,* **4,** 259–264.

Natarajan, A. T., Van Buul, P. P. W., and Raposa, T. (1978). An evaluation of the use of peripheral blood lymphocyte systems for assessing cytological effects induced *in vivo* by chemical mutagens. *In* "Mutagen-Induced Chromosome Damage in Man" (H. J. Evans and D. C. Lloyd, eds.), pp. 268–274. Edinburgh University Press, Edinburgh, Scotland.

Nevstad, N. P. (1977). Sister chromatid exchanges and chromosome aberrations induced in human lymphocytes by the cytostatic adriamycin *in vivo* and *in vitro*. Abstract, Helsinki Chromosome Conference, 1977.

Nordenson, I., Beckman, G., Beckman, L., and Nordenson, S. (1978a). Occupational and environmental risk in and around a smelter in northern Sweden. II. Chromosomal aberrations in workers exposed to arsenic. *Hereditas* **88,** 47–50.

Nordenson, I., Beckman, G., Beckman, L., and Nordström, S. (1978b). Occupational and environmental risk in and around a smelter in northern Sweden. IV. Chromosomal aberrations in workers exposed to lead. *Hereditas* **88**, 263–267.

Nordenson, I., Beckman, G., Beckman, L., Rosenthal, N. and Stjenberg, N. (1980a). Is exposure to dioxide clastogen? Chromosomal aberrations among workers at a sulphite pulp factory. *Hereditas* **93**, 161–164.

Nordenson, I., Beckman, G., Beckman, L., and Lemperg, R. (1980b). Chromosomal aberrations in children exposed to diagnostic X-rays. *Hereditas* **93**, 177–179.

Nordström, S., Beckman, L., and Nordenson, I. (1978a). Occupational and environmental risk in and around a smelter in northern Sweden. I. Variations in birth weight. *Hereditas* **88**, 43–46.

Nordström, S., Beckman, L., and Nordenson, I. (1978b). Occupational and environmental risk in and around a smelter in northern Sweden. III. Frequencies of spontaneous abortions. *Hereditas* **88**, 51–54.

Nordström, S., Beckman, G., and Nordenson, I. (1979a). Occupational and environmental risk in and around a smelter in northern Sweden. V. Spontaneous abortion among female employees and decreased birth weight in their offspring. *Hereditas* **90**, 291–296.

Nordström, S., Beckman, L., and Nordenson, I. (1979b). Occupational and environmental risk in and around a smelter in northern Sweden. VI. Congenital malformations. *Hereditas* **90**, 297–302.

Norppa, H., Sorsa, M., Pfäffli, P., and Vainio, H. (1980). Styrene and styrene oxide induced SCEs and are metabolised in human lymphocyte cultures. *Carcinogenesis* **1**, 357–361.

Perry, P., and Evans, H. J. (1975). Cytological detection of mutagen-carcinogen exposure by sister chromatid exchange. *Nature (London)* **258**, 121–125.

Picciano, D. (1979). Cytogenetic investigation of occupational exposure to epichlorhydrin. *Mutat. Res.* **66**, 169–173.

Pilinskaja, M. A. (1970). Chromosome aberrations in the persons contacted with Ziram (in Russian). *Genetika* **6**, 157–163.

Pilinskaja, M. A. (1974). Results of cytogenetic examination of persons occupationally contacted with fungicide Zineb (in Russian). *Genetika* **5**, 140–146.

Purchase, I. F. H. (1978). Chromosomal analysis of exposed populations: A review of industrial problems. In "Mutagen-Induced Chromosome Damage in Man" (H. J. Evans and D. C. Lloyd, eds.), pp. 258–268. Edinburgh University Press, Edinburgh, Scotland.

Purchase, I. F. H., Richardson, C. R., and Anderson, D. (1976). Chromosomal effects in peripheral lymphocytes. *Proc. R. Soc. Med.* **69**, 290–291.

Purchase, I. F. H., Richardson, C. R., Anderson, D., Paddle, G. M., and Adams, W. G. F. (1978). Chromosomal analysis in vinyl chloride exposed workers. *Mutat. Res.* **57**, 325–334.

Rapoport, I. S. (1948). Effect of ethylene oxide, glycidol, and glycols on gene mutations (in Russian). *Bull. Acad. Sci. USSR* **60**, 469–472.

Scott, D., and Lyons, C. Y. (1979). Homogeneous sensitivity of human peripheral blood lymphocytes to radiation-induced chromosome damage. *Nature (London)* **278**, 756–758.

Seabright, M. (1971). A rapid banding technique for human chromosomes. *Lancet* **I**, 971–972.

Sofuni, T., Shimba, H., Okhtaki, K., and Awa, A. A. (1978). A cytogenetic study of Hiroshima atomic-bomb survivors. In "Mutagen-Induced Chromosome Damage in

Man" (H. J. Evans and D. C. Lloyd, eds.), pp. 108–115. Edinburgh University Press, Edinburgh, Scotland.

Sorsa, M. (1979). Cytogenetic monitoring of persons at risk. Abstract, Finnish-Soviet Symposium on Longterm Effects in Occupational Health, Helsinki, 1979.

Sorsa, M. (1981). Chromosomal aberrations as a monitoring method for possible genotoxic risks. In "Biological Monitoring and Health Surveillance of Workers Exposed to Chemicals" (A. Aition, V. Riihimäbi and H. Vainio, eds.), Hemisphere Publ. Co., Washington D.C.

Sorsa, M., Hyvönen, M., Järventaus, H., and Vainio, H. (1979). Chromosomal aberrations and sister chromatid exchange in children of women reinforced plastic industry. Abstract, Eur. Toxicol. Symp., Roma, 1979.

Sumner, A. T., Evans, H. J., and Buckland, A. A. (1971). New technique for distinguishing between human chromosomes. Nature (London) 232, 31–32.

Šrám, R., Zudová, Z., and Kuleshov, N. P. (1980). Cytogenetic analysis of peripheral lymphocytes in workers occupationally exposed to epichlorhydrin. Mutat. Res. 70, 115–120.

Tough, I. M., and Court-Brown, W. M. (1965). Chromosome aberrations and exposure to ambient benzene. Lancet I, 684.

Tough, I. M., Buckton, K. E., Baikie, A. G., and Court-Brown, W. M. (1960). X-ray induced chromosome damage in man. Lancet II, 849.

Uggla, A. H., Andersson, H. C., Tranberg, E. A., and Zetterberg, L. G. (1980). Correlation between exposure to styrene and the frequency of chromosomal aberrations and sister chromatid exchanges in lymphocytes of workers in a plastic boat factory. Abstract, EEMS 9th Annual Meeting, Tučepi-Makarska, 1979. Mutat. Res. 74, 199.

Viola, P. L., Biogetti, A., and Caputo, A. (1971). Oncogenic response of rat skin, lungs, and bones to vinyl chloride. Cancer Res. 31, 516–522.

Vogel, F. (1978). Genetic aspects of induced mutation. Hum. Genet. Suppl. 1, 141–147.

Wolff, S., and Perry, P. (1974). New Giemsa method for the differential staining of sister chromatids. Nature (London) 251, 156–158.

Yoder, J., Watson, M., and Benson, W. W. (1973). Lymphocyte chromosome analysis of agricultural workers during extensive occupational exposure to pesticides. Mutat. Res. 21, 335–340.

Zavala, C., Arroyo, P., Lisker, R., Carnevale, A., Salamanca, F., Navarrete, J. I., Jiménez, F. M., Blanco, B., Vázquez, V., Sánchez, J., and Canún, S. (1979). Variability between and within laboratories in the analysis of structural chromosomal abnormalities. Clin. Genet. 15, 377–381.

Chapter 10

The Rationale and Methodology for Quantifying Sister Chromatid Exchange in Humans

Anthony V. Carrano and Dan H. Moore II

267

MUTAGENICITY:
NEW HORIZONS IN GENETIC TOXICOLOGY

I. RATIONALE FOR THE ANALYSIS OF SISTER CHROMATID EXCHANGE IN HUMANS

Man has continually been exposed to a variety of natural or synthetic substances. These have taken the form of physical agents such as ultraviolet light (UV), cosmic rays, and heat, as well as chemicals present in the diet or in the environment. We are well aware today of the ability of such agents to modify the genetic material, and thus it is reasonable to expect that they played a significant role in the induction of mutation and the process of evolution. In addition, the growth of technologies, industrialization, and the desire for a more comfortable lifestyle have been associated with the production, use, and exposure to a wide spectrum of new substances. It is obvious that there is a need both to identify any potential genetic toxicity of these substances and to estimate their biological impact on man. Identification is a role well suited to *in vitro* and animal bioassays while risk estimation has been traditionally accomplished by epidemiological methods. The two methods should not be mutually exclusive. If we accept the premise that *in vitro* or animal bioassays cannot accurately predict the unique response of an individual to an environmental insult, an optimal solution would be to couple multiple human bioassays with the longer term epidemiology. This chapter will focus on one of these human bioassays, the analysis of sister chromatid exchange (SCE) in peripheral blood lymphocytes.

A. The Significance of Sister Chromatid Exchange

A sister chromatid exchange is the cytologiocal manifestation of a four-strand exchange in the DNA. This has been clearly demonstrated by Taylor using tritium-labeled DNA (1958) and later by the higher resolution fluorescence plus Giemsa (FPG) technique (Perry and Wolff, 1974) in *Vicia faba* (Kihlman and Kronberg, 1975), Chinese hamster (Wolff and Perry, 1975), and human (Wolff *et al.*, 1975) cells. Thus the formation of an SCE involves breakage of the genetic material and subsequent recombination of the four DNA strands. Mechanisms of SCE have been postulated which involve either a type of recombinational repair (Bender *et al.*, 1974), altered replication of DNA past damaged bases (Shafer, 1977), or asynchrony in DNA synthesis between adjacent replicon clusters (Painter, 1980a), but experimental evidence to conclusively demonstrate any specific mechanism is lacking. Even though the mechanism(s) of SCE formation has not yet been established, the importance of SCE as a measure of exposure to potential mutagens or carcinogens or as an indicator of abnormalities in the maintenance of cellular homeostasis should not be diminished. Elevated SCE frequencies can therefore be produced in response to environmentally induced DNA damage or can arise "spontaneously" (i.e., in the absence of an external inducing agent) as occurs in cells from patients with Bloom's syndrome (Chaganti et al., 1974). These two possibilities should be clearly distinguished since the mechanisms involved and the biological significance may be different. We will concern ourselves here only with SCEs induced directly or indirectly by environmental agents.

The evidence implicating the SCE as having biological significance derives from in vitro studies with known mutagens and carcinogens. Perry and Evans (1975) clearly demonstrated in Chinese hamster ovary cells (CHO) that SCEs are induced by many mutagens and that the induction is dose dependent. This work was amplified by several additional investigations using a variety of physical and chemical agents (Latt, 1974; Abe and Sasaki, 1977; Banerjee and Benedict, 1979; see also reviews by Wolff, 1977; Perry, 1980). Together these studies have provided information as to the relative efficiency of different DNA lesions for producing SCEs. In general, those substances that form covalent adducts to the DNA or otherwise distort the DNA bases are relatively good inducers of SCEs. Ultraviolet light and mono- and bifunctional alkylating agents such as the alkyl sulfates, nitrosoureas, as well as the reactive forms of the larger aromatic hydrocarbons are examples. Substances that do not induce SCE as efficiently include those agents that are capable of directly breaking the DNA backbone. Ionizing radiation, heavy metals, and other substances fall into this category. This group of agents is generally efficient at producing a different cytogenetic end point, namely, chromosomal aberrations. Because it might not always be possible to predict whether a chemical will be more efficient as an inducer of SCEs or chromosomal aberrations in humans, the analysis of SCE should not be used as the only cytogenetic end point in a monitoring program but rather should be coupled to the measurement of chromosomal aberrations in the same individuals.

Recent investigations to determine the biological significance of SCE have suggested a formal relation between SCE and single gene mutation for a variety of inducing agents. Carrano et al. (1978) treated Chinese hamster cells in vitro with four mutagens, each of which formed different types of DNA lesions and demonstrated a linear relation between induced SCE and induced gene mutation. The relationship was also found to hold when other chemicals, ionizing radiation, and UV light were studied (Carrano et al., 1979; Carrano and Thompson, 1981). It is important to note that the ratio of the number of SCEs to gene mutations induced was not the same for all substances but was dependent on the agent, i.e., the nature of the lesion. Other investigators have reported similar results in different cell systems and with additional chemicals (Sirianni and Huang, 1980; Okinaka et al., 1981). Qualifications to this relationship have been suggested (Bradley et al., 1979; Connell, 1979) and were further discussed in terms of alternative interpretations (Carrano and Thompson, 1981).

The correlation between induced SCE and mutation must be carefully interpreted. The presence of an SCE does not necessarily imply the presence of a mutation. In fact, the above studies suggest that the mutagenic lesions may be a subset of the many types of lesions that elicit SCEs. Further, the ratio of SCE to mutation can be modified by the duration of exposure, intervening repair time between exposure and analysis of the two endpoints, and cellular processing of the DNA damage. At the risk of generalization, it should be pointed out that with few, if any exceptions, the ability of an agent to induce mutations in an in vitro system is accompanied by the ability of the agent to induce SCEs in the same system. It may be possible, however, to elicit elevated SCEs in response to an agent without an accompanying increase in mutation. This might be attributed to the insensitivity of the particular mutagenicity assay for the agent and/or the ability of SCEs to respond to a broader spectrum of lesions.

B. The Lymphocyte as a Cell for Sister Chromatid Exchange Analysis

A cell type that is used for the analysis of sister chromatid exchange in human populations should be (1) readily accessible; (2) capable of undergoing proliferation in vitro in order that the DNA can be substituted with bromodeoxyuridine (BUdR) for visualization of the SCEs; and (3) representative for exposure over any portion of the body. No single cell type can fulfill all these criteria but the circulating peripheral blood lymphocyte offers some unique advantages. Small amounts of blood can be easily and repeatedly obtained by venipuncture from test individuals and the methods for the growth of lymphocytes following mitogen stimulation in vitro are routine. As the carriers of nutrients, metabolites, and other chemicals to and from the cells of the body, the peripheral blood circulates to every organ. Exposure of the blood cells to a toxic agent localized in the initially exposed cells is therefore dependent on and potentially limited by the transport of the agent itself or its metabolites from the target cells to the blood. The pharmacology of many chemicals is known and for these it is possible to estimate blood versus tissue distributions and hence doses. Often it is possible to directly measure blood levels of the toxic chemical. Thus, with the possible exception of external exposures to the skin, the peripheral blood lymphocyte could serve as a useful indicator of exposure. For studies designed to measure SCEs in individuals over long periods of time, the presence of long-lived lymphocyte subpopulations can be an advantage. Buckton *et al.* (1978) have measured the loss of dicentric chromosomes in lymphocytes of patients after radiation therapy. The kinetics of dicentric loss were biphasic; a moderate initial rate (approximately 43% per year for the first 3 or 4 years) was followed by a slow rate of loss (approximately 14% per year up to 20 years). This result suggests a very long-lived component of the lymphocyte population. If lesions that are capable of forming SCEs can persist in the nondividing lymphocyte *in vivo*, then the long-lived lymphocyte may also be able to accumulate damage following long-term chronic or repeated exposures or may serve as an indicator of acute exposure for some time after the damaging event.

There is evidence to support the concept of persistent SCE-forming lesions both from animal and human exposures. When rabbits are given a single intraperitoneal injection of a known mutagen, the frequency of SCE observed in peripheral blood lymphocytes, following *in vitro* culture, is increased within 24 hours. Subsequently, the SCE frequency decreases and gradually returns to normal (Stetka and Wolff, 1976). This increased SCE frequency observed in lymphocytes shortly after *in vivo* exposure is attributed to the induction of lesions *in vivo*, persistence of the lesions in the lymphocyte at least up to the first S phase in culture and their ability to cause the formation of SCEs *in vitro*. The observation of SCEs *in vitro* after *in vivo* exposure should, however, be viewed with some caution. As pointed out by Littlefield *et al.* (1980) and Dufrain *et al.* (1980), it is possible that the chemical administered *in vivo* may be present in the serum of the animal or human and carried into the lymphocyte culture. If this were the case, then the chemical could act on the DNA to form SCEs when the lymphocytes reach S phase *in vitro* and lead to an erroneous interpretation that genetic damage was induced *in vivo*. On the other hand, it might be envisioned that some chemicals produce lesions which can be efficiently repaired in the mitogen stimulated lymphocyte *in vitro* but not in the circulating lymphocyte. In this case, a small or even no increase in the level of SCEs may be observed leading to the

erroneous conclusion that no DNA damage occurred *in vivo*. Littlefield
et al. (1979a,b) have suggested that some lesions can be repaired prior
to S phase following treatment of the lymphocyte *in vitro*. If the possibility of such events exists, appropriate *in vitro* experiments can and
should be designed to assist in the interpretation of the *in vivo* results.

The loss of SCE-forming lesions in the circulating lymphocyte *in
vivo* may be attributed to at least two factors: (1) the repair of the
lesions; and/or (2) dilution of the damaged cell population through
cell death or by recruitment of new lymphocytes into the circulating
pool. The increase in SCEs and the subsequent loss of SCE-forming
lesions in circulating lymphocytes following *in vivo* exposure has also
been observed in patients undergoing chemotherapy (Lambert *et al.*,
1978a; Raposa, 1978; Nevstad, 1978; Musilova *et al.*, 1979; Otter *et
al.*, 1979; Ohtsuru *et al.*, 1980). These studies demonstrate that the
rate of decrease in SCE-forming lesions in the circulating lymphocyte
seems to be dependent on the inducing agent, the method of patient
treatment (i.e., single or multiple exposures), and the unique response
of each patient. The persistence of the lesions can be highly variable.
For example, for two patients treated with a single dose of mitomycin
C, the frequency of SCE in the peripheral lymphocytes was observed
to return to normal within two days (Ohtsuru *et al.*, 1980), however,
increased SCE frequencies have been observed up to 4 months after
treatment with cyclophosphamide (Musilova *et al.*, 1979) or 1-(2-chlo-
roethyl)-3-cyclohexyl-l-nitrosourea (Lambert *et al.*, 1979). The implication of these findings for human population studies is obvious. In
the case of known exposures to suspect mutagens or DNA-damaging
agents, it is important to obtain the blood sample as soon as possible
following exposure in order to optimize the detection of a positive
SCE response. Failure to achieve an increased SCE frequency in children treated with chemotherapy for malignant lymphoma has been
attributed to the long time interval between cessation of therapy and
acquisition of blood for cytogenetic analysis (Haglund *et al.*, 1980).
In this case, the DNA-damaging effect of the therapeutic agents could
be documented by the increased frequency of stable aberrations, i.e.,
translocations, in the same children.

Because the long-lived lymphocyte may be able to integrate damage
over its lifetime, we asked whether repeated treatment of rabbits with
low doses of mitomycin C would produce a measureable persistence
of SCE-forming lesions in the circulating lymphocytes (Stetka *et al.*,
1978). Measureable increases in SCE frequency could be observed 6
months after the last mitomycin-C treatment when the experiment was
terminated. Although the mean SCE frequency for each animal was

elevated, more significant was the distribution of SCEs among the lymphocytes examined. As the time interval from the last injection increased, the proportion of cells with an increased SCE frequency gradually diminished and the distribution became dominated by cells with a normal SCE frequency. About 5–10% of the lymphocytes, however, still possessed a high SCE frequency at 6 months posttreatment. This suggested that an important parameter to quantify after exposure may be the proportion of cells with a high SCE frequency, i.e., high-frequency cells (HFCs). Whether these HFCs are due to accumulated persistent damage in the long-lived lymphocyte is only conjectural but their presence has been noted in humans as well (Lezana et al., 1977; Lambert et al., 1978b; Raposa, 1978; Kowalczyk, 1980). This chapter details our efforts to quantify the frequency of HFCs in nonexposed individuals and in cigarette smokers, to describe them statistically, and to justify their consideration in conjunction with the overall mean SCE frequency for human population studies.

II. METHODOLOGICAL ASPECTS OF SISTER CHROMATID EXCHANGE ANALYSIS IN HUMAN POPULATIONS

Before we can interpret the significance of induced SCE in human populations, an understanding of normal human variation is essential. The factors influencing the measurement of a baseline SCE level fall into two categories: (1) inherent factors associated with the growth and culture of lymphocytes; and (2) external factors associated with the genotype, lifestyle, or physiological state of the individual. Although the distinction is often nebulous, variation associated with the former is, to a limited extent, controllable while variation associated with the latter generally is not. The objective then is to quantify SCE induction above an existing noisy baseline.

A. Variation in Sister Chromatid Exchange Frequency Associated with the Lymphocyte Culture

The reported baseline frequency of SCE in peripheral blood lymphocytes of presumably nonexposed individuals ranges from about 1.4 to 45 SCE per cell (Lambert et al., 1976; Alhadeff and Cohen, 1976; Crossen et al., 1977; Galloway, 1977). Even within the same laboratory the frequency of SCE for nonexposed individuals may vary by a factor

of two or more. We recently completed a study to determine the preparative and analytical factors that might influence the measurement
of SCE in normal individuals (Carrano et al., 1980). Several sources
of variation were quantified and a few of these are summarized below.

The major source of inherent variation in baseline SCE frequencies
can be attributed to the amount of BUdR present in the culture medium
relative to the number of lymphocytes initially added. The frequency
of SCE was demonstrated to increase from about 6 to 9 per cell over
the BUdR concentration range of 10–160 μM. The ability of BUdR to
induce SCEs has been well documented in human and other cell
systems (Wolff and Perry, 1974; Lambert et al., 1976; Latt and Juergens,
1977; Mazrimas and Stetka, 1978), and it was further shown that an
increased SCE frequency can also be achieved by decreasing the cell
number while maintaining a constant concentration of BUdR (Stetka
and Carrano, 1977). The number of lymphocytes that respond to mitogen and eventually incorporate the analogue will be highly variable
and generally not controllable, and therefore, the effect of BUdR can
best be minimized by standardizing the concentration of BUdR and
white blood cell or lymphocyte number when the culture is initiated.
Since the increases above the baseline SCE frequency following most
chronic human exposure might be expected to be small (i.e., of the
order of the BUdR effect), careful consideration of the BUdR effect is
warranted. The variation associated with the analysis of replicate cultures, replicate slides, and with cell-to-cell differences within the same
individual was also quantified but these demonstrated less variability
than is attributed to the BUdR effect. In fact, the cell-to-cell variation
can be reduced by one-half if the number of cells scored per person
is increased from 20 to 80.

Other inherent factors have also been considered as possibly influencing the measurement of SCEs in human lymphocytes. A concern
of most investigators is the delineation of an optimal time to harvest
the cells for chromosome analysis. At least three investigations have
concluded that the baseline frequency of SCE is not dependent on the
time of harvest. Carrano et al. (1980) found that for any individual
the SCE frequency could vary as much as 30% as a function of culture
duration but that the pattern of variation was not consistent nor was
it demonstrated for all individuals. Beek and Obe (1979), Becher et
al. (1979), and Morimoto and Wolff (1980) also did not observe any
significant alteration in the baseline SCE frequency as a function of
culture time. This finding did not hold, however, for SCEs induced
by the alkylating agent trenimon (Beek and Obe, 1979; Riedel and Obe,
1980). Snope and Rary (1979) reported that, for four individuals, the

baseline SCE frequency observed in 58-hour cultures was about 30% less than that in 70-hour cultures. They considered this difference to be significant. One explanation that has been proposed to account for the change in SCE frequency is the presence of lymphocyte subpopulations with different sensitivities to mutagens. Alternatively, the subpopulations might respond to mitogen stimulation at different times or possess different cell cycle times such that they would be observed at different culture times. Although there have been no definitive experiments to resolve these possibilities, there is some evidence that the T and B lymphocytes differ in the baseline SCE frequency (Santesson et al., 1979). These investigators demonstrated that for several concentrations of BUdR B lymphocytes have a significantly lower SCE frequency relative to the T lymphocytes. Lezana et al. (1977) suggested that lymphocyte subpopulations might also account for their observed bimodal distribution of SCEs among cells of normal and Down's syndrome individuals. Further work will be needed to identify these populations and to determine whether the culture time differences correspond to T, B, or other lymphocyte classes, or are a sampling artifact.

Two additional factors have been demonstrated to influence the baseline frequency of SCE: the sera used in the culture medium and the culture temperature. Using sera from four different sources, McFee and Sherrill (1981) showed that the baseline SCE frequency of swine lymphocytes was different for each serum source and that for two sera sources the SCE frequency increased up to 50% with increasing serum in the medium. The authors concluded that some commercially available sera contain a factor that causes SCE. Kato and Sandberg (1977) also demonstrated an effect of sera on the production of SCE. The frequency of SCE has been shown to increase almost fivefold with increasing culture temperature for cells of the amphibian, *Xenopus laevis* (Speit, 1980). For Chinese hamster cells the situation is more complex. Both lower (33°C) and higher (40°C) temperatures increased the frequency of SCEs approximately 60% for several concentrations of BUdR. Because the altered temperature was only effective in altering SCE frequencies while BUdR was being incorporated, the author concluded that the SCEs might arise via a temperature-dependent disturbance of DNA replication.

Taken together, the experience accumulated from the use of lymphocyte cultures suggests that there are many inherent factors that could influence the frequency of SCE. This suggests that in order to maximize the sensitivity of this technique, the use of standardized methods is essential.

B. Other Factors Affecting the Measurement of Sister Chromatid Exchanges in Human Lymphocytes

In contrast to the preparative and cell culture factors above, conditions unique to the individual being sampled may influence the SCE results. The effect of sex, age, diet, genotype, medication, and smoking potentially all play a role in the induction or expression of SCEs. Some evidence has been obtained relating to each of these.

Several investigators have compared the baseline frequency of SCEs in male versus female donors and have not observed significant differences (Galloway and Evans, 1975; Alhadeff and Cohen, 1976; Crossen et al., 1977; Latt and Juergens, 1977; Morgan and Crossen, 1977; Cheng et al., 1979). There is no a priori reason to expect a difference between sexes in normal individuals unless the cyclic hormonal variations in females influence the response of the lymphocyte to mitogen stimulation or to uptake of BUdR. A rigorously controlled study to ascertain such effects might be useful. A subset of our own data illustrating the SCE frequencies in nonexposed males and females is shown in Fig. 1. Statistical analysis indicates that the group means, variances, and distribution of means are not significantly different. The data do, however, emphasize the range of SCE frequencies found

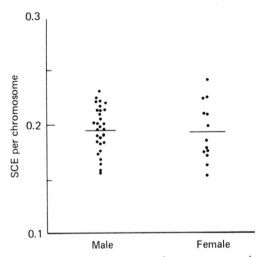

Fig. 1. The SCE frequency per chromosome for 29 nonexposed males and 13 nonexposed females. Each dot represents one individual and the group means are plotted as horizontal lines.

among normal individuals. The absence of clear documentation of a sex difference for baseline SCE frequencies suggests that this factor may not have to be matched in a population study.

Another variable of concern in population studies is the age of the test individual since it is not always possible to obtain age matched donors. The results shown in Fig. 2 indicate that the baseline SCE frequency appears to be independent of age, at least for ages 21–63 years. The lack of an age effect in lymphocytes of presumably nonexposed individuals has also been noted by other investigators (Galloway and Evans, 1975; Morgan and Crossen, 1977; Hollander et al., 1978; Cheng et al., 1979; Lambert and Lindblad, 1980), but in patients with tumors and chronic myelogenous leukemia SCE frequency declined with age (Cheng et al., 1979). The response of aging individuals to mutagen exposure may be more complex. Kato and Stich (1976) reported that the baseline SCE frequency in human diploid fibroblasts in vitro increases as the cells approach senescence, but this was not confirmed in a later study by Schneider and Monticone (1978). More significant was the finding in this latter study that the SCE frequency induced by three different alkylating agents was decreased in older relative to the younger cultures of human diploid fibroblasts. The same phenomenon, i.e., a decrease in mutagen-induced SCEs in older cell populations, was also observed in human skin fibroblasts established from young and old donors (Schneider and Gilman, 1979) suggesting that diminution of mutagen-induced SCEs with age occurs in vivo as well as in vitro. Further evidence of an aging effect in vivo was demonstrated in mouse and rat bone marrow cells (Kram et al., 1978). At moderate to high mitomycin C doses, induced-SCE frequencies were significantly reduced in bone marrow of the older animals. Whether a similar phenomenon will be found in lymphocytes of exposed humans remains to be seen but anyone conducting population studies involving older individuals should be aware of this possibility.

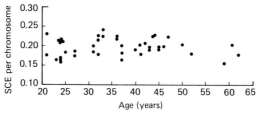

Fig. 2. The SCE frequency as a function of age for the 42 nonexposed individuals from Fig. 1. Each dot represents a single individual.

Genetic factors have been suggested by many investigators as a potential source of the variation in baseline SCE frequencies of nondiseased humans. In a study designed to identify genetic determinants for baseline SCE frequencies, Pedersen et al. (1979) examined the cultured lymphocytes from 11 monozygotic and 9 dizygotic same-sexed twin pairs. They could find no significant difference either in the total number of SCEs or their intrachromosomal distribution in monozygotic versus dizygotic twins. No statistically significant variation was found between individuals either. In fact, for single chromosomes, the variation in SCE distribution between cells was of the same order of magnitude as the variation between individuals suggesting that, if genetic factors do contribute to variation in baseline SCEs, they probably represent only a small component of the total variation. As with the studies on aging, it is not known whether the same conclusions will apply to mutagen exposed individuals. In a related study, baseline SCE frequencies were measured in normal and high-cancer-risk individuals (Cheng et al., 1979). Those individuals from cancer-prone families who were at high risk did not have increased SCEs compared to normal individuals. Thus if a genetic component for the induction of cancer exists, it does not appear to be reflected in the SCE frequency of the high-risk individual.

Other factors are worth considering as potentially affecting the measurement of human SCEs. Because a large fraction of women use oral contraceptives, care must be taken to ensure that this does not confound the results of exposure to other agents. A recent study of oral contraceptive users demonstrated a significant increase in lymphocyte SCEs associated with the daily intake of d-norgestrel or ethenyl estradiol over a period of 6–24 months (Balakrishna Murthy and Prema, 1979). Additional studies are warranted to confirm and extend these results. A study conducted by Pero et al. (1976) is very interesting. Their findings suggest that individuals with high blood pressure may be at increased risk to genetic damage following exposure to known mutagens. When lymphocytes were exposed in vitro, the amount of unscheduled DNA synthesis, the frequency of chromosomal aberrations, and the amount of carcinogen binding to DNA were greater in the lymphocytes from individuals with high blood pressure. Because hypertension is such a common disease, it would be worthwhile to pursue this finding to determine if SCE frequencies are also related to increased blood pressure.

Despite our knowledge of natural mutagens in food and the role of food processing in creating mutagens (Sugimura et al., 1977), little attention has been given to the diet as a factor in causing baseline SCE variation. Such a population study might be difficult to properly control, but religious groups with known dietary restrictions may serve as an appropriate starting point. At one extreme, it has been reported that severe protein malnutrition in children is associated with increased lymphocyte SCE frequencies (Balakrishna Murthy et al., 1980). Following nutritional rehabilitation, SCE frequencies were observed to decrease. Subsequently, it has been demonstrated that mice placed on restricted protein diets have an increased level of SCEs in the bone marrow cells (Balakrishna Murthy and Srikantia, 1981). The SCE frequency increased from 3.3 per cell at the control level (22% dietary protein) to 4.9 per cell at 6% protein, 8.5 per cell at 4% protein, and 10.0 per cell at 2% protein. The effect was seen in essentially all the animals in each group and might well be mediated through interference with DNA metabolism.

For all practical purposes, most human studies will not be this extreme. Of more concern are the potential effects associated, for example, with the intake of caffeine, saccharin, other additives, or well-cooked foods. The evidence supporting the ability of these chemicals to induce SCEs in vitro is equivocal or the observed effect is small. The attainment of conclusive data in man himself would be difficult, if not impossible. Abe and Sasaki (1977) reported a slight increase in SCE frequency in Chinese hamster cells treated with sodium saccharin that did not show a dose dependence. A dose-dependent increase was observed both in Chinese hamster cells and in human lymphocytes treated in vitro in a study by Wolff and Rodin (1978), but Saxholm et al. (1979) did not confirm the increase in human lymphocytes. Caffeine alone or in combination with other agents has been reported to produce a multitude of cytotoxic, DNA replicative, and cytogenetic effects. In a recent study, Painter (1980b) has shown that caffeine can alter the conformation of intracellular chromatin in such a way as to modify the effects of DNA damaging agents. This might explain the variety of effects observed when caffeine is given before, after, or during exposure to other substances. Chinese hamsters given caffeine directly by intubation were found to have about a 50% increase in SCEs in their bone marrow cells at the highest doses (Basler et al., 1979). Thus caffeine alone might be expected to induce SCEs in humans but the increase would probably be small. In combination with other substances, however, the response could be complex and should stimulate those doing population studies not to overlook this confounding factor.

Taken together, the preparative and external factors that can influence the baseline SCE frequency in lymphocytes are numerous, and it is not surprising that considerable variation exists among individuals. We have not yet considered the effects of a major variable, cigarette smoking, because this will be dealt with later in the chapter.

C. Methodology for Sister Chromatid Exchange Analysis

The procedures for SCE analysis in human populations will depend on the specific circumstances of the study. We present here a generalized outline of a procedure that, based on our own experience, we have found useful.

In order to identify any of the factors that may confound the analysis of SCEs, it is important to obtain relevant information on the background of each individual in the study population. This information is best obtained by means of a questionnaire. The purpose of the proposed study and the necessity for the questionnaire can be explained to several individuals at the same time. This will help to gain their confidence in the study and ensure their continued motivation. In addition to the obvious questions related to personal and familial cancer incidence, known genetic disorders, and current health, there are specific questions that relate to the interpretation of SCEs. These include (1) smoking history; (2) medication and drug use; (3) known or suspected exposures to physical or chemical agents occurring at work, at home, or during the pursuit of various hobbies or recreation; and (4) nutrition, particularly the adherence to special diets or use of caffeine and artificial sweetners. Because a complete list of agents that may directly or indirectly induce SCEs is by no means established, it is better to collect too much information rather than not enough. An abbreviated questionnaire is useful at the time the blood sample is taken if some time has elapsed since the first questionnaire was completed. As a routine practice we utilize a one-page questionnaire at the time of the blood sample to identify any medication, drug, caffeine, or smoking exposure during the previous 24 hours. Because the amount of information accumulated for each individual can be substantial, a computerized data-base management system is helpful and, for large studies, almost essential. We currently utilize the PDP 11/34A system (Digital Equipment Corp., Maynard, Mass.) for storage, sorting, and retrieval of questionnaire data.

At the time the blood is collected, it should be mixed with heparin, acid-citrate dextrose, or other appropriate anticoagulant. Our previous results (Carrano et al., 1980) suggested this mixture could be stored at 4°–37°C for several days prior to culture but this is not desirable. Lymphocyte loss, reduction of mitogenic response, and possible selective loss of SCE-forming lesions may occur during storage. Therefore, if possible, cultures should be established within 24 hours. A total white-blood-cell count and a slide for a differential count should be made from the blood sample just prior to establishing the culture. These end points can serve as indicators of abnormally high or low lymphocyte counts, factors that may have some bearing on the observed frequency of SCEs.

It is important for the analysis of SCEs that a standard lymphocyte culture procedure be employed. The exact procedure may differ among laboratories but there are some common requirements. Because BUdR can induce SCEs in human lymphocytes and, at doses normally employed for these studies, this induction can be dependent on both the BUdR concentration and the number of lymphocytes potentially incorporating the drug, these two factors should be held as constant as possible (Stetka and Carrano, 1977; Mazrimas and Stetka, 1978). In order to ensure this in our own laboratory, we have established a dose response for BUdR-induced SCEs for a sampling of eight individuals (Carrano et al., 1980). In general, this induction curve increases rapidly at low ratios of BUdR concentration to lymphocyte number and then appears to plateau or increases much more slowly at higher ratios. A combination of the appropriate number of lymphocytes and BUdR concentration that was in the middle of the plateau region of the induction curve was selected as standard; the plateau region being defined as that range of BUdR concentration per lymphocyte for which the slope of the dose–response curve is minimal. What cannot be anticipated even under these standard conditions is the individual response to the mitogen. Although the number of lymphocytes added to each culture is kept fairly constant, each individual or even the same individual at different times may have a different fraction of lymphocytes stimulated to enter the cell cycle. This may be a source of variability that will not be easily controlled but, if the plateau region is sufficiently broad, selecting a BUdR and lymphocyte concentration in the middle of this region may allow for some flexibility if the stimulated fraction varies. Selection of culture medium, serum type and concentration, whole blood versus purified lymphocyte cultures, and duration of culture is, to some extent, arbitrary and should be based on historical experience. The serum lot should be pretested for its potential to support lymphocyte growth and the reproducibility of baseline SCE frequencies. If feasible, the same serum lot should be used for each population study. The duration of the culture should be such that it yields a high percentage of second division metaphases so that the SCE frequency is representative of a major fraction of the lymphocytes. Needless to say, exposure of the cultures to other than yellow or red light should be avoided. Replicate cultures should be establishied for each individual and, when a large number of individuals is processed, the inclusion of blood cultures from people sampled longitudinally is useful as a negative control. Occasional addition of a positive control, e.g., an individual on appropriate chemotherapy, ensures the investigator of the continued responsiveness of the methods.

At the appropriate culture time, the cells can be arrested in mitosis by a brief treatment with Colcemid, colchicine, or other suitable agent. Several slides should be prepared from each culture and the remaining cell pellet stored in absolute methanol:glacial acetic acid (3:1) at 4°C or lower for future reference. The slides can be stained for analysis by fluorescence alone (Latt. 1973), fluorescence plus Giemsa (Perry and Wolff, 1974), or a reproducible variation thereof. Once a staining procecure has been standardized and adopted, it should be used at least throughout the course of the study.

Well differentiated metaphases should be accepted for scoring. The coded slides can be scanned under low magnification ($100–200 \times$) and selected for scoring on the basis of good staining and chromosome number. We found that it is not necessary to select for SCE analysis only those cells that have a complete complement of chromosomes. As a general rule, cells with 41 or more chromosomes are acceptable. The SCEs are counted for each chromosome in the cell with the final results tabulated as the number of SCEs per chromosome or converted to the number of SCEs per diploid cell. Although the precise number of cells to be scored per individual will ultimately depend on the desired sensitivity, we believe a minimum of 50 cells should be scored, 25 cells from each replicate culture.

The accumulation of a large amount of data occurs very rapidly in any population study and the need to repeatedly update historical or concurrent controls necessitates continual statistical treatment. Much of this work can be simplified by automation. Data acquisition is amenable to the application of microprocessor-based technology as is the statistical treatment of the data. The use of microprocessors as opposed to computer systems is cost effective for most small laboratories. Figure 3 illustrates such a microprocessor-based system used in our laboratory for SCE analysis. It consists of an Apple II Plus microprocessor with 48K of RAM (Apple Computer Inc., Cupertino, Calif.), video monitor, data entry keypad, dual $5\frac{1}{4}$ inch floppy disks, and a printer. The operator enters the number of SCE for each chromosome as it is visualized

Fig. 3. A microprocessor-based data acquisition and analysis system used for the analysis of SCEs. Using the data entry keypad, the scorer enters the number of SCEs for each chromosome as it is observed in the microscope. Simultaneous and subsequent data analysis is performed on the same system. The system is compact, portable, and relatively inexpensive. (Photo courtesy of G. Watchmaker.)

in the microscope and after completion of each cell, the data can be stored on the disk in addition to being printed as hard copy. Statistical analysis can be performed on the cells from each individual and also pooled for many individuals. All the statistics described in the next section are performed on this system.

III. STATISTICAL CONSIDERATIONS

If the sister chromatid exchange method is to be used as an assay for genetic damage it is important to understand the distribution of

SCEs in normal individuals. In this section we describe how statistical methods can be applied to aid in characterizing the distribution of SCEs within cells of the same person and among cells from different persons. We will show how to test the SCEs from a single person for Poisson distribution, how to compare one individual to a group of individuals, and how to compare two cohorts of individuals.

A. SCE Distribution within Individuals

What is the distribution of SCEs in unexposed individuals? If it is assumed that the occurrence of an SCE on a chromosome is similar to that of the decay of a radioactive particle, then a reasonable model for the distribution of SCEs is the Poisson distribution. To define the problem mathematically, let S_{ij} = number of SCEs on the jth chromosome ($j = 1, \ldots, C_i$) from the ith cell ($i = 1, \ldots, n$), C_i = number of chromosomes scored in the ith cell ($1 \le C_i \le 46$), and n = number of cells scored in the experiment. Under a Poisson model the probability that the number of SCEs $S_{ij} = k$ is given by

$$Pr\{S_{ij} = k\} = \frac{e^{-\lambda_{ij}}\lambda_{ij}^{k}}{k!} \quad \text{for } k = 0, 1, 2, \ldots, \tag{1}$$

where λ_{ij} is the Poisson density parameter and is equal to the expected number of SCEs on the jth chromosome from the ith cell. In this model we allow the Poisson parameter λ_{ij} to vary from chromosome to chromosome and from cell to cell. We are primarily interested in the distribution of the sum of the SCEs in each cell,

$$S_i = \Sigma S_{ij} \tag{2}$$

where the sum is taken over $j = 1, \ldots, C_i$, the number of chromosomes scored in the ith cell. One nice feature of the Poisson distribution is that the sum of a set of Poisson random variables also has a Poisson distribution with parameter equal to the sum of the individual parameters. In our model S_i will have a Poisson distribution with parameter

$$\Lambda_i = \Sigma \lambda_{ij} \tag{3}$$

However, in order for Λ_i to be the same in every cell it is necessary that the sum be taken over the same set of chromosomes in each cell. This will be true whenever all 46 chromosomes are scored in each cell. When varying numbers of chromosomes are scored in each cell, as is usually the case due to overlapping and missing chromosomes, the raw SCE counts, S_i, will have varying expected values, Λ_i, depending on which chromosomes are scored and which are omitted.

This means that in order to test the Poisson distribution assumption the raw counts S_i must be adjusted to take into account varying expected values Λ_i.

The simplest adjustment is to divide the raw count, S_i, by the number of chromosomes scored, C_i. Let $X_i = S_i/C_i$ denote this quantity. What can we say about the distribution of X_i? First, X_i, just like S_i, is expected to vary from cell to cell depending on which chromosomes were scored. Second, if the number of chromosomes omitted in each cell is not very large (say five or fewer), then the true Poisson parameter Λ_i will not vary significantly from cell to cell. Therefore, we can assume that the number of SCEs per cell has expected value λC_i, where λ is the average SCE per chromosome. Note that this model assumes that λ is the same for all chromosomes, i.e., the SCE per chromosome for the No. 1 chromosome is the same as that for the No. 21 chromosome. (We know that this is not true but when 41 or more of the 46 chromosomes are scored, the difference between λC_i and the true value Λ_i will be less than 10%.) Whether or not this model is valid for an observed set of data is easily tested by comparing the observed variance of the X_i,

$$V = \frac{\Sigma(X_i - \bar{X})^2}{n - 1} \tag{4}$$

with its expected value

$$EV = \frac{\lambda\Sigma(1/C_i)}{n} \tag{5}$$

where λ is estimated by $\bar{X}$, the mean of the X_i. [A better estimate of λ is provided by $\Sigma S_i/\Sigma C_i$, the total sum of SCEs divided by the total number of chromosomes taken over all cells scored. This has a smaller variance than $X = \Sigma(S_i/C_i)/n$.] The ratio V/EV, multiplied by $(n - 1)$, has a χ^2 distribution with $(n - 1)$ degrees of freedom. Large values of χ^2 suggest lack of fit between the data and the Poisson model.

B. Application of the Poisson Test to Cells from a Single Individual

We applied the χ^2 test to SCE counts, expressed as SCE per chromosome, based on scoring 80 cells from each of 42 healthy persons. Frequency histograms for each of the 42 persons are shown in Fig. 4. The means and standard deviations for these distributions are shown in Table I. The column labeled "P value for Poisson" in Table I shows that 23 of the 42 distributions fail the Poisson test at the 5% level and

TABLE I

Summary Statistics for SCE per Chromosome

Individual	Mean	SD	Skew	Kurtosis	P value for Poisson
43	0.199	0.068	0.26	2.85	0.54
45	0.158	0.067	0.16	2.84	0.14
46	0.185	0.074	0.83	4.71	0.10
48	0.219	0.096	1.00	3.74	0.00
50	0.165	0.068	0.71	4.35	0.54
52	0.175	0.077	0.53	2.74	0.00
53	0.168	0.060	0.00	2.64	0.65
54	0.187	0.084	0.83	4.11	0.00
55	0.198	0.080	0.95	3.55	0.00
56	0.216	0.080	0.51	2.79	0.03
57	0.183	0.081	1.27	6.06	0.04
58	0.154	0.059	0.40	2.85	0.42
59	0.178	0.084	1.42	8.08	0.00
83	0.222	0.091	0.60	2.96	0.00
84	0.224	0.083	0.39	2.75	0.00
85	0.241	0.102	1.50	7.67	0.00
92	0.224	0.102	1.08	5.07	0.00
93	0.176	0.063	0.73	3.75	0.41
94	0.162	0.066	0.54	3.19	0.08
95	0.179	0.067	0.55	3.40	0.17
97	0.177	0.077	0.70	2.80	0.00
100	0.216	0.079	0.37	2.86	0.03
101	0.197	0.075	0.89	4.18	0.03
103	0.157	0.068	0.66	4.12	0.02
106	0.189	0.081	0.32	2.81	0.00
108	0.220	0.067	−0.20	2.79	0.69
111	0.188	0.084	1.04	4.58	0.00
114	0.231	0.080	0.69	3.51	0.06
115	0.216	0.078	0.61	3.36	0.05
119	0.211	0.075	0.13	2.58	0.11
122	0.187	0.079	0.93	4.13	0.00
123	0.172	0.070	0.33	2.46	0.05
125	0.185	0.070	0.59	3.01	0.12
134	0.208	0.077	0.18	2.50	0.05
135	0.174	0.072	1.01	4.23	0.03
137	0.197	0.063	0.91	3.81	0.71
138	0.222	0.092	0.73	3.12	0.00
139	0.206	0.075	0.51	2.99	0.06
140	0.224	0.088	0.57	4.18	0.00
142	0.201	0.076	0.67	3.44	0.04
143	0.200	0.089	0.85	4.22	0.00
148	0.202	0.068	0.32	3.16	0.37

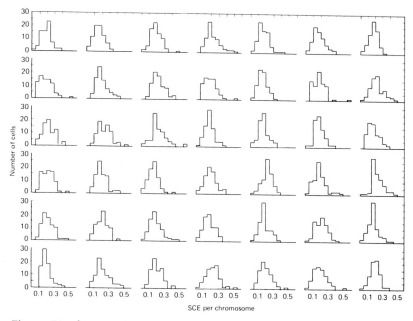

Fig. 4. Distributions of SCE frequencies for 42 unexposed persons based on scoring 80 cells per person. These histograms show that the distributions of SCEs varies widely from person to person. Most of the distributions cannot be fit by a Poisson distribution due to the presence of high-frequency cells (HFCs). The HFCs appear as "tails" to the right of the main peak in these histograms.

that 16 failed at the 1% level of significance. These results can also be visualized graphically in Fig. 5 where those distributions failing the Poisson test are seen to have variances much larger than those predicted by the Poisson model. A comparison between the histograms of those that fail the Poisson test and those that pass reveals that in most cases the failures are due to high SCE frequencies (HFCs). For example, the sixth histogram in the second row of Fig. 4 has two HFCs, one at 0.372 and one at 0.580. The cause for these values is not known but their presence must be taken into account when setting up criteria for baseline SCE frequencies.

Since the Poisson model does not appear to fit our SCE per chromosome data, is there any other distributional form that can be used? In Fig. 6 we have plotted β_2, the ratio of the fourth moment to the square of the second moment, against β_1, the ratio of the square of the third moment to the cube of the second moment. Plots such as these are helpful in deciding what kind of distribution can be used to fit

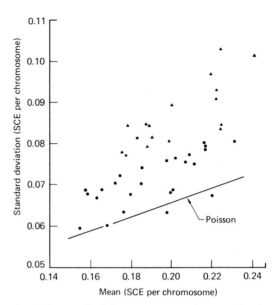

Fig. 5. This plot of the standard deviation versus the mean for the 42 distributions of Fig. 4 shows that the Poisson distribution fails to fit the majority of the distributions. Those which fail the Poisson distribution test at the 1% level are plotted as triangles, those which fail at the 5% level are plotted as squares, while those which pass are plotted as circles.

a set of data (Johnson and Kotz, 1970). The two lines drawn on this plot represent families of Poisson and log-normal distributions. The normal distribution lies at the single point $\beta_1 = 0, \beta_2 = 3$ on this diagram. Our data are quite scattered, some points are reasonably close to the Poisson line while others are nearer to the log-normal line. However, the wide scatter among the points suggests that no single family of distributions can be used as a model for the SCE per chromosome distribution; no matter which distribution is chosen it will fail to fit a significant number of our data sets. We have even tried fitting negative binomials, mixtures of Poissons, normals, and log-normals to our data, but all result in unsatisfactory fits to some of the data sets.

C. Distribution of Sister Chromatid Exchanges in the Same Individual over Time

Figure 7 shows the distribution of mean SCE per chromosome for one person who has been followed for over 2 years. The variability

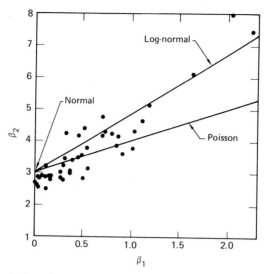

Fig. 6. Kurtosis (β_2) plotted against square of skew (β_1) for distributions of Fig. 4. Normal distributions have skew = 0 and kurtosis = 3.0 so that samples from the normal family will fall near the point (0,3). Poisson distributions are characterized by the relationship $\beta_2 = 3 + \beta_1$ so that samples from this family will fall near the Poisson line. Log-normal distributions are characterized by the relationship $\beta_2 = 3 + 1.83\,\beta_1$. Samples from this family are expected to fall near the log-normal line.

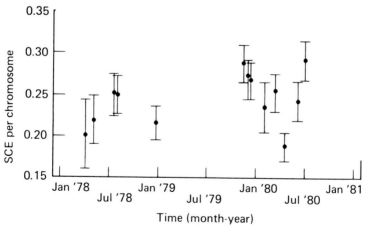

Fig. 7. This plot shows the wide variation in SCE frequencies in the same person over a 28-month period. The mean SCE frequency is plotted as a circle with one SD indicated by the vertical lines. The variation among the means is significantly greater than the variation among the cells at each time point but not as large as the variation among persons.

of these measurements over time can be compared with the cell-to-cell variability for each person. If we let X_i denote the mean and SD_i the standard deviation of the SCE per chromosome based on scoring n_i cells at the ith time point, the variation among time points is estimated by

$$S_a^2 = \frac{\Sigma n_i (X_i - \bar{X})^2}{(I - 1)} \tag{6}$$

where I is the number of time points and $\overline{X}$ is the mean of the X_i. A pooled estimate for the within person cell-to-cell variation is given by

$$S_w^2 = \frac{\Sigma (n_i - 1) SD_i^2}{\Sigma (n_i - 1)} \tag{7}$$

The F ratio given by

$$F = S_a^2 / S_w^2 \tag{8}$$

can be used to test whether the variation over time is significantly greater than the cell-to-cell variation by comparing the F ratio with tables of the F distribution with $(I - 1)$ and $\Sigma(n_i - 1)$ degrees of freedom. In Table II we summarize the results of applying this test to six individuals for whom we had multiple time-point measurements of SCE per chromosome. The table shows that variation over time is significantly greater than cell-to-cell variation in four of the six individuals. In the next section we will show that this variation is smaller than the variation among different individuals.

D. Distribution of Sister Chromatid Exchanges among Individuals

It is evident from the histograms shown in Fig. 4, from the means and standard deviations shown in Fig. 5, and from the β_2 versus β_1 plot of Fig. 6 that the distribution of SCEs varies widely from person to person, even among healthy, genetically normal persons. Analysis of variance can also be used to compare the variation among the means from different persons with the cell-to-cell variation within persons. The same F ratio described in the preceding section is used for comparing the variance among the SCE means (S_a^2) with the within individual cell-to-cell variance (S_w^2). The only change is that X_i and SD_i now denote the mean and standard deviation, respectively, for the ith individual rather than the ith time point. The standard deviation for among person variability is 0.202 while the pooled within person standard deviation is 0.078. This leads to an F ratio of 6.70 which is

TABLE II

Variability of SCE Frequency with Time

Individual	Time period (months)	Number of time points	SD Over time	Cell-to-cell	F	P
19	27	13	0.2584	0.0981	6.94	<0.001
43	24	7	0.1328	0.0748	3.15	0.005
46	25	10	0.1672	0.0742	5.08	<0.001
47	28	6	0.2171	0.1623	1.79	0.113
48	29	6	0.1136	0.0833	1.86	0.100
50	26	6	0.2632	0.0785	11.24	<0.001

significant at the 0.01% level. As indicated previously several investigators have reported finding significant variability among the SCE means from different healthy individuals.

E. Distribution of Individual Sister Chromatid Exchange Means

If we look at the distribution of means, shown in Table I, we see that they range from a low of 0.155 SCE per chromosome (7.1 SCE per cell) to a high of 0.241 (11.1 SCE per cell). The mean of the means is 0.196 SCE per chromosome (9.0 SCE per cell) and the standard error of the means is 0.023. (This standard error is equal to the among person standard deviation divided by the square root of the number of cells scored per person.) Furthermore, the distribution of these means can be fit reasonably well by a normal distribution. If we rank these means in numerical order from largest to smallest and compare these ranked values with the expected median values, M_i, for the order statistics from a standard normal distribution, given by

$$M_i = \Phi^{-1}\{m_i\} \text{ where } m_i = \begin{cases} 1 - (.5)^{1/n} & i = 1 \\ \dfrac{i - .3175}{(n + .365)} & i = 2, \ldots, n-1 \\ (.5)^{1/n} & = n \end{cases} \quad (9)$$

where Φ^{-1} is the inverse cumulative normal function (Zelen and Severo, 1964) we find a correlation coefficient of 0.989 which is above the 10% value reported by Filliben (1975). This suggests that the population or cohort mean SCE frequency per chromosome, based on scoring at least 41 chromosomes in each of 80 cells, is reasonably normally distributed.

F. Comparing a Single Individual to a Group of Individuals

Now we can use these measurements on 42 healthy persons to construct normal baseline values for SCE frequencies. This baseline can be based on SCE frequencies for individual cells or on mean SCE frequencies based on scoring n cells from each individual. Since the means based on scoring $n = 80$ cells per person are reasonably normally distributed we can use tabled values (such as those found in Owen, 1962) to find an upper 95% tolerance bound for SCE means. In our case this upper bound is equal to 0.243, found by multiplying 2.111 (interpolated value from the table for a sample size of 42) by the standard error (0.023) and adding the result to the overall mean (0.196). The interpretation of this bound is that we expect (with 95% confidence) fewer than 5% of future SCE means from healthy individuals to exceed this bound.

Alternatively, we can use the pooled distribution of all 3360 SCE measurements (shown in Fig. 8), the result of pooling 80 cells from each of our sample of 42 unexposed persons to establish a baseline for cellular SCEs based on the HFC concept. A nonparametric procedure, described by Walsh (1962), is required since the cellular SCEs

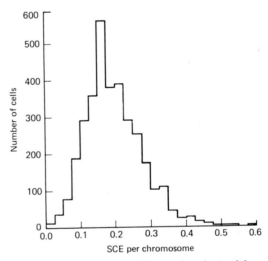

Fig. 8. Pooled distribution of 3360 SCE frequencies obtained from scoring 80 cells from each of 42 unexposed persons. The shoulder to the right of the main peak contains the subpopulation of high-frequency cells (HFCs).

are clearly not normally distributed. Under this procedure we must determine a number k large enough so that 95% of the SCEs will fall below the kth largest SCE in our sample. In our case where the total number of SCEs is large k can be found from the formula

$$k = n(1 - p) + 0.5 + \sqrt{1.645\ np(1 - p)} \qquad (10)$$

In our case $n = 3360$ and $p = 0.95$ so that $k = 148$. The 148th largest SCE frequency in our sample is 0.348 SCE per chromosome. Thus, we define an HFC as a cell with more than 0.348 SCE per chromosome or, equivalently, more than 16 SCEs per cell. In unexposed individuals we expect (with 95% confidence) that no more than 5% of their cells will exceed 0.348 (i.e., there should be no more than 5% HFCs among their cells). When we score 80 cells per person we expect, on the average, 4 HFCs (5% of 80) but we need an upper bound for this number in order to determine whether or not an individual has an "abnormal" number of HFCs. In unexposed individuals the number of HFCs should be binomially distributed so that when 80 cells are scored, the probability that m HFCs are present is given by

$$\binom{80}{m}(0.05)^{m}(0.95)^{80-m} \qquad (11)$$

The upper 95% bound for this distribution is at $m = 7$ HFCs so that we will conclude that an individual does not conform to our baseline if 8 or more of his cells are HFCs.

Although this criterion for normal SCE frequencies is conceptually more difficult to understand than the one for SCE means, it has two advantages. First, it is nonparametric and requires no assumption of normal distribution of SCE frequencies. Second, the results of computer simulation suggest that this test is more powerful than the one based on means in detecting departures from the baseline. We will see that this is true for some data on smokers in the next section.

G. Comparing Two Groups of Individuals

The methods described in the preceding section for comparing single individuals to a group of individuals can be extended to compare two groups of persons. First, we count the number of individuals in each group who have SCE per chromosome values falling outside our previously established 95% tolerance bounds. Letting M_i be this number for the ith group ($i = 1,2$) when N_i individuals are scored we form the following 2 × 2 tabulation:

	No. above bound	No. below bound	Totals
Group 1	M_1	$N_1 - M_1$	N_1
Group 2	M_2	$N_2 - M_2$	N_2
Totals	M	$N - M$	N

The χ^2 statistic

$$\chi^2 = \frac{[M_1(N_2 - M_2) - M_2(N_1 - M_1)]^2 N}{N_1 N_2 M(N - M)} \qquad (12)$$

can be used to test whether or not the groups differ. The M_i can refer either to the number of persons whose mean SCE frequency is above 0.243 or to the number of persons who have 8 or more HFCs (out of 80).

As an example we applied these tests to a group of smokers to compare them with our controls. Table III and Fig. 9 summarize the results of these tests. The tests strongly reject the hypothesis that SCE frequencies are identical in the two populations. In our samples there

TABLE III

Comparisons between Smokers and Nonsmokers Using Two Different Tests

Group	Individual means		
	No. above bound	No. below bound	Total
Nonsmokers	0	42	42
Smokers	5	10	15
Total	5	52	57

$$\chi^2 = 15.35, \ p < .001$$

Group	HFCs per individual		
	No. with 8 or more outliers	No. with 7 or less outliers	Total
Nonsmokers	4	38	42
Smokers	10	5	15
Total	14	43	57

$$\chi^2 = 19.45, \ p < .001$$

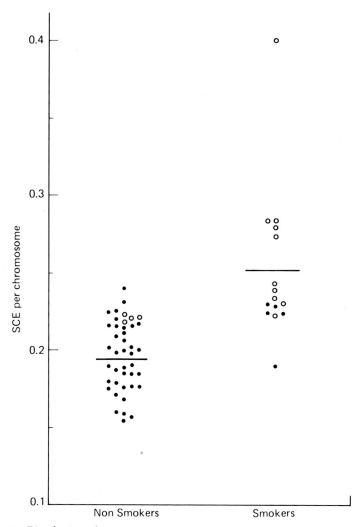

Fig. 9. Distribution of mean SCE frequencies for two groups of persons: one group of 42 nonsmokers and one group of 15 smokers. Persons with 8 or more HFCs are plotted as open circles; those with 7 or fewer are plotted as solid circles. Ten of the 15 smokers, but only 4 of the 42 nonsmokers have a significant number of HFCs. This difference between the two groups is significant at the 0.001 level.

is good evidence that smokers are at greater risk of having elevated SCE frequencies in a large proportion of their cells. Other investigators have also shown an increased SCE frequency associated with cigarette smoking (Lambert et al., 1978b; Balakrishna Murthy, 1979; Mäki-Paakkanen et al., 1980; Lambert and Lindblad, 1980; Ardito et al., 1980), but no difference was observed in two additional studies (Hollander et al., 1978; Crossen and Morgan, 1980).

H. Sample Size Considerations

In previous sections we have described statistical tests that can be used to determine whether or not an individual differs from a standard pool of individuals and have shown how the results of these tests can be summarized in a 2 × 2 tabulation to compare a group of individuals with our standard pool. In this section we will discuss the relationship between sample size and ability of these tests to detect differences.

In order to do this we must model the distribution of SCEs and SCE means in the two groups to be compared. We have already shown that no single statistical distribution can adequately characterize all of the observed data; however, the negative binomial distribution came the closest to fitting most of the distributions and it provided an acceptable fit to the distribution of those cells in which all 46 chromosomes were scored. We also found that the individual mean SCE frequencies, based on scoring 80 cells per person, were reasonably normally distributed. Thus, we will use a negative binomial distribution to model SCEs in individual cells and a normal distribution to model the behavior of SCE means based on scoring n cells per person.

Table IV shows the probabilities of detecting a single individual with an elevated SCE frequency for various degrees of elevation. This

TABLE IV

Power of SCE Tests: Probabilities of Detecting an Elevated SCE Frequency

Sample size	Test	Percentage elevation above unexposed				
		10	15	20	25	30
20	Mean	0.12	0.21	0.29	0.35	0.40
	HFC	0.14	0.31	0.52	0.69	0.81
40	Mean	0.14	0.25	0.35	0.41	0.47
	HFC	0.26	0.55	0.80	0.93	0.98
80	Mean	0.19	0.33	0.43	0.51	0.56
	HFC	0.42	0.80	0.97	0.99	0.99

TABLE V

Sample Sizes Required for 90% Probability of Detecting an Increased SCE Frequency in a Test Group

Number of cells per sample	Test	Percentage elevation in SCE above unexposed					
		10	12	14	16	18	20
20	Mean	151	33	11	9	8	7
	HFC	596	402	283	42	20	11
40	Mean	45	20	9	8	7	6
	HFC	290	33	12	7	7	9
80	Mean	11	9	7	6	5	5
	HFC	20	8	8	5	7	9

table shows that the test based on individual SCE frequencies is more powerful than the one based on SCE means. When 80 cells are scored per individual we are reasonably certain (i.e., have high probability) of detecting a 15% increase in the frequency of SCE per chromosome when using the HFC test. If the test is based on the SCE means, the probability of detecting a significant increase is much lower.

Table V shows the number of persons required to detect (with 90% assurance) an elevated mean in a group of persons who are to be compared with our standard population. This time the test based on SCE means is more powerful than the one based on HFCs. This apparent anomaly is explained by the observation that none of the control individuals had significantly elevated means so that the appearance of a single SCE mean above the bound is highly significant in the 2 × 2 summary table. In contrast, four members of our control population had significantly high numbers of HFCs so that a larger number of individuals from a population with elevated SCE frequencies is required for significance in the 2 × 2 summary table. We expect that as our pool of normal individuals increases the upper bound for the means test will decrease (assuming that future individuals are similar to those already in the pool). This will make the means test more powerful for individuals and, perhaps, less powerful for groups. Thus, the two test methods will become more alike.

IV. APPLICATION TO HUMAN STUDIES

From the foregoing information, it is evident that we need more quantitative information on the extent of variation both within and among individuals. The reason for this is that the SCE increases measured in exposed humans will likely be small. A large control population will increase our sensitivity to detect either abnormal SCE means or individuals with HFCs. It should also be feasible to establish several control cohorts based, for example, on age, sex, medication use, or smoking history, in order to appropriately match an exposed cohort. This is particularly relevant for occupational or accidental exposures when appropriate concurrent controls may not always be available.

Several human studies already conducted support the statement that the SCE increases associated with occupational exposure will be small or not detectable. These studies are summarized in Table VI. It is clear

TABLE VI

SCE Frequencies Associated with Human Occupational Exposures

Agent	SCE per cell	References
Organic solvents		
Controls	13.1	Funes-Cravioto et al. (1977)
Exposed	19.7	Funes-Cravioto et al. (1977)
Controls	14.0	Lambert and Lindblad (1980)
Exposed	19.9	Lambert and Lindblad (1980)
Ethylene oxide		
Controls	14.0	Lambert and Lindblad (1980)
Exposed	16.1	Lambert and Lindblad (1980)
Controls	6.4	Garry et al. (1979)
Exposed	10.3	Garry et al. (1979)
Toluene		
Controls	8.0	Mäki-Paakkanen et al. (1980)
Exposed	7.9[a]	Mäki-Paakkanen et al. (1980)
Tetrachloroethylene		
Controls	8.2	Ikeda et al. (1980)
Exposed	8.6[a]	Ikeda et al. (1980)
Styrene		
Controls	7.5	Andersson et al. (1980)
Exposed	8.4	Andersson et al. (1980)
Epoxy resins		
Controls	8.7	Mitelman et al. (1980)
Exposed	9.0[a]	Mitelman et al. (1980)
Vinyl chloride		
Controls	9.4	Kucerova et al. (1979)
Exposed	13.8	Kucerova et al. (1979)
Pesticides/herbicides	5/57 individuals were elevated	Crossen et al. (1978)

[a] Not significantly different from controls.

that the observed frequencies are only about 10 to 60% greater than the control values. These increases are of the same magnitude as those attributable to either the preparative or other confounding factors previously discussed. Future human studies, therefore, might do well to identify and quantify potentially confounding variables, to minimize them, and to develop more sensitive methods of data analysis. We have attempted to emphasize these goals in our own human studies and stress their significance in this chapter. Although their attainment will not be easy, the end point—a sensitive method to monitor environmental exposure directly in humans—should be well worth the effort.

ACKNOWLEDGMENT

This work was supported by the U.S. Department of Energy under contract number W-7405-Eng-48 by the Lawrence Livermore National Laboratory.

REFERENCES

Abe, S., and Sasaki, M. (1977). Chromosome aberrations and sister chromatid exchanges in Chinese hamster cells exposed to various chemicals. *J. Natl. Cancer Inst.* **58,** 1635–1641.

Alhadeff, B., and Cohen, M. M. (1976). Frequency and distribution of sister chromatid exchanges in human peripheral lymphocytes. *Isr. J. Med. Sci.* **12,** 1440–1447.

Andersson, H. C., Tranberg, E. A., Uggla, A. H., and Zetterberg, G. (1980). Chromosomal aberrations and sister chromatid exchanges in lymphocytes of men occupationally exposed to styrene in a plastic-boat factory. *Mutat. Res.* **73,** 387–401.

Ardito, G., Lamberti, L., Ansaldi, E., and Ponzetto, P. (1980). Sister chromatid exchanges in cigarette-smoking human females and their newborns. *Mutat. Res.* **78,** 209–212.

Balakrishna Murthy, P. (1979). Frequency of sister chromatid exchanges in cigarette smokers. *Hum. Genet.* **52,** 343–345.

Balakrishna Murthy, P., and Prema, K. (1979). Sister chromatid exchanges in oral contraceptive users. *Mutat. Res.* **68,** 149–152.

Balakrishna Murthy, P., and Srikantia, S. G. (1981). SCE frequency in malnourished mice. *Metabolism* **30,** 1–2.

Balakrishna Murthy, P., Bhaskaram, P., and Srikantia, S. G. (1980). Sister chromatid exchanges in protein-energy malnutrition. *Hum. Genet.* **55,** 405–406.

Banerjee, A., and Benedict, W. F. (1979). Production of sister chromatid exchanges by various cancer chemotherapeutic agents. *Cancer Res.* **39,** 797–799.

Basler, A., Bachmann, U., Roszinsky-Köcher, G., and Röhrborn, G. (1979). Effects of caffeine on sister chromatid exchanges (SCE) *in vivo. Mutat. Res.* **59,** 209–214.

Becher, R., Schmidt, C. G., Theis, G., and Hossfeld, D. K. (1979). The rate of sister chromatid exchange in normal human bone marrow cells. *Hum. Genet.* **50,** 213–216.

Beek, B., and Obe, G. (1979). Sister chromatid exchanges in human leukocyte chromosomes: Spontaneous and induced frequencies in early- and late-proliferating cells *in vitro. Hum. Genet.* **49,** 51–61.

Bender, M. A., Griggs, H. G., and Bedford, J. S. (1974). Recombinational DNA repair and sister chromatid exchanges. *Mutat. Res.* **24**, 117–123.

Bradley, M. O., Hsu, I. C., and Harris, C. C. (1979). Relationship between sister chromatid exchange and mutagenicity, toxicity, and DNA damage. *Nature (London)* **282**, 318–320.

Buckton, K. E., Hamilton, G. E., Paton, L., and Langlands, A. O. (1978). Chromosome aberrations in irradiated ankylosing spondylitis patients. In "Mutagen-Induced Chromosome Damage in Man" (H. J. Evans and D. C. Lloyd, eds.), pp. 142–150. Edinburgh University Press, Edinburgh, Scotland.

Carrano, A. V., and Thompson, L. H. (1981). Sister chromatid exchange and single-gene mutation. In "Sister Chromatid Exchange" (S. Wolff, ed.), Wiley, New York (in press).

Carrano, A. V., Thompson, L. H., Lindl, P. A., and Minkler, J. L. (1978). Sister chromatid exchanges as an indicator of mutagenesis. *Nature (London)* **271**, 551–553.

Carrano, A. V., Thompson, L. H., Stetka, D. G., Minkler, J. L., Mazrimas, J. R., and Fong, S. (1979). DNA crosslinking, sister chromatid exchange, and specific locus mutations. *Mutat. Res.* **63**, 175–188.

Carrano, A. V., Minkler, J. L., Stetka, D. G., and Moore, D. H., II (1980). Variation in the baseline sister chromatid exchange frequency in human lymphocytes. *Environ. Mut.* **2**, 325–337.

Chaganti, R. S. K., Schonberg, S., and German, J. (1974). A manyfold increase in sister chromatid exchanges in Bloom's syndrome lymphocytes. *Proc. Natl. Acad. Sci. U.S.A.* **71**, 4508–4512.

Cheng, W., Mulvihill, J. J., Greene, M. H., Pickle, L. W., Tsai, S., and Whang-Peng, J. (1979). Sister chromatid exchanges and chromosomes in chronic myelogenous leukemia and cancer families. *Int. J. Cancer* **23**, 8–13.

Connell, J. R. (1979). The relationship between sister chromatid exchange, chromosome aberration, and gene mutation induction by several reactive polycyclic hydrocarbon metabolites in cultured mammalian cells. *Int. J. Cancer* **24**, 485–489.

Crossen, P. E., and Morgan, W. F. (1980). Sister chromatid exchange in cigarette smokers. *Hum. Genet.* **53**, 425–426.

Crossen, P. E., Drets, M. E., Arrighi, F. E., and Johnston, D. A. (1977). Analysis of the frequency and distribution of sister chromatid exchanges in cultured human lymphocytes. *Hum. Genet.* **35**, 345–352.

Crossen, P. E., Morgan, W. F., Horan, J. J., and Stewart, J. (1978). Cytogenetic studies of pesticide and herbicide sprayers. *N. Z. Med. J.* **88**, 192–195.

Dufrain, R. J., Littlefield, L. G., and Wilmer, J. L. (1980). The effect of washing lymphocytes after *in vivo* treatment with streptonigrin on the yield of chromosome and chromatid aberrations in blood cultures. *Mutat. Res.* **69**, 101–105.

Filliben, J. (1975). The probability plot correlation coefficient test for normality. *Technometrics* **17**, 111–117.

Funes-Cravioto, F., Kolmondin-Hedman, B., Lindsten, J., Nordenskjöld, M., Zapata-Gayon, C., Lambert, B., Norberg, E., Olin, R., and Swensson, A. (1977). Chromosome aberrations and sister chromatid exchange in workers in chemical laboratories and a photoprinting factory and in children of women laboratory workers. *Lancet* **2**, 322–325.

Galloway, S. M. (1977). Ataxia telangiectasia: The effects of chemical mutagens and X-rays on sister chromatid exchanges in blood lymphocytes. *Mutat. Res.* **45**, 343–349.

Galloway, S. M., and Evans, H. J. (1975). Sister chromatid exchange in human chromosomes from normal individuals and patients with ataxia telangiectasia. *Cytogenet. Cell Genet.* **15**, 17–29.

Garry, V. F., Hozier, J., Jacobs, D., Wade, R. L., and Gray, D. G. (1979). Ethylene oxide: Evidence of human chromosomal effects. *Environ. Mut.* **1**, 375–382.

Haglund, U., Hayder, S., and Zech, L. (1980). Sister chromatid exchanges and chromosome aberrations in children after treatment for malignant lymphoma. *Cancer Res.* **40**, 4786–4790.

Hollander, D. H., Tockman, M. S., Liang, Y. W., Borgaonkar, D. S., and Frost, J. K. (1978). Sister chromatid exchanges in the peripheral blood of cigarette smokers and in lung cancer patients; and the effect of chemotherapy. *Hum. Genet.* **44**, 165–171.

Ikeda, M., Koizumi, A., Watanabe, T., Endo, A., and Sato, K. (1980). Cytogenetic and cytokinetic investigations on lymphocytes from workers occupationally exposed to tetrachloroethylene. *Toxicol. Lett.* **5**, 251–256.

Johnson, N. L., and Kotz, S. (1970). "Continuous Univariate Distributions-1," pp. 74–75. Houghton, Boston, Massachusetts.

Kato, H., and Sandberg, A. A. (1977). The effect of sera on sister chromatid exchanges *in vitro. Exp. Cell Res.* **109**, 445–448.

Kato, H., and Stich, H. F. (1976). Sister chromatid exchanges in aging and repair deficient human fibroblasts. *Nature (London)* **260**, 447–448.

Kihlman, B. A., and Kronberg, D. (1975). Sister chromatid exchanges in *Vicia faba.* I. Demonstration by a modified fluorescent plus Giemsa (FPG) technique. *Chromosoma* **56**, 101–109.

Kowalczyk, J. (1980). Sister chromatid exchanges in children treated with nalidixic acid. *Mutat. Res.* **77**, 371–375.

Kram, D., Schneider, E. L., Tice, R. R., and Gianas, P. (1978). Aging and sister chromatid exchange. I. The effect of aging on mitomycin-C induced sister chromatid exchange frequencies in mouse and rat bone marrow cells *in vivo. Exp. Cell Res.* **114**, 471–475.

Kucerova, M., Pilivkova, Z., and Batora, J. (1979). Comparative evaluation of the frequency of chromosomal aberrations and the SCE numbers in peripheral lymphocytes of workers occupationally exposed to vinyl chloride monomer. *Mutat. Res.* **67**, 97–100.

Lambert, B., and Lindblad, A. (1980). Sister chromatid exchange and chromosome aberrations in lymphocytes of laboratory personnel. *J. Toxicol. Environ. Health* **6**, 1237–1243.

Lambert, B., Hansson, K., Lindsten, J., Sten, M., and Werelius, B. (1976). Bromodeoxyuridine-induced sister chromatid exchanges in human lymphocytes. *Hereditas* **82**, 163–174.

Lambert, B., Ringborg, U., Harper, E., and Lindblad, A. (1978a). Sister chromatid exchanges in lymphocyte cultures of patients receiving chemotherapy for malignant disorders. *Cancer Treat. Rep.* **62**, 1413–1419.

Lambert, B., Lindblad, A., Nordenskjöld, M., and Werelius, B. (1978b). Increased frequency of sister chromatid exchanges in cigarette smokers. *Hereditas* **88**, 147–149.

Lambert, B., Ringborg, U., and Lindblad, A. (1979). Prolonged increase of sister chromatid exchanges in lymphocytes of melanoma patients after CCNU treatment. *Mutat. Res.* **59**, 295–300.

Latt, S. (1973). Microfluorometric detection of deoxyribonucleic acid replication in human metaphase chromosomes. *Proc. Natl. Acad. Sci. U.S.A.* **70**, 3395–3399.

Latt, S. A. (1974). Sister chromatid exchanges, indices of human chromosome damage and repair: Detection by fluorescence and induction by mitomycin-C. *Proc. Natl. Acad. Sci. U.S.A.* **71**, 3162–3166.

Latt, S. A., and Juergens, L. A. (1977). Determinants of sister chromatid exchange frequencies in human chromosomes. In "Population Cytogenetics: Studies in Humans" (E. B. Hook and I. H. Porter, eds.), pp. 217–236. Academic Press, New York.

Lezana, E. A., Bianchi, N. O., Bianchi, M. S., and Zabala-Suarez, J. E. (1977). Sister chromatid exchanges in Down syndromes and normal human beings. *Mutat. Res.* **45**, 85–90.

Littlefield, L. G., Colyer, S. P., Joiner, E. E., and Dufrain, R. J. (1979a). Sister chromatid exchanges in human lymphocytes exposed to ionizing radiation during G_0. *Radiat. Res.* **78**, 514–521.

Littlefield, L. G., Colyer, S. P., Sayer, A. M., and Dufrain, R. J. (1979b). Sister chromatid exchanges in human lymphocytes exposed during G_0 to four classes of DNA-damaging chemicals. *Mutat. Res.* **67**, 259–269.

Littlefield, L. G., Colyer, S. P., and Dufrain, R. J. (1980). Comparison of sister chromatid exchanges in human lymphocytes after G_0 exposure to mitomycin *in vivo* vs. *in vitro*. *Mutat. Res.* **69**, 191–197.

McFee, A. F., and Sherrill, M. N. (1981). Mitotic response and sister chromatid exchanges in lymphocytes cultured in sera from different sources. *Experientia* **37**, 27–29.

Mäki-Paakkanen, J., Husgafvel-Pursiainen, K., Kalliomäki, P. L., Tuominen, J., and Sorsa, M. (1980). Toluene-exposed workers and chromosome aberrations. *J. Toxicol. Environ. Health* **6**, 775–781.

Mazrimas, J. A., and Stetka, D. G. (1978). Direct evidence for the role of incorporated BUdR in the induction of sister chromatid exchanges. *Exp. Cell Res.* **117**, 23–30.

Mitelman, F., Fregert, S., Hedner, K., and Hillbertz-Nilsson, K. (1980). Occupational exposure to epoxy resins has no cytogenetic effect. *Mutat. Res.* **77**, 345–348.

Morgan, W. F., and Crossen, P. E. (1977). The incidence of sister chromatid exchanges in cultured human lymphocytes. *Mutat. Res.* **42**, 305–312.

Morimoto, K., and Wolff, S. (1980). Increase of sister chromatid exchanges and perturbations of cell division kinetics in human lymphocytes by benzene metabolites. *Cancer Res.* **40**, 1189–1193.

Musilova, J., Michalova, K., and Urban, J. (1979). Sister chromatid exchanges and chromosomal breakage in patients treated with cytostatics. *Mutat. Res.* **67**, 289–294.

Nevstad, N. P. (1978). Sister chromatid exchanges and chromosmal aberrations induced in human lymphocytes by the cytostatic drug adriamycin *in vivo* and *in vitro*. *Mutat. Res.* **57**, 253–258.

Ohtsuru, M., Ishii, Y., Takai, S., Higashi, H., and Kosaki, G. (1980). Sister chromatid exchanges in lymphocytes of cancer patients receiving mitomycin-C treatment. *Cancer Res.* **40**, 477–480.

Okinaka, R. T., Barnhart, B. J., and Chen, D. J. (1981). Comparison between sister chromatid exchange and mutagenicity following exogenous metabolic activation of promutagens. *Mutat. Res.* **91**, 57–61.

Otter, M., Palmer, C., and Baehner, R. L. (1979). Sister chromatid exchange in lymphocytes from patients with acute lymphoblastic leukemia. *Hum. Genet.* **52**, 185–192.

Owen, D. B. (1962). "Handbook of Statistical Tables," p. 126. Addison-Wesley, Reading, Massachusetts.

Painter, R. B. (1980a). A replication model for sister chromatid exchange. *Mutat. Res.* **70**, 337–341.

Painter, R. B. (1980b). Effect of caffeine on DNA synthesis in irradiated and unirradiated mammalian cells. *J. Mol. Biol.* **143**, 289–301.

Pedersen, C., Olah, E., and Merrild, U. (1979). Sister chromatid exchanges in cultured peripheral lymphocytes from twins. *Hum. Genet.* **52**, 281–294.

Pero, R. W., Bryngelsson, C., Mitelman, F., Thulin, T., and Norden, A. (1976). High blood pressure related to carcinogen-induced unscheduled DNA synthesis, DNA–carcinogen binding, and chromosomal aberrations in human lymphocytes. *Proc. Natl. Acad. Sci. U.S.A.* **73**, 2496–2500.

Perry, P. (1980). Chemical mutagens and sister chromatid exchange. *Chem. Mutagens* **6**, 1–39.

Perry, P., and Evans, H. J. (1975). Cytological detection of mutagen–carcinogen exposure by sister chromatid exchange. *Nature (London)* **258**, 121–125.

Perry, P., and Wolff, S. (1974). New Giemsa method for the differential staining of sister chromatids. *Nature (London)* **251**, 156–158.

Raposa, T. (1978). Sister chromatid exchange studies for monitoring DNA damage and repair capacity after cytostatics *in vitro* and in lymphocytes of leukemic patients under cytostatic therapy. *Mutat. Res.* **57**, 241–251.

Riedel, L., and Obe, G. (1980). Trenimon-induced SCEs and structural chromosomal aberrations in early- and late-dividing lymphocytes. *Mutat. Res.* **73**, 125–131.

Santesson, B., Lindahl-Kiessling, K., and Mattsson, A. (1979). SCE in B and T lymphocytes: Possible implications for Bloom's syndrome. *Clin. Genet.* **16**, 133–135.

Saxholm, H. J. K., Iverson, O. H., Reith, A., and Brogger, A. (1979). Carcinogenesis testing of saccharin. No transformation or increased sister chromatid exchange observed in two mammalian cell systems. *Eur. J. Cancer* **15**, 509–513.

Schneider, E. L., and Gilman, B. (1979). Sister chromatid exchanges and aging. III. The effect of donor age on mutagen-induced sister chromatid exchange in human diploid fibroblasts. *Hum. Genet.* **46**, 57–63.

Schneider, E. L, and Monticone, R. E. (1978). Aging and sister chromatid exchange. II. The effect of the *in vitro* passage level of human fetal lung fibroblasts on baseline and mutagen-induced sister chromatid exchange frequencies. *Exp. Cell Res.* **115**, 269–276.

Shafer, D. A. (1977). Replication bypass model of sister chromatid exchanges and implications for Bloom's syndrome and Fanconi's anemia. *Hum. Genet.* **39**, 177–199.

Sirianni, S. R., and Huang, C. C. (1980). Comparison of induction of sister chromatid exchange, 8-azaguanine- and ouabain-resistant mutants by cyclophosphasmide, ifosfamide and 1-(pyridyl-3)-3,3-dimethyltriazene in Chinese hamster cells cultured in diffusion chambers in mice. *Carcinogenesis* **1**, 353–355.

Snope, A. J., and Rary, J. M. (1979). Cell-cycle duration and sister chromatid exchange frequency in cultured human lymphocytes. *Mutat. Res.* **63**, 345–349.

Speit, G. (1980). Effects of temperature on sister chromatid exchanges. *Hum. Genet.* **55**, 333–336.

Stetka, D. G., and Carrano, A. V. (1977). The interaction of Hoechst 33258 and BrdU substituted DNA in the formation of sister chromatid exchanges. *Chromosoma* **36**, 21–31.

Stetka, D. G., and Wolff, S. (1976). Sister chromatid exchange as an assay for genetic damage induced by mutagen-carcinogens. Part I. *In vivo* test for compounds requiring metabolic activation. *Mutat. Res.* **41**, 333–342.

Stetka, D. G., Minkler, J., and Carrano, A. V. (1978). Induction of long-lived chromosome damage, as manifested by sister chromatid exchange, in lymphocytes of animals exposed to mitomycin C. *Mutat. Res.* **51**, 383–396.

Sugimura, T., Nagao, M., Kawachi, T., Honda, M., Yahagi, T., Seino, Y., Sato, S., Matsukura, N., Matsushima, T., Shirai, A., Sawamura, M., and Matsumoto, H. (1977). Mutagen–carcinogens in food, with special reference to highly mutagenic pyrolytic products in broiled foods. In "Origins of Human Cancer" (H. H. Hiatt, J. D. Watson, and J. A. Winsten, eds.), pp. 1561–1577. Cold Spring Harbor Lab., Cold Spring Harbor, New York.

Taylor, J. H. (1958). Sister chromatid exchanges in tritium-labeled chromosomes. Genetics **43**, 515–529.

Walsh, J. E. (1962). Non-parametric confidence intervals and tolerance regions. In "Contributions to Order Statistics" (A. Sarhan and B. Greenbery, eds.), pp. 136–143. Wiley, New York.

Wolff, S. (1977). Sister chromatid exchange. Annu. Rev. Genet. **11**, 183–201.

Wolff, S., and Perry, P. (1974). Differential Giemsa staining of sister chromatids and the study of sister chromatid exchanges without autoradiography. Chromosoma **48**, 341–353.

Wolff, S., and Perry, P. (1975). Insights on chromosome structure from sister chromatid exchanges and the lack of both isolabelling and heterolabelling as determined by the FPG technique. Exp. Cell Res. **93**, 23–30.

Wolff, S., and Rodin, B. (1978). Saccharin-induced sister chromatid exchanges in Chinese hamster and human cells. Science **200**, 543–545.

Wolff, S., Bodycote, J., Thomas, G. H., and Cleaver, J. E. (1975). Sister chromatid exchange in xeroderma pigmentosum cells that are defective in DNA excision repair or post-replication repair. Genetics **81**, 349–355.

Zelen, M., and Severo, N. (1964). Probability functions. In "Handbook of Mathematical Functions" (M. Abramowitz and I. A. Stegun, eds.), p. 933. U.S. Dept. of Commerce, National Bureau of Standards, Applied Math Series #55, Washington, D.C.

Chapter 11

The 6-Thioguanine-Resistant Peripheral Blood Lymphocyte Assay for Direct Mutagenicity Testing in Humans

Richard J. Albertini, David L. Sylwester, and Elizabeth F. Allen

305

MUTAGENICITY:
NEW HORIZONS IN GENETIC TOXICOLOGY

I. INTRODUCTION

A. Testing for Environmental Mutagens and Carcinogens

Many tests are now in use for identifying environmental mutagens, some of which may also identify environmental carcinogens (Brusick, 1977; McCann and Ames, 1977). They range from relatively simple, rapid assays in prokaryotes, to complex, time-consuming, and expensive tests in intact animals (cf. Waters, 1977; Kilbey et al., 1979). Intermediate in terms of time and expense are tests using mammalian cells in tissue culture (cf. Hsie et al., 1979). Results here are thought to be especially relevant for extrapolation to man because of the similarity among animal cells of "genetic targets."

There is general agreement that some approach must be made to the task of assessing, in terms of their threat to human health, the thousands of environmental agents with potential for genotoxic damage to germinal or somatic cells. Added to these thousands of individual agents are the virtually unlimited combinations in which they may occur, with unknown capacity for synergy of action.

A sequential or tier approach to mutagenicity testing has been suggested (Bridges, 1974; Waters, in press; EPA, 1979). In this scheme, initial evaluation of any agent or combination is made with simple, relatively inexpensive tests. These provide qualitative results as regards mutagenicity and have a low potential for producing false negatives, but may produce false positives. More complex tests, including those using mammalian cells in vitro, are reserved for agents considered to be positive in first tier tests. These confirm potential mutagenicity and determine possible carcinogenicity of the agents.

Final testing is difficult and is reserved for agents deemed clearly hazardous on the basis of lower level testing. Currently, this level of evaluation requires whole-animal studies. It is designed to confirm the genotoxicity of the agent under study and to attempt quantitation of the risk involved in terms of human health.

Thus, most current tests of mutagenicity are concerned only with the identification of mutagens (Newcombe, 1978), because most test systems yield qualitative information in terms of human health risks. Even results obtained from whole animal studies require extrapolation to be relevant to man. Furthermore, extrapolation must assume homogeneity of susceptibility in man—a situation that is clearly not so (Arlett and Lenman, 1978; Swift and Chase, 1979). Decisions regarding the health of humans must consider individual as well as average risk factors.

The test system described here attempts to detect directly somatic cell mutation occurring *in vivo* in man. It may be a prototype for laboratory mutagenicity tests with relevance for making human health risk assessments.

B. Direct Mutagenicity Testing

The term "direct mutagenicity testing" (DeMars, 1979) as used here refers to all tests that detect genetic damage occurring *in vivo* in an intact animal or man. The damage may be somatic or germinal and is detected in single cells. Damage may be at the chromosome or gene level.

Because direct mutagenicity tests detect events that occur *in vivo*, *in vitro* growth of cells being studied either does not occur or is limited to the extent necessary to allow recognition of the variant phenotype. Clearly the variant phenotype must be detectable at the single cell level and result from gene mutation. It is convenient if the cells under study are easily obtainable. It is required that they be polyclonal in origin with the potential for reflecting independent genetic events.

Several systems currently in use are direct mutagenicity tests. Cytogenetic tests have long been used as such (Brogger, 1979). Measurements of sister chromatid exchanges (SCEs) likewise score for genetic events occurring *in vivo* (Carrano et al., 1978, 1980). Both detect real or presumed genetic damage occurring at the chromosome level.

Because of the obvious utility of direct mutagenicity tests, novel systems are under development. The detection of sperm head morphological abnormalities (Wyrobek and Bruce, 1978) or the presence of double Y bodies in mammalian sperm (Kapp et al., 1979) are examples. The recently described detection of structurally mutant hemoglobins (Hb) (Stamatoyannopoulos et al., 1980) within red blood cells (RBCs) allows detection of rare variant (presumably mutant) RBCs arising *in vivo*. The Hb system, and the one described here, are designed to detect specific locus mutations occurring in somatic cells in man.

The utility of direct tests, particularly those which detect specific locus mutations, for linking mutagenicity testing and risk assessment will depend on the following: (1) specific locus somatic cell mutations must occur in vivo in man; (2) mutant somatic cells arising in vivo must be quantitatively detectable; and (3) frequencies of mutant cells arising in vivo must (a) correlate with exogenous or endogenous genotoxic influences, (b) be quantitative indicators of genetic damage occurring in vivo in somatic or germinal cells, and (c) be quantitative predictors of human health risks.

C. Purine Analogue Resistance in Human Somatic Cells

1. The Lesch–Nyhan Mutation in Human Fibroblasts

The development of the variant lymphocyte assay for direct mutagenicity testing in man was a direct outgrowth of earlier studies of purine analogue resistance in cultured diploid human fibroblasts (Albertini and DeMars, 1973). Its rationale is, therefore, the same as that given for the fibroblast mutational system. Both systems are based on detecting damage to the X-chromosomal gene controlling the enzyme hypoxanthine-guanine phosphorybosyltransferase (HPRT).

The naturally occurring Lesch–Nyhan (LN) mutation (Lesch and Nyhan, 1964; Seegmiller et al., 1967), which specifies an unambiguous phenotype at the single cell level, served as the prototype mutation in developing the fibroblast system (Albertini and DeMars, 1970, 1973). Variant LN-like fibroblasts arising in vitro in populations of non-LN fibroblasts are recognized by their resistance to the cytotoxic effects of some purine analogues (8-azaguanine resistant = AG^r, 6-thioguanine resistant = TG^r). Variant resistant fibroblasts are mutagen inducible in populations of normal fibroblasts, presumably by somatic cell mutation in vitro of the relevant X-linked gene. The fibroblast system has been used to study the mutagenicity of X-rays, ultraviolet (UV) light, and several chemical agents in normal and repair-deficient human cells (Albertini and DeMars, 1973; Maher et al., 1976; Burger and Simons, 1979; Jacobs and DeMars, 1979). Induced-mutant clones have been recovered and characterized as to HPRT deficiency, degree of purine analogue resistance, and ability to utilize hypoxanthine, the normal substrate of HPRT, for growth (Jacobs and Demars, 1979). The variant phenotype in fibroblasts is heritable over the somatic line. Thus, recovered mutants satisfy several of the criteria necessary to define them as genetic mutants (Chu and Powell, 1976).

2. *The LN Mutation in Human Peripheral Blood Lymphocytes*

The LN prototype mutation was used also to develop the variant TG^r peripheral blood lymphocyte (PBL) assay, as it is expressed in PBLs as well as in fibroblasts. As with fibroblasts, LN PBLs are AG^r and TG^r whereas normal PBLs are AG and TG sensitive (AG^s, TG^s).

Purine-analogue resistance in PBLs is determined by their resistance to inhibition of phytohemagglutinin (PHA)-stimulated tritiated-thymidine ([^{3}H]thymidine) incorporation *in vitro* by AG or TG (Albertini and DeMars, 1974). Analogue-resistant LN PBLs may be distinguished from analogue-sensitive normal PBLs by various methods. Initial studies (Albertini and DeMars, 1974) quantitated [^{3}H]thymidine incorporation by scintillation spectrometry. With this method, as few as 1% LN PBLs could be detected in artificial mixtures with normal cells, and LN heterozygous females were diagnosed by detecting mosaicism of their PBLs. However, these early studies showed also that scintillation spectrometry is too insensitive to detect rare LN-like PBLs in non-LN individuals.

Subsequently, an autoradiographic method sensitive enough to detect rare TG^r PBLs in normal non-LN individuals was developed to enumerate variant TG^r PBLs (Strauss and Albertini, 1977, 1979). Using this method, several studies in humans assessed the effects of various potentially mutagenic therapies on the LN-like TG^r PBL variant frequencies (V_f). In several instances, exposure of humans to known or suspected mutagens was correlated with elevated median TG^r PBL V_f values as compared to normals (Strauss and Albertini, 1977, 1979; Strauss et al., 1979). Based on these results, as well as on earlier results with fibroblasts, we have proposed that TG^r PBLs result from LN-like mutations occurring *in vivo* and that their detection allows quantitative estimation of somatic cell mutation occurring *in vivo* in man.

3. *A Problem of Phenocopies*

Recently, we reported that mononuclear cells (MNCs) obtained from fresh blood samples contain infrequent TG^s cells that appear as labeled TG^r cells under our original conditions of assay (Albertini and Allen, in press; Albertini et al., in press). We first suspected the presence of these cells from the wide range of apparent TG^r PBL V_f values determined by careful testing of fresh blood samples from normal, non-mutagen exposed individuals. Furthermore, with fresh blood samples, intraindividual variations in apparent TG^r PBL V_f values are as great as interindividual variations. Subsequent studies suggested that the cells responsible for this phenomenon arise, at least in part, from MNCs that are spontaneously cycling *in vivo* and/or during the early periods of cell culture *in vitro*. We have referred to these presumably nonmutant TG^s cells which appear as mutant TG^r PBLs as "phenocopies."

An original working assumption in developing the TG^r PBL assay was that all PBLs are in an arrested G_0 state of their cell cycle in the peripheral blood prior to testing. However, we have since shown that fresh samples of MNCs (which contain the PBLs) are heterogenous mixtures containing infrequent cycling cells (Albertini *et al.*, in press). The exact frequency of cycling cells is variable and perhaps influenced by a variety of factors. Nonetheless, these cells probably are not reliably inhibited from initiating DNA synthesis *in vitro* during their short exposure to TG. Apparently more time is required for TG to inhibit TG^s cycling cells than is required for it to inhibit TG^s resting G_0 cells. We had noted previously that dividing human TG^s fibroblasts undergo several division cycles *in vitro* in the presence of TG before finally being arrested. Similarly, a fraction of TG^s cycling MNCs may appear as TG^r cells.

Thus, analysis of the cell culture conditions underlying the originally proposed TG^r PBL V_f method suggests the possible origin of phenocopies. DNA synthesis in the assay is induced *in vitro* in PBLs by PHA stimulation. Our culture interval is brief (see Section II,C): 30 hours of culture plus 12 hours of label. This shortened interval was chosen deliberately so that no PBL which enters the cell cycle following PHA stimulation *in vitro* can complete its initial cycle, replicate, and be in its second period of DNA synthesis when labeled with [^{3}H]thymidine (Strauss and Albertini, 1979). Any PBL which becomes labeled does so during its first round of DNA synthesis *in vitro*. Purine-analogue resistance, therefore, antedates the *in vitro* assay rather than being produced by it and TG^r PBLs detected by the method must necessarily arise *in vivo*. This, of course, is critical for direct mutagenicity testing.

The solution to this phenocopy problem then is either to eliminate cycling cells or to interfere with their ability to synthesize DNA during the critical interval of assay. Fortuitously, at approximately the same time as we recognized the problem, we also noted lower and consistent TG^r PBL V_f values in normal individuals when MNC samples that had been cryopreserved in liquid nitrogen were tested. A systematic investigation of this phenomenon indicated that freezing interferes with the [^{3}H]thymidine incorporation of phenocopies in TG during the critical interval of assay (Albertini *et al.*, in press). Although the mechanisms are unknown, cryopreservation and, to a certain extent, re-frigeration at 4°C appears to reduce or eliminate phenocopies at the current level of sensitivity of the method. Thus, for test purposes, we now use only cryopreserved MNCs or samples which have been held at 4°C for at least 12 hours (Albertini et al., in press). Employing these methods to eliminate phenocopies, we still find that humans exposed to known mutagens have elevated TG^r PBL V_f values as compared to normals. We feel these TG^r PBLs are in large part LN-like mutants.

II. THE THIOGUANINE-RESISTANT (TG^r) PBL ASSAY METHOD

A. Cell Preparation

Whole blood is obtained by venipuncture into syringes coated with 0.1 ml heparin (beef lung, 1000 U/ml, benzylalcohol preservative; Upjohn) per 10 ml blood. Syringes are inverted to mix contents and may be left up to 18 hours at room temperature before separating MNCs.

MNCs are separated from whole blood by the Ficoll–Hypaque method (Boyum, 1968). Materials are prewarmed to 37°C. Blood samples greater than 5 ml require one 50-ml, sterile glass, round-bottom, centrifuge tube for separation of each 15 ml of blood. To each tube is added 15 ml of sterile Ficoll–Hypaque (specific gravity 1.077). Blood samples of 5 ml or less are separated in smaller tubes, to which is added Ficoll–Hypaque in a volume equal to that of blood. Blood is gently layered over the Ficoll–Hypaque and centrifuged at room temperature at 600 g for 30 minutes. After centrifugation, the upper layer in tubes is plasma, the next visible layer is the MNC layer, and at the bottom are the other components of the blood.

The plasma is gently aspirated with a plugged pipette to approximately 5 mm above the MNC layer and may be centrifuged at 1000 g for 10 minutes and used immediately as a medium supplement or frozen for future use. The MNC layer is gently aspirated using a plugged pipette. Mononuclear cells are transferred to point-bottom centrifuge tubes which are then filled with sterile phosphate-buffered saline (PBS). Tubes are centrifuged at room temperature at 150 g for 10 minutes, the supernatant removed, and the PBS wash repeated. Following another centrifugation, the supernatant is removed leaving the MNC pellet. If fresh cells are to be tested, the MNCs are diluted in complete medium (see Section II,C) after counting. Otherwise, cells are refrigerated overnight or frozen as described further on.

On average, $1-2 \times 10^6$ MNCs are recovered from each ml of blood drawn from an adult.

B. Cryopreservation

For freezing, the MNC pellet, after washing with PBS, is resuspended in medium RPMI 1640 with additives (25 mM HEPES, 2 mM glutamine, 100 U/ml penicillin, 100 µg/ml streptomycin) supplemented with 10% human AB serum or autologous plasma and 7.5% dimethyl sulfoxide (DMSO) so as to achieve a final cell density of 10^7/ml. The procedure is done rapidly because DMSO is toxic to cells at room temperature. The MNC-freezing mixture is aliquoted as 1-ml samples into 1-ml Nunc ampuls. The ampuls are then frozen in a Union Carbide Biological Freezer for controlled cooling according to manufacturer's instructions. Frozen cells are stored in the vapor phase in liquid-nitrogen containers. Storage of cells may be for many months.

If cells are to be refrigerated at 4°C for 12 hours rather than frozen in liquid nitrogen, they are resuspended in complete medium (see Section II,C) following the PBS wash indicated above.

C. Cell Culture

If fresh or refrigerated cells are tested, a very small aliquot of cells in PBS or medium RPMI 1640 alone is diluted with trypan blue for viability counts. The remainder of the cells is resuspended in complete medium (CM) which consists of medium RPMI 1640, supplemented with 25 mM HEPES, 2 mM glutamine, penicillin/streptomycin (100 U/ml and 100 µg/ml, respectively, GIBCO), and 30% autologous plasma or 20% AB human serum (Biobee) and counted. After counting, cells are resuspended in CM at a final density of 1.4×10^6 cells/ml.

If frozen cells are tested, ampuls of cells prepared as described in Section II,B are rapidly thawed in a 37°C water bath, added drop by drop to prewarmed (37°C) medium RPMI 1640 to a final dilution of at least 1 to 10, centrifuged at room temperature at 600 g for 10 minutes, and the resultant pellet resuspended in a small volume of medium RPMI 1640. An aliquot is removed for viability counts, and the remainder resuspended in CM and counted and diluted to 1.4×10^6 cells/ml as above. Cryopreserved cell recovery varies from 30 to 100%. Viability should be in excess of 90%.

Culture tubes are prepared to receive control (without TG) or test (with TG) cultures. Phytohemagglutinin (PHA-P, 4 µg/0.2 ml; Burroughs Wellcome), is added to several screw-top, point-bottom glass culture tubes. For each person tested, two or more tubes are prepared to receive control cultures and several tubes are prepared to receive test cultures. For control cultures, tubes receive, in addition to the PHA-P, medium RPMI 1640 that has been adjusted to the pH of the stock TG solution. For TG cultures, tubes receive, in addition to the PHA-P, stock 2×10^{-3} M TG solution. For both, additions are in 0.1-ml volumes per 1 ml final culture. Finally, the MNC suspension prepared as described above (1.4×10^6 cells/ml) is added to all tubes at 0.7 ml per 1 ml final culture volume. Thus, all cultures are made to contain 1×10^6 or 2×10^6 MNCs in 1- or 2-ml final culture volumes, respectively.

Inoculated culture tubes with loosely applied screw tops are incubated in a humidified, 5% CO_2 atmosphere at 37°C for 30 hours. At 30 hours, 5 μCi [^{3}H]thymidine (New England Nuclear, S.A. 6.7 Ci/mM) is added per ml culture. Cultures are incubated for an additional 12 hours and terminated.

D. Termination, Coverslip Preparation, and Autoradiography

1. Termination

Cultures are terminated by adding 4 ml of 0.1 M citric acid to each tube. After mixing, tubes are centrifuged at room temperature at 600 g for 10 minutes. The supernatant is aspirated and put into radioactive waste. The remaining pellet contains free nuclei and cytoplasmic fragments. Four milliliters of freshly made fixative containing 7 parts methanol to 1.5 parts glacial acetic acid are added to each tube and the nuclei resuspended. Tubes again are centrifuged at room temperature at 600 g for 10 minutes, supernatant again aspirated for radioactive waste, and the pellets resuspended in 0.2 ml of fixative. Tightly capped tubes are refrigerated at 4°C for at least 1 hour for fixation. Samples may be kept for several days at this stage.

2. Coverslip Preparation

Nuclei in fixative suspension are counted with the Coulter Counter (ZBI Model). We count 10 μl of the suspension and add the remainder, in a carefully measured volume, to 18 × 18 mm glass coverslips previously fixed with permount to glass slides. The suspension is allowed to spread evenly and care is taken that sample does not spill over the coverslip. The total number of nuclei added to a coverslip is calculated from the density of nuclei per microliter (Coulter count) times the number of microliters added to the coverslip.

Slides so prepared are air dried, dipped in 1% filtered aceto-orcein stain for 1 minute, dipped in distilled water, rinsed in cold running tap water, and again air dried.

3. *Autoradiography*

In complete darkness, dried stained slides in holders are dipped for 10–15 seconds in autoradiographic emulsion (NTB2; Kodak) which is stored at 4°C away from radioactive material and prewarmed in a 50°C oven 5–6 hours prior to use. After draining to remove excess emulsion, slides in holders are placed in a light tight holding box for exposure. The box is closed, wrapped in black plastic bags, and sealed with tape. The box is placed in a refrigerator (4°C) or freezer (-20°C) for a minimum of 24 hours (but may be kept in the freezer for several weeks before developing).

After exposure, slides in holders are removed from the holding box, again in total darkness, and dipped for 4 minutes in developer D-19 (Kodak), drained, placed in a stop bath for 10–30 seconds, removed, again drained, and placed in fixer (Kodak) for 5 minutes. Finally, slides (no longer light sensitive) are washed briefly in cold water and allowed to dry.

E. Enumeration of TG^r PBLs: Calculation of TG^r PBL Variant Frequency (V_f) Values

Coverslips with nuclei from cultures containing PHA-P and pH adjusted medium RPMI 1640 without TG (controls) are viewed under high-power microscopy ($970\times$). Twenty-five hundred nuclei chosen at random are counted on each coverslip (2 coverslips = 5000 nuclei counted) and the incidence of autoradiographically labeled (positive) cells determined. (All nuclei containing silver granules above background levels are considered positive.) The labeling index (*LI*) of control cultures is determined:

LI (control cultures) = No. labeled nuclei in 5000 nuclei/5000.

Coverslips prepared from test cultures containing PHA-P and TG are viewed at low power ($160\times$) so that all nuclei are scanned. All labeled nuclei on each coverslip are counted. Autoradiographically positive nuclei detected at low power are confirmed at high power ($970\times$). The *LI* of test cultures is determined by dividing the number of all labeled nuclei on all test coverslips by the total number of nuclei on the coverslips, which is obtained from Coulter counts as indicated above (Section II,D,2). Thus,

LI (test cultures) = No. labeled nuclei from all test cultures/total No. nuclei recovered from test cultures.

The TG^r PBL V_f is calculated as the ratio of the control and test LI values:

$$TG^r \text{ PBL } V_f = LI \text{ (test cultures)}/LI \text{ (control cultures)}.$$

III. SAMPLE RESULTS

A. TG^r PBL V_f Assay: Appearance of Slides

A portion of a coverslip containing nuclei from a control culture (without TG) is shown in photomicrographs made at low power ($160\times$; Fig. 1A) and high-power oil-immersion microscopy ($1000\times$; Fig. 1B). The restricted field shown in Fig. 1A has an LI of 11%. When 5000 nuclei from the control culture were examined (see Section II,E) the LI was 13%. Positive nuclei on slides made from test cultures have this same appearance. All positive nuclei on test slides are confirmed at high-power microscopy under oil immersion ($1000\times$).

B. TG^r PBL V_f Values: Normal Nonmutagen Exposed Individuals: Fresh versus Cryopreserved MNCs

Table I gives TG^r PBL V_f values determined for 10 healthy non-LN individuals. After blood samples were obtained and the MNC fractions separated, an aliquot of each sample was cryopreserved in liquid nitrogen. The remainder of the MNC fraction of each sample was tested fresh.

As seen from the table, V_f values determined on the fresh MNCs range from 2.2×10^{-5} to 2.3×10^{-3}, with a median value of 3.4×10^{-4} for the 10 individuals. By contrast, TG^r PBL V_f values determined on cryopreserved MNCs derived from the same original blood samples were considerably lower ranging from 0 to 2.7×10^{-5} (median $V_f = 1.4 \times 10^{-5}$). As shown in Table I, in every instance, the TG^r PBL V_f determined for the cryopreserved sample was lower than that determined for the paired fresh sample obtained at the same time from the same individual. This has been observed repeatedly when testing fresh versus cryopreserved samples (see Section I,C,3). Freezing and thawing appear to prevent the scoring of phenocopies which inflate V_f values obtained with fresh cells. Assuming no real difference in

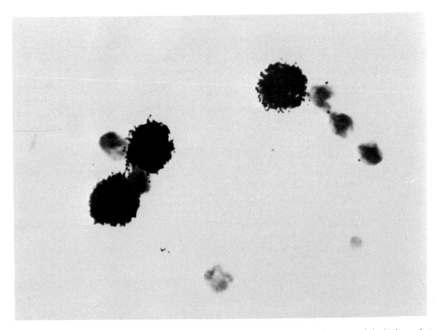

Fig. 1. (A) Appearance of nuclei from a control culture showing 11% labeled nuclei (160×). (B) Appearance of nuclei from a control culture showing three labeled nuclei and six unlabeled ones (1000×).

TABLE I

TG⁻ PBL V_f: Normal Healthy Controls: Fresh and Frozen MNCs

	Fresh MNCs				Frozen MNCs				
Subject	No. labeled nuclei[a]	No. nuclei/ slide × 10^6[b]	LI[c]	$V_f \times 10^{-6}$[d]	No. labeled nuclei[a]	No. nuclei/ slide × 10^6[b]	LI[c]	$V_f \times 10^{-6}$[d]	95% Confidence interval $V_f \times 10^{-6}$
1 F	5	0.059	0.43	280	0	0.300	0.36	4.5	(0.449, 25.13)
	9	0.056			1	0.319			
2 M	9	0.082	0.29	390	3	0.319	0.16	27	(6.81, 69.4)
	13	0.112			1	0.599			
3 F		—	0.29	450	0	0.268	0.19	0	(0, 34.90)
	25	0.190			0	0.290			
4 F	16	0.140	0.34	490	1	0.246	0.20	21	(2.07, 74.53)
	26	0.109			1	0.237			
5 F	8	0.104	0.35	170	0	0.327	0.20	14	(1.39, 50.27)
	12	0.232			2	0.389			
6 M	7	0.164	0.27	83	1	0.374	0.14	11	(1.13, 63.09)
	2	0.238			0	0.260			
7 M	1	0.115	0.33	22	0	0.139	0.17	0	(0, 73.53)
	1	0.166			0	0.157			
8 M	21	0.159	0.41	320	0	0.143	0.20	0	(0, 48.30)
	13	0.101			0	0.240			
9 F	22	0.084	0.33	820	0	0.125	0.18	19	(1.90, 106.18)
	39	0.140			1	0.168			
10 M	186	0.140	0.42	2300	1	0.328	0.25	7.1	(0.71, 39.93)
	187	0.252			0	0.233			

[a] Obtained by microscopy count of entire slide.

[b] Obtained by Coulter count of suspension of nuclei in fixative.

[c] Labeling index, control culture.

[d] $LI \times$ No. nuclei/slide $\times 10^6$.

mean values, the probability that the TG^r PBL V_f value of the frozen–thawed member of each paired sample is the lower of the two is quite small $[p = (\tfrac{1}{2})^{10} \cong 0.001]$.

The TG^r PBL V_f values determined on cryopreserved MNCs as given in Table I have wide 95% confidence intervals (CIs; see Section IV,C). This is due in large part to the small numbers of evaluatable nuclei available for each individual tested. Larger experiments are required to allow meaningful quantitation of individual V_f values and assessment of differences among individuals.

C. TG^r PBL V_f Values: Normal versus Cancer Patients Exposed to Chemotherapy

Because of the wide 95% CIs for TG^r PBL V_f values determined in experiments with small numbers of evaluatable nuclei, larger experiments designed to evaluate at least 5×10^5 nuclei per individual tested were performed. Table II gives the results of two such studies where MNCs from three normal individuals and four cancer patients were tested. Data from one of these studies have been published elsewhere (Albertini et al., in press).

As seen from Table II, blood samples from the three control individuals gave TG^r PBL V_f values of 8.3×10^{-6}, 2.4×10^{-6}, and 4.8×10^{-6}. These estimates were made on more than 5×10^5 evaluatable PBLs per individual tested, and have 95% CIs that allow useful comparisons of these values with each other, and with those determined for cancer patients receiving chemotherapy. TG^r PBL V_f values for the four cancer patients were 1.3×10^{-5}, 5.35×10^{-5}, 5.2×10^{-5}, and 1.7×10^{-5}, all of which are significantly greater than values determined for the controls.

All MNC samples tested in these experiments were cryopreserved as described except for cells from cancer patient No. 4. The MNCs from patient 4 were precultured at 37°C in mitogen free medium for 4 days as an alternative method for removing phenocopies. Although we do not have sufficient data to evaluate this method, the values obtained are comparable to those obtained with freezing and thawing and are included for illustrative purposes.

Comparisons of TG^r PBL V_f values may be made among all individuals tested in these two experiments. Values for some treated cancer patients are significantly higher than those for others and higher also than the V_f values determined for the normal controls (see Section IV and Table III).

IV. STATISTICAL ANALYSIS METHODS

A. Notation and Basic Assumptions

In this section we introduce the basic statistical assumptions, notation, and approach to the problem of finding point estimates, confidence intervals and tests of hypothesis regarding variant frequencies. In Section IV,B, we describe a general approach using logarithims to transform products and ratios of random variables into sums and differences with terms that are approximately normally distributed. In Section IV,C we show by several numerical examples that this simple approach yields results which are very close to other more exact but less flexible methods. We also present alternative methods of drawing confidence intervals when numbers of variants counted are small. Section IV,D discusses methods for determining necessary sample sizes to achieve desired precision for confidence intervals.

Let the true variant frequency be denoted by v. The estimate of v is simply the ratio of the number M of variants observed to the N evaluatable cells from TG cultures. $N = T \cdot LI$ is the product of the number, T, of cells from TG cultures, multiplied by the labeling index, LI (obtained from control cultures).

$$V_f = \frac{M}{N} = \frac{M}{T \cdot LI} \tag{1}$$

For the labeling index, LI, let C be the number of cells scored from control cultures and let B denote the number of labeled cells from control cultures. Then

$$LI = B/C \tag{2}$$

We assume that T and C are measured without (substantial) error. They will be treated as known constants in our development. For a given subject under prescribed assay conditions, the number of variants M and the number of labeled cells B are treated as random variables. The M is assumed to have a Poisson distribution with parameter μ, while B is assumed to have a binomal distribution with parameter C equal to the number of independent trials (which is simply the number of cells scored) and p equal to the probability that a cell is labeled. With this notation we can represent v, the "true" variant frequency by an expression that parallels the sample version in Eq. (1):

$$v = \frac{\mu}{TP} \tag{3}$$

TABLE II

TG^r PBL V_f Values^a

	Normal individual						Cancer patients: Multiple agent chemotherapy							
	(1)		(2)		(3)		(1)		(2)		(3)		(4)	
	a^b	b^c	a	b	a	b	a	b	a	b	a	b	a	b
LI^d	0.21		0.24		0.20		0.17		0.21		0.16		0.32	
Culture No.														
1	3	1.004	0	1.104	—	—	1	0.809	12	0.705	6	0.801	0	0.242
2	1	1.315	1	1.116	0	0.256	0	1.015	8	0.877	9	0.704	2	0.451
3	3	1.373	1	1.141	0	0.387	5	1.124	12	1.264	6	1.23	1	0.188
4	5	1.394	3	1.136	0	0.851	4	1.176			9	0.708	0	0.168
5	4	1.376	—	—	1	0.702	1	1.165					3	0.149
6	1	1.366	0	1.418	0	0.874	4	0.917					3	0.251
7	4	1.375	0	1.239	0	0.583	2	1.020					2	0.438
8	1	1.292	0	1.662	3	0.554	2	0.848					6	0.316
9	0	1.677	1	1.426	—	—							1	0.304
10	2	1.304	1	1.395									1	0.451
11													4	0.795

12							0	0.586
13							2	0.332
14							1	0.184
15							1	0.245
Sum of labeled nuclei	24	7	4	19	32	30	27	
Sum of nuclei recovered	13.476	11.637	4.207	8.074	2.846	3.510	4.900	
Evaluatable PBLs[e]	2.830	2.793	0.841	1.373	0.598	0.562	1.570	
V_f[f]	8.3×10^{-6}	2.4×10^{-6}	4.8×10^{-6}	1.3×10^{-5}	5.35×10^{-5}	5.2×10^{-5}	1.7×10^{-5}	
95% Confidence interval $\times 10^6$	(5.54, 12.43)	(1.14, 5.04)	(1.80, 12.81)	(8.26, 20.5)	(36.66, 73.8)	(36.2, 74.8)	(11.6, 24.8)	

[a] Large experiments: multiple cultures, MNCs from normals and treated cancer patients; cryopreserved MNCs – 2×10^6 MNCs/ml culture; 2-ml cultures.

[b] a = No. labeled nuclei; obtained by microscopy count of entire slide.

[c] b = No. nuclei recovered per culture; obtained by Coulter count of suspension of nuclei in fixative, $\times 10^6$.

[d] Labeling index, control culture.

[e] Evaluatable PBLs = $LI \times b$, $\times 10^6$.

[f] Variant frequency = $\dfrac{a/b}{LI} = \dfrac{a}{b \times LI} = \dfrac{a}{\text{evaluatable PBL}}$.

The variables introduced here possess several important properties. A Poisson variable with parameter μ has that value as both mean and variance: $\mu = E(M) = Var(M)$. The observed count M is the "best" estimate of the parameter μ, where "best" includes the properties of being the maximum likelihood estimator, being unbiased, and having the smallest variance of all unbiased estimates; M is approximately normal for moderate or large values of μ (Hogg and Craig, 1970).

Because B is assumed to be binomial with parameter p, then the labeling index $LI = B/C$ can be shown to be the best estimate of p just as M is the best estimate of μ. Because $E(LI) = p$, LI is unbiased and is approximately normally distributed with mean equal to p and variance equal to pq/C, where $q = 1 - p$.

The maximum likelihood estimate of the true variant frequency is then the observed variant frequency $V_f = M/(T \cdot LI)$. Table II gives some numerical values obtained from our studies. For example, on cancer patient number 2, we see that $M = 32$, $T = 2.846 \times 10^6$, $LI = 0.21$ and so the estimate of v is $V_f = 5.35 \times 10^{-5}$.

Because M and LI are random variables, the quantity V_f will also be a random variable. We are interested in finding the standard deviation of V_f so that we can assess the degree of precision with which V_f is an estimate of v. Similarly, we wish to carry out a variety of tests of hypothesis regarding v or more usually, several v values from different subjects.

B. Approximate Confidence Interval Using the Transformation $Y = \ln(X)$

We shall make extensive use of the transformation $Y = \ln(X)$, where ln denotes the natural logarithm. Using Taylor series expansions of $\ln(X)$ around the point $x = E(X)$, we have the approximate relationship

$$\ln(X) \doteq \ln[E(X)] + [X - E(X)]/E(X) \tag{4}$$

For X representing a Poisson variable or a sample mean, it can be shown that $\ln(X)$ is approximately normally distributed with mean $E[\ln(X)] \doteq \ln[E(X)]$ and variance $Var[\ln(X)] \doteq Var(X)/[E(X)]^2$. This general result will be applied to find a confidence interval (CI) for $E(X)$. The essential probability statement for a normal variate Y is

$$Prob(-1.96 \leqslant [Y - E(Y)]/\sigma_Y \leqslant 1.96) = 0.95 \tag{5}$$

For X a Poisson variate and $Y = \ln(X)$, we have $Var[\ln(X)] \doteq 1/E(X)$, which we can estimate by $1/X$. We then obtain an approximate 95% CI for $\ln[E(X)]$:

$$[\ln(X) - 1.96/\sqrt{X}, \ln(X) + 1.96/\sqrt{X}] \qquad (6)$$

Taking antilogs gives a 95% CI for $E(X)$:

$$(X \cdot e^{-1.96/\sqrt{X}}, X \cdot e^{1.96/\sqrt{X}}) \qquad (7)$$

For example, for cancer patient number 2, we obtain the 95% CI (22.63, 45.25) which is close to (21.8, 45.1) the exact CI obtained from tables (Beyer, 1966; Pearson and Hartley, 1976).

As a second example, let X be the observed fraction of successes of C independent trials where p denotes the probability of success on each trial. Then $E(X) = p$ and $Var(X) = pq/C$. Applying Eq. (4) we have $E[\ln(X) \doteq \ln(p)$ and $Var(\ln(X)] \doteq q/pC$. If $C = 5000$ and the observed fraction is $X = 0.21$, then the estimated standard deviation of $\ln(X)$ is

$$\sigma \doteq \sqrt{(0.79)/(0.21)(5000)} = 0.0274$$

Then the approximate 95% CI for p is

$$[(0.21)e^{(-1.96)(0.0274)}, (0.21)e^{(1.96)(0.0274)}] \qquad (8)$$

which yields (0.199, 0.222) while the usual CI, based on the normality of a binomial fraction, is (0.199, 0.221).

The above approximations may be used to find CIs for variant frequencies and ratios of variant frequencies. In the TG^r PBL assay recall that $V_f = M/(T \cdot LI)$, where M, the number of variants counted, is assumed to be a Poisson variate with parameter μ; T is the number of cells recovered (TG culture) and is assumed to be counted with small error. The labeling index, LI, is the fraction of the C cells (control cultures) that are labeled. The transformation to natural logs gives

$$\ln(V_f) = \ln(M) - \ln(LI) - \ln(T) \qquad (9)$$

Using Taylor series expansions for the terms containing M and LI gives the approximation

$$\ln(V_f) \doteq \ln(\mu) + \frac{1}{\mu}(M - \mu) - \ln(p) - \frac{1}{p}(LI - p) - \ln(T) \quad (10)$$

The approximate mean of $\ln(V_f)$ is then $\ln(\mu) - \ln(p) - \ln(T) = \ln(v)$ and the approximate variance is

$$Var[\ln(V_f)] \doteq \frac{1}{\mu^2} Var(M) + \frac{1}{p^2} Var(LI) = \frac{1}{\mu} + \frac{q}{pC} \qquad (11)$$

A convenient estimate of the approximate variance is

$$s^2 = \frac{1}{M} + \frac{1 - LI}{C \cdot LI} \tag{12}$$

A CI for v is determined in the same way as the CI for μ above. The final form is

$$(V_f e^{-1.96s}, \ V_f e^{1.96s}) \tag{13}$$

We illustrate the use of Eq. (13) for normal subject number 1 in Table II for which $M = 24$, $T = 13.476 \times 10^6$, $LI = 0.21$, $C = 5000$, and $V_{f1} = 8.3 \times 10^{-6}$. Then the approximate standard deviation for $\ln(V_f)$ is $s = 0.206$ so that the CI is $(5.54 \times 10^{-6}, 12.43 \times 10^{-6})$. For cancer patient number 2, the approximate standard deviation of $\ln(V_{f2})$ is $s = 0.179$ and the CI for v is $(37.68 \times 10^{-6}, 76.03 \times 10^{-6})$.

To compare two variant frequencies, V_{f1} and V_{f2}, we simply note that $\ln(V_{f2}/V_{f1}) = \ln(V_{f2}) - \ln(V_{f1})$ and apply the preceding methods. Since V_{f1} and V_{f2} are assumed independent, the variance for the difference in logarithms is simply the sum of the two variances. Using the data for normal control number 1 and cancer patient number 2 discussed above, the estimated ratio is $V_{f2}/V_{f1} = 6.27$. The approximate variance of $\ln(V_{f2}/V_{f1})$ is $s_1^2 + s_2^2 = .042 + .032 = .074$, and the approximate CI for v_2/v_1 is (3.69, 10.69). Since this CI excludes the particular value $v_2/v_1 = 1$ we conclude that v_1 and v_2 are significantly different from each other at $P \leq 0.05$.

Table III gives the 95% CIs using the individual V_f values and using ratios of the V_f values presented in Table II. For example, comparing cancer patient number 4 with control number 1, we see that the CI for the ratio is (0.853, 2.87). Since the interval contains 1, we cannot conclude that the two true variant frequencies are significantly different from each other.

Table III presents 28 separate 95% CIs. Although it would be more appropriate to use simultaneous statistical procedures for these comparisons, we have not done so in order to better illustrate the method. Relevant simultaneous test procedures are discussed in Miller (1966).

The approximate normality of the log transformation can also be used when more than two variant frequencies are to be compared. For example, a set of estimated variant frequency rates can be tested for equality by a χ^2 statistic.

$$\chi^2 = \sum_{i=1}^{n} \frac{[\ln(V_{fi}) - \overline{\ln(V_f)}]^2}{\ln(V_f)} \tag{14}$$

TABLE III

Estimates and 95% Confidence Intervals for Variant Frequencies and Ratios of Variant Frequencies

	Normal controls			Cancer patients			
	1	2	3	4	5	6	7
$V_f \times 10^{6a}$	8.3	2.4	4.8	13.0	53.5	52.0	170.0
$Var[\ln(V_f)]^b$	0.0424	0.1435	0.2508	0.0536	0.0320	0.0344	0.0375
L^c	5.54	1.14	1.80	8.26	37.68	36.15	116.30
U^d	12.43	5.04	12.81	20.47	76.03	74.79	248.43
V_{fi}/V_{f1}		0.29	0.58	1.56	6.45	6.27	20.48
L		0.12	0.20	0.853	3.78	3.64	11.77
U		0.68	1.67	2.87	11.00	10.79	35.64
V_{fi}/V_{f2}	3.46		2.00	5.42	22.29	21.67	70.83
L	1.48		0.58	2.27	9.80	9.48	30.77
U	8.05		6.85	12.93	50.66	49.52	163.06
V_{fi}/V_{f3}	1.73	0.50		2.71	11.14	10.83	35.42
L	0.60	0.15		0.92	3.93	3.80	12.36
U	5.00	1.72		7.99	31.60	30.86	101.45
V_{fi}/V_{f4}	0.64	0.18	0.37		4.12	4.00	13.08
L	0.35	0.08	0.13		2.32	2.24	7.24
U	1.18	0.44	1.09		7.30	7.15	23.63
V_{fi}/V_{f5}	0.16	0.04	0.09	0.24		1.03	3.17
L	0.09	0.02	0.03	0.14		0.62	1.89
U	0.26	0.10	0.25	0.43		1.70	5.33
V_{fi}/V_{f6}	0.16	0.05	0.09	0.25	0.97		3.27
L	0.09	0.02	0.03	0.14	0.59		1.93
U	0.27	0.11	0.26	0.45	1.61		5.53
V_{fi}/V_{f7}	0.05	0.01	0.03	0.08	0.31	0.31	
L	0.03	0.006	0.01	0.04	0.19	0.18	
U	0.08	0.03	0.08	0.14	0.53	0.52	

[a] Variant frequency.
[b] Variance $[\ln(V_f)]$.
[c] Lower 95% boundary.
[d] Upper 95% boundary.

which will tend to be large in case at least one of the $\ln(V_{fi})$ values is quite different from the average value $\overline{\ln(V_f)}$. Similarly, one can investigate possible relationships between a set of variant frequencies and other variables (such as degree of exposure to potential mutagenic agents) by weighted regression analysis of the $\ln(V_{fi})$ using weights that are inversely proportional to the estimated variances. Most statistical computer packages include weighted regression routines.

C. Alternative Methods for Obtaining Confidence Intervals

In this section we compare the log transformation with alternate approaches to finding CIs. This includes methods that can be used when M is in the range of 0 to 10 where the log transformation is not well approximated by the normal distribution. Methods presented here will be used in Section IV,D to determine approximate sample sizes to achieve desired precision.

Again, recall that the estimated variant frequency is $V_f = M/N = M/(T \cdot LI)$, where M is a Poisson variate and the labeling index LI is a binomial fraction. Tables of exact CIs for Poisson variates can be consulted to obtain a CI for $E(M) = \mu$. For example, if $M = 32$, the exact 95% CI is $(a_M, b_M) = (21.8, 45.1)$. In regard to the denominator of V_f, the number N of cells evaluated is a binomial random variable with parameters T and p where p is itself estimated from the sample of C control cells. The mean and variance of N is calculated by using conditional means and variances (Parzen, 1960). The final result is

$$E(N) = T \cdot p \tag{15}$$

$$Var(N) = \frac{T^2 pq}{C} \left(1 + \frac{C-1}{T}\right) \tag{16}$$

For our assay, C is much smaller than T so that $(C-1)/T \doteq 0$ in Eq. (16). Thus the approximate standard deviation of N is $s_N = T(pq/C)^{1/2}$. Since N is approximately normal we can easily find a 95% CI for the actual number of cells evaluated. For example for cancer patient number 2 we obtain $T \cdot LI = N = 0.598 \times 10^6$ and $s_N = 16.47 \times 10^3$ so that a 95% CI is $N \pm (1.96) s_N = (a_N, b_N) = (.566 \times 10^6, .630 \times 10^6)$.

We can now form two different CIs for $\mu/E(N)$. The narrower CI is to treat N in the denominator as a constant and divide that value into the end points of the CI for the numerator to obtain $(a_M/N, b_M/N)$. Using this approach with the data given above we obtain $(36.45 \times 10^{-6}, 75.42 \times 10^{-6})$. The wider CI is to divide the lower confidence interval bound in the numerator by the upper bound in the denominator to obtain a_M/b_N, the smallest "possible" value for the lower bound. Using the data above we find $(a_M/b_N, b_M/a_N) = (34.60 \times 10^{-6}, 79.68 \times 10^{-6})$. Recall that the approximate CI obtained using the log transformation is $(37.68 \times 10^{-6}, 76.03 \times 10^{-6})$ which is close to the two found here.

The latter approach for determining a CI for v values is especially relevant when the observed M is very small so that $\ln(M)$ is not approximately normal. The data in Table I are illustrative of this situation. For frozen PBLs we note that the number of labeled nuclei ranges from 0 to 4 for the ten subjects. Using Poisson tables to obtain exact CIs for the numerators and dividing those bounds by the estimated number of evaluatable cells, we obtain the narrowest bounds. For example, for subject number 1, we have $M = 1$ which gives a 95% CI for μ of $(0.1, 5.6)$. Dividing $N = 0.619 \times 10^6$ into the confidence bounds gives the 95% CI for v of $(0.449 \times 10^{-6}, 25.13 \times 10^{-6})$. The true CI is a little wider than this approximation, so that it is clear that experiments in which the count M is small cannot yield very precise estimates of true variant frequency rates.

Finding a CI for the ratio of variant frequencies when M_1 and M_2 are small is somewhat more complicated. As shown above, most of the variability in V_f is due to the variability in M and little is due to the variability in N, the number of evaluatable cells. Accordingly, when comparing two V_f values we shall assume that the respective N_1 and N_2 are fixed and known. It then follows that M_1 is a Poisson variate with parameter $\mu_1 = N_1 v_1$, and similarly for M_2. It can be shown that the conditional distribution of M_2 given the total $T = M_1 + M_2$ is a binomial variable with parameters T and

$$p = \frac{v_2 N_2}{v_1 N_1 + v_2 N_2} \tag{17}$$

One can then use tables (Diem, 1962) to get an exact CI for p from which an exact CI for v_2/v_1 can be obtained (by algebra) since Eq. (17) is equivalent to

$$\frac{v_2}{v_1} = \frac{N_1}{N_2(1/p - 1)} \tag{18}$$

For example, comparing v_3 to v_2 for the third and second control subjects in Table II, we have $T = M_2 + M_3 = 7 + 4 = 11$ and the 95% CI for p is $(0.11, 0.69)$. Using N_3, N_2, and the two CI end points for p we find from Eq. (18) the CI for v_3/v_2 is $(0.407, 7.465)$. The interval is (disappointingly) wide but that must be expected since small values of M will always lead to relatively wide CIs. Bross (1954) gives further details and some modifications of this approach. The procedure can be extended to the simultaneous comparison of sets of data using the multinomial distribution which is a generalization of the binomial. However, the statistical power of the approach will be low because of the large variability in the original counts, M_i, $i = 1, 2, \ldots, K$.

D. Sample Size Determinations

It is useful to have at least rough estimates of required sample sizes to obtain preassigned precision in CIs for individual variant frequencies and for ratios of variant frequencies. Tables of exact CIs for Poisson variables will suggest desired approximate sample sizes for a single variant frequency. For example, if $M = 35$, the 95% CI for μ is (24.3, 48.7). Note that the upper limit is about twice as large as the lower limit and each of the limits are within about 30% of the estimated value $M = 35$. If such precision is adequate for the study purposes, then one needs to evaluate enough cells to achieve (approximately) $M = 35$.

One method of achieving this is to collect sufficient evaluatable cells, N, until exactly $M = 35$ labeled cells are scored. In that case M is treated as a preassigned constant and N is the variable. It can then be shown (Parzen 1960) that $W = 2\nu N$ is a χ^2 variable with $2M$ degrees of freedom. This leads to a CI for ν based on the χ^2 distribution. For example, if the data for cancer patient number 3 had been gathered in this manner so $M = 30$ (fixed), then $2(0.562 \times 10^6)$ would be the value of a χ^2 variable with $2 \times 30 = 60$ degrees of freedom. The corresponding 95% CI for ν is $(36.0 \times 10^{-6}, 74.1 \times 10^{-6})$ which is very close to the interval $(36.2 \times 10^{-6}, 74.8 \times 10^{-6})$ found in Table III when M was treated as the variable.

The approximate sample sizes required to detect a doubling in variant frequencies will depend on the *relative number of evaluatable cells* tested in the two cases being compared. For example, if the experiment is designed so that $N_1 = N_2$, then Eqs. (17) and (18) simplify to

$$p = \frac{\nu_2}{\nu_1 + \nu_2} \tag{19}$$

$$\frac{\nu_2}{\nu_1} = \frac{1}{1/p - 1} = \frac{p}{q} \tag{20}$$

Suppose we seek necessary sample sizes to ensure that the CI for ν_2/ν_1 excludes the specific value 1 (which implies that $\nu_2 = \nu_1$). If the ratio of sample variant frequencies equals 2, then the sample estimate of p is 0.667. If the ratio of the sample variant frequencies equals one, then the sample estimate of $p = 0.500$. Our problem thus becomes that of finding the necessary values of $T = M_1 + M_2$ so that the CI for p includes 0.667 but excludes 0.500. By considering several values for binomial proportions, we find that $T = 42$, $M_2 = 28$, and $M_1 = 14$ yields the 95% CI for p of (0.5045, 0.8043). Thus, if the true ratio of variant frequencies equals 2, a study with sufficient evaluatable cells to achieve $M_2 = 28$ and $M_1 = 14$ will yield a CI for the ratio that does not include the special value one. A similar calculation shows that to detect a tripling requires $T = 20$ with $M_2 = 15$ and $M_1 = 5$.

If the experiment is carried out by increasing the counts N_1 and N_2 of evaluatable cells until one has achieved preassigned values M_1 and M_2 then the CI for v_2/v_1 is based on N_1 and N_2 as the random variables. Since $2vN$ is a χ^2 variable with $2M$ degrees of freedom, we can base the CI on the F statistic which has $2M_1$ and $2M_2$ degrees of freedom:

$$F = F(2M_1, 2M_2) = \frac{v_1 N_1/M_1}{v_2 N_2/M_2} \tag{21}$$

For example, suppose the data for cancer patients number 1 and number 3 had been gathered *in this manner* so that $M_3 = 30$ and $M_1 = 19$ are considered fixed. The random variables are the numbers of evaluatable cells with values $N_3 = 0.562 \times 10^6$ and $N_1 = 1.373 \times 10^6$. Then a 95% CI for the ratio v_2/v_1 is based on the F statistic

$$F = \frac{v_1 N_3/M_3}{v_3 N_1/M_1} = \frac{v_3(0.562)/30}{v_1(1.373)/19} \tag{22}$$

From F tables with $2 \times 30 = 60$ and $2 \times 19 = 38$ degrees of freedom, we find a 95% CI for v_3/v_1 to be (2.19, 7.06), which is very close to the CI in Table III (2.24, 7.15), obtained when the counts M were variables and N was fixed.

This approach can be extended to testing for the equality of K rates: $v_1 = v_2 = \cdots = v_K$. If the prespecified M values are equal, then the distribution of the maximum of the F statistics is tabled (Pearson and Hartley, 1976). If the M values are not all equal, approximations must be made.

V. DISCUSSION

Our belief that at least some TG^r PBLs are specific locus mutants that arise *in vivo* rests, at present, by analogy with the human fibroblast system. Since the LN mutation is expressed as clearly at the single cell level in PBLs as it is in fibroblasts, this appears valid and the LN mutation again provides a prototype somatic cell mutation for system development. Women heterozygous for this mutation present an ideal model for the development of a direct mutagenicity assay.

Our early demonstration of PBL mosaicism in LN heterozygotes (Albertini and DeMars, 1974), based on single X-chromosome inactivation, recapitulated the earlier experience with fibroblasts (Rosenbloom et al., 1967; Salzmann et al., 1968). However, because of negative selection in vivo, LN TG^r PBLs are in a decided minority in the peripheral blood of heterozygotes, occurring at frequencies of 10^{-3} to 5×10^{-2} (Strauss et al., 1980). This makes the heterozygote model particularly relevant. For direct mutagenicity testing, rare mutant cells arising in individuals in vivo must be detected in vitro in large majority populations of normal analogous cells. Mutant TG^r PBLs in LN heterozygotes unquestionably arise in vivo. Even though infrequent, these mutant cells are detectable in vitro in majority populations of normal autologous cells. The shortened life-span of mutant TG^r PBLs, indicated by the unequal mosaicism in heterozygotes, suggested that an additional advantage for direct mutagenicity testing would be temporal correlation between the induction of LN-like mutant PBLs and increases in their frequencies.

Rare LN-like TG^r PBLs were then found in blood samples from normal, non-LN individuals in low frequencies (Strauss and Albertini, 1977, 1979). As expected from the LN-heterozygote model, variant cells did not increase in frequency with the age of the individual tested. However, individuals exposed to mutagens showed clear increases in variant frequencies. There appeared to be a dosage effect in humans receiving X-irradiation therapy (Albertini, 1980).

Two sorts of experiments then suggested that the variant TG^r phenotype is heritable in PBLs. Human patients receiving chronic therapy with azathioprine, a purine analogue similar to TG, showed markedly elevated TG^r PBL V_f values (Strauss and Albertini, 1979). We interpreted these elevations to result, in part, from selection in vivo against the normal TG^s phenotype with the enrichment of mutants. Second, a single EB virus-induced lymphoblastoid cell line was initiated in TG in vitro using peripheral blood MNCs from an azathioprine treated patient with a high TG^r PBL V_f (Albertini, 1979, 1980). This could not be accomplished using MNCs from normal individuals. Similarly, there was a short interval of propagation (growth factor maintained) in TG of PBLs from an azathioprine treated patient, again with a high TG^r PBL V_f (Albertini, 1979, 1980). Taken together, these results suggest heritability in PBLs in vivo and in vitro of the LN-like phenotype.

Criteria have evolved which allow phenotypically variant cells arising in vitro to be defined as to be mutant somatic cells. (Chu and Powell, 1976). Judged by these, our evidence concerning the mutant nature of TG^r PBLs is admittedly indirect. Rigorous demonstration of the mutational basis of TG^r in PBLs will require methods which allow propagation of the variant cells *in vitro* under conditions that restrict analysis to only those altered phenotypes that arise *in vivo*. Methods available for these purposes are presently being explored.

Currently, when cryopreserved MNCs are tested, we find TG^r PBL V_f values in normal nonexposed individuals to be of the order of 10^{-6} when experiments of sufficient size are performed. Humans therapeutically exposed to mutagens, such as cancer patients, usually show elevated V_f values when compared to normals. Thus, mutagenic influences in man, at least of this magnitude, are detectable. As outlined in Section IV,D, experiments of sufficient size should be able to detect a doubling of the normal V_f.

Values of TG^r PBL V_f are *frequencies* and not *rates*. The rate of somatic cell mutation to TG^r resistance in cultured human fibroblast has been found in several studies to be of the order of 10^{-6} per cell generation (DeMars and Held, 1972; Buchwald, 1977; Gupta and Siminovitch, 1978). Furthermore, fluctuation tests measuring mutation rates *in vitro* in lymphoblastoid cells using TG^r and HLA antigen loss as markers also give values of the order of 10^{-6} per cell generation (Pious and Soderland, 1977; Thilly *et al.*, 1980). The analogous spontaneous mutation rate in PBLs *in vivo* is unknown. However, background frequencies of the order of 10^{-6} to 10^{-5} seem reasonable.

There are several potential sources of error in the TG^r PBL V_f assay. Phenocopies of one sort have been considered in Section I,C,3. We know now that fresh peripheral blood MNCs contain cells that are cycling *in vivo*. These cells are rare, but even at very low frequencies their numbers may exceed those of mutant TG^r PBLs. At best this results in insensitivity of the method. At worst, however, the frequency of cycling cells may not be constant but may be influenced by a variety of physiological or pathological conditions. Such cells may increase in frequency *in vivo* in individuals as a function of disease or mutagen exposure and introduce systematic errors. An analogous situation of mutagen-induced phenocopies was encountered when fetal hemoglobin (HbF)-containing mature red blood cells (RBCs) were studied as indicators of somatic cell mutation occurring *in vivo* and resulted in abandonment of that system (Papayonnopoulou *et al.*, 1977). Clearly, cycling cells must be removed or rendered unable to incorporate [^{3}H]thymidine in TG under the conditions of assay. We feel that cryopreservation as outlined above accomplishes this in large measure.

There are several technical sources of error with the method, all of which operate to effectively decrease the duration or degree of exposure of cells to TG. Because this kind of error most likely will cause interexperimental variation, it now seems wise to include a technical control in each mutagenicity test. We have shown that cryopreserved MNCs give a reproducible TG^r PBL V_f when tested repeatedly over time (greater than 60 days). The inclusion of such a sample of known V_f will serve to mitigate this source of technical error.

There are, in addition, considerations of error due to sampling. As noted above, a MNC sample represents a polyclonal mixture of cells. A constant relationship between mutants (TG^r PBLs) and mutations depends, in part, on maintenance of polyclonality and homogeneity of clonal expansion with cell division *in vivo*. However, PBLs are not neutral cells but participate by proliferation *in vivo* in various physiological and pathological conditions. Certainly, the clonality of immunocompetent cells proliferating in response to any given immunological stimulus (antigen) may be more restricted than the polyclonality of all PBLs in the peripheral blood. An increase in mutant TG^r PBL V_f might therefore result from a single mutant immunocompetent TG^r PBL undergoing clonal expansion that is greater than that of nonmutant TG^s PBLs. The magnitude of errors of this sort is not known at present and will require experiments in animals for assessment.

Despite these problems, we feel that the TG^r PBL method is useful and has the potential to answer basic questions regarding somatic cell mutation *in vivo*. Unlike the variant hemoglobin methods in RBCs, nucleated PBLs provide genomes that may be propagated *in vitro* allowing investigation of the mutational basis of the phenotypic variation.

The development of direct mutagenicity tests seems well worth the effort. Such tests have clear, qualitative advantages over other sorts of mutagenicity testing. They go beyond identification of mutagens and, indeed, will not have their greatest utility in that regard. Rather, the advantages of these tests are in terms of their metabolic and pharmacokinetic realism, their ability to assess the actual genotoxic effects of environmental mixtures of importance to man, and their ability to detect individual differences as regards human susceptibility to genotoxic agents.

Finally, we contend that direct mutagenicity testing can close the gap between laboratory mutagenicity testing and quantitative human health risk assessment. Direct testing is capable of providing data on individuals or groups that may then be correlated with clinical and epidemiological data as regards health outcomes. Thus, the possibility exists of validating direct mutagenicity tests in terms of utility as predictors of human disease.

We have discussed elsewhere situations where such validation may be attempted (Albertini, 1980; Albertini and Allen, in press). Accidental human exposures to genotoxic agents, unfortunately, still occur. Blood samples from exposed individuals should be obtained for direct mutagenicity testing. These individuals should be tested repeatedly and followed medically for extended periods of time, both to render necessary medical care for those who show late genotoxic illnesses and to correlate the occurrences of these late illnesses with early evidence in tests of somatic cell mutation. Follow-up and repetitive testing may allow identification of susceptible individuals for more anticipatory medical treatment, and, at the same time, evaluate the mutagenicity test or tests used as meaningful quantitative predictors of clinically relevant genotoxicity in man.

We have suggested that cancer patients receiving necessary treatments with mutagenic anticancer agents be monitored in this manner (Albertini and Allen, in press). Some of these patient groups are known to have relatively high frequencies of second malignancies, particularly of acute leukemia (Casciato and Scott, 1979). Longitudinal monitoring during and after treatment may allow validation of direct mutagenicity tests as quantitative predictors of this health risk.

Direct mutagenicity testing in man will achieve its ultimate value, then, when it allows reliable, quantitative test results to replace diseased individuals as realistic end points of human health hazards. We present the TG^r PBL method in its present state as a possible step in this direction.

ACKNOWLEDGMENTS

This work was supported by National Institutes of Health Grant No. PHS R01 25035, Cancer Center Support Grant CA 22435, Genetics Center Support Grant MCH-46-1001-02-0 and the Surgical Associates Foundation, Inc. E. F. Allen was supported by the National Institutes of Health National Research Service Award No. 5-F 32 CA 05918.

REFERENCES

Albertini, R. J. (1979). Direct mutagenicity testing with peripheral blood lymphocytes. Banbury Rep. **2**, 359–373.
Albertini, R. J. (1980). Drug resistant lymphocytes in man as indicators of somatic cell mutation. Teratog. Carcinog. Mutagen. **1**, 25–48.

Albertini, R. J., and Allen, E. F. (1981). Direct mutagenicity testing in man. In "Health Risk Analysis" (P. J. Walsh, C. R. Richmond, and E. D. Copenhaver, eds.), Franklin Institute Press, (in press).

Albertini, R. J., and DeMars, R. (1970). Diploid azaguanine-resistant mutants of cultured fibroblasts. *Science* **169**, 481.

Albertini, R. J., and DeMars, R. (1973). Detection and quantification of X-ray induced mutation in cultured, diploid human fibroblasts, *Mutat. Res.* **18**, 199.

Albertini, R. J., and DeMars, R. (1974). Mosaicism of peripheral blood lymphocyte populations in females heterozygous for the Lesch-Nyhan mutation. *Biochem Genet.* **11**, 397–411.

Albertini, R. J., Allen, E. F., Quinn, A. S., and Albertini, M. R. (in press). Human somatic cell mutation: In vivo variant lymphocyte frequencies as determined by 6-thioguanine resistance. In "Birth Defects Institute Symposium XI" Academic Press, New York (in press).

Arlett, C. F., and Lenman, A. R. (1978). Human disorders showing increased sensitivity to the induction of genetic damage. *Am. Rev. Genet.* **12**, 95–115.

Beyer, W. H., ed. (1966). "CRC Handbook of Tables for Probability and Statistics." Chem. Rubber Publ. Co., Cleveland, Ohio.

Boyum, A. (1968). Separation of leukocytes from blood and bone marrow. *Scand. J. Clin. Lab. Invest. Suppl.* 97, **21**, 51–76.

Bridges, B. A. (1974). The three-tier approach to mutagenicity screening and the concept of radiation-equivalent dose. *Mutat. Res.* **26**, 335–340.

Brogger, A. (1979). Chromosome damage in human mitotic cells after in vivo and in vitro exposure to mutagens. In "Genetic Damage in Man Caused by Environmental Agents" (K. Berg, ed.), pp. 87–99. Academic Press, New York.

Bross, I. (1954). A confidence interval for a percentage increase. *Biometrics* **10**, 245–250.

Brusick, D. J. (1977). In vitro mutagenesis assays as predictors of chemical carcinogenesis in mammals. *Clin. Toxicol.* **10**, 79–109.

Buchwald, M. (1977). Mutagenesis at the ouabain resistance locus in human diploid fibroblasts. *Mutat. Res.* **44**, 401–412.

Burger, P. M., and Simons, J. W. I. M. (1979). Mutagenicity of 8-methoxypsoralen and long-wave ultraviolet irradiation in diploid human skin fibroblasts: An improved risk estimate in photochemotherapy. *Mutat. Res.* **63**, 371–380.

Carrano, A. V., Thomspon, L. H., Lindl, P. A., and Minkler, J. L. (1978). Sister chromatid exchanges as an indicator of mutagenesis. *Nature (London)* **271**, 551–553.

Carrano, A. V., Minkler, J. M., Stetka, D. G., Moore D. H., II, (1980). Variation in the baseline of sister chromatic exchanges frequencies in human lymphocytes, *Environmental Mutagenesis,* **2**, 325–337.

Casciato, D. A., and Scott, J. L. (1979). Acute leukemia following prolonged cytotoxic agent therapy. *Medicine (Baltimore)* **58**, 32–47.

Chu, E. H. Y., and Powell, S. S. (1976). Selective systems in somatic cell genetics. *Adv. Hum. Genet.* **7**, 189–259.

DeMars, R. (1979). Suggestions for increasing the scope of direct testing for mutagens and carcinogens in intact humans and animals. *Banbury Rep.* **2**.

DeMars, R., and Held, K. (1972). The spontaneous azaguanine-resistant mutants of diploid human fibroblasts. *Humangenetik* **16**, 87–110.

Diem, K., ed. (1962). "Documenta Geigy: Mathematical Tables." Geigy Pharmaceuticals, Ardsley, New York.

EPA (1979). Environmental assessment: Short-term tests for carcinogens, mutagens and other genotoxic agents. Environmentl Protection Agency Document 625/9-79-003, July, 1979.

Gupta, R., and Siminovitch, L. (1978). Isolation and characterization of mutants of human diploid fibroblasts resistant to diphtheria toxin. *Proc. Natl. Acad. Sci. U.S.A.* **75**, 3337–3340.

Hsie, A. W., O'Neill, J. P., and McElheny, V. K. eds. (1979). "Mammalian Cell Mutagenesis: The Maturation of Test Systems," pp. 1–497. Cold Spring Harbor Lab., Cold Spring Harbor, New York, Banbury Report 2.

Hogg, R. V., and Craig, A. T. (1970). "Introduction to Mathematical Statistics." Macmillan, New York.

Jacobs, L., and DeMars, R. (1979). Chemical mutagenesis with diploid human fibroblasts. *In* "Handbook of Mutagenicity Test Procedures" (B. J. Kilbey, M. Legator, W. Nichols, and C. Ramel, eds.), pp. 193–220. Elsevier, Amsterdam.

Kapp, R. W., Jr., Picciano, D. J., and Jackson, C. B. (1979). Y-Chromosomal nondisjunction in dibromochloropropane-exposed workmen. *Mutat. Res.* **64**, 47–51.

Kilbey, B. J., Legator, M., Nichols, W., and Ramel, C. eds. (1979). "Handbook of Mutagenicity Test Procedures." Elsevier, Amsterdam.

Lesch, M., and Nyhan, W. L. (1964). A familial disorder of uric acid metabolism and central nervous system function. *Am. J. Med.* **36**, 561–570.

McCann, J., and Ames, B. N. (1977). The *Salmonella*/microsome mutagenicity test: Predictive value for animal carcinogenicity. *In* "Origins of Human Cancer" (H. H. Hiatt, J. D. Watson, and J. A. Winsten, eds.) Book C, pp. 1431–1450. Cold Spring Harbor Lab., Cold Spring Harbor, New York.

Maher, V. M., Ouellette, L. M., Curren, R. D., and McCormick, J. J. (1976). Frequency of ultraviolet light-induced mutations is higher in xeroderma pigmentosum variant cells than in normal human cells. *Nature (London)* **271**, 593.

Miller, R. (1966). "Simultaneous Statistical Inference." McGraw-Hill, New York.

Newcombe, H. B. (1978). Problems of assessing the genetic impact of mutagens on man. *Can. J. Genet. Cytol.* **20**, 459–470.

Papayonnopoulou, T. H., Brice, M., and Stamatoyannopoulos, G. (1977). Hemoglobin F synthesis *in vitro*: Evidence for control at the level of primitive erythroid stem cells. *Proc. Natl. Acad. Sci. U.S.A.* **74**, 2923–2927.

Parzen, E. (1960). "Modern Probability Theory and Its Applications." Wiley, New York.

Pearson, E. S., and Hartley, H. O., eds. (1976). "Biometrika Tables for Statisticians." Charles Griffin and Co., Ltd., London.

Pious, D., and Soderland, C. (1977). HLA variants of cultured human lymphoid cells: Evidence for mutational original and estimation of mutation rate. *Science* **197**, 769–771.

Rosenbloom, F. M., Kelley, W. N., Henderson, J. F., and Seegmiller, J. E. (1967). Lyon hypothesis and X-linked disease. *Lancet* **ii**, 305–306.

Salzmann, J., DeMars, R., and Benke, L. (1968). Single allele expression at an X-linked hyperuricimia locus in heterozygous human cells. *Proc. Natl. Acad. Sci. U.S.A.* **60**, 545–552.

Seegmiller, J. E., Rosenbloom, F. M., and Kelley, W. N. (1967). Enzyme defect associated with a sex-linked human neurological disorder and excessive purine synthesis. *Science* **155**, 1682.

Stamatoyannopoulos, G., Nute, P. E., Papayannopoulou, Th., McGuire, T., Lim, G., Bunn, H. F., and Rucknagel, D. (1980). Development of a somatic mutation screening system using Hb mutants. IV. Successful detection of red cells containing the human frameshift mutants Hb Wayne and Hb Cranston using nonspecific fluorescent antibodies. *Am. J. Hum. Genet.* **32,** 484–496.

Strauss, G. H., and Albertini, R. J. (1977). 6-Thioguanine resistant lymphocytes in human peripheral blood. *In* "Progress in Genetic Toxicology" (D. Scott, B. A. Bridges, F. H. Sobels, eds.), pp. 327–336. Elsevier, Amsterdam.

Strauss, G. H., and Albertini, R. J. (1979). Enumeration of 6-thioguanine resistant peripheral blood lymphocytes in man as a potential test for somatic cell mutation arising *in vivo. Mutat. Res.* **61,** 353–379.

Strauss, G. H., Albertini, R. J., Krusinski, P. A., and Baughman, R. D. (1979). 6-Thioguanine resistant peripheral blood lymphocytes in humans following psoralen long-wave UV light therapy. *J. Invest. Dermatol.* **73,** 211–216.

Strauss, G. H., Allen, E. F., and Albertini, R. J. (1980). An enumerative assay of purine analogue resistant lymphocytes in women heterozygous for the Lesch-Nyhan mutation. *Biochem. Genet.* **18,** 529–547.

Swift, M., and Chase, C. (1979). Cancer in families with xeroderma pigmentosum. *J. Natl. Cancer Inst.* **62,** 1415–1421.

Thilly, W. G., De Luca, J. G., Furth, E. F., Hoppe IV, H., Krolewski, J. J., Kaden, D. A., Liber, H. L., Skopek, T. R., Slapikoff, S. A., Tizard, R. J., and Penman, B. W. (1980). Gene-locus mutation assays in diploid human lymphoblast lines. *Chem. Mutagens* **6,** 331–364.

Waters, M. D. (in press). Monitoring the environment. *In* "Toxicity Testing *In Vitro*" (M. Nardone, ed.). Academic Press, New York.

Wyrobek, A. J., and Bruce, W. R. (1978). *Chem. Mutagens* **5,** 257–285.

Chapter 12

Sperm Assays as Indicators of Chemically Induced Germ-Cell Damage in Man

A. J. Wyrobek

I. INTRODUCTION

The studies of sperm in men exposed to dibromochloropropane (DBCP) illustrated that exposure to an occupational toxin can have profound antispermatogenic effects (Whorton *et al.*, 1977; Sandifer *et*

337

MUTAGENICITY:
NEW HORIZONS IN GENETIC TOXICOLOGY

al., 1979). However, DBCP is by no means the only chemical agent known to affect human spermatogenesis. Approximately 35 different chemical exposures have been shown to have antispermatogenic effects in man (Wyrobek *et al.*, 1982a).

Sperm assays have a long history in the diagnosis of infertility in man and domestic animals. It was therefore not surprising that early attempts to assess chemically altered human spermatogenic function used sperm parameters common to fertility diagnosis, i.e., sperm number (counts), motility, and morphology. Many animal and human studies have shown that sperm anomalies can be used as indicators and, in certain cases, dosimeters of induced antispermatogenic effects (for review, see Wyrobek *et al.*, 1982a,b). Induced sperm anomalies are clearly linked to testicular damage. However, their relationship to fertility and induced, heritable, genetic defects is not yet clear; it is an active research area. Sperm assays have received much attention by those who monitor human exposure to mutagens and carcinogens because (1) sperm are easy cells to obtain, (2) they can reflect damage to the gonads, and (3) they can be studied in humans and model animals.

II. DESCRIPTION OF AVAILABLE HUMAN SPERM ASSAYS AND THEIR RELATIVE SENSITIVITIES

A. Available Human Sperm Assays

Four human sperm parameters are available: number (count), motility, morphology, and the new YFF test which is thought to test Y-chromosomal nondisjunction during human spermatogenesis (Amelar 1966; Wyrobek *et al.*, 1982a). With the exception of YFF, the ranges of "normal" for these sperm parameters are primarily based on early studies with patients from infertility, prenatal, and vasectomy clinics and with prison volunteers.

1. Sperm Count

Sperm count is usually determined by hemocytometer and reported as number of sperm per milliliter of ejaculate. Automated counters can be used. Interpretation of results may be confounded by a number of factors such as: variable continence time before obtaining specimen, sample-to-sample variation in the same individual, and collection of incomplete ejaculate.

2. Motility

The swimming ability of the sperm, motility is expressed either as a percentage of motile sperm or on a graduated scale of 1 to 4. Although motility may be one of the best measures of the relation of testicular function and fertility, it is also the most subjective parameter and is very sensitive to time and temperature after collection. Thus, motility is a difficult if not impossible parameter to measure in a large-scale cross-sectional study, especially when samples are collected at home.

3. Sperm Morphology

As done in the past, sperm morphology has been relatively easy to conduct, but is subject to much interlaboratory and interscorer variability. Recent approaches to improving the use of the sperm morphology assay class sperm into shape categories (MacLeod, 1974) and use standard slides to assure quantitative and reproducible measurements (Wyrobek, 1981). Determination of chemically induced effects on morphology are made by statistically comparing groups of samples from exposed and unexposed men or by comparing samples taken before and after exposure to an individual. This assay is performed with air-dried smears that are fixed and stained with a modified Papanicolao method and scored by a trained technician. Normal ranges have been established for several unexposed populations (Wyrobek, 1981).

4. YFF

The YFF test scores the frequency of sperm with more than one fluorescent body after staining with the fluorescent dye, quinacrine mustard or dihydrochloride. However, the method is new and its link to chromosomal nondisjunction controversial. Very few populations of exposed men have been analyzed yet normal ranges have been tentatively established (Kapp and Jacobson, 1980).

B. Relative Statistical and Biological Sensitivities

Table I compares the relative statistical sensitivities of the sperm assays for counts, morphology, and YFF. These data were obtained from the control group in one of our occupational studies (Wyrobek et al., 1981a). Of the three assays, that for morphology requires the smallest sample size to detect with confidence a 25% change from the

TABLE I

The Relative Statistical Sensitivities of Three Assays of Human Sperm

	Counts	Morphology[a]	YFF[a]
Assumed distribution	Log-normal	Normal	Normal[b]
Mean[c]	132×10^6/ml	41.9%	0.8%
SD	160×10^6/ml	12.4%	0.7%
25% change in mean (SD units)	0.2	0.8	0.3
Sample size for 5% level test with 80% power[d]	214	26	41

[a] Using the criteria of Wyrobek et al. (1981a).
[b] Square-root transformation (Snedecor and Cochran, 1978).
[c] Analysis is based on 35 unexposed new hires in a pesticide plant (Wyrobek et al., 1981a).
[d] Method of Owen (1962).

mean value of the control group. However, it is important to realize that the statistical sensitivities may be independent of the relative biological sensitivities of these assays. For example, Lancranjan et al. (1975) found that sperm morphology was at least as sensitive to lead exposure as sperm counts or motility. However, in a study of DBCP workers, the analyses of sperm morphology suggest no DBCP-related differences, although counts showed major differences (Sandifer et al., 1979). We have recently completed a study of human sperm in which sperm morphology was the only assay that yielded positive findings in a group of pesticide production workers scored for changes in sperm counts, morphology, and YFF (Wyrobek et al., 1981a). Since the data are still insufficient to predict which assay would be biologically most sensitive to a given agent, the conservative approach to assessing changes in human testicular function should include assays for sperm count, morphology, YFF, and, whenever practical, motility.

III. REVIEW OF THE USES OF SPERM ASSAYS IN CHEMICALLY EXPOSED MEN

A recent survey of the literature (Wyrobek et al., 1982a) showed that human sperm assays have been more widely used than was generally suspected. More than 100 papers, covering approximately 70 different chemical exposures, were found on the use of semen assays in assessing chemically induced changes in human testicular function. About 85% of the exposures were to experimental or therapeutic drugs, about 10% were occupational exposures, and about 5% involved personal drug use. The studies reviewed 42 single agents, 11 complex mixtures, and 17 sets of multiple agents. Tables II and III list the single agents and complex mixtures categorized by (1) those agents that were found to significantly reduce the sperm quality (reduction in sperm counts, decrease in sperm motility, increase in morphologically abnormal forms, or increase in the percentage of YFF sperm) of exposed men, (2) those that gave suggestive but inconclusive evidence of reduced sperm quality, and (3) those that showed no effect. Of the 42

TABLE II

Single Agents Studied with Human Sperm Assays[a]

Agents causing a reduction in sperm quality[b]	Agents with data suggestive of effects	Agents showing no effects
Acridinyl anisidine	Centchroman	CIBA-32644 Ba
Adriamycin	Colchicine	Lysine
Aspartic acid	Diethylstilbestrol	Methyltestosterone
Chlorambucil	Hydroxymethyl	Ornithine
Cyclophosphamide	nitrofurantoin	WIN 59491
Cyproterone acetate	Methadone	
Enovid	Methotrexate	
Gossypol	Metronidazole	
6-Medroxyprogesterone	Nitrofurantoin	
Metanedienone	Trimeprazine	
Nilevar		
Norlutin		
Prednisolone		
Progesterone		
Salicylazosulfapyridine		
Testosterone		
Testosterone cylopentylpropionate		
Testosterone enanthate		
Testosterone propionate		
WIN 13099		
WIN 17416		
WIN 18446		

[a] For specific data, proper chemical names and literature sources for specific agents see Wyrobek et al., 1982a.

[b] Assignments to columns are based on studies of sperm counts, motility, morphology, and YFF; assignment into specific columns may change as data for more studies become available.

TABLE III

Complex Mixtures Studied with Human Sperm Assays[a]

Mixtures causing a reduction in sperm quality[b]	Mixtures showing suggestive effects	Mixtures showing no effects
Chronic alcoholism	Carbaryl[c]	Anesthetic gases[c]
Carbon disulfide[c]	Tobacco smoke	Epichlorohydrin[c]
		Glycerine
Dibromochloropropane[c]		compounds[c]
Lead[c]		Polybrominated
Marijuana		biphenyls[c]

[a] For specific data on mixtures and literature sources see Wyrobek et al., 1982a; each entry represents the most likely agent causing the effect observed.

[b] Assignments to specific columns are based on studies of sperm counts, motility, morphology, and YFF.

[c] Occupational exposures.

single agents studied, 22 reduced sperm quality, 9 gave suggestive but inconclusive evidence of induced reductions, and 5 showed no changes. Six agents not shown in Table II (clomiphene citrate, coenzyme Q7, fluoxymesterone, kallikrein, luteinizing hormone releasing factor, and vitamin B_{12} showed some evidence, often of marginal statistical significance, of increased sperm counts and/or motility in some of the infertile patients studied. Vitamin B_{12} was the only agent found to improve sperm morphology based on studies in selected infertile patients (see Wyrobek et al., 1982a, for details).

Eleven complex mixtures were studied using the sperm assays. As shown in Table III, 5 appeared to reduce sperm quality, 2 showed suggestive but inconclusive effects, and 4 showed no effects. Most of these studies involve occupational exposures in which single active agents have been implicated (DBCP, lead, etc.).

Sperm assays have also been used in men exposed to at least 17 sets of 2 or more agents in consort. Of these, 10 combinations caused reductions in semen quality (e.g., cyclophosphamide plus prednisone, danazol plus testosterone enanthate, and MVPP cancer chemotherapy).

Of the 70 different human exposures evaluated, including all single agents, multiple agents, and complex mixtures, 97% of the studies used sperm counts as one of the parameters studied; 25% of the studies used counts as the only parameter measured. Overall, 58% of the studies used motility, 48% used morphology, but only 7% used YFF.

As shown in Table IV, 27 agents or groups of agents were tested for their ability to induce morphologically abnormal sperm. Fifteen were

TABLE IV

Agents Tested in Man for Induction of Morphologically Abnormal Sperm

	Agents causing significant increase	Agents causing inconclusive effects	Agents showing no response
Single agents and complex mixtures	Acridinyl anisidide (1)[a] Carbaryl (1) Carbon disulfide (1) Cyproterone acetate (2) Gossypol (1) Lead (1) Marijuana (1) 6-Medroxyprogesterone (1) Metanedienone (1) Metronidazole (1) Salicylazosulfapyridine (1) Tobacco smoke (1) WIN 18446 (1)	Clomiphene citrate (2) Cyclophosphamide (1) Diethylstilbestrol (2) Testosterone enanthate (3)	Anesthetic gases (1) Fluoxymesterone (1) Nitrofurantoin (1) CIBA-32644-Ba (1) Methyltestosterone (1) Polybrominated biphenyls (2) Glycerine production compounds (1)
Multiple agents	Danazol + methyl testosterone (1) Danazol + testosterone enanthate (1)	Cyclophosphamide + prednisone (1)	

[a] Number in parentheses is the number of studies evaluated (see Wyrobek et al., 1982a, for details).

associated with significant decreases in numbers of normally shaped sperm in the ejaculate, 5 showed inconclusive effects, and 7 showed no changes in sperm morphology.

IV. RATIONALE, STRATEGIES, AND PROBLEMS ENCOUNTERED IN THE USE OF HUMAN SPERM ASSAYS IN EXPOSED POPULATIONS

Human sperm studies are considerably more complex than animal sperm studies. Although other approaches for assessing induced testicular dysfunction in man have been investigated (e.g., biopsy and blood gonadotropin level), sperm assays are the only feasible, direct method of assessing chemical effects on the male germ cells. In general, sperm studies in man are warranted if there are (1) sufficient animal data available that suggest testicular effects; or, (2) questionnaire data or case reports that suggest a problem with infertility or pregnancy outcome linked to the male.

In cases where specific chemical agents are under scrutiny, both animal and preliminary human data are desirable before major studies are to be initiated. Recent comprehensive surveys of the applications of animal and human sperm assays (Wyrobek et al., 1982a,b) suggested that there is good agreement between results from animal sperm assays and human sperm assays. Mouse sperm morphology is the most widely used animal sperm assay. Because it is simple, fast, inexpensive, and highly quantitative, its use in the very early stages of a study is warranted if there are (1) significant human exposures to an agent or mixture that can be isolated, (2) insufficient data on testicular toxicity from other animal studies, and (3) adequate quantities of chemical available for testing.

A strategy for the application of human sperm assays might best be described in the context of a scenario where there is concern over a group that has been exposed to an agent that may affect fertility, reproductive outcome, or both. The first level of response might be to survey and compile all human and animal studies of testicular dysfunction with the suspected agent. If no animal or human data exist, if there is sufficient concern about the human exposures, and if the suspected agent is available in sufficient quantities, an animal study such as the mouse sperm morphology assay may be done to assess the likelihood of antispermatogenic effects. The second level of activity might be a screening questionnaire to determine if there are major changes in human fertility. If there continues to be sufficient evidence for concern and if the population is sufficiently large, this may be followed by a more detailed questionnaire to determine to what extent the infertility may be male or female related. It is at this point, either concurrent with or in place of the more detailed questionnaire, that human sperm studies might be considered.

The following is an outline of the key steps of a human sperm study:

1. Identify a population of men exposed to a potential testicular toxin.
2. Arrange access to the study population.
3. Identify a control group.
4. Develop an appropriate study design.
5. Organize a referral strategy.
6. Recruit individual volunteers.
7. Collect exposure, medical, and relevant personal data by questionnaire.
8. Collect semen samples.
9. Analyze samples in laboratory.
10. Analyze the data statistically.
11. Report the results.

For a description of each step see Wyrobek (1981). The first four steps can be exceedingly time consuming, bureaucratic, and political. They usually occur before any contact has been made with the sperm donors. Steps 5, 6, 7, and 11 are best done with physicians so that the physician–patient relationship may be used to advantage in the interactions with the donors. Step 9, the laboratory analyses, is a relatively short aspect of the entire process. Sperm counts, motility, morphology, and YFF can be determined for each sample. Steps 4 and 10 require epidemiological and biostatistical input. Distributions of sperm data from control and exposed men can be analyzed by the t test or Kolmogorov–Smirnov test, as was done in the DBCP study of Whorton et al. (1977) or by analyses of the proportion of men with semen pathologies, as in the lead study (Lancranjan et al., 1975). The investigator must consider relevant confounding factors, generate dose–response curves if possible, and consider postexposure time-related effects to (1) determine the agents or environmental factors that may be responsible for any sperm defects found, and (2) assess the magnitude and reversibility of the effect.

When designing a study, it is important to know the number of men exposed to the agent as well as the likelihood of high exposures. Because between-male variability in semen characteristics is high even among fertile and presumably healthy men, rather large numbers of subjects are required to establish differences between control and exposed groups in cross-sectional studies in which each individual is usually sampled only once (See Table I). Since some men will not participate because of vasectomies or personal beliefs, even larger numbers of men must be identified. Cross-sectional studies are therefore typically large, e.g., 170 in the study of smoking effects (Viczian, 1969) and 200 in the study of lead effects (Lancranjan *et al.*, 1975).

Longitudinal study designs may be more appropriate when fewer men are available for sampling. In this study design, repeated semen samples from an individual are compared to assess chemically induced sperm defects. Since variation of sperm morphology within an individual is considerably less than variation among individuals (MacLeod 1965, 1974; Sherins *et al.*, 1977), fewer people are required to detect induced changes. These studies, however, have several constraints: (a) no precedent exists for such studies in the workplace, although studies have been successfully conducted on men exposed to X-rays and men receiving drugs (MacLeod, 1965; Heller *et al.*, 1965); (b) repeated samplings during a period of months and perhaps years are required; (c) samples before exposure are needed (or within days of an acute exposure, that is, before an induced semen defect can be seen); and (d) the number of men needed for an effective study is unknown.

V. THE IMPLICATIONS OF CHEMICALLY INDUCED SPERM CHANGES

A. Evidence from Human Studies

Although it is generally agreed that major reduction in sperm counts, motility, and normal sperm forms are linked to reduced fertility, it remains unclear as to which, if any, of the sperm parameters indicates embryonic failure or heritable genetic abnormalities. Human data on this question are very limited. In one study, fathers of 201 spontaneous abortions showed significantly higher sperm abnormalities and lower sperm counts than 116 fathers of normal pregnancies (Furuhjelm *et al.*, 1962); this evidence suggests a link between poor semen quality and frequency of spontaneous abortions. Several studies with habitually aborting women support this observation (e.g., Czeizel *et al.*, 1976). More human studies are needed to compare exposure of the male parent, induced sperm defects, and reproductive outcome.

B. Murine Studies of Sperm Morphology

Most studies on chemically induced sperm abnormalities have been conducted in the mouse (Wyrobek and Bruce, 1978; Bruce and Heddle, 1979; Topham, 1980). Several lines of evidence link induction of abnormal sperm and heritable genetic abnormalities (for review, see Wyrobek et al., 1982b). First, it is clear that sperm shaping and the production of abnormal sperm are polygenically controlled by autosomal as well as sex-linked genes. Second, there seems to be good agreement between the mutagenicity of a chemical and its ability to induce abnormal sperm. A survey of the literature indicated that there is a greater than 80% concordance between the activity of compounds in the mouse sperm morphology test and their activities in one or more of the following tests: the morphological specific locus, F_1 sperm morphology, heritable translocation, and/or dominant lethal tests. Third, in several studies with agents that induce sperm abnormalities in mice, abnormalities were transmitted to the male offspring of the exposed parent. Therefore, the mouse sperm morphology test may be a useful screen for compounds that constitute a potential genetic hazard. Spindle poisons that may cause nondisjunction in germ cells have also been identified by the mouse morphology test (Topham, 1980).

Correlations between sperm changes and carcinogenicity have been determined with data from the mouse sperm morphology test. As part of the International Program for the Evaluation of Short-Term Tests for Carcinogenesis, 6 carcinogen/noncarcinogen pairs and 5 unpaired carcinogens were surveyed as unknowns, using the mouse sperm abnormality assay (Wyrobek et al., 1981b). No false positive responses were found, suggesting that the sperm assay has a high specificity for carcinogens. However, several false negatives were obtained, suggesting that not all carcinogens induce sperm abnormalities in mice. A positive result on the sperm morphology assay may be strong evidence that an agent is a carcinogen while no conclusions can be drawn from negative results. Similar findings were made in the review prepared for the Gene-tox Program (Wyrobek et al., 1982b). The detailed comparisons of the sperm abnormality assay and the other short-term mutagenicity and carcinogenicity assays are still in progress.

C. Interpretation

Because of our poor understanding of the genetic mechanisms underlying various induced sperm anomalies and their relationship to carcinogenesis, at present the main information that can be gained from human sperm assays is whether human spermatogenesis is affected by exposure to a chemical agent or mixture of agents. These data, together with results of other short-term assays for mutagenicity, may indicate which of these agents may be a potential germ cell mutagen. Clearly more research is needed to develop sensitive sperm assays with defined mutational endpoints so that risks of heritable damage may be assessed directly using human sperm.

VI. CONCLUSIONS

The human sperm assays have numerous advantages: they can be used to monitor effects directly in exposed men, and large numbers of sperm can be easily examined. The changes in sperm parameters probably arise from interference by the test substance with the genetically controlled differentiation of the sperm cell, and therefore these assays are intrinsically relevant to safety evaluation and assessment of potential effects of the agent on male fertility and possibly reproductive outcome. The laboratory methods are generally rapid, straightforward, and quantitative. Sperm assays have major advantages over other approaches for assessing induced changes in testicular function. Testicular biopsies are impractical, traumatic, invasive, and may themselves affect testicular function. Relying solely on epidemiological surveys of reproductive function, using questionnaires, requires large sample sizes and may not be very sensitive. Analyses of blood levels of gonadotropins are expensive and generally insensitive to small changes in testicular function. Compared with these methods, sperm assays are noninvasive, inexpensive, require small sample sizes for effective analyses, and are sensitive to small changes.

The major disadvantages of sperm assays are that (1) the heritability of the induced damage is not yet clearly understood; (2) limited sperm sampling times and dosage regimens may reduce the sensitivity of the assay (e.g., agents that exert only transient effects may be missed using single sampling times); (3) other factors such as ischemia, marijuana use, high fever, medications taken, and testicular trauma may produce spurious false positive responses (questionnaire data are of vital importance to identify these confounding factors); and (4) there may be difficulties in obtaining samples especially in environmental and occupational studies.

ACKNOWLEDGMENTS

I wish to thank Dr. Lowry Dobson, Laurie Gordon, and George Watchmaker for help in the preparation of this manuscript.

Work was performed under the auspices of the U.S. Department of Energy by the Lawrence Livermore National Laboratory under contract number W-7405-Eng-48 and by funding from the United States Environmental Protection Agency.

REFERENCES

Amelar, R. D. (1966). The semen analysis. In "Infertility in Men: Diagnosis and Treatment," pp. 30–53. Davis, Philadelphia, Pennsylvania.

Bruce, W. R., and Heddle, J. A. (1979). The mutagenic activity of 61 agents as determined by the micronucleus, *Salmonella*, and sperm abnormality assays. *Can. J. Genet. Cytol.* **21**, 319–334.

Czeizel, E., Hancsok, M., and Viczian, M. (1976). Examination of the semen of husbands of habitually aborting women. *Orv. Hetil.* **108**, 1591–1595.

Furuhjelm, M., Jonson, B., and Lagergren, C. G. (1962). The quality of human semen in spontaneous abortion. *Int. J. Fertil.* **7**, 17–21.

Heller, C. G., Wootton, P., Rowley, M. J., Lalli, M. F., and Brusca, D. R. (1965). Action of radiation upon human spermatogenesis. *Int. Congr. Ser. Excerpta Med. No. 112*, 408–410.

Kapp, R. W., and Jacobson, C. B. (1980). Analysis of human spermatozoa for Y chromosomal nondisjunction. *Teratogen. Carcinogen. Mutagenesis* **1**, 193–211.

Lancranjan, I., Popescu, H. I., Gavanescu, O., Klepsch, I., and Serbanescu, M. (1975). Reproductive ability of workmen occupationally exposed to lead. *Arch. Environ. Health* **30**, 396–401.

MacLeod, J. (1965). Human seminal cytology following the administration of certain antispermatogenic compounds. In "A Symposium on Agents Affecting Fertility" (C. R. Austin and J. S. Perry, eds.), pp. 93–123. Little, Brown, Boston, Massachusetts.

MacLeod, J. (1974). Effects of environmental factors and antispermatogenic compounds on the human testis as reflected in seminal cytology. *Proc. Serono Symp.* **5**, pp. 123–148. Academic Press, New York.

Owen, D. B. (1962). "Handbook of Statistical Tables," Addison-Wesley, Reading, Mass.

Sandifer, S. H., Wilkins, R. I., Loadholt, C. B., Lane, L. G., and Eldridge, J. C. (1979). Spermatogenesis in agricultural workers exposed to dibromochloropropane (DBCP). *Bull. Environ. Contam. Toxicol.* **23**, 703–710.

Sherins, R. J., Brightwell, D., and Sternthal, P. M. (1977). Longitudinal analysis of semen of fertile and infertile men. In "The Testis in Normal and Infertile Men" (P. Troen and H. R. Nankin, eds.), pp. 473–488. Raven, New York.

Snedecor, G. W., and Cochran, W. G. (1978). "Statistical Methods," pp. 325–327. Iowa State Univ. Press, Ames, Iowa.

Topham, J. (1980). The detection of carcinogen-induced sperm head abnormalities in mice. *Mutat. Res.* **69**, 149–155.

Viczian, M. (1969). Ergebnisse von Spermauntersuchungen bei Zigarettenrauchern. *Z. Haut Geschlechtsk.* **44**, 183–187.

Whorton, D., Krauss, R. M., Marshall, S., and Milby, T. H. (1977). Infertility in male pesticide workers. *Lancet* **2**, 1259–1261.

Wyrobek, A. J., and Bruce, W. R. (1978). The induction of sperm-shape abnormalities in mice and humans. *In* "Chemical Mutagens" (A. Hollaender and F. de Serres, eds.) Volume 5, pp. 257–285. Plenum Publishing Corporation, New York.

Wyrobek, A. J. (1981). Methods for human and murine sperm assays. *In* "Short-Term Tests for Chemical Carcinogens" (H. F. Stich and R. M. C. San, eds.), pp. 408–419. Chapter 36. Springer-Verlag, Berlin and New York.

Wyrobek, A. J., Burkhart, J. G., Francis, M. C., Gordon, L., Kapp, R. W., Letz. G., Malling, H. V., Topham, J. C., and Whorton, M. D. (1982a). Chemically induced alternations of spermatogenic function in man as measured by semen analyses parameters: A report for the Genetox program. *Mutat. Res. Rev. Genet. Toxicol.* (in press).

Wyrobek, A. J., Burkhart, J. G., Francis, M. C., Gordon, L., Kapp, R. W., Letz, G., Malling, H. V., Topham, J. C., and Whorton, M. D. (1982b). An evaluation of the mouse sperm-morphology assay and sperm assays in other non-human mammals: A report for the Genetox program. *Mutat. Res. Rev. Genet. Toxicol.* (in press).

Wyrobek, A. J., Watchmaker, G., Gordon, L., Wong, K., Moore, D., II, and Whorton, D. (1981a). Sperm shape abnormalities in carbaryl exposed employees. *Environ. Health Perspect.* **40**, 255–265.

Wyrobek, A. J., Gordon, L., and Watchmaker, G. (1981b). The effects of 17 chemical agents including 6 carcinogen/noncarcinogen pairs on sperm shape abnormalities in mice. *In* "Evaluation of Short-Term Tests for Carcinogens" (F. de Serres and J. Ashby, eds.) pp. 712–717. Elsevier North-Holland, New York.

Chapter 13

Cytogenetic Events *in Vivo*

J. G. Brewen and D. G. Stetka

I. INTRODUCTION

A highly significant proportion of the total human genetically based disease burden can be attributed to structural and numerical chromosome anomalies. It is estimated that one in every hundred live births carries an abnormal chromosome constitution and that at least 25% of all conceptions possess serious chromosomal defects. The great majority of these chromosomal disorders are lethal and result in early abortion, and therefore they do not represent a truly genetic, or heritable, threat. They do, however, generate emotional crises and, in some instances, life-threatening situations for the mother.

MUTAGENICITY:
NEW HORIZONS IN GENETIC TOXICOLOGY

Many chromosomal disorders are viable and cause very severe phenotypic effects. Some of the better known numerical anomalies are Klinefelter's and Turner's syndromes, and one form of Down's syndrome. Structural aberrations that are purportedly responsible for genetic disorders include a simple deletion in Wilm's tumor, and unbalanced segregation products from balanced translocations in some cases of Down's syndrome and Cri-du-Chat syndrome.

Because many chemical and physical agents are capable of inducing chromosomal aberrations, it stands to reason that any program in mutagenicity assessment should include assays that screen for the ability of the substance under test to produce effects at the chromosome level. Furthermore, because the principal concern is the potential effect in humans, such assays should be done using an appropriate whole-animal test system. The remainder of this chapter will deal with the present state-of-the-art of in vivo cytogenetics and discuss approaches that appear to hold promise for the future. Because there are several excellent reviews of the techniques utilized for in vivo cytogenetic studies and of the data accumulated, to date, on various chemical substances (Adler and Brewen, 1981; Brewen and Preston, 1982), this chapter will emphasize the advantages and disadvantages of currently employed procedures and will attempt to critically substantiate the need for a more futuristic approach to solving the problems encountered in determining the cytogenetic effects of chemicals. Recent developments in the field of sister chromatid exchange (SCE) analysis will be included, because SCE induction in vivo has proved to be a very sensitive indicator of exposure to chemical mutagens.

II. STRUCTURAL CHROMOSOME ABERRATIONS

A. Mechanisms of Aberration Formation

The proper design, and eventual interpretation, of any cytogenetic study that purportedly evaluates the clastogenic activity of a chemical substance must take into account the general aspects of the mechanism by which the chemical produces its effect. It has become increasingly clear that chemical mutagens have a wide spectrum of mechanisms by which they exert their mutagenic action. This is manifested not only by the various types of lesions that are suspected of being responsible for the end point under consideration, but also by such general phenomena as the involvement of many normal cellular repair processes. Due to space limitations and the volume of relevant literature, it is beyond the scope of this chapter to present an in-depth review of the evidence that is suggestive of mechanisms of aberration formation and to deal also with the other aspects of *in vivo* cytogenetics that should be discussed. Instead, a general review of some of the current concepts of aberration formation will be provided.

1. Specificity of Clastogen Activity

It was determined very early in the study of chemical mutagenicity that many substances produced aberrations by a so-called "delayed effect" process whereas others seemed to mimic ionizing radiation in that aberrations were observed in cells shortly after treatment with the chemical, ergo a "nondelayed effect" (Kihlman, 1966). Those chemicals that exhibit a delayed effect produce only chromatid-type aberrations. Although initial studies on nondelayed effect chemicals reported predominantly chromatid-type aberrations (Kihlman, 1955, 1964; Cohen *et al.*, 1963; Brewen, 1965), subsequent studies showed at least three of them to be capable of producing chromosome-type aberrations in early G_1 cells (Brewen and Christie, 1967; Dresp *et al.*, 1978; DuFrain *et al.*, 1980). The general feeling of most investigators was that those chemicals that exhibited a delayed effect produced the aberrations in late G_1 and S and that an intervening round of DNA synthesis was required. Those chemicals that did not exhibit a delayed effect were thought to either interfere with "normal" DNA metabolism (Ahnstrom and Natarajan, 1966) or to produce DNA strand breaks directly (Bender *et al.*, 1974b), as does ionizing radiation. From these two general observations evolved the concept of S-dependent and radiomimetic chemical clastogens.

a. S-Dependent Agents. Although the concept of the need for an intervening round of DNA replication as an explanation of the action of delayed effect chemicals had been considered for some time, it was the work of Evans and Scott (1964, 1969) that unequivocally demonstrated this phenomenon. They combined treatment of *Vicia faba* roots with maleic hydrazide and nitrogen mustard with exposure to tritiated thymidine, and demonstrated that the first cells to reach mitosis with aberrations had also incorporated the radioisotope into their DNA. Furthermore, their data clearly showed that the G_1 cells, as ascertained by the lack of radioisotope incorporation, also contained only chromatid-type aberrations. This latter observation argued for the persistence of the lesion in the chromatin so that when the cell eventually replicated its DNA, the lesion was translated into an aberration.

Subsequent studies indicate that the great majority of chemical chromosome-damaging agents act in this fashion. This certainly is true for alkylating agents, nitroso compounds, some antibiotics, and DNA base analogues. The persistence of G_1-produced lesions until DNA replication varies from chemical to chemical, and is determined to a large degree by the cell's repair competence, as will be discussed later.

b. Radiomimetic Agents. The early work of Sax (1938, 1940, 1941) demonstrated that ionizing radiation resulted in the formation of structural chromosome aberrations very shortly after the exposure of the target cell. As was previously mentioned, this is not the case with most of the clastogenic chemicals studied to date. There are, however, some chemicals that do in fact produce structural aberrations within a short time interval after treatment of the target cells. Because of this similarity with ionizing radiation these chemicals are frequently referred to as being radiomimetic. The most thoroughly studied examples of these are streptonigrin (Cohen *et al.*, 1963; Kihlman, 1964; Dufrain *et al.*, 1980), bleomycin (Dresp *et al.*, 1978; Vig and Lewis, 1978), cytosine arabinoside (Brewen and Christie, 1967), adriamycin (Vig, 1977; Au and Hsu, 1980), and 8-ethoxycaffeine (Kihlman, 1955; Scott and Evans, 1964). The precise mechanism of action of these compounds is not clear but it appears that they are capable of either producing strand breaks in the DNA, or, as is the apparent case for cytosine arabinoside, effectively inhibiting the repair of spontaneously arising DNA strand breaks.

In addition to those chemicals that are either S-dependent or radiomimetic, there are some that apparently possess both capabilities. These agents are weakly radiomimetic in that they produce some chromosome-type aberrations in G_1 cells, or act on G_2 cells, but are much more effective when late G_1 and S phase cells are treated. The best example of this class is the polyfunctional alkylating agent triethylenemelamine (TEM) (Caine and Lyon, 1977; Brewen and Payne, 1978; Luippold *et al.*, 1978).

2. *Models of Aberration Formation*

It sometimes appears that there have been as many models proposed to explain the mechanism of aberration formation following chemical treatment as there are chemicals that have been studied. The great majority of these models have common basic components, and although none of them can adequately explain all the experimental observations, most of them focus on several fundamental theses and facts: (1) chromosomal DNA is the principal target molecule; (2) DNA strand breaks are either the initial lesion, or are generated, in the nascent strand, during DNA replication, opposite a lesion; and (3) some type of recombination occurs between strands of DNA in different chromosomes or different regions of the same chromosome.

a. DNA as the Target Molecule. Several lines of evidence exist that suggest DNA is the principal target for the induction of lesions that ultimately give rise to structural chromosome aberrations. In the case of ionizing radiation there is an excellent correlation between the production of double-strand breaks in DNA and the formation of structural aberrations. Furthermore, Brewen and Peacock (1969) demonstrated that reunion of chromatids following X-irradiation was strictly controlled by the unidirectional polarity conferred by the 5' and 3' termini of the single strands of the DNA duplex. More recent work by Hiss and Preston (1977) and Preston (1980) substantiate these conclusions.

The initial evidence that DNA is the primary target for UV-induced aberrations was provided by the observations of Stadler and Uber (1942) on *Zea mays*, and Kirby-Smith and Craig (1957) on *Tradescantia*, that the wavelength of 2500–2600 Å (which DNA absorbs) is much more efficient than any other at producing aberrations. This observation was extended to mammalian cells by Chu (1965). The ultimate confirmation that DNA was the primary target for UV-induced aberrations was provided by the experiments of Griggs and Bender (1973). They demonstrated that an amphibian cell line capable of photoreactivating pyrimidine dimers was able to remove the lesions responsible for chromosome aberration formation when exposed to photoreactivating light immediately after UV exposure. Because enzymatic photoreactivation does nothing but split cyclobutane dimers, the conclusion is that dimers formed between adjacent pyrimidines are the lesions that are ultimately translated into structural aberrations.

The fact that most chemical clastogens require replicative DNA synthesis either during, or subsequent to, cell treatment in order to form structural aberrations is in itself very suggestive that DNA is the principal target molecule in the production of the aberrations.

Chemical studies supporting the proposition that DNA is the critical target for aberration induction are of four types. First, reactions of many chemical carcinogens with DNA have been thoroughly documented (Lutz, 1979, Miller and Miller, 1971). Second, the careful consideration of chemical reaction kinetics (Osterman-Golkar et al., 1970) has strongly correlated mutagenicity of chemical agents to their reactivity with DNA targets. Third, the effects of altered mutagen structure and reaction rate, which lead to subtle changes in the pattern of potential DNA lesions, have been shown (Vogel and Natarajan, 1979) to give changes in mutation spectra which are predictable from knowledge of chemical studies. Fourth, the effect of storage of chemically damaged sperm cells on the relative frequencies of recovered mutations is consistent only with a time-dependent change in the DNA.

Studies of chemical reactions between mutagens and DNA *in vivo* have greatly advanced our understanding of true dose versus response kinetics of mutagenesis as well as carcinogenesis. For example, Lutz (1979), in reviewing the available literature on the binding of a wide array of carcinogens (many of which require activation) to rodent cell DNA, found a "crude semiquantitative correlation" between the amount of DNA-bound carcinogen in liver and frequency of hepatocarcinoma (i.e., more bound carcinogen, more cancer). More direct is the demonstration that induction of point mutations by EMS in *Drosophila* sperm is proportional to the amount of ethyl groups attached to DNA (Aaron and Lee, 1978).

Most mutagens that have been carefully studied have been shown to introduce a variety of chemical groups into DNA. Correlations between the nature and location of the various DNA adducts formed and the structure of the mutagen have been rigorously performed for relatively few chemicals (Sun and Singer, 1975); but, the kinetic properties of many more mutagens have been studied. Linear free-energy relationships such as that of Swain and Scott (1953) have been described for many series of chemicals, and have been shown by Ehrenberg and his co-workers (Osterman-Golkar *et al.*, 1970) to provide a means of correlating chemical reactivity *in vitro* with genetic and other effects of alkylating agents. For example, using the Swain–Scott correlation it was possible to attribute the toxic effects of several alkylating chemicals to reaction with proteinaceous targets, while genetic effects appear to be due to DNA alkylation. Vogel and Natarajan (1979) showed that the correlation of the mutational spectrum with chemical reaction pattern extends to germline effects in *Drosophila melanogaster*. In essence, what they showed was that at equal levels of point mutational damage, the relative yield of chromosome aberrations is inversely correlated with the propensity of a mutagen to alkylate oxygen atoms of DNA. For example, ethylnitrosourea (ENU) gave very high frequencies of point mutations, but no chromosome aberrations (II-III translocations); ENU gives relatively high levels of oxygen alkylation. Methylmethane sulfonate (MMS), on the other hand, does not alkylate oxygen sites very efficiently, but produces high frequencies of chromosome aberrations.

In addition, when treated sperm cells are stored in the spermathecae of *Drosophila melanogaster* female, an increase is observed in the proportion of chromosome breaks among all mutations. Such storage experiments with alkylating agents (Schalet, 1955; Abrahamson et al., 1969; Slizynska, 1973) provide clear evidence that some factors responsible for chromosome aberrations are subject to a time-dependent maturing process. It is known that events such as alkylation on the N-7 position of guanine lead to instability and loss of the base, and presumably to strand scission. Such events may ultimately be reflected as breakage of the chromosome.

b. The Role of DNA Strand Breaks. Whether the aberration is an apparent simple deletion, or an exchange of material between chromosomes, the premise is that DNA strand breaks play a major role in aberration formation. This derives initially from the early studies on radiations of varying LET. Densely ionizing radiation is much more efficient at producing double-strand breaks than is sparsely ionizing radiation and is also more efficient at producing structural aberrations. These observations lead to the conclusion that it is the double-strand break that is principally involved in aberration formation. This hypothesis receives support from the observation that the only chemicals capable of producing aberrations in all stages of the cell cycle are those that produce double-strand breaks. Recent data by Preston (1980), to be discussed later, cloud this simple assumption because they show that accumulation of single-strand breaks enhances the yield of aberrations.

If double-strand breaks are required for aberration formation, the question can be asked as to how they arise following UV and chemical treatment. This entire subject has been eloquently discussed by Bender and colleagues (Bender et al., 1973a,b, 1974a,b) in a series of articles dealing with models of aberration formation, by Evans (1977), and by Scott (1980). In essence, the suggestion is made that the lesion placed into DNA by the treatment, if not repaired, results in a gap in the nascent strand opposite the lesion at the time of DNA replication. If this gap is not repaired by a postreplication repair process the single-stranded region in the parental strand is susceptible to attack by single-strand endonucleases, thus generating double-strand breaks. These double-strand breaks, in turn, interact to form the aberration.

There are several lines of evidence that support this contention, the most prominent being the effect of caffeine. Caffeine has been shown to interfere with postreplication repair in mammalian cells (Cleaver and Thomas, 1969; Fujiwara and Kondo, 1972; Trosko and Chu, 1973) and it greatly enhances the clastogenic effect of many chemicals when given as a postchemical exposure treatment (Kihlman, 1977). It is suggested that the inhibition of the filling in of the gap in the nascent strand opposite the lesion increases the probability of the parental strand undergoing enzymatic degradation with the result of the formation of a double-strand break. Scott (1980) reports that a Yoshida sarcoma cell line deficient in postreplication repair has enhanced sensitivity to chemical induction of aberrations and is relatively insensitive to effects of caffeine when compared to a repair proficient line.

All these data argue strongly for double-strand breaks as the primary requisite for aberration formation but they do not conclusively prove it.

c. **The Role of Recombinational Events.** The mere formation of an aberration involving two different chromosomes means that recombination between those chromosomes had to take place. The question is therefore not whether a recombinational process is involved but rather if the critical step of recombination involves single- or double-strand breaks. The preponderance of data cited at this point supports the concept of the necessity for a double-strand break. This is not to say, however, that single-strand breaks cannot lead to aberrations, particularly in certain instances, e.g., during DNA replication and on rare occasions in G_1 and G_2 cells. Normal excision repair of lesions probably does not lead to chromosome aberrations. This is based on two observations: (1) cells with diminished excision repair capabilities, such as Xeroderma pigmentosum, do not show altered sensitivity to the production of aberrations in G_1 by either X-rays, or chemicals; (2) aberrations are not formed in excision repair competent G_1 cells following chemical treatment.

Evans (1977), however, presents perfectly logical models for aberration formation during DNA replication. One involves the exchange of Okasaki fragments of two daughter strands followed by excision of the lesion and gap filling that would complete the exchange process to include the duplex molecule. The other model, in simple terms, involves heteroduplex formation between parental strands during DNA replication (S). These heteroduplexes can theoretically arise as the result of large gaps in the nascent strands opposite lesions. There is some evidence (Rommelaere and Miller-Faures, 1975; Moore and Holliday, 1976) for the formation of heteroduplexes after treatment with alkylating agents. Presumably these heteroduplex structures occur in regions of different chromosomes that contain base sequence homology, such as repetitive sequences that are dispersed throughout the genome. Once the heteroduplex is formed the aberration could be finalized by incision and ligation to the daughter strands (see Evans, 1977, for details). Although no firm evidence exists to support these models they should be considered as possibilities until ruled out.

3. The Role of Molecular Repair in Aberration Formation

Any discussion of repair mechanisms involved in chromosome aberration formation must include the classical work of Wolff and colleagues (Wolff and Luippold, 1958; Wolff, 1964). These studies showed quite clearly that oxidative phosphorylation and protein synthesis were required for repair to occur and that interference with nucleotide biosynthesis had no effect on repair. This latter observation was interpreted as indicating that DNA was not the principal molecule involved in aberration formation.

In recent years considerable data have been generated that strongly implicate DNA-repair processes as contributing factors in chromosomal aberration formation. The data do not delineate the precise mechanisms involved but they do indicate that aberration frequencies can be modified by the absence, or inhibition, of normal repair processes. Some examples of this are the results of Griggs and Bender (1973), mentioned earlier and more recently the data of Sasaki (1980). Sasaki's data on the cytogenetic effects of chemical mutagens in Xeroderma pigmentosum (XP) and Fanconi's anemia (FA) lymphocytes illustrate the role of repair processes in moderating a cell's response to clastogenic agents. Xeroderma pigmentosum cells are unable to excise the DNA damage produced by 4-nitroquinoline oxide (4NQO) and decarbomyl mitomycin C (DCCMC), and FA cells are unable to repair the DNA damage produced by mitomycin C (MMC). Treatment of normal G_0, or G_1, lymphocytes with these compounds results in very little, if any, cytogenetic effort. Treatment of G_0, or G_1, XP lymphocytes with 4NQO or DCMMC leads to a very high frequency of chromatid aberrations and treatment of G_1 FA lymphocytes with MMC produces a high frequency of chromatid aberrations. Similar results are obtained with ataxia telangiectasia (AT) lymphocytes and ionizing radiation in that presumably unrepaired DNA base damage (Patterson *et al.*, 1976) in irradiated G_0 AT lymphocytes generates chromatid aberrations (Bender, 1980). Although these data do not lend themselves to constructing a model of how the aberrations are actually formed, they do allow a semiprecise identification of the lesions responsible for the aberrations.

Experiments that provide more insight, but not the definitive answer, into the mechanism of aberration formation are those of Holmberg and Jonasson (1974), J. G. Brewen and R. J. Preston (unpublished), and Preston (1980). Holmberg and Jonasson and Brewen and Preston demonstrated that although UV irradiation of normal human G_0 lymphocytes resulted in very few chromosome-type aberrations, UV irradiation did act synergistically with low LET ionizing radiation. The interpretation of these results was that the excision of UV-induced pyrimidine dimers resulted in a few single-strand gaps in DNA at any one moment—too few to generate many aberrations by interacting with one another. The ionizing radiation, on the other hand, generated a large number of strand breaks that persisted for a moderately long period of time. Consequently, the gaps generated by normal excision of UV-induced lesions had strand breaks available to interact with as they were produced in the normal course of excision repair.

Preston (1980) carried these experiments one step farther when he studied the effect of cytosine arabinoside (ara-C) when applied after treatment of G_1 lymphocytes with MMS, 4NQO, and X-rays. Cytosine arabinoside had been shown (Hiss and Preston, 1977) to inhibit the ligation step in excision repair and consequently it leads to the accumulation of single-strand gaps when applied after mutagen treatment. Preston's data show that ara-C enhances the yield of chromosome-type aberrations following treatment of G_1 lymphocytes with MMS and 4NQO. This latter observation is particularly interesting in that neither of these chemicals produces chromosome-type aberrations in normal repair-proficient lymphocytes because they are S-dependent clastogens.

The interpretation that comes immediately to mind is that ara-C treatment results in the accumulation of single-strand breaks by inhibiting the ligation step of excision repair of base damage, and these single-strand breaks in turn interact to form the aberrations. It is conceivable, however, that the intact polynucleotide strand opposite the gap is cleaved by an endonuclease resulting in double-strand breaks that interact to form the aberrations. The enhancement of the X-ray effect does not diverge from either of these models because it is possible that some radiation-induced aberrations result from single-strand breaks and ara-C simply increases the number of these single-strand breaks by blocking the final step of base damage repair.

The precise mechanism by which structural aberrations are formed is as yet not known. There are, however, compelling data to suggest that the first step is the production of a discontinuity in either the single, or double, polynucleotide strand followed by an erroneous recombinational step.

B. Somatic-Cell Aberrations

Perhaps the two easiest *in vivo* cytogenetic procedures to perform involve the use of somatic cells. These procedures employ the use of bone marrow or circulating immature lymphocytes. The use of peripheral lymphocytes has the unique advantage that results obtained in laboratory animals can be compared, when the occasions occur, to results obtained in instances of human exposure. This is because samples of venous blood can be obtained from human subjects with little discomfort whereas other tissues involve much more complex sampling procedures and routine sampling is very impractical. To date, several dozen mammalian species have been used to perform cytogenetic analysis on peripheral lymphocytes. The more commonly used ones are man, marmoset, rhesus monkey, mouse, rat, rabbit, and domestic swine. In addition to the ease in obtaining blood samples, the use of peripheral lymphocytes also presents the opportunity of taking repeat samples from the same animal. Repetitive sampling of other suitable tissue from the same animal is often impossible, and when it is possible (e.g., bone marrow) it is a tedious procedure.

The peripheral lymphocyte system has, however, one major disadvantage for studying chemical clastogens. It was stated earlier that, with few exceptions, chemical clastogens require replicative DNA synthesis in order to have the lesions they induce translated into structural aberrations. It has been determined that in excess of 99% of the circulating lymphocytes are in what is called a G_0 stage of the cell cycle, and do not replicate their DNA until they are induced to do so in tissue culture by treatment with an appropriate mitogenic agent. This fact suggests that most chemicals will not produce structural aberrations when they are administered to the animal. If they do, however, produce a long-lived lesion that persists until the cells are stimulated to divide in tissue culture it is conceivable that aberrations will be formed at this time.

At this time very few data exist that lend themselves to interpreting the utility, and efficacy, of using structural aberrations in peripheral lymphocytes for detecting *in vivo* exposures to chemical clastogens. Schnizel and Schmid (1976) report a very large study in which they analyzed peripheral lymphocytes from patients undergoing chemotherapy. In this study some individuals had excessive levels of chromosome-type aberrations but these individuals all had a history of either radiotherapy, chemotherapy with radiomimetic compounds, or an extended treatment with S-dependent chemicals at very high dose levels. In this latter instance, the aberrations observed were probably surviving events produced in hemopoietic precursor cells. There were only a few instances in which the frequency of chromatid-type aberrations was elevated above the control level and these involved high doses of known S-dependent compounds. The results of this study led the authors to conclude, "this testing system is therefore judged to be inadequate for monitoring weak or questionable mutagens in exposed populations." This is an understatement because the system is also probably inadequate for detecting strong mutagens except in those instances where the chemicals are radiomimetic or massive exposures are involved as in chemotherapy.

With this in mind, the wisdom of monitoring human populations purportedly exposed to mutagenic substances for structural aberrations must be seriously questioned. This is particularly true until appropriate animal model experiments are done that can verify the sensitivity, or insensitivity, of the assay. These experiments are straightforward and easy to perform. One parameter that must be studied is the cell's ability to repair the lesions introduced into the DNA. Normal human lymphocytes can repair most DNA lesions and if a substance is S-dependent there is an excellent probability the lesions will be repaired before the lymphocyte replicates its DNA *in vitro*.

The use of bone marrow, or other normally proliferating cells, for *in vivo* cytogenetic analysis circumvents the previously discussed problem because proliferating cells routinely pass through an S phase. Hence, in routine testing for possible clastogenic effects of unknown substances, it is advisable that a proliferating tissue be used. There are many tissues that can be used and some thought should be given to which one is chosen. In instances where a compound might be detoxified by gut bacteria, and where the principal route of exposure is anticipated to be oral, duodenal epithelium is a suitable tissue to study. If a compound is expected to be metabolized to an active, or inactive, form by the liver, regenerating liver might be considered. Instances will surely arise when substantial concern about effects in the fetus must be accounted for. In these cases the investigator should consider looking for cytogenetic effects in the fetus.

There exists a considerable literature on the cytogenetic effects of agents in various somatic tissues and no attempt will be made to make a compendium of that literature. It should be noted, however, that in those instances when fetal studies are conducted care should be taken to treat the females after implantation. This is because preliminary data generated in our laboratory indicate that no, or minimal, effects are observed in preimplantation embryos if treatment occurs after fertilization. More research is needed in this area before the assay can be used as a routine screening procedure for detecting potential fetal effects.

1. Types of Aberrations

Structural chromosome aberrations are generally regarded as occurring at three levels of chromosome organization. Those aberrations that occur at the same locus of both chromatids are called chromosome-type aberrations and have been shown to be formed in G_1 phase of the cell cycle when the chromosome is functionally a single-stranded structure. Chromosome-type aberrations can also be observed as a result of scoring second division cells that originally contained chromatid-type aberrations. Chromatid-type aberrations usually involve only one chromatid of the metaphase chromosome and are produced from late G_1 through the S and G_2 phases of the cell cycle when the chromosome is functionally a double-stranded structure. The third, and least important type of aberration with respect to the ensuing discussion, is the subchromatid type. These are generally formed during prophase when the chromosomes are condensing for cell division. It is still not clear if they represent aberrations involving chromatid subunits, the super-coiled major component of the chromatid, or simply stickiness.

Both chromosome- and chromatid-type aberrations can be further classified with regards to whether they involve one, two, or more chromosomes and whether they are symmetrical or asymmetrical. Symmetrical aberrations result in all the chromosome material being involved with a centromere whereas asymmetrical aberrations generate an acentric fragment. Figure 1 presents diagrammatic examples of these various types of aberrations.

2. Fate and Consequences of Structural Aberrations

It is a commonly held belief that centromeres of different chromosomes segregate at random with respect to each other at mitosis and that acentric fragments will usually, but not always, be lost. With this in mind it seems appropriate to consider what happens to the structural aberrations commonly analyzed at mitotic metaphase.

Loss of large segments of genetic material through chromosome deletion will impair the cell's ability to survive if it does not kill the cell outright. Deletion aberrations are expected, therefore, to be lost at a

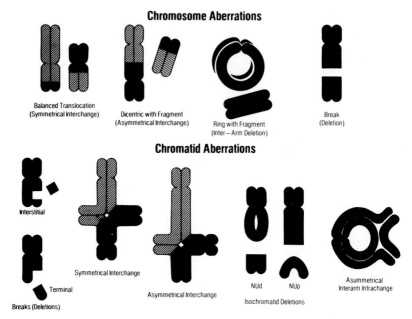

Fig. 1. Examples of chromosome and chromatid aberrations as they appear in metaphase.

moderately high rate after they are induced. Not all deletions need be cell lethal as there are the odd examples that are not lethal to the whole organism. It is usually accepted, however, that they are lethal in excess of 90% of the time. For this reason alone it is imperative that cells be analyzed in their first posttreatment cell division to insure determination of the maximum effect.

The same precautions apply to other types of aberrations. For example, all asymmetrical inter- and intrachanges generate acentric fragments, and in the case of interchanges dicentric chromosomes or chromatids. Hence, there are two mechanisms by which they can be lost, namely, mechanical difficulties at anaphase due to bridge formation and loss of the acentric fragment. Although symmetrical chromatid-type interchanges do not generate either acentric fragments or dicentric chromatids they only have a 25% probability of being recovered in subsequent cell generations. This is because duplication-deficient genotypes for the exchanged material are expected to be produced by random segregation 50% of the time and the normal and translocated chromatids, in balanced form, recovered the other 50% of the time. Only the symmetrical chromosome-type aberrations are transmitted to future progeny cells with a frequency of 1.0.

To date there have been no thorough studies that compare the rate of production of all types of aberrations, as determined by analysis of first posttreatment division, to the frequency of eventual recovery in future cell generations. If the human peripheral lymphocyte system is to ever be used as a monitoring tool with any degree of confidence in its ability to detect clastogenic effects, these kinds of studies must be done.

The foregoing discussion illuminates the need for careful experimental design when investigating the possible clastogenic effects of any chemical. Care must be taken to ensure the analysis of first posttreatment cell divisions as well as the sampling of cells from all stages of the cell cycle in order to detect cell stage specific effects.

C. Germ-Cell Aberrations

Other than the fact that chromosome aberrations produced in germ cells can be transmitted to future progeny and thus are truly heritable, there are only a few basic differences between germ and somatic cells that must be considered when discussing germ-cell cytogenetics. First, whereas balanced chromosome-type translocations are very difficult to detect in somatic cells they are detected quite easily in meiotic cells due to homologous pairing and subsequent multivalent formation. This allows for the easy detection of one of the two major types of heritable chromosome aberrations. Second, because of a haploid chromosome complement is recovered in each gamete after meiosis the probability of recovering a chromatid interchange in the balanced form is the square of that after mitosis, or one in 16, and the probability of recovering a chromosome translocation is one in four. This applies to both oocytes and primary spermatocytes. Third, since there are no cell divisions subsequent to meiosis any cytogenetic effects produced postmeiotically must be studied in the early zygote, preferably at the first cleavage division.

Other than for the preceding considerations the types, fates, and consequences of aberrations produced in germ cells are the same as those in somatic cells. Generally speaking, aberrations that normally confer lethality at the cellular level produce that lethality in the zygote when they are produced in the spermatocytes or oocytes.

In the past 10–15 years, there has been a considerable literature generated in the area of cytogenetic effects of chemicals in mammalian germ cells and some of the observations merit some discussion because they appear to be phenomenologically inconsistent. For example, with a few exceptions (Tates and Natarajan, 1976; Luippold *et al.*, 1978; van Buul and Goudzwaard, 1980; Adler, 1980; Au and Hsu, 1980) studies designed to detect reciprocal translocations in primary spermatocytes following spermatogonial treatment have yielded negative results. When positive results have been observed the yields of aberrations have been low and the chemicals were radiomimetic. These negative results were obtained in spite of the fact that the chemicals were shown to be clastogenic in other systems. To date there is no proved explanation for these results although several hypotheses have been proposed (see Adler and Brewen, 1981, for review). It may well be that at high doses of S-dependent chemicals, the sensitive and susceptible cells are killed but at lower doses they would survive and contain structural aberrations.

There have been reports on the cytogenetic effects of S-dependent chemicals on postmeiotic male germ cells ascertained by analysis of early cleavage embryos (Brewen et al., 1975; Matter and Jaeger, 1975; Burki and Sheridan, 1978; Katoh and Tonaka, 1979; Generoso et al., 1979). In theory only chromatid-type aberrations are expected to be observed due to the S dependency of the chemicals. The principal aberration type reported in these studies was the chromatid-type, or derived chromosome-type, except in two instances. Burki and Sheridan report ring chromosomes at first cleavage after TEM treatment and Katoh and Tonaka report dicentric chromosomes in first cleavage embryos following MMS treatment. These studies should be repeated because if chromosome-type aberrations are being produced by S-dependent chemicals in mature spermatozoa when they do not have such an effect in other cells there may be something unique about the stability of the lesion produced in sperm as seems to be the case in *Drosophila*. Also, similar studies should be done with known radiomimetic compounds.

In recent years the oocyte has become amenable to cytogenetic studies. The published data indicate that S-dependent compounds produce chromosomal aberrations only during pronuclear DNA synthesis in the embryo after treatment of the female followed by mating (Brewen and Preston, 1982). As in the case of postmeiotic male germ cells there are only a few thorough studies (Jagiello, 1967; Brewen and Payne, 1976, 1978; Caine and Lyon, 1977) and more research is called for. One area of research that deserves more attention is that which deals with the fate of aberrations induced by chemical treatment of the female as the embryo develops. One such study has been done (DuFrain et al., 1981) and more are called for. Studies of this nature could answer questions such as: Are some large deletions viable up to late embryonic development? Is there preferential inclusion of blastomeres not carrying aberrations into the formation of the inner cell mass? These are all points that will enhance our knowledge on the role of structural aberrations in fetal loss and malformation.

III. SISTER CHROMATID EXCHANGES (SCEs)

A. Historical Perspectives

Sister chromatid exchanges (SCEs), which represent reciprocal exchanges of homologous (or nearly homologous) chromatid segments between sister chromatids, were first demonstrated by Taylor (1958) in autoradiographic studies of plant chromosomes. One of the most significant subsequent developments was the discovery of staining procedures, such as the fluorescence plus Giemsa (FPG) technique of Wolff and Perry (1974), that provide permanently stained, "harlequinized" chromosomes in which SCEs can be easily and accurately scored. These new procedures, due to their elegant simplicity, led directly to the proliferation of SCE studies in the recent literature. One such work is that of Perry and Evans (1975) which in retrospect probably did as much as any other to establish SCE induction as a valid indicator of mutagenic potential. This and many later publications demonstrated that essentially all mutagenic carcinogens will induce SCEs under the appropriate conditions, and that in general the SCE assay is more sensitive than other mammalian cell assays for mutagenicity.

B. Mechanisms of Sister Chromatid Exchange Formation

Despite the fact that SCEs have been studied intensively, the mechanism of their formation remains unknown, as does the nature of the causative DNA lesion(s). Several models for SCE formation have been proposed, many of which suggest that SCE is a manifestation of DNA repair processes such as "postreplication repair" (cf. Bender *et al.*, 1974a; Kato, 1977). Attempts to correlate SCE induction with DNA repair, however, have been largely unsuccessful. For example, although cells are known to be defective in excision repair or postreplication repair, they are more sensitive to SCE induction by light UV and certain chemicals than are normal cells (Wolff *et al.*, 1977) providing good evidence that DNA repair systems are not actively involved in SCE formation. deWeird-Kastelein *et al.* (1977) found no good correlations between excision repair capacity and SCE production; and, they also found that XP variant cells (with a presumed defect in postreplication repair) do not differ from normal cells in terms of SCE frequency. Furthermore, Bloom's syndrome cells, which display exceptionally high spontaneous SCE levels, have proved to be normal in terms of their capacity to carry out known DNA-repair processes (XP and Bloom's syndrome will be discussed below in more detail). Another model for SCE formation, involving not DNA repair but rather "replication bypass" of DNA cross-links (Shafer, 1977), has been questioned on theoretical grounds (Stetka, 1979) and refuted in part by experimental test results (Carrano et al., 1979). The most promising model to date is that of Painter (1980), which is based on the premise that double-strand breaks are generated at the junctions between completely duplicated replicon clusters and partially duplicated replicon clusters. Sister chromatid exchange is "initiated when daughter strands of a duplicated cluster recombine with the parental strands of partially replicated cluster." When the latter cluster completes replication, the SCE is completed. This model predicts that agents which block DNA fork displacement should be efficient SCE inducers, as indeed they are (e.g., mitomycin-C, Carrano et al., 1978; and uv light, Kato, 1973). It also accurately predicts that agents such as X-rays, which induce DNA strand breaks and hence inhibit replicon initiation (Povirk and Painter, 1976), should be relatively inefficient inducers of SCE (cf. Perry and Evans, 1975). The model might also account for the high SCE frequency observed in Bloom's syndrome cells, as Hand and German (1975) have demonstrated that the fork-displacement rate is 30% below normal in these cells. Recent experiments have confirmed the model's prediction that pretreatment with X-rays should, by delaying the onset of replicon cluster replication, thereby decrease the frequency of SCEs induced by an agent that blocks chain elongation (R. B. Painter, personal communication). Further confirmation comes from the recently published results of Kato (1980) which demonstrate that fluorodeoxyuridine (FUdR) treatment, which slows DNA fork-displacement, enhances SCE induction by UV light in bromodeoxyuridine (BUdR)-substituted chromosomes. Thus, the model by Painter is clearly the best yet developed; but, unfortunately, there is still no absolute proof that it is correct.

An area of SCE research that has provided some mechanistic insights is that concerned with SCE levels in cells from human subjects with various genetic disease syndromes. (This subject is also covered in Chapter 10.) These studies have utilized a variety of cells, including lymphocytes from whole blood, bone marrow cells, and fibroblasts from skin biopsies; and, they have helped to characterize a number of human diseases including Fanconi's anemia, Bloom's syndrome, Down's syndrome, ataxia telangiectasia, and xeroderma pigmentosum. The first four share the common feature of chromosome fragility, while all five share a predisposition for the development of neoplasia. Of the five, only Bloom's syndrome exhibits a marked elevation in baseline SCE frequency (cf. Chaganti *et al.*, 1974; Wolff *et al.*, 1975; Yu and Borgaonkar, 1977), suggesting that SCE analysis will not be developed as a clinical tool for the detection of such rare genetic disorders. However, various studies have demonstrated that cells which carry these genetic defects can respond quite differently, in terms of SCE frequency, to treatment with known mutagens. Latt and co-workers (1975), for example, observed an approximately half-normal increase in SCE frequency in FA lymphocytes after *in vitro* exposure to MMC, a cross-linking agent. As FA lymphocytes remove (repair) DNA cross-links at greatly reduced rates, it would be tempting to speculate that this repair defect is responsible for the abnormally high SCE frequencies. Unfortunately, Latt's work also indicated that FA fibroblasts show an almost normal response to MMC, and so no firm mechanistic conclusions can be drawn from these studies. Studies with XP cells have shown that they are more sensitive to UV light (Bartram *et al.*, 1976) and to both UV- and radio-mimetic chemicals (Wolff *et al.*, 1977), in terms of SCE induction, than are normal human cells. Although most XP cell lines are deficient in the repair of pyrimidine dimers, it does not appear that this deficiency is responsible for their susceptibility to SCE induction (cf. Wolff *et al.*, 1977; deWeird-Kastelein *et al.*, 1977). The fact that radio-mimetic chemicals (whose damage XP cells seem able to repair) are so effective at SCE induction suggests that although most DNA damage is repaired, there may be some small but long-lived component(s) of damage that leads to SCE formation (Wolff *et al.*, 1977). Speculation has centered on alkylation at the O-6 position of guanine, based primarily on work with ENU, because (1) this mutagen alkylates the O-6 position very efficiently, and such base damage is excised more slowly from XP cells than from normal cells (Goth-Goldstein, 1977); and (2) ENU treatment results in elevated SCE levels that persist longer in XP cells than in normal cells. [Chinese hamster ovary (CHO) cells are also deficient at removal of O^6-alkylguanine (Goth-Goldstein, 1980) and these cells also show long-lived elevations in SCE frequency following ENU treatment (D. G. Stetka, unpublished results)]. The primary objection to this speculation is the lack of sufficient information concerning other forms of DNA damage that might also be excised slowly in cells where elevated SCE levels persist longer than normal. Perhaps the most interesting human condition yet discovered with regard to SCE is Bloom's syndrome. Lymphocytes from Bloom's syndrome patients, when cultured in the presence of BUdR, display at least 10-fold increases in SCE frequency over normal cells (e.g., Chaganti et al., 1974). The defect is clearly genetic in nature, but the molecular basis of the problem is unknown. German and co-workers (1977) have reported the coexistence, in individuals, of cells with both high and normal SCE levels, suggesting that the defect is regulatory. Tice and co-workers (1978), on the other hand, concluded from cocultivation experiments that diffusible, SCE-inducing substances are produced by Bloom's syndrome cells, so that all cells in culture with abnormal Bloom's cells should show elevated SCE frequencies. In conflict with the evidence from Tice et al. are the results derived from hybrids formed by fusion of Bloom's with normal human cells (Bryant et al., 1979). Such hybrids show complete reversion to normal SCE levels, consistent with the idea that high SCE frequencies in Bloom's syndrome result from an intrinsic genetic defect (a defect overcome by fusion with normal cells). Perhaps the deficiency involves enzymes required for DNA replication, which would explain the slower than normal fork-displacement rate (Hand and German, 1975) and also, possibly, the elevation of SCE frequency (which would be expected if the model of Painter, referred to above, is correct). Two other human conditions, Down's syndrome and ataxia telangiectasia, have proved uninteresting in SCE studies, because both exhibit normal SCE frequencies (Yu and Borgaonkar, 1977; Galloway and Evans, 1975).

C. Biological Significance of Sister Chromatid Exchange

Although the mechanism of SCE production remains to be determined, there is little doubt that SCE induction indicates a biologically significant and potentially deleterious event at the chromosomal level. In vitro tests have shown conclusively that mutagenic carcinogens nearly always induce SCEs, often at doses well below those required to show positive results in other mammalian cell assays (cf. Perry and Evans, 1975; Abe and Sasaki, 1977; Carrano et al., 1978). In addition, it is reasonably clear that SCEs are caused by agents that attack DNA and are thus potentially mutagenic, although the nature of the DNA lesions leading to SCE remains unknown (see Wolff, 1977, for a thorough discussion of SCE-inducing lesions).

Attempts to correlate SCE, mutation, and chromosome-aberration induction have yielded a confusing array of sometimes contradictory results. For example, Connell (1979), using metabolites of benzo[a]pyrene and bromoethylbenz[a]anthracene as inducers, found that SCE induction was not directly related to either aberration or mutation induction; and Nichols *et al.* (1978) presented data suggesting that SCE and aberrations induction by Simian virus 40 have different viral mechanisms. Carrano *et al.* (1978), using four mutagens, observed linear relationships between induced SCEs and induced mutations (at the HPRT locus), but the slopes were different for each agent. Their data suggest that each mutagen produces a spectrum of lesions, some of which are more readily converted to one form of damage (e.g., SCEs or mutations) than to another. The findings of Swenson *et al.* (1980) also indicate that SCE induction does not necessarily result from a single specific DNA lesion. Despite the lack of good correlations between specific DNA lesions and SCE induction, or between SCEs and mutations or aberrations, an objective analysis of published data indicates that SCE inducers are typically mutagens, and often clastogens as well.

Overall, it seems reasonable to conclude that many forms of primary DNA damage can lead ultimately to SCE when the affected cell traverses S phase (Wolff *et al.*, 1974), and that SCEs may result from processes that are basically different from those that cause mutations or aberrations. It seems equally clear, however, that SCE induction is at least a qualitative indicator of the mutagenic and clastogenic potential of the inducing agent.

D. Sister Chromatid Exchanges *in Vivo:* Background Information

Accepting the premise that SCEs represent potential genetic damage, the ability to detect SCEs induced *in vivo* acquires a certain significance. *In vitro* systems, including those that incorporate certain aspects of mammalian metabolism via the addition of liver enzymes (Stetka and Wolff, 1976a; Natarajan *et al.*, 1976), cannot accurately or sufficiently duplicate the metabolic pathways and physiological conditions that exist in an animal. Therefore, if one wishes to assess potential genetic risk to man, based on SCE results obtained with a suspected mutagen, then *in vivo* testing is required. Animals must be exposed to test agents, and cells from various tissues must be analyzed for SCE frequency determination.

One technical problem, unique to SCE analysis, must be dealt with in all SCE test systems. Recently developed staining procedures have facilitated the visualization and enumeration of SCEs (compares with autoradiographic methods), but these new procedures work only when cells have completed two successive rounds of DNA replication in the presence of 5-bromodeoxyuridine (a thymidine analogue). The BUdR is incorporated into nascent DNA and so after two rounds of replication one chromatid of each chromosome contains DNA that is bifilarly substituted with BUdR, while its sister chromatid is unifilarly substituted. Such chromatids can be differentially stained using techniques developed by, among others, Zakharov and Egolina (1972), Latt (1973), and Wolff and Perry (1974). Differential staining permits detection of SCE, and, as it requires prior BUdR incorporation, all SCE assays must in some way provide BUdR to the dividing cells of interest. This is accomplished *in vivo* using a variety of techniques, as discussed below.

E. Sister Chromatid Exchanges *in Vivo:* Systems Amenable to Study

Many cell systems are amenable to *in vivo* SCE analysis, including selected tissues of plants (Kihlman and Kronberg, 1975), birds (Bloom and Hsu, 1975), fish (Kligerman and Bloom, 1976), and many mammalian tissues. The present discussion is limited to mammalian systems, where SCEs can be analyzed in cells from testes (spermatogonial), various fetal tissues, regenerating liver, salivary glands, bone marrow, lung (alveolar macrophages), blood (peripheral lymphocytes), and potentially, embryos. Bromodeoxyuridine is in some cases provided to cells *in vivo* (and *in situ*), either by multiple intraperitoneal injections (Allen and Latt, 1976; Vogel and Bauknecht, 1976); continuous infusion, subcutaneous (Pera and Mattias, 1976) or intravenous (Schneider *et al.*, 1976); subcutaneous implantation of BUdR tablets (Allen *et al.*, 1977); or intraperitoneal injection of BUdR absorbed to activated charcoal (Russev and Tsanev, 1973; Ramirez, 1980). Where *in vivo* BUdR provision is employed, as it is in most of these assays, the assays also require *in vivo* administration of colchicine or Colcemid to arrest dividing cells in metaphase, followed by harvest of the cells (tissues) of interest after animals have been sacrificed. An alternate approach involves exposure of experimental animals to test agents, followed by removal and subsequent culture of selected cells *in vitro* with BUdR. This technique is employed in the peripheral lymphocyte assay and can also be used for the analysis of SCEs in embryos and bone marrow (see below); and, such assays may therefore be referred to as "*in vivo–in vitro* assays."

Experimental test results from several *in vivo* assays will serve to illustrate the relative merits of each system. Allen and Latt (1976) developed the technique for detection of SCEs formed in spermatogonial cells of mice. Fourteen hourly injections (IP) of BUdR were followed by IP exposure to various concentrations of MMC and subsequent exposure to Colcemid. After 3–5 hours, spermatogonial chromosomes were prepared from isolated seminiferous tubules and then analyzed for SCE frequency. The SCE levels increased with dose of MMC, while aberration frequencies failed to do so, demonstrating the sensitivity of SCE versus aberration induction *in vivo*. Equally important is the fact that this assay measures genetic damage in germ cells, and as such constitutes a relevant system for the study of genetic risk associated with potential mutagens. Another system concerned with congenital genetic damage, but not necessarily with genetic risk in the strictest sense, is one that detects SCE in various tissues of developing fetuses *in utero* (Kram *et al.*, 1980). In one experiment, BUdR and test mutagens were administered by intravenous infusion, and SCE frequencies were subsequently determined in fetal liver, lung, and gut (Kram *et al.*, 1980). Each tissue proved equally sensitive to SCE induction by direct-acting mutagens (MMC and daunomycin), while the activation-requiring compound cyclophosphamide induced many more SCEs in the liver than in the other organs. Thus, this system can be used to determine which fetal organs are sensitive to particular mutagens at specific times in development. *In vivo* SCE induction in mature liver can be studied following partial hepatectomy, which leads to regeneration via cell division. Schreck and Latt (1980), for example, treated two strains of partially hepatectomized mice with 3-methylcholanthrene (MC) to induce liver enzymes, and then administered benzo[a]pyrene (BP) by ip injection and BUdR by subcutaneous implantation of pellets. They found that inducibility of the liver enzyme B[a]P hydroxylase varied significantly between the two strains but that the induction of SCE by BP showed little correlation with such inducibility. Similarly, their data from bone marrow cells suggest little correlation between enzyme induction (by MC or phenobarbital) and cyclophosphamide-induced SCEs in the same two mouse strains. These results serve to illustrate several important points: (1) mammalian species/strains differ in terms of their ability to metabolize potential mutagens into active, mutagenic chromosome-damaging agents; (2) enzyme induction does not necessarily correlate in any meaningful way with genetic damage; and (3) there is at present no adequate substitute for direct genetic (e.g., cytogenetic) assays for the evaluation of potential genetic damage *in vivo*. The most often employed *in vivo* SCE assays are the various bone marrow tests (which differ only in the means of supplying BUdR to the animal). To cite just a few results, mouse bone marrow SCEs are induced by cyclophosphamide and triaziquone (Vogel and Bauknecht, 1976), MMC

(Kram *et al.*, 1978), and benzene (Tice *et al.*, 1980). Bone marrow assays have one clear advantage over other *in vivo* tests, that being their relative simplicity. It is quite easy to harvest bone marrow and to then prepare metaphase chromosomes on microslides for subsequent SCE analysis. One additional *in vivo* system that merits attention is the salivary gland assay of Ramirez (1980) in which BUdR is supplied via intraperitoneal injection of suspensions of activated charcoal to which BUdR is absorbed, and in which synchronized division of salivary gland cells is induced by injections of isoproterenol-HCl. Exposure of the animals to cyclophosphamide produced high SCE frequencies in the dividing cells. This system represents a breakthrough in that SCEs can now be studied in a tissue where cells do not actively divide under normal circumstances, and, unlike the liver assays, surgery is not required to stimulate cell division.

Nearly all currently available *in vivo* tests share certain advantages and disadvantages. The advantages are obvious. Genetic damage is assayed following exposure of whole animals to known or suspect mutagens, and so the many facets of mammalian physiology, metabolism, etc., are automatically incorporated into these tests. When compared to available *in vitro* tests, which at best incorporate only limited aspects of mammalian promutagen metabolism (e.g., Ames *et al.*, 1975; Stetka and Wolff, 1976a; Natarajan *et al.*, 1976), *in vivo* tests are obviously superior if the purpose of the test is to estimate potential genetic risk to man. Another advantage is that chromosome damage can be studied in more than one tissue from the same animal; thus, relative tissue sensitivities can be compared. The disadvantages of these various tests are not so obvious, but two especially important ones do exist. First, the test animals are often exposed to test mutagens at some time after the first exposure to BUdR, and so nuclear DNA is substituted with this thymidine analogue prior to attack by chemical agents. It has been demonstrated (Stetka and Carrano, 1977) that SCE induction by an otherwise weak inducer can become quite significant if the DNA is substituted with BUdR, a finding that at least throws into question any quantitative interpretation of data derived from *in vivo* studies where BUdR was available to cells prior to mutagen administration. Of course this problem is easily circumvented by exposing the animals to mutagens prior to BUdR treatment, assuming the chemical is rapidly cleared from the body, but this approach is not generally used. The second problem is that *in vivo* assays developed to date generally involve sacrifice of the animals in order to obtain the tissue (cells) under study. Longitudinal studies therefore require many animals, especially when multiple doses are employed, and such studies may thus become expensive and time consuming. One solution to this problem (an assay where repeated samples can be obtained from a single animal) is discussed below.

So-called *in vivo–in vitro* SCE assays are those in which tissues are removed from animals and then cultured (*in vitro*) in the presence of BUdR. One such assay, for which results have been published, was developed by Stetka and Wolff (1976b). It required only minimal modifications of procedures for removal of blood samples for marginal ear veins of rabbits, followed by culture of the blood samples in the presence of BUdR for subsequent SCE analysis in phytohemagglutinin-stimulated lymphocytes. In the original work animals were bled, then acutely exposed to either EMS, MMS, or cyclophosphamide, and then bled at various times after exposure. Analysis of the blood samples showed that SCE frequencies increased significantly within 1 day of exposure and then dropped to normal, preexposure levels within 1–2 weeks. This system clearly possessed certain advantages over other whole animal genetic assays: (1) the procedures were simple and inexpensive; (2) the sensitivity appeared to exceed that of other *in vivo* assays (e.g., cyclophosphamide effects were detectable at 20 mg/kg); and (3) perhaps most importantly, the time course of events could be followed in each animal, before and after exposure, and in this way each animal served as its own control and also provided results obtainable only with far more animals had sacrifice been required. Because human blood cells can also be cultured and analyzed for SCE levels, these rabbit-blood results led to speculation that SCE analysis in human lymphocytes might serve as a sensitive monitor for human exposure to mutagens. It was pointed out, however, that the transient nature of the SCE-frequency increase following mutagen exposure might preclude the use of this system unless subjects were available for testing within a few days of a suspected incident. With this potential drawback in mind, subsequent rabbit studies were conducted in which the animals were exposed subchronically (weekly) to low levels of known mutagens. These studies produced the somewhat surprising result that repeated exposure to MMC (Stetka *et al.,* 1978), benzo[a]pyrene, or 3-methylcholanthrene (D. G. Stetka, unpublished results) induced increases in peripheral lymphocyte SCE frequency that persisted for several months after termination of the exposure period. Thus, it appears that SCE-inducing DNA lesions are long-lived in peripheral lymphocytes, and that lymphocyte SCE analysis might serve as a very sensitive assay for human exposure to chemical mutagens (see below). (Other *in vivo–in vitro* systems, which remain to be validated, involve SCE analysis in bone marrow and embryo cells from exposed animals following *in vitro* culture of these tissues with BUdR. These systems will be discussed in the last section).

F. Sister Chromatid Exchanges in Man

Rabbit lymphocyte test results suggested that human exposure to mutagens might be detectable by SCE analysis, and results from human subjects tend to support this hypothesis. Exposure to mutagenic agents for medical/chemotherapeutic purposes, for example, induces SCEs in human lymphocytes. Positive results have been obtained following cancer therapy with a variety of cytostatic agents including adriamycin (Perry and Evans, 1975; Nevsted, 1978), CCNU {1-[2-chloroethyl-3-(4-methylcyclohexyl)]-1-nitrosourea} and melphalan (Gebhart *et al.*, 1980; Lambert *et al.*, 1978), and cyclophosphamide (Raposa, 1978), and also following UV phototherapy for neonatal hyperbilirubinemia (Goyanes-Villaescusa *et al.*, 1977). Occupational exposure to potential mutagens has also been shown to induce lymphocyte SCEs in workers using pesticides and herbicides (Crossen *et al.*, 1978), and in other workers exposed to a variety of toxic chemicals (Funes-Cravioto *et al.*, 1977). Considered together, the results of human and animal studies indicate that lymphocyte SCE analysis should be included in future cytogenetic studies of subjects with exceptional exposure to chemicals.

The efficacy of the SCE assay for human population studies depends heavily on the ability of the assay to detect small differences in SCE frequency between (1) lymphocytes sampled at various times from an individual subject, for longitudinal studies, and (2) cohort populations whose lymphocytes were sampled at similar times (one population being the matched control group). A recent study by Carrano and co-workers (described in Chapter 10) involved repeated determination of lymphocyte SCE frequencies in many individuals in order to determine the sensitivity of this assay. Their results indicate that a 30% difference in SCE frequency between two cohort populations can be detected with 95% probability at a 5% level of significance when 11 individuals per cohort are studied. Also, for longitudinal studies of single individuals, a 50% increase in SCE frequency can be detected with 95% probability at a 5% level of significance when only 25 cells per sample are analyzed. Thus, it appears feasible and practical to apply the SCE assay to humans as a measure of genetic damage from environmental agents. There is evidence, however, that while most clastogens also induce SCEs, there are some that do not; and so since SCEs and chromosomal aberrations might represent manifestations of different molecular events (Gebhart *et al.*, 1980), the SCE method cannot completely supplant classical cytogenetic analysis.

G. Areas for Future Research

The assignment of greater significance to SCE induction awaits determination of the primary DNA lesions as well as the mechanisms responsible for SCE formation. Progress in these fields will most likely result from *in vitro* studies. The present chapter, however, deals with *in vivo* events, and so this discussion will be limited to research needs in this area.

Finally, the sensitivity of the SCE system should be exploited in assays for heritable genetic damage. Midgestation embryos can be extracted and then cultured with BUdR for SCE analysis (Galloway *et al.*, 1980) and cells from fetal tissues can also be analyzed (Kram *et al.*, 1980), but studies to date have employed either maternal exposure (postfertilization) or *in vitro* exposure of cultured embryos. What remains to be accomplished is SCE analysis in early embryos following precopulation exposure of either parent. Experiments to develop and validate such a system are currently underway in this and other laboratories. Success in this area will provide a meaningful and rapid assay for genetic risk associated with chemical mutagens.

REFERENCES

Aaron, C. S., and Lee, W. R. (1978). Molecular dosimetry of the mutagen ethyl methanesulfonate in *Drosophila* melanogaster spermatozoa: Linear relation of DNA alkylation per sperm cell (dose) to sex-linked recessive lethals. *Mutat. Res.* **49**, 27–44.

Abe, S., and Sasaki, M. (1977). Chromosome aberrations and sister chromatid exchanges in Chinese hamster cells exposed to various chemicals. *J. Natl. Cancer Inst.* **58**, 1635–1641.

Abrahamson, S., Kiriazis, W. C., and Sabol, E. M. (1969). A storage effect of ethyl methanesulfonate (EMS) on the induction of translocations in *Drosophila* sperm. *D.I.S.* **44**, 110.

Adler, I. D. (1980). *In* "Cytogenetic Test Systems for Environmental Chemicals" (T. C. Hsu, ed.), Allenheld Publ. Co., (in press).

Adler, I. D., and Brewen, J. G. (1981). Effects of chemicals on chromosome abberation production in male and female germ cells. *Chem. Mutagens* (in press).

Ahnstrom, G., and Natarajan, A. T. (1966). Mechanisms of chromosome breakage—A new theory. *Hereditas* **54**, 379–388.

Allen, J. W., and Latt, S. A. (1976). Analysis of sister chromatid exchange formation *in vivo* in mouse spermatogonia as a new test system for environmental mutagens. *Nature (London)* **260**, 449–451.

Allen, J. W., Shuler, C. F., Mendes, R. W., and Latt, S. A. (1977). A simplified technique for *in vivo* analysis of sister-chromatid exchanges using 5-bromodeoxyuridine tablets. *Cytogenet. Cell Genet.* **18**, 231–237.

Ames, B. N., McCann, J., and Yamasaki, E. (1975). Methods for detecting carcinogens and mutagens with the *Salmonella*/mammalian-microsome mutagenicity test. *Mutat. Res.* **31**, 1–15.

Au, W. W., and Hsu, T. C.(1980). The genotoxic effects of adriamycin in somatic and germinal cells of the mouse. *Mutat. Res.* **79**, 351–361.

Bartram, C. R., Koske-Westphal, T., and Passarge, E. (1976). Chromatid exchanges in ataxia telangiectasia, Bloom's syndrome, Werner syndrome, and xeroderma pigmentosum. *Ann. Hum. Genet.* **40**, 79–86.

Bender, M. A. (1980). Relationship of DNA lesions and their repair to chromosomal aberration production. In "DNA Repair and Mutagenesis in Eukaryotes" (W. M. Generoso, M. D. Shelby, and F. J. de Serres, eds.), Vol. 15, 245–266. Plenum, New York.

Bender, M. A., Griggs, H. G., and Walker, P. L. (1973a). Mechanism of chromosomal aberration production. I. Aberration induction by ultraviolet light. *Mutat. Res.* **20**, 387–402.

Bender, M. A., Bedford, J. S., and Mitchell, J. B. (1973b). Mechanisms of chromosomal aberration prdouction. II. Aberrations induced by 5-bromodeoxyuridine and visible light. *Mutat. Res.* **20**, 403–416.

Bender, M. A., Griggs, H. G., and Bedford, J. S. (1974a). Recombinational DNA repair and sister chromatid exchanges. *Mutat. Res.* **24**, 117–123.

Bender, M. A., Griggs, H. G., and Bedford, J. S. (1974b). Mechanisms of chromosomal aberration production. III. Chemicals and ionizing radiation. *Mutat. Res.* **23**, 197–212.

Bloom, S. E., and Hsu, T. C. (1975). Differential fluorescence of sister chromatids in chicken embryos exposed to 5-bromodeoxyuridine. *Chromosoma* **51**, 261–267.

Brewen, J. G. (1965). The induction of chromatid lesions by cytosine arabinoside in post-DNA-synthetic human leukocytes. *Cytogenetics* **4**, 28–36.

Brewen, J. G., and Christie, N. T. (1967). Studies on the induction of chromosomal aberrations in human leukocytes by cytosine arabinoside, *Exp. Cell Res.* **46**, 276–291.

Brewen, J. G., and Peacock, W. J. (1969). Restricted rejoining of chromosomal subunits in aberration formation: A test for subunit dissimilarity. *Proc. Natl. Acad. Sci. U.S.A.* **62**, 389–394.

Brewen, J. G., and Payne, H. S. (1976). Studies on chemically induced dominant lethality in maturing dictyate mouse oocytes. *Mutat. Res.* **37**, 77–82.

Brewen, J. G., and Payne, H. S. (1978). Studies on chemically induced dominant lethality. III. Cytogenetic analysis of TEM effects on maturing dictyate mouse oocytes. *Mutat. Res.* **50**, 85–92.

Brewen, J. G., and Preston, R. J. (1982). Cytogenetic analysis of mammalian oocytes in mutagenicity studies. In "Handbook of Cytogenetic Assay Systems for Environmental Mutagens" (T. C. Hsu, ed.), Allenheld Publ. Co.

Brewen, J. G., Payne, H. S., Jones, K. P., and Preston, R. J. (1975). Studies on chemically induced dominant lethality. I. The cytogenetic basis of MMS-induced dominant lethality in post-meiotic male germ cells. *Mutat. Res.* **33**, 239–246.

Bryant, E. M., Holger, H., and Martin, G. M. (1979). Normalization of sister chromatid exchange frequencies in Bloom's syndrome by euploid cell hybridisation. *Nature* (London) **279**, 795–796.

Burki, K., and Sheridan, W. (1978). Expression of TEM-induced damage to post-meiotic stages of spermatogenesis of the mouse during early embryogenesis. II. Cytological investigations. *Mutat. Res.* **52**, 107–115.

Caine, A., and Lyon, M. F. (1977). The induction of chromosome aberrations in mouse dictyate oocytes by X-rays and chemical mutagens. *Mutat. Res.* **45**, 325–331.

Carrano, A. V., Thompson, L. H., Lindl, P. A., and Minkler, J. L. (1978). Sister chromatid exchange as an indicator of mutagenesis. *Nature* (London) **271**, 551–553.

Carrano, A. V., Thompson, L. H., Stetka, D. G., Minkler, J. L., Masrimas, J. A., and Fong, S. (1979). DNA crosslinking, sister chromatid exchange and specific-locus mutations. *Mutat. Res.* **63,** 175–188.

Chaganti, R. S. K., Schonberg, S., and German, J. (1974). A manyfold increase in sister chromatid exchanges in Bloom's syndrome lymphocytes. *Proc. Natl. Acad. Sci. U.S.A.* **71,** 4508–4512.

Chu, E. H. Y. (1965). Effects of ultraviolet radiation on mammalian cells. I. Induction of chromsome aberrations. *Mutat. Res.* **2,** 75–94.

Cleaver, J. E., and Thomas, G. H. (1969). Single strand interruptions in DNA and the effects of caffeine in Chinese hamster cells irradiated with ultraviolet light. *Biochem. Biophys. Res. Commun.* **36,** 203.

Cohen, M. M., Shaw, M. W., and Craig, A. P. (1963). The effects of streptonigrin on cultural human leukocytes. *Proc. Natl. Acad. Sci. U.S.A.* **50,** 16–24.

Connell, J. R. (1979). The relationship between sister chromatid exchange, chromosome aberration, and gene mutation induction by several reactive polycyclic hydrocarbon metabolites in cultured manmalian cells. *Int. J. Cancer* **24,** 485–489.

Crossen, P. E., Morgan, W. F., Horan, J. J., and Stewart, J. (1978). Cytogenetic studies of pesticide and herbicide sprayers. *N. Z. Med. J.* **88,** 192–195.

deWeird-Kastelein, E. A., Keijizer, W., Rainaldi, G., and Bootsma, D. (1977). Induction of sister chromatid exchanges in xeroderma pigmentosum cells following exposure to ultraviolet light. *Mutat Res.* **45,** 253–261.

Dresp, J., Schmid, E., and Bauchinger, M. (1978). The cytogenetic effect of bleomycin as human peripheral lymphocytes *in vitro* and in vivo. *Mutat. Res.* **56,** 341–353.

DuFrain, R. J., Littlefield, L. G., and Wilmer, J. L. (1980). The effect of washing lymphocytes after *in vivo* treatment with streptonigrin on the yield of chromosome and chromatid aberrations in blood cultures. *Mutat. Res.* **69,** 101–105.

DuFrain, R. J., Littlefield, L. G., and Wilmer, J. L. (1981). Evaluation of chemically induced cytogenetic lesions in rabbit oocytes. I. The test system and the effects of streptonigrin. *Mutat. Res.* (in press).

Evans, H. J. (1977). Molecular mechanisms in the induction of chromosome aberrations. *Dev. Toxicol. Environ. Sci.* **2,** 57–76.

Evans, H. J., and Scott, D. (1964). Influence on DNA synthesis on the production of cbromatid aberrations by X-rays and maleic hydrozide in *Vicia faba. Genetics* **49,** 17–38.

Evans, H. J., and Scott, D. (1969). The induction of chromosome aberrations by nitrogen mustard and its dependence on DNA synthesis. *Proc. R. Soc. London Ser. B* **173,** 491–512.

Fujiwara, Y., and Kondo, T. (1972). Caffeine-sensitive repair of ultraviolet light-damaged DNA of mouse L-cells. *Biochem. Biophys. Res. Commun.* **47,** 557.

Funes-Cravioto, F., Kolmodin-Hedman, B., Lindsten, J., et al. (1977). Chromsome aberrations and sister-chromatid exchange in workers in chemical laboratories and a rotoprinting factory and in children of woman laboratory workers. *Lancet* **8033,** 322–325.

Galloway, S. M., and Evans, H. J. (1975). Sister chromatid exchange in human chromosomes from normal individuals and patients with ataxia telangiectasia. *Cytogenet. Cell Genet.* **15,** 17–29.

Galloway, S. M., Perry, P. E., Meneses, J., Nebert, D. W., and Pedersen, R. A. (1980). Cultured mouse embryos metabolize benzo[a]pyrene during early gestation: Genetic differences detectable by sister chromatid exchange. *Proc. Natl. Acad. Sci. U.S.A.* **77,** 3524–3528.

Gebhart, E., Windolph, B., and Wopfner, F. (1980). Chromsome studies on lymphocytes of patients under cytostatic therapy. II. Studies using the BUdR-labelling technique in cytostatic interval therapy. *Hum. Genet.* **56**, 157–167.

Generoso, W. M., Cain, K. T., Krishna, M., and Huff, S. W. (1979). Genetic lesions induced by chemicals in spermatozoa and spermatids of mice are repaired in the egg. *Proc. Natl. Acad. Sci. U.S.A.* **76**, 435–437.

German, J., Schonberg, S., Louis, E., and Chaganti, R. (1977). Bloom's syndrome. IV. Sister-chromatid exchanges in lymphocytes. *Am. J. Hum. Genet.* **29**, 248–255.

Goth-Goldstein, R. (1977). Repair of DNA damaged by alkylating carcinogens is defective in xeroderma pigmentosum-derived fibroblasts. *Nature (London)* **267**, 81–82.

Goth-Goldstein, R. (1980). Inability of Chinese hamster ovary cells to excise O^6-alkyl guanine. *Cancer Res.* **40**, 2623–2624.

Goyanes-Villaescusa, V. J., Ugarte, M., and Vazquez, A. (1977). Sister chromatid exchange in babies treated by photo-therapy. *Lancet* **8047**, 1084–1085.

Griggs, H. G., and Bender, M. A. (1973). Photoreactivation of ultraviolet-induced chromosomal aberrations. *Science* **179**, 86–88,

Hand, R., and German, J. (1975). A retarded rate of DNA chain growth in Bloom's syndrome. *Proc. Natl. Acad. Sci. U.S.A.* **72**, 758–762.

Hiss, E. A., and Preston, R. J. (1977). The effect of cytosine arabinoside on the frequency of single-strand breaks in DNA of mammalian cells following irradiation or chemical treatment. *Biochem. Biophys. Acta.* **478**, 1–8.

Holmberg, M., and Jonasson, J. (1974). Synergistic effect of X-ray and UV irradiation on the frequency of chromsome breakage in human lymphocytes. *Mutat. Res.* **23**, 213–221.

Jagiello, G. (1967). Streptonigrin: Effect on the first meiotic metaphase of the mouse egg. *Science* **157**, 453–454.

Kato, H. (1973). Induction of sister chromatid exchanges by UV light and its inhibition by caffeine. *Exp. Cell Res.* **82**, 383–390.

Kato, H. (1977). Mechanisms for sister chromatid exchanges and their relation to the production of chromsomal aberrations. *Chromosoma* **59**, 179–191.

Kato, H. (1980). Evidence that the replication point is the site of sister chromatid exchange. *Cancer Genet. Cytogenet.* **2**, 69–77.

Katoh, M., and Tonaka, N. (1979). Relationship between chromosome aberrations in the first cleavage metaphases and unscheduled DNA synthesis following paternal MMS treatment. *Jpn. J. Genet.* **55**, 55–65.

Kihlman, B. A. (1964). The production of chromosomal aberrations by streptonigrin in *Vicia faba*. *Mutat. Res.* **1**, 54–62.

Kihlman, B. A. (1955). Chromsome breakage in *Alluim* by 8-ethoxy-caffeine and X-rays. *Exp. Cell Res.* **8**, 345–368.

Kihlman, B. A. (1966). "Actions of Chemicals on Dividing Cells." Prentice-Hall, Englewood Cliffs, New Jersey.

Kihlman, B. A. (1977). "Caffeine and Chromsomes." Elsevier, Amsterdam.

Kihlman, B. A., and Kronberg, B. (1975). Sister chromatid exchanges in *Vicia faba*. I. Demonstrations by a modified fluorescent plus Giemsa (FPG) technique. *Chromosoma* **51**, 1–10.

Kirby-Smith, J. S., and Craig, D. L. (1957). The induction of chromosome aberrations in *Tradescantia* by ultraviolet light. *Genetics* **42**, 176–187.

Kligerman, A. D., and Bloom, S. E. (1976). Sister chromatid differentiation and exchanges in adult mudminnows (Umbra limi) after *in vivo* exposure to 5-bromodeoxyuridine. *Chromosoma* **56**, 101–109.

Kram, D., Bynum, G., Senula, G., Bickings, C., and Schneider, E. (1980). *In utero* analysis of sister chromatid exchange: Differential sensitivity of fetal tissues to mutagenic damage. Abstract, 11th Annual Environ. Mutagen Soc. Meeting, Nashville, TN.

Kram, D., Schneider, E. L., Singer, L., and Martin, G. R. (1978). The effects of high and low fluoride diets on the frequencies of sister chromatid exchanges. *Mutat. Res.* **57,** 51–55.

Lambert, B., Ringborg, U., Harper, E., and Lindblad, A. (1978). Sister chromatid exchanges in lymphocyte cultures of patients receiving chemotherapy for malignant disorders. *Cancer Treat. Rep.* **62,** 1413–1419.

Latt, S. A., Stetten, G., Juergens, L. A., Buchanan, G. R., and Gerald, P. S. (1975). Induction by alkylating agents of sister chromatid exchanges and chromatid breaks in Tanconi's anemia. *Proc. Natl. Acad. Sci. U.S.A.* **72,** 4066–4070.

Latt, S. A. (1973). Microfluorometric detection of deoxyribonucleic acid replication in human metaphase chromosomes. *Proc. Natl. Acad. Sci. U.S.A.* **70,** 3395–3399.

Luippold, H. E., Gooch, P. C., and Brewen, J. G. (1978). The production of chromsome aberrations in various mammalian cells by triethylene-malamine. *Genetics* **88,** 317–326.

Lutz, W. K. (1979). *In vivo* covalent binding of organic chemicals to DNA as a quantitative indicator in the process of chemical carcinogenesis. *Mutat. Res.* **65,** 289–356.

Matter, B. E., and Jaeger, I. (1975). Premature chromosome condensation, structural chromosome aberrations, and micronuclei in early mouse embryos after treatment of paternal postmeiotic germ cells with triethylene-melamine, possible mechanisms for chemically induced dominant-lethal mutations. *Mutat. Res.* **33,** 251–260.

Miller, J. A., and Miller, E. C. (1971). Chemical carcinogenesis: Mechanisms and approaches to its control. *J. Natl. Cancer Inst.* **47,** 15–14.

Moore, P. D., and Holliday, R. (1976). Evidence for the formation of hybrid DNA during mitotic recombination in Chinese hamster cells. *Cell* **8,** 573–579.

Natarajan, A. T., Tates, A. D., van Buul, P. P. W., Meijers, M., and deVogel, N. (1976). Cytogenetic effects of mutagens/carcinogens after activation in a microsomal system *in vitro*. I. Induction of chromosome aberrations and sister chromatid exchanges by diethylnitrosamine (DMN) and dimethylnitrosamine (DMN) in CHO cells in the presence of rat-liver microsomes. *Mutat. Res.* **37,** 83–90.

Nevsted, N. P. (1978). Sister chromatid exchanges and chromosomal aberrations induced in human lymphocytes by the cytostatic drug adraimycin *in vivo* and *in vitro*. *Mutat. Res.* **57,** 253–258.

Nichols, W. W., Bradt, C. I., Toji, L. H., Godley, M., and Segawa, M. (1978). Induction of sister chromatid exchanges by transformation with Simian virus 40. *Cancer Res.* **38,** 960–964.

Osterman-Golkar, S., Ehrenberg, L., and Washtmeister, C. A. (1970). Reaction kinetics and biological action in barley of monofunctional methanesulfonic esters. *Radiat. Biol.* **10,** 303–327.

Painter, R. B. (1980). A replication model for sister-chromatid exchange. *Mutat. Res.* **70,** 337–341.

Patterson, M. C., Smith, B. P., Lohman, P. H. M., Anderson, A. K., and Fishman, L. (1976). Defective excision repair of X-ray damaged DNA in human (ataxia telangiectasia) fibroblasts. *Nature (London)* **260,** 444–447.

Pera, F., and Mattias, P. (1976). Labelling of DNA and differential sister chromatid staining of BrdU treatment *in vivo*. *Chromosoma* **57,** 13–18.

Perry, P., and Evans, H. J. (1975). Cytological detection of mutagen-carcinogen exposure by sister chromatid exchange. *Nature (London)* **258,** 121–125.

Povirk, L. F., and Painter, R. B. (1976). The effect of 313 nanometer light on initiation of replicons in mammalian cell DNA containing bromodeoxyuridine. *Biochem. Biophys. Acta* **432**, 267–272.

Preston, R. J. (1980). DNA repair and chromosome aberrations: The effect of cytosine arabinoside on the frequency of chromosome aberrations induced by radiation and chemicals. *Teratogen. Carcinogen. Mutagen.* **1**, 147–159.

Ramirez, P. M. (1980). Analysis *in vivo* of sister chromatid exchange in mouse bone-marrow and salivary gland cells. *Mutat. Res.* **74**, 61–69.

Raposa, T. (1978). Sister chromatid exchange studies for monitoring DNA damage and repair capacity after cytostatics *in vitro* and in lymphocytes of leukaemic patients under cytostatic therapy. *Mutat. Res.* **57**, 241–251.

Rommelaere, J., and Miller-Faures, A. (1975). Detection by density equilibrium centrifugation of recombinant-like DNA molecules in somatic mammalian cells. *J. Miol. Biol.* **98**, 195–218.

Russev, G. C., and Tsanev, R. G. (1973). Continuous labeling of mammalian DNA *in vivo*. *Anal. Biochem.* **54**, 115–119.

Sasaki, M. S. (1980). Chromosome aberration formation and sister chromatid exchange in relation to DNA repair in human cells. *In* "DNA Repair and Mutagenesis in Eukaryotes" (W. M. Generoso, M. D. Shelby, and F. J. deSerres, eds.), Vol. 15, 285–314. Plenum, New York.

Sax, K. (1938). Induction by X-rays of chromosome aberrations in Tradescantia microspores. *Genetics* **23**, 494.

Sax, K. (1940). An analysis of X-ray induced chromosomal aberrations in Tradescantia. *Genetics* **25**, 41–68.

Sax, K. (1941). Types and frequencies of chromosomal aberrations induced by X-rays. *Cold Spring Harbor Symp. Quant. Biol.* **9**, 93–101.

Schalet, A. (1955). The relationship between the frequency of nitrogen mustard induced translocations in mature sperm of *Drosophila* and utilization of sperm by females. *Genetics* **40**, 594.

Schneider, E. L., Chaillet, J. R., and Tice, R. R. (1976). *In vivo* BUdR labeling of mammalian chromsomes. *Exp. Cell Res.* **100**, 396–399.

Schnizel, A., and Schmid, W. (1976). Lymphocyte chromosome studies in humans exposed to chemical mutagens. The validity of the method in 67 patients under cytostatic therapy. *Mutat. Res.* **40**, 139–166.

Schreck, R. R., and Latt, S. A. (1980). Influence of metabolic potential over benzo[a]pyrene induced SCE. Abstract, 11th Annual Environ. Mutagen Soc. Meeting, Nashville, TN.

Scott, D. (1980). Molecular mechanisms of chromosome structural changes. *Dev. Toxicol. Environ. Sci.* **7**, 101–114.

Scott, D., and Evans, H. J. (1964). On the requirement of DNA synthesis in the production of chromosome aberrations by 8-ethoxycaffeine. *Mutat. Res.* **1**, 146–156.

Shafer, D. A. (1977). Replication bypass model of sister chromatid exchanges and implications for Bloom's syndrome and Fanconi's anemia. *Hum. Genet.* **39**, 177–190.

Slizynska, H. (1973). Cytological analysis of storage effects on various types of complete and mosaic changes induced in *Drosophila* chromosomes by some chemical mutagens. *Mutat. Res.* **19**, 199–213.

Stadler, L. J., and Uber, F. H. (1942). Genetic effects of ultraviolet radiation in maize. IV. Comparison of monochromatic radiations. *Genetics* **27**, 84–118.

Stetka, D. G. (1979). Further analysis of the replication bypass model for sister chromatid exchange. *Hum. Genet.* **49**, 63–69.

Stetka, D. G., and Carrano, A. V. (1977). The interaction of Hoechst 33258 and BUdR substituted DNA in the formation of sister chromatid exchanges. *Chromosoma* **63**, 21–31.

Stetka, D. G., and Wolff, S. (1976a). Sister chromatid exchange as an assay for genetic damage induced by mutagen-carcinogens. II. *In vitro* test for compounds requiring metabolic activation. *Mutat. Res.* **41**, 343–350.

Stetka, D. G., and Wolff, S. (1976b). Sister chromatid exchange as an assay for genetic damage induced by mutagen-carcinogens. I. *In vivo* test for compounds requiring metabolic activation. *Mutat. Res.* **41**, 333–342.

Stetka, D. G., Minkler, J., and Carrano, A. V. (1978). Induction of long-lived chromosome damage, as manifested by sister chromatid exchange, in lymphocytes of animals exposed to mitomycin-C. *Mutat. Res.* **51**, 383–396.

Sun, L., and Singer, B. (1975). The specificity of different classes of ethylating agents toward various sites of HeLa cell DNA *in vitro* and *in vivo*. *Biochem.* **14**, 1795.

Swain, C. G., and Scott, C. B. (1953). Quantitative correlation of relative rates. Comparison of hydroxide ion with other nucleophilic reagents toward alkyl esters, epoxides and acyl halides. *J. Am. Chem. Soc.* **75**, 141–147.

Swenson, D. H., Harbach, P. R., and Trzos, R. J. (1980). The relationship between alkylation of specific DNA bases and induction of sister chromatid exchange. *Carcinogenesis* **1**, 931–936.

Tates, A. D., and Natarajan, A. T. (1976). Correlative study on genetic damage induced by chemical mutagens in bone marrow and psermatogonia of mice. I. CNU-ethanol. *Mutat. Res.* **37**, 267–278.

Taylor, J. H. (1958). Sister chromatid exchanges in tritium-labeled chromosome. *Genetics* **43**, 515–529.

Tice, R., Rary, J. M., and Windler, G. (1978). Effect of co-cultivation on SCE frequencies in Bloom's syndrome and normal fibroblast cells. *Nature (London)* **273**, 538–540.

Tice, R. R., Costa, D. L., and Drew, R. T. (1980). Cytogenetic effects of inhaled benzene in murine bone marrow: Induction of sister chromatid exchanges, chromosomal aberrations, and cellular proliferation inhibition in DBA/2 mice. *Proc. Natl. Acad. Sci. U.S.A.* **77**, 2148–2152.

Trosko, J. E., and Chu, E. H. Y. (1973). Inhibition of repair of UV-damaged DNA by caffeine and mutation induction in Chinese hamster cells. *Chem. Biol. Interact.* **6**, 317–332.

van Buul, P. P. W., and Goudzwaard, J. H. (1980). Belomycin induced structural chromosome aberrations in spermatogonia and bone marrow cells of mice. *Mutat. Res.* **69**, 319–324.

Vig, B. (1977). Genetic toxicology of mitomycin-C, actinomycin, duanomycin, and adriamycin. *Mutat. Res.* **49**, 189–238.

Vig, B. K., and Lewis, R. (1978). Genetic toxicology of bleomycin. *Mutat. Res.* **55**, 121–145.

Vogel, W., and Bauknecht, T. (1976). Differential chromatid staining by *in vivo* treatment as a mutagenicity test system. *Nature (London)* **260**, 448–449.

Vogel, E., and Natarajan, A. T. (1979). The relation between reaction kinetics and mutagenic action of monofunctional akylating agents in higher eukaryotic system. I. Recessive lethal mutations and translocations in *Drosophila*. *Mutation Res.* **62**, 51–100.

Wolff, S. (1964). Mechanisms of dose-rate effects: Insights obtained from intensity and fractionation studies on chromosome aberration induction. *Jpn. J. Genet.* **40**, 38–48.

Wolff, S. (1977). Sister chromatid exchange. *Am. Rev. Genet.* **11**, 183–201.

Wolff, S., and Luippold, H. (1958). Modification of chromosomal aberration yield by postirradiation treatment. *Genetics* **43**, 493–501.

Wolff, S., and Perry, P. (1974). Differential Giemsa staining of sister chromatids and the study of sister chromatid exchanges without autoradiography. *Chromosoma* **48**, 341–353.

Wolff, S., Bodycote, J., and Painter, R. B. (1974). Sister chromatid exchanges induced in Chinese hamster cells by UV irradiation of different stages of the cell cycle: The necessity for cells to pass through S. *Mutat. Res.* **25**, 73–81.

Wolff, S., Bodycote, J., Thomas, G. H., and Cleaver, J. E. (1975). Sister chromatid exchange in xeroderma pigmentosum cells that are defective in DNA excision repair or post-replication repair. *Genetics* **81**, 349–355.

Wolff, S., Rodin, B., and Cleaver, J. E. (1977). Sister chromatid exchanges induced by mutagenic carcinogens in normal and xeroderma pigmentosum cells. *Nature (London)* **265**, 347–348.

Yu, C. W., and Borgaonkar, D. S. (1977). Normal rate of sister chromatid exchange in Down's syndrome. *Clin. Genet.* **11**, 397–401.

Zakharov, A. F., and Egolina, N. A. (1972). Differential spiralization along mammalian mitotic chromosomes. I. BUdR-revealed differentiation in Chinese hamster chromosomes. *Chromosoma* **38**, 341–365.

Chapter 14

Dominant Skeletal Mutations: Applications in Mutagenicity Testing and Risk Estimation

Paul B. Selby

385

MUTAGENICITY:
NEW HORIZONS IN GENETIC TOXICOLOGY

I. INTRODUCTION

Mutation frequency data, expressed per locus, have provided most of what we know about the biological, physical, and chemical factors affecting the mutation rate in mammals for gene mutations and small deficiencies. Most of these data have been collected using the specific-locus method developed in the mouse by W. L. Russell (1951). Mutation frequencies measured per locus are not suitable, however, for estimating overall genetic damage. Some of the reasons for this are illustrated by a discussion of a once used method for using specific-locus data in estimating genetic risk in the male (UNSCEAR Committee, 1972). The specific-locus mutation frequency expressed per locus per rad for mouse spermatogonia was multiplied by an estimate of the number of genetic loci in humans, which was 30,000. The product was then multiplied by the estimate that 2–5% of induced recessive mutations adversely affect heterozygotes, thus yielding an estimate of first-generation genetic damage. One obvious difficulty with this approach is uncertainty as to how typical the seven loci studied in the mouse are of the 30,000 loci in humans. The estimate of heterozygous effects was based on *Drosophila* data, and it was stressed that the adverse effects studied in the flies were difficult to relate to humans. Because an estimate of the number of mutations induced per million gametes is almost meaningless unless it is known whether most, some, or very few cause serious genetic effects in heterozygotes, the lack of an accepted measure of induced damage in a mammal severely hampered risk estimation.

Specific-locus data are still used in estimating genetic risk in the doubling-dose approach (UNSCEAR Committee, 1977; BEIR III Committee, 1980; Ehling, 1980). One of the limitations of this commonly used approach is the necessity of assuming that the likelihood of causing a serious handicap is about the same for both spontaneous and induced mutations. No one knows if this assumption is valid.

Some of the many strengths of the specific-locus method, to be discussed later, make it an excellent technique for studying certain questions about genetic risk. However, in view of its limitations, W. L. Russell had always realized the need for complementing the specific-locus approach with ways of measuring overall induced damage. One approach was his collaboration, which he has detailed elsewhere (W. L. Russell, 1981), with Ehling in setting up an attempt to find skeletal variants in the offspring of male mice exposed to X-irradiation. Ehling's (1966) pioneering studies suggested strongly that high-dose irradiation induces a rather high frequency of dominant mutations that cause malformations in the mouse skeleton. With the exception of one small experiment (Ehling, 1970), mice in his experiments were killed for skeletal preparation at 4 weeks of age, thus making breeding tests to check on transmissibility of likely mutants impossible. For this reason, Ehling's conclusion was based almost entirely on presumed mutations. Probably for this reason, his data were not used by committees in making estimates of overall genetic risk.

II. Dominant Skeletal Mutation Rate Experiment Using Breeding Tests

A. Summary of Procedure and Results

Ehling's conclusion was shown to be correct in an experiment (Selby and Selby, 1977, 1978a,b) that was almost identical to one of Ehling's but that included breeding tests. Male mice from strain 101 (the same one used by Ehling) were exposed to 100 R + 500 R of 60 R/min γ-radiation, with a 24-hour interval between exposures. This exposure was chosen because the same exposure with X-rays had yielded the highest presumed mutation frequency in Ehling's experiments, and it was thought that it would thus provide a large number of presumed mutations to test. Only offspring derived from germ cells irradiated as spermatogonia were observed. As in Ehling's experiments, irradiated males were mated with C3H females. Cleared and alizarin-stained skeleton preparations were examined under a dissecting microscope.

The important new feature of this experiment was that F_1's were permitted to breed instead of being killed when 4 weeks old. Thus, when an F_1 was found to have skeletal malformations that made it likely that a dominant mutation was present, a sample of his progeny was usually present that could be examined to determine whether the anomalies were transmitted. As a result, it was possible to show without question that many presumed mutants were indeed mutants and to learn exactly which abnormalities were caused by each such mutation.

Thirty-seven dominant skeletal mutations were found in our sample of 2646 F_1 progeny. Thirty-one of these were proved to be mutations by breeding tests (Selby and Selby, 1978b) and 6, for which there were no offspring, were concluded to be mutations based on presumed-mutation criteria supported by the data (Selby and Selby, 1978a). Most of the mutations cause multiple effects. If the number of different abnormalities is counted for each of the 31 proved mutations, 5 caused only one anomaly each, 19 caused 2–5 anomalies, 5 caused 6–10, and 2 caused 11–13. Effects occurred in almost all regions of the skeleton, and individual mutations often affect widely separated parts of the skeleton. Most abnormalities were (1) fusions of bones or other changes in the number of individual bones, (2) shifts in the relative positions of bones, or (3) gross changes in the shapes of bones. Photographs of some of the malformations have been published (Selby and Selby, 1978b).

Judging from the 31 proved mutations, it seems to be a valid generalization that dominant mutations have incomplete penetrance for some or all of their effects. At least 9 of the 31 have incomplete penetrance for every effect that they are known to cause. Another, more standard, way of expressing the above is to say that at least 9 of the mutations have incomplete penetrance and that all of the 31 mutations show variations in expressivity. The former way of stating this situation seems preferable in relating the results to humans because it puts the emphasis on the individual effects. After all, mutations in humans that have low penetrance for severe effects and complete penetrance for innocuous effects may go undetected in most carriers if those effects with complete penetrance are not detected clinically.

Very few of the dominant skeletal mutations cause any effect that is externally visible, and for those that do, most such effects occur in only a small fraction of carriers. A small fraction of the dominant skeletal mutations, including at least 3 of them, are reciprocal translocations (Selby, 1979). Balanced carriers exhibit the effects of these mutations. Eight of the 37 mutations found have been studied in much more detail. All 8 are homozygous lethal, 7 killing between implantation and birth, and 1 killing before adulthood (Selby, 1979). At least 2 of these 8 have incomplete penetrance for low levels of dominant lethality (Selby, 1979).

B. Use of These Data in Risk Estimation

The experiment just described formed the basis for the direct method of risk estimation used by the UNSCEAR Committee (1977) and the BEIR III Committee (1980). The importance for risk estimation of having an empirical measure of the extent and nature of induced damage in a mammal is made clear by the prominence given this experiment even though it involves an acute and fractionated exposure. Because risk estimates are generally made for small protracted exposures, it was necessary, in order to make use of the estimate of induced skeletal damage in the first generation, to assume that the fractionation and dose-rate factors based on specific-locus data could be used validly for extrapolation downward to the expected results under conditions of small protracted exposures. Mutation-rate results for the much rarer dominant visible mutations in mice gave some support for the assumption that such extrapolation based on specific-locus data was valid (Searle, 1974).

Accordingly, both the UNSCEAR (1977) and the BEIR III (1980) Committees divided the mutation frequency of 37 dominant skeletal mutations in 2646 offspring by 600, the exposure in roentgens; by 3, the dose-rate effect (Russell, 1972); and by 1.9, the fractionation effect (Russell, 1964a) to give the expected induced mutation frequency of 4×10^{-6} dominant skeletal mutations per rem per gamete for low-level irradiation of spermatogonia. The mutation frequency of 37/2646 in the skeletal experiment had been assumed to be an approximation of the induced mutation frequency (Selby and Selby, 1977) largely on the basis of the conclusion from Ehling's earlier work (Ehling, 1966) that the spontaneous mutation frequency must be very low. It was thought that in view of the probably very low spontaneous mutation frequency any error in overestimating the induced rate made by taking all mutations as induced would probably be more than counterbalanced by the failure to detect some mutations because of viability effects or very low penetrance.

In order to use this induced-mutation frequency estimated for 1 rem of protracted exposure in estimating overall damage in the first generation, it is necessary to expand it to all body systems and to restrict it to only that damage that would cause a serious handicap if it occurred in humans. It seems obvious that some of the abnormalities would be innocuous or would blend in with normal variation. Following a detailed discussion between V. A. McKusick and myself of the 37 individual dominant skeletal mutations, it seemed that about one-half of them would, if similar effects occurred in humans, result in a serious handicap. The 1977 UNSCEAR Committee, which was also familiar with the descriptions of the 37 mutations that have since been published (Selby and Selby, 1978a,b), accepted this estimate. One of the most important features of the dominant skeletal mutation approach is that it does provide the possibility of evaluating the severity of induced damage. The health implications of a given mutation frequency would be infinitely different if all of the mutations seemed to be innocuous, on the one hand, or serious, on the other. Per locus mutation-rate data by themselves provide no way of addressing this vital question.

In order to expand a measure of induced damage to the skeleton to all body systems, the UNSCEAR Committee (1977) multiplied the frequency of dominant skeletal mutations by 10, a factor arrived at in the following way. According to McKusick's (1975) tabulation of monogenic disorders in humans, 74 of 328 clinically important dominant disorders (about one-fifth of them) involve the skeleton. This suggests that the number of skeletal mutations should be multiplied by about five in order to get the total number of dominant mutations. However, skeletal malformations, like those affecting a few other body systems (for example, the senses), are clearly more easily diagnosed than the average genetic disorder, so it is likely that one-fifth is too large a fraction. It is unlikely, though, that the true fraction is much smaller than one-fifth because many dominant mutations in humans and in other animals exhibit pleiotropism and affect more than one body system. In view of the above considerations and of the opinions of the human geneticists, C. O. Carter and V. A. McKusick (see UNSCEAR Committee, 1977), it was decided that the factor of 10 was reasonable. It should be stressed that the value of one-fifth derived from McKusick's catalog is not more than one clue in trying to guess the size of the extrapolation factor. A precise derivation of this factor would require, for example, complete information on the individual incidences (not given in McKusick's catalog) of all disorders with dominant inheritance, including not just monogenic ones but also those that are irregularly inherited (the bulk of our genetic load) or that are gross chromosomal aberrations. In addition, complete information on pleiotropy would have to be considered. In view of the thousands of genetic disorders, it would require extreme optimism to expect to obtain, even in the far off future, a precise extrapolation factor. However, such precision is not needed in risk estimation. Adoption of a reasonably accurate factor at least makes possible a rough estimate of total induced damage and the comparison of total risk from different mutagens. Furthermore, it makes possible the comparison of risk estimates based on induced damage to the skeleton with risk estimates based on induced damage to other body systems for which appropriate extrapolation factors would have to be derived.

Applying the extrapolation factors of one-half and 10 and multiplying by one million, which is the number of liveborn offspring in which the incidence is to be estimated, the risk estimate was calculated to be 20 (UNSCEAR Committee, 1977). This means that there would be about 20 serious induced dominant disorders per million liveborn offspring in the first generation following exposure of all fathers to 1

R of X- or γ-irradiation in a period of 30 years. This is 20 in addition to the current incidence of serious genetic disorders, which is estimated to be roughly 107,100 per 1,000,000 liveborn humans (BEIR III Committee, 1980). The genetic disorders included in this estimate are all of those that cause a serious handicap at some time during life. The risk estimate based on the skeletal data specifically applies to those conditions in humans that are (1) autosomal dominants, (2) irregularly inherited disorders that are caused by dominants with incomplete penetrance, or (3) chromosomal aberration disorders that have dominant inheritance. The UNSCEAR Committee (1977) made a direct estimate only for paternal irradiation, but it stated that maternal risk would be expected to be low.

Rather than make a point estimate of risk, the BEIR III Committee (1980) chose to express risk as a range encompassing what it felt to be a reasonable degree of uncertainty. It assumed that from one-quarter to three-quarters of the dominant skeletal mutations cause a serious defect, as had been suggested in the first publication of the skeletal experiment (Selby and Selby, 1977). Furthermore, it assumed that the number of skeletal mutations should be multiplied by the range of 5 to 15 to encompass uncertainty as to the number of mutations affecting all body systems. As a result of using these ranges, the BEIR III first-generation estimate, based on the skeletal data, is 5–45 serious disorders per 1,000,000 liveborn for 1 rem of paternal irradiation per generation. Based mainly on a detailed reanalysis of specific-locus data by Russell (1977), the BEIR III Committee (1980) assumed that genetic risk from protracted maternal irradiation could be anything from negligible to 44% of that in the male. As a result, the first-generation risk estimate for exposure of both sexes to 1 rem per generation is 5–65 serious genetic disorders per 1,000,000 liveborn offspring.

III. THE NEW SENSITIVE-INDICATOR METHOD

A. Origin and Description of Method

The procedure described in Section II, to be referred to briefly as the breeding-test method, requires considerable time and a high level of training on the part of the observer. These requirements make it likely that this method will be applied in chemical mutagenesis only for those chemicals for which a detailed quantitative estimate of genetic risk is needed. Some of the applications for which the breeding-test method would be useful will be described later.

Recently, we have developed a much simpler and much more rapid application of the dominant skeletal approach that shows great promise of being a powerful new tool in mutagenesis research. This method, termed the sensitive-indicator method, stems from the observation in our earlier experiment that a few specific abnormalties appear to be excellent indicators of the presence of a mutation. For example, of the 2646 F_1 males examined in that experiment, 3 had a triangular-pisiform (2 bones in the wrist) fusion, 3 had an extra rib or rib pair at the posterior end of the rib cage, 2 had a rib on the seventh cervical vertebra, and 2 had the vertebra prominens (a large spikelike projection) shifted from the second to the third thoracic vertebra. Most of these 10 mice had additional malformations, and all 10 were shown to be mutants by breeding tests. Thus, every time that any of these effects were found they were caused by a dominant skeletal mutation.

Two conclusions seemed clear from that observation. First, a large number of genes must be able to mutate to alleles causing these particular malformations in order for them to occur repeatedly in a sample of this size. (Of course, some hot spots where mutations easily occur that cause sensitive-indicator malformations might, in part, explain the repeated occurrences.) Second, all these effects are extremely rare as nonmutant variants in our strain. There is no question but that some or all the sensitive-indicator traits that are so useful in our strain would be useless in some others as the result of their being common nonmutant variants in them. It would be expected that any strain would have its own array of sensitive indicators. To date, we have only determined these for (C3H × 101)F_1 hybrids.

In view of these conclusions, it seemed reasonable to attempt to screen for such sensitive indicators alone as a much more rapid application of the dominant skeletal approach. In order to make the procedure more efficient, eight other malformations that seemed likely to be sensitive indicators, based on the earlier experiment, were also included.

In a sensitive-indicator experiment, mice of the same strains as in the earlier skeletal mutation experiments are used. Randomized males are treated in one of various ways or left untreated. All F_1 offspring that live until the age of weaning (3–4 weeks) are coded, prepared for skeletal examination at 6–8 weeks of age (or when they die), examined, classified, and uncoded. The number of mice that have a sensitive indicator is divided by the number of mice in the sample to obtain the mutation frequency.

Four things contribute to make the method much more rapid than the breeding-test method: (1) only the part of the skeleton ahead of the hips is prepared, (2) only parts of the skeleton preparation are examined, thus decreasing examination time by roughly a factor of 5,

(3) no breeding test is done, and (4) a technician can easily be taught to examine skeletons because only a rather small number of specific, easily seen, malformations needs to be checked for. This fourth difference would make it much easier to scale up experiments to very fast-flow rates than would be possible using the breeding-test approach.

One additional feature incorporated in the procedure to make it more efficient is the inclusion of dominant skeletal mutations detected on the basis of the presumed-mutation criteria (Selby and Selby, 1978a) developed in the course of our earlier experiment. Briefly stated, either the presence of two or more different abnormalities that are rare or the presence of a single bilateral malformation that is extremely rare is sufficient reason to consider a mouse to be a presumed mutant. The sum of the mice with sensitive indicators and of the mice concluded to be presumed mutants based on presumed-mutation criteria is referred to as the number of probable mutants. Some probable mutants have both a sensitive-indicator malformation and meet the presumed-mutation criteria. The inclusion of mutants detected using presumed-mutation criteria has little effect on the time required for the experiment. A technician simply sets aside those very few skeletons noticed to have one or more unusual malformations in the course of examining them for sensitive indicators. The remaining parts of these skeletons are then examined by me to see if there is enough evidence to conclude that the malformation was caused by a dominant mutation. Although some subjectivity is involved in making decisions as to which mice meet presumed-mutation criteria, the decisions are unbiased because the mice are not uncoded until after they have been classified.

B. Results for Probable Mutants and Discussion

A number of experiments are being used to test the sensitive-indicator method and to permit a preview of likely relative mutation frequencies that could be expected in later more detailed studies using the breeding-test method. Table I shows the treatments that have already yielded data. All offspring are derived from spermatogonia of treated mice.

To date, 14 mice with sensitive indicators have been found and 16 mice have met presumed-mutation criteria. Four of the mice meeting presumed-mutation criteria had within the syndrome a sensitive indicator, so the total number of probable mutants is 26. Table I shows how these are distributed among the five experimental groups and the control. Point estimates and 95% confidence limits of the mutation

TABLE I

Frequencies of Probable Skeletal Mutations after Various Treatments of Spermatogonia

| | Number of | | Mutation frequency per gamete (%) | | |
| | Probable mutants | F$_1$ offspring examined | Point estimate | 95% Confidence limits | |
Treatment				Lower	Upper
Control	2	1223	0.2	0.03	0.55
600 R acute[a] X-rays	8	631	1.3[b]	0.52	2.36
100 R + 500 R acute[a] X-rays	7	415	1.7[c]	0.79	3.32
300 R acute[a] X-rays	0	200	0	0	1.64
600 R chronic[d] γ-rays	2	155	1.3	0.23	4.31
Ethylnitrosourea 150 mg/kg	7	309	2.3[e]	1.06	4.45

[a] Dose rate of 85–93 R/min.
[b] Significantly higher than that of control, $P < 0.004$.
[c] Significantly higher than that of control, $P < 0.002$.
[d] Dose rate of 0.005 R/min.
[e] Significantly higher than that of control, $P < 0.0004$.

frequencies are shown. As would be expected from specific-locus results, the point estimate for the fractionated 600-R exposure is higher than that for the single 600-R exposure; however, the difference is not yet statistically significant. Few data have been collected in the 300-R experiment, so the zero mutation frequency has little meaning. It is somewhat surprising that in the low-dose-rate experiment 2 mutations have already been found. However, in view of the very wide confidence limits, it is too soon to worry that this may indicate a dose-rate effect for dominant skeletal mutations markedly different from the factor of 3 assumed when making the direct risk estimates based on skeletal mutations.

Of the 26 probable mutations found to date, only 2 occurred in the control even though 42% of the offspring were in the control. The present low control frequency could easily be an overestimate because only one of the 16 mutants meeting presumed-mutation criteria was on the borderline as to whether it should be considered a mutation, and that one, when uncoded, was found to be in the control. The remaining control mutation caused severe effects.

The very low control frequency for probable mutants supports our earlier assumption that the true frequency must be very low. That assumption, it should be noted, was based on Ehling's much earlier control frequency, for which the criteria used were very different. The mutation frequencies for the 600-R acute, 600-R fractionated, and 150 mg/kg ethylnitrosourea (ENU) experiments are all highly significantly higher than that of the control. The mutation frequency for the protracted 600-R exposure is on the borderline of being significantly higher than that of the control, $P = 0.065$.

The mutation frequencies in the 600-R acute and the 100 R + 500 R fractionated acute experiments are so far running somewhat higher than expected in view of the frequency of 1.4% found in our earlier experiment using the breeding-test method following a 100 R + 500 R fractionated acute exposure. It seems unlikely that the probable mutation frequencies are inflated much at all by the presence of non-mutant variants. One factor that would increase the relative mutation frequency found in a sensitive-indicator experiment over that found in our earlier experiment is that mice are killed earlier in the sensitive-indicator experiments. All mice living to weaning age (3–4 weeks of age) are killed instead of only those living to at least 9 weeks of age, and it is known that some mutants die early. In contrast, factors tending to decrease the mutation frequency in the sensitive-indicator experiments relative to that found in the earlier experiment are that part of the skeleton is not prepared (thus some effects cannot be seen) and no breeding tests are performed (thus those mutations causing only one effect go undetected unless that one effect is either a sensitive indicator or extremely rare and bilateral). It seems unlikely, however, that any of these factors would change mutation rates found to a large extent. Should the probable mutation frequencies stay as high as they are at present, it will be interesting to determine the reason why.

C. Ethylnitrosourea Experiment

When the supermutagen ENU was found to yield many mutations that were intermediate alleles (Russell et al., 1979), the question arose as to whether the mutations induced by ENU might have fewer deleterious effects in heterozygotes than those induced by radiation. One way of testing this question was to see if ENU was as much of a supermutagen in inducing dominant skeletal mutations as it is in inducing specific-locus mutations. Our finding of a frequency of 2.3%

shows that ENU is clearly a supermutagen for dominant skeletal mutations, too. When the point estimates for ENU for both dominant skeletal mutations and for specific-locus mutations (W. L. Russell, personal communication) are compared to those for 600-R acute X-irradiation in the single and fractionated exposures, it appears that the proportion of induced mutations that has serious effects may be only about one-half as high for ENU as it is for radiation; however, the confidence limits are still too wide to exclude there being equal severity for these two classes of mutations. At least for this first chemical tested in both the specific-locus and sensitive-indicator methods, the specific-locus data were correct in predicting that many serious mutations would be induced by ENU. It will, of course, be important to collect many more data for ENU and for other chemicals to determine how well the specific-locus method serves as a predictor of the amount of dominant genetic damage across an array of different types of mutagens.

It has long been known that some mutations detected using the standard specific-locus test have some dominant deleterious effects (W. L. Russell, 1951; L. B. Russell, 1971), so it comes as no great surprise to find a suggestion of a reasonable correlation between such data and induced dominant damage. Should any highly efficient biochemical specific-locus techniques become available for widespread mutagenesis testing, it would be of vital importance to determine if there is a good correlation between such frequencies and induced dominant damage.

A major assumption often made in risk estimation (for example, in the doubling-dose approach) is that the likelihood that a mutation will cause a handicap is about the same for both spontaneous and induced mutations. This assumption seems questionable in view of the differences in the spectra of specific-locus mutations, and in the proportions of such mutations that are lethal in homozygous condition, for different chemical (Ehling, 1980) and radiation (Russell, 1971) treatments. Such differences also have been reported for different germ-cell stages given the same treatment (W. L. Russell, 1964b; L. B. Russell, 1971; Ehling, 1980). However, because it is not known if there is any relationship between the spectrum and homozygous lethality, on the one hand, and the frequency of induced handicaps, on the other, and, furthermore, because the seven loci may not be representative of the rest of the genome, a measure of induced dominant damage is needed to test the assumption. It is worth pointing out that the sensitive-indicator experiments in progress (and those to be done using other chemicals) provide a way of testing this assumption. Thus, for example, if the ratio of the induced specific-locus mutation frequency between a given ENU exposure and a given radiation exposure turns out to be clearly different from the ratio found for serious dominant skeletal mutations, it will be apparent that the mutations induced by these two treatments do not have the same average severity. Such a finding would contradict the major assumption being considered because it would show that at least one of these mutagens does not induce mutations having the same average severity as spontaneous mutations.

D. Results for Sensitive Indicators Alone and Discussion

Table II presents the results for sensitive indicators only. The mutation frequencies for 600 R acute, 100 R + 500 R acute and fractionated, and for 150 mg/kg of ENU are again highly statistically significantly higher than that of the control. As for probable mutants, these frequencies are in the same relative order predicted from specific-locus mutations. It is of interest that the 14 sensitive-indicator mice

TABLE II

Frequencies of Sensitive Indicators after Various Treatments of Spermatogonia

Treatment	Number of F₁ offspring with a sensitive indicator	Number of F₁ offspring examined	Point estimate	95% Confidence limits Lower	95% Confidence limits Upper
Control	0	1223	0	0	0.27
600 R acute[a] X-rays	6	631	0.95[b]	0.41	2.03
100 R + 500 R acute[a] X-rays	4	415	0.96[c]	0.33	2.31
300 R acute[a] X-rays	0	200	0	0	1.64
600 R chronic[d] γ-rays	0	155	0	0	2.12
Ethylnitrosourea 150 mg/kg	4	309	1.29[e]	0.44	3.11

[a] Dose rate of 85–93 R/min.
[b] Significantly higher than that of control, $P < 0.002$.
[c] Significantly higher than that of control, $P < 0.005$.
[d] Dose rate of 0.005 R/min.
[e] Significantly higher than that of control, $P < 0.002$.

have been distributed among the experimental groups with none being found in the control. This finding supports our earlier conclusion that sensitive indicators are, at most, rare as nonmutant variants.

Of the 12 different malformations selected as sensitive indicators, 3 have occurred once, 4 have been found twice, and 1 has occurred 3 times. The repeated occurrence already of 5 of these sensitive indicators supports our earlier conclusion that either many loci must be able to mutate to cause sensitive indicators or else there are hot spots.

E. Comparison of Sensitive-Indicator and Breeding-Test Methods

It seems useful to compare some of the relative merits of the sensitive-indicator method and of the breeding-test method, which includes study of the entire skeleton. It seems that the sensitive-indicator method is at least 10 times faster than the breeding-test approach for a highly trained investigator. This contrast becomes much more extreme when the much greater ease of establishing a team of assistants to carry out a sensitive-indicator experiment is considered. Thus, a sensitive-indicator experiment could much more easily be scaled up to a high rate of sample collection, and it could be much more easily used by a group with little prior experience. The breeding-test method has the important merits of permitting proof of transmissibility, of giving an estimate of induced dominant damage to the entire skeleton, and of permitting the setting up of mutant stocks of the many fascinating mutations that are found.

The sensitive-indicator method, although it provides no way of proving that individual probable mutations are actual mutations, does provide an estimate of the induced-mutation frequency for use in comparing treatments because the control reveals the combined frequency with which spontaneous mutations and nonmutant variants occur. By subtracting the frequency of probable mutations in the control from the frequency in an experimental group, one is left with the induced mutation frequency for the experimental group for all of the types of skeletal damage that can be picked up using the sensitive-indicator method. Although an exact determination of the proportion of total damage to the skeleton that is identified by the sensitive-indicator method must await more experimentation using the breeding-test method, it seems likely that the induced probable mutation frequency represents roughly three-quarters of the total damage to the skeleton.

The sensitive-indicator method can be used to test whether or not a chemical induces dominant skeletal mutations and, at the same time, it provides a rough approximation of overall dominant damage to the skeleton. The results thus provide a quick preview of results that might be expected using the breeding-test method, which could then be selectively applied only in those cases in which it is most needed. When the correction factor for converting probable mutation frequencies into estimates of overall damage has been determined, as discussed in the prior paragraph, probable mutation frequencies will acquire much more importance.

The sensitive-indicator method could also be used for determining doubling doses, should the doubling-dose approach continue to be used in risk estimation, especially for genetic equilibrium estimates. Such an estimate of the doubling dose would be very useful, especially if it is restricted to serious mutations alone and thus becomes free of the assumption that the likelihood of causing a serious handicap is about the same for induced and spontaneous mutations. Use of the sensitive-indicator method in this way, however, must await determination of the spontaneous mutation frequency and of the frequency of nonmutant variants in future experiments using the breeding-test method. At the present time, it should be stressed that doubling doses calculated from sensitive-indicator, or probable mutation, frequencies would be too high unless there are no nonmutant variants. As an illustration, if the frequency of nonmutant variants is the same as the spontaneous mutation frequency, then a doubling dose calculated from the data would be twice as large as the true value, and, as a result, risk estimates obtained by applying it would only be one-half as high as they should be.

F. Absolute Estimate of Cost in Time of Sensitive-Indicator Method

Comparisons of the relative rates of sample collection for the two methods for studying dominant skeletal mutations were made above, but an absolute measure is needed to understand approximate costs of applying the sensitive-indicator method to chemical mutagenesis testing. From our experience, it appears that the cost in time of a sample of 100 offspring, including all facets of the experiment, is about 1 work week of 40 hours for one person. Accordingly, the results reported here for this method would have required slightly more than 3 months of labor for two persons involved only in sensitive-indicator work. Ethylnitrosourea was known to be a mutagen in this system by the time we uncoded a sample of 98 offspring in that experiment. In other words, that discovery required 1 person-week of labor.

IV. OTHER MAJOR QUESTIONS ABOUT GENETIC RISK NOW AMENABLE TO STUDY

The dominant skeletal approach has been used so far only in estimating genetic risk in the first generation. This approach could be extended to later generations by using multigeneration exposures, either with radiation or a potent chemical mutagen. Although multigeneration experiments in the past, performed using radiation, have given mostly negative results (Green, 1968), there is every reason to expect that an accumulation of damage could be detected and quantified using dominant skeletal mutations (Selby, 1981).

A basic assumption in risk estimation is that it is valid to extrapolate from mice to humans. While the considerable similarities between the skeletons of these two species suggest that extrapolation based on induced damage may be reasonably safe, it must be borne in mind that the only estimate of induced dominant skeletal damage obtained to date is for spermatogonia of 101 strain male mice, and almost all our knowledge about the induction of gene mutations in mammals, for radiation or chemicals, is for the germ cells of F_1 hybrids between the 101 and C3H strains. If comparisons of induced dominant skeletal damage in mice and distantly related mammals show reasonable agreement, there would be much less reason to worry about the validity of extrapolation between mice and humans.

V. STUDY OF INDUCED DOMINANT DAMAGE TO OTHER BODY SYSTEMS

The ways in which the UNSCEAR (1977) and BEIR III (1980) Committees expanded from dominant damage to the skeleton to total dominant damage were described earlier. It would be useful to have measures of dominant damage for other body systems in order to judge the reasonableness of such an expansion. Data have recently become available on the induction of dominant cataract mutations in the mouse (see Kratochvilova, 1981). It is hoped that approaches capable of detecting many types of damage to some other body systems will be attempted.

VI. EFFECTS OF DECREASED VIABILITY AND INCOMPLETE PENETRANCE ON EXPERIMENTAL FREQUENCIES OF INDUCED DOMINANT DAMAGE

As has been discussed earlier (Selby and Selby, 1977), both decreased viability and incomplete penetrance would lead to an underestimation of the true frequency of induction of dominant damage. It is worthwhile to discuss in more detail the extent to which these factors would influence risk estimation based on dominant damage.

There is no question but that some mice with dominant disorders die before adulthood or even weaning age, with the result that the observed induced frequency is lower than the true frequency of induction. However, risk estimates cannot pretend to have absolute precision anyway, and this error would be serious only if it were large. An obvious way to judge the extent of the underestimate would be to measure precisely the mutation frequency at a very early age. This would not be accomplished easily. Indeed, one attempt has been made (Bartsch–Sandhoff, 1974), but it yielded no useful approximation of the extent of underestimation. A simpler, but crude, approach to this problem is to turn the question around and ask, "What proportion of the 107,100 per 1,000,000 liveborn humans that will someday have a serious genetic disorder lives to be teenagers?" No precise figure is published, but it seems clear that a large proportion survives this long, and this gives some support for feeling that dominant skeletal mutation frequencies as measured are not seriously in error because of reduced viability.

It should be mentioned that the UNSCEAR Committee (1977) and BEIR III Committee (1980) risk estimates actually have a built in, but unstated, compensation for some loss owing to decreased viability. This is because they include a separate first-generation risk estimate for chromosomal aberrations even though the estimates based on dominant skeletal mutations would include such disorders, too, unless mice affected by unbalanced chromosomal disorders almost always die when very young.

Turning now to the question of the effect of incomplete penetrance, it should be noted that this could lead to an underestimation in two ways. First, an F_1 descendant of the exposed parent may not show the effect of its mutation. While this occurrence would result in an underestimation of the true mutation frequency, it would not lead to an underestimation of the frequency of mice actually having dominant disorders in the first generation and, accordingly, it causes no error in the estimation of first-generation induced damage. Second, the F_1 may show the dominant damage, but, owing to incomplete penetrance, too few of his progeny may show the effect to permit demonstration of transmissibility. He would thus not be considered a mutant even though he really is one. Such an error would clearly be important for risk estimation. It is worthy of mention, however, that measurements of induced damage based on a whole body system would be less adversely affected by penetrance in this way than those based on a restricted type of damage to one body system. This is easily understood by recalling that dominant skeletal mutations usually cause multiple effects. (Of 37 mutations, 32 caused more than one effect.) For the majority of these mutations, one or more effect has full penetrance and can easily be used to establish proof of transmissibility of a mutation having incomplete or low penetrance for other effects. In contrast, only cataract mutations that have high or complete penetrance could easily be shown to be transmissible.

The greater susceptibility of cataract experiments (or of any other experiments based on a restricted type of damage) to underestimation error resulting from incomplete penetrance could, of course, be in part compensated for by the examination of many more offspring in the breeding test than are examined in skeletal experiments. Just as was the case with decreased viability, the underestimation of the frequency of F_1 offspring with serious effects because of incomplete penetrance probably does not introduce a serious error in risk estimates based on dominant skeletal mutations.

VII. RELATIVE MERITS OF SENSITIVE-INDICATOR AND STANDARD SPECIFIC-LOCUS METHODS

Having had considerable experience in using the standard specific-locus method, it seems fitting for me to compare that method with the sensitive-indicator method. These two methods were designed to measure different aspects of mutagenesis, so it is not surprising that each has merits not found in the other. Merits of the specific-locus method not found in the sensitive-indicator method include the extreme ease and speed of examining individual F_1 offspring for the presence of mutations, the certainty of knowing that only definite mutations are being counted, the extensive body of results obtained using this method with which results can be compared, and the ease of setting up stocks carrying the mutations found, in order to permit additional study of them. Merits of the sensitive-indicator method not found in the specific-locus method are that it detects mutations at a presumably very large number of genes, that it deals with abnormalities that are easy to relate to human genetic disorders, and that it detects only mutations that cause phenotypic changes in heterozygotes, and, thus, it is easily related to genetic damage in early generations following exposure. Furthermore, the sensitive-indicator method gives a rough estimate of damage to the skeleton, and the frequencies of induction are much higher, with the result that much smaller sample sizes are needed for drawing conclusions.

VIII. SPECIAL CASE IN WHICH RISK ESTIMATION IS GREATLY SIMPLIFIED BY USING SENSITIVE-INDICATOR AND SPECIFIC-LOCUS METHODS

Russell (1979) demonstrated that if the exposure to a mouse is massive compared to human exposure levels and if (in view of pharmacokinetic considerations, the extent of spermatogonial killing, and so on) it appears valid to extrapolate from massive mouse exposures to small human exposures, questions of genetic risk can be resolved in very small and inexpensive specific-locus experiments. This possibility seems especially attractive, for at least some chemicals, in view of the great need for mutagenesis data collected for mammalian germ cells. The need for such data has been amply demonstrated by the poor record that short-term tests in lower organisms have had in accurately predicting the mutational response in germinal cells in the mouse (Ehling, 1980; W. L. Russell, 1981; L. B. Russell et al., 1981).

Russell's suggestion of how to say much about risk with few data could also be applied to the sensitive-indicator method. As an example, assume that a reliable estimate of the spontaneous mutation frequency of serious probable mutants turns out to be 0.1% and that in 1 week of labor for 1 person using the sensitive-indicator method for chemical X, no serious probable mutations are found in the sample of 100 F_1 skeletons examined. Assume that the parents of the F_1 offspring were exposed to 2000 times the exposure level in humans. By using the binomial distribution, it can be shown that the probability is less than 0.05 that the experimental frequency is as high as 3%. On the basis of this, it seems unlikely that the induced frequency is any higher than 29 times the spontaneous mutation frequency of serious probable mutations [that is, (3.0% − 0.1%) divided by 0.1%]. In view of this and the exposure level 2000 times that in humans, and if it appears valid to extrapolate downward to the much lower human exposure level, it could be stated with reasonable confidence, following 1 person-week of labor, that the genetic risk from chemical X to humans is unlikely to be any more than 1.4% of the spontaneous rate.

A problem faced by companies developing chemicals is what to do if they get a positive result in a short-term test. If the prime focus in regulation is carcinogenesis, perhaps the extreme expense of necessary long-term animal tests might justify termination of the development of such a chemical. However, if the primary focus is on genetic disorders (a view that may become more common as more and more cancers are prevented or become curable), the availability of relatively inexpensive specific-locus and sensitive-indicator tests (*in the application being discussed*) to evaluate genetic risk might make it reasonable to proceed in the development of certain chemicals in spite of positive results in certain short-term tests. Surely, if a chemical showed potential of being useful, a company might find it economically justifiable to invest considerably more than 1 person-week of labor in a sensitive-indicator test of the chemical. The data so obtained, because of the ease of application to the question of genetic hazard, could be very useful in making decisions as to the wisdom of developing and marketing the chemical.

At least in this special situation of much lower exposure in humans than in mice, the sensitive-indicator and specific-locus methods should not be considered only as top-tier genetic tests. The money spent on usual tier-1 and tier-2 tests might be better spent doing pharmacokinetic studies to sharpen the accuracy of extrapolation. The much greater relevance of specific-locus data and, especially, of dominant skeletal mutation data to genetic risk in humans provides important justification for applying these approaches for studying gene mutation to other types of chemicals as well.

ACKNOWLEDGMENTS

Research sponsored by the Office of Health and Environmental Research, United States Department of Energy under contract W-7405-eng-26 with the Union Carbide Corporation.

REFERENCES

Bartsch-Sandhoff, M. (1974). Skeletal abnormalities in mouse embryos after irradiation of the sire. *Humangenetik* **25**, 93–100.

BEIR III Committee (Advisory Committee on the Biological Effects of Ionizing Radiation of the United States National Academy of Sciences) (1980). "The Effects on Populations of Exposure to Low Levels of Ionizing Radiation," pp. 91–180 in typescript ed. and pp. 71–134 in printed ed. Nat. Acad. Press, Washington, D.C.

Ehling, U. H. (1966). Dominant mutations affecting the skeleton in offspring of X-irradiated male mice. Genetics 54, 1381–1389.

Ehling, U. H. (1970). Evaluation of presumed dominant skeletal mutations. In "Chemical Mutagenesis in Mammals and Man" (F. Vogel and G. Rohrborn, eds.), pp. 162–166. Springer-Verlag, Berlin and New York.

Ehling, U. H. (1980). Induction of gene mutations in germ cells of the mouse. Arch. Toxicol. 46, 123–138.

Green, E. L. (1968). Genetic effects of radiation on mammalian populations. Annu. Rev. Genet. 2, 87–120.

Kratochvilova, J. (1981). Dominant cataract mutations detected in offspring of gamma-irradiated male mice. J. Hered. 72, 302–307.

McKusick, V. A. (1975). "Mendelian Inheritance in Man: Catalogs of Autosomal Dominant, Autosomal Recessive and X-linked Phenotypes." 4th ed., Johns Hopkins University Press, Baltimore.

Russell, L. B. (1971). Definition of functional units in a small chromosomal segment of the mouse and its use in interpreting the nature of radiation-induced mutations. Mutat. Res. 11, 107–123.

Russell, L. B., Selby, P. B., Von Halle, E., Sheridan, W., and Valcovic, L. (1981). The mouse specific-locus test with agents other than radiations: Interpretation of data and recommendations for future work. Mutat. Res. 86, 329–354.

Russell, W. L. (1951). X-ray-induced mutations in mice. Cold Spring Harbor Symp. Quant. Biol. 16, 327–336.

Russell, W. L. (1964a). Effect of radiation dose fractionation on mutation frequency in mouse spermatogonia. Genetics 50, 282.

Russell, W. L. (1964b). Evidence from mice concerning the nature of the mutation process. Genet. Today, Proc. Int. Congr., 11th, 1963 2, 257–264.

Russell, W. L. (1972). The genetic effects of radiation. Peaceful Uses At. Energy Proc. Int. Conf., 4th 1971 13, 487–500.

Russell, W. L. (1977). Mutation frequencies in female mice and the estimation of genetic hazards of radiation in women. Proc. Natl. Acad. Sci. U.S.A. 74, 3523–3527.

Russell, W. L. (1979). Comments on mutagenesis risk estimation. Genetics, May Suppl. 92, s187–s194.

Russell, W. L. (1981). Problems and solutions in the estimation of genetic risks from radiation and chemicals. In "Measurement of Risks, Proceedings of Rochester International Conference on Environmental Toxicity, 13th, 1980." (G. G. Berg and H. D. Maillie, eds.) pp. 361–384. Plenum, New York.

Russell, W. L., Kelly, E. M., Hunsicker, P. R., Bangham, J. W., Maddux, S. C., and Phipps, E. L. (1979). Specific-locus test shows ethylnitrosourea to be the most potent mutagen in the mouse. Proc. Natl. Acad. Sci. U.S.A. 76, 5818–5819.

Searle, A. G. (1974). Mutation induction in mice. Adv. Radiat. Biol. 4, 131–207.

Selby, P. B. (1979). Radiation-induced dominant skeletal mutations in mice: Mutation rate, characteristics, and usefulness in estimating genetic hazard to humans from radiation. Radiat. Res. Proc. 6th Int. Cong. pp. 537–544.

Selby, P. B. (1981). Radiation genetics. In "The Mouse in Biomedical Research: History, Genetics and Wild Mouse" (H. L. Foster, J. D. Small, and J. G. Fox, eds.) pp. 263–283, Vol. I, Academic Press, New York.

Selby, P. B., and Selby, P. R. (1977). Gamma-ray-induced dominant mutations that cause skeletal abnormalities in mice. I. Plan, summary of results and discussion. *Mutat. Res.* **43**, 357–375.

Selby, P. B., and Selby, P. R. (1978a). Gamma-ray-induced dominant mutations that cause skeletal abnormalities in mice. III. Description of presumed mutations. *Mutat. Res.* **50**, 341–351.

Selby, P. B., and Selby, P. R. (1978b). Gamma-ray-induced dominant mutations that cause skeletal abnormalities in mice. II. Description of proved mutations. *Mutat. Res.* **51**, 199–236.

UNSCEAR Committee (United Nations Scientific Committee on the Effects of Atomic Radiation) (1972). Ionizing radiation: Levels and effects, Vol. II: Effects. Report to the General Assembly, with Annexes, pp. 199–302. United Nations, New York, Sales No. E.72.IX.18.

UNSCEAR Committee (United Nations Scientific Committee on the Effects of Atomic Radiation) (1977). Sources and effects of ionizing radiation. Report to the General Assembly, with Annexes, pp. 425–564. United Nations, New York, Sales No. E.77.IX.1.

Chapter 15

Plants as Sensitive *in Situ* Detectors of Atmospheric Mutagens

William F. Grant and K. D. Zura

MUTAGENICITY:
NEW HORIZONS IN GENETIC TOXICOLOGY

Whatever the reasons may be, the use of plants in monitoring and screening systems for mutagens has not been widely accepted. The recent successful introduction of the use of *Tradescantia* staminal hairs to detect airborne mutagens and carcinogens may be the beginning of the recognition of various plant assays which are inexpensive, easy to handle and applicable for indoor as well as outdoor detection of environmental mutagens (Stich and San, 1980).

I. INTRODUCTION

The notion that chemicals in the environment can bring about changes in the human hereditary process has gained acceptance only in the last decade (Ames, 1979). Accumulating epidemiological evidence suggests that cancer has a large environmental component or may even be considered an environmental disease (Heath, 1978). However, it is well known that man has long been exposed to toxic substances naturally present in the environment. One of the first atmospheric pollutants to be identified as early as the eighteenth century was soot from burning fires which caused scrotal cancer in chimney sweepers. To our knowledge, the first recorded observation of a correlation between reduction in plant fertility and cytological abnormality dates back to 1931 when Kostoff (1931) observed greatly reduced seed set in tobacco plants that had been fumigated with nicotine sulfate; an examination of meiosis revealed many chromosome irregularities which he considered to be the cause of the partial sterility of the plants.

Since World War II, chemicals entering the environment have increased manyfold (Anon, 1975). Today, a wide spectrum of toxic agents that are mutagens or potential mutagens are entering the atmosphere both from natural and anthropogenic sources, such as emissions of volcanic gases and smoke produced in natural fires, and from vehicle emissions, and industrial use categories and their degradation products (Maugh, 1978).

Recently, Hughes *et al.* (1980) have reviewed sources and types of air pollution, the collection, fractionation and chemical analysis of air samples, and problems such as sample size and the complexity of the air sample (both vapor phase and particulate). Fishbein (1976) has surveyed a number of atmospheric mutagens which include sulfur dioxide, bisulfite, nitrous oxide, nitrous acid, polynuclear aromatic hydrocarbons (e.g., benzo[a]pyrene, benz[a]anthracene, dibenzanthracene), peroxyacyl nitrates, peroxides, ozone, halogenated hydrocarbons (e.g., di- and trichlorofluoromethane, vinyl chloride, tri- and tetrachloroethylene, carbon tetrachloride, chloroprene, ethylene dibromide, ethylene chlorohydrin), sodium fluoride, pesticides (DDT, dichlorvos), polychlorinated biphenyls, formaldehyde, ethylene oxide, and inorganic and organic derivatives of lead and mercury.

Many plant species act as natural bioconcentrators of atmospheric pollution by accumulating pollutants on or in their tissues. Air pollutants may be assimilated by the plant through foliar or root absorption from the soil directly (Cataldo and Wildung, 1978), or from plant detritus particles derived from plant decomposition (Odum and Driftmeyer, 1978).

In view of the increasing levels of potentially mutagenic substances entering the environment, there is a need for methods of evaluating specific environments for mutagenic hazards. Many higher plant mutagen assay systems have been developed for the detection of environmental mutagens (de Serres, 1978). At present, only a few are suitable for gaseous compounds, but a number of mutagen assay systems are being developed for airborne pollutants.

The detection of airborne pollutants falls into two general classes: (1) *in situ*, or on-site methods, and (2) laboratory, or off-site methods. Plants offer unique test systems for mutagen *in situ* monitoring, but progress in their development and refinement has been slow. The latter may be accounted for partly by the lack of general recognition of the high resolution that higher plant test systems provide for mutagenic testing and monitoring and the consequent lack of support for this area of research (Nilan, 1978).

In situ environmental monitoring has the advantage that actual conditions, which may involve multiple chemicals, are reflected in the mutagenicity assays. Such conditions may be difficult to duplicate in the laboratory. Unlike laboratory studies, *in situ* monitoring may also be used wherever chemical pollution is suspected in the environment.

This chapter outlines the major *in situ* higher plant assays currently available for detecting and monitoring mutagenic environmental pollutants and indicates some systems that have been used successfully for determining the mutagenicity of gaseous compounds under laboratory conditions. Chemicals that are normally gaseous or volatile but were incorporated into solutions for test purposes have been excluded from this review.

II. *Tradescantia* as an *in Situ* System for the Detection of Atmospheric Mutagens

A. Genetic Systems

Of all the higher plants, *Tradescantia paludosa* Anders. has features which, at the present time, make this species the most ideal for the detection and monitoring of atmospheric mutagens (Table I). A number of *Tradescantia* genetic systems have been developed by the late Arnold Sparrow and colleagues at the Brookhaven National Laboratory. Although originally intended for studying the effects of ionizing radiation, these systems have several genetic end points that can be used for assays: mutations in flower petals and stamen hairs, meiotic chromosome aberrations, micronucleus frequency in quartet (tetrad) cells, and mitotic chromosome aberrations in microspores. This species has already been used successfully to detect airborne mutagens at the Brookhaven National Laboratory following an accidental exposure to chemicals that were exhausted from a fume hood duct on a roof and carried by air currents into the greenhouse intake vents (Sparrow and Schairer, 1971).

Clones of *Tradescantia paludosa*, heterozygous for flower color, are used for the analysis of somatic mutations in flower petals and stamen hairs. Clones 02, 0106, and 4430, the most commonly used, are diploid ($2n = 12$). Clones are easily reproduced by vegetative propagation and will flower continuously throughout the year if maintained in growth chambers (Underbrink *et al.*, 1973). The clones are hybrids between pink- and blue-flowering parents, with blue dominant to pink (Mericle and Mericle, 1967; Emmerling-Thompson and Nawrocky, 1980). The visually discernible marker used in the test system is the phenotypic change in pigmentation from blue to pink (or colorless cells) in either petals or stamen hairs. Illustrations of mutant petal sectors and stamen hair cells are given by Schairer *et al.* (1978a).

B. Petal System

Although mutant homozygous recessive pink sectors occur spontaneously in the petals, their frequency greatly increases following exposure to mutagens. In young developing floral tissue, mutations are expressed as isolated pink (or colorless) cells which appear within 5–18 days after exposure to a mutagen. In the mature flower, they appear as sectors or groups of cells (Mericle and Mericle, 1967).

Significant increases in mutation rates in flower color have been obtained after exposures to 3 ppm of gaseous ethylmethane sulfonate or 1,2-dibromoethane (Sparrow and Schairer, 1974). Likewise, significant increases in mutation rate above spontaneous levels were observed for ozone, SO_2, and N_2O but these agents were considered at best weak mutagens in this system since the maximum response obtained was only slightly more than twice the spontaneous rate (Sparrow and Schairer, 1974).

C. Stamen Hair System

The stamen hair system is extremely sensitive to both radiation and gaseous compounds. A linear relationship between mutation frequency and dosage is found over a wide range of radiation levels with no evidence of a threshold effect even at levels as low as 250 mrad of X-rays (Schairer et al., 1978a).

The procedure for testing for mutagenicity with the stamen hair assay is relatively easy. Cuttings containing young flower buds are exposed to the atmosphere to be tested. In 1–3 weeks mutant pink and colorless cells (also giant and dwarf pink cells) can be seen in the mature stamen hairs. For facility, only pink cells are scored routinely (Schairer et al., 1978a). Detailed procedures for growing, scoring and statistical treatment are given in the papers by Underbrink et al. (1973) and Grant et al. (1981).

In the case of gaseous compounds, a comparison of chemical exposure concentration and tissue dose for various exposure periods indicated that the penetration of 1,2-dibromoethane (ethylene dibromide: DBE) through the outer sepal and petal tissues was both rapid and uniform and the compound readily reaches the stamen hair cells (Nauman et al., 1976, 1979). As with radiation dosages, data on exposure to gaseous ethylmethane sulfonate (EMS) and DBE indicate that a linear relationship exists between response and duration of treatment for higher concentrations (15, 20, and 45 ppm) and exposures of up to 6 or 8 hours (Sparrow et al., 1974; Sparrow and Schairer, 1976; Nauman et al., 1976). In the case of DBE, a linear relationship was found over the range from 3.6 to 148.2 ppm for an exposure of 6 hours (Sparrow et al., 1974). The mutation response frequency after a 6-day exposure at the low dose of 2 ppm of DBE was 5 events per 100 stamen hairs or the equivalent to that produced by about 50 rad of X-ray (Sparrow and Schairer, 1976). The sensitivity of the *Tradescantia* stamen hair test system, with significant responses at < 0.1 ppm for DBE, has demonstrated that gaseous chemicals can be as, or more, mutagenic than X-rays, with EMS being somewhat more mutagenic than DBE. Clone 4430 showed a higher sensitivity to chemical mutagens than to X-rays (Nauman et al., 1976).

Lower et al. (1978) used clone 4430 to test for mutagenicity at three sites (0.3, 1.7, and 11.4 km from a lead smelter plant) in southeastern Missouri where the major effluents were lead, cadmium, copper, zinc, and sulfur. There was a significant difference between the mutational frequency at the site 1.7 km from the smelter and the control, whereas the increase in mutation frequencies at the 0.3- and 11.4-km sites, while elevated above controls, were not significantly different from

TABLE I

Studies Using the *Tradescantia* Test Systems

Agent	Stage treated	Target tissue	Range of exposures	References
1,2-Dibromoethane (DBE)	Cuttings bearing inflorescences	Stamen hairs for pink mutations and meiotic chromosomes for micronuclei in tetrads	3.6–148.2 ppm for 6 hours	Ma et al., 1978
DBE	Cuttings bearing inflorescences	Developing stamen hairs	3.6–148.2 ppm for 6 hours	Sparrow et al., 1974
[³H]DBE	Cuttings bearing inflorescences	Developing stamen hairs	4.8–88.2 ppm for 0.25–6 hours	Nauman et al., 1979
Ethylmethane sulfonate (EMS)	Cuttings bearing inflorescences	Developing stamen hairs	5–250 ppm for 1–19 hours (mostly 6 hours)	Schairer et al., 1973; Nauman et al., 1974, 1975, 1976; Sparrow et al., 1974; Villalobos-Pietrini et al., 1974; Sparrow and Schairer, 1974, 1976; Nauman and Grant, 1982

EMS	Cuttings bearing inflorescences	Meiotic chromosomes for micronuclei in tetrads	1000 ppm for 6 hours	Ma, 1979
Ethylene oxide	Mature pollen grains cultured	Chromosomes at pollen mitosis	5–7 ml for 5 minutes	Smith and Lofty, 1954
Hydrazoic acid	Cuttings bearing inflorescences	Meiotic chromosomes for micronuclei in tetrads	136 ppm for 6 hours	Ma, 1979
Ketene	Mature pollen grains cultured	Chromosomes at pollen tube mitosis	Saturated atmosphere for 5–15 seconds	Smith and Lofty, 1954
Methyl chloride	Mature pollen grains cultured	Chromosomes at pollen tube mitosis	6–7 ml for 5 minutes	Smith and Lofty, 1954
N_2O, ozone, SO_2	Cuttings bearing inflorescences	Petals or stamen hairs	Data not given	Sparrow and Schairer, 1974
SO_2	Pollen cultured on microslides	Chromatid aberrations in pollen mitosis	6.05–0.075 ppm for 30 minutes	Ma et al., 1973; Ma and Khan, 1976
Vinyl chloride	Cuttings bearing inflorescences	Developing stamen hairs	10–150 ppm for 6 hours	Nauman and Grant, 1982

control levels. However, when analyses of soil samples from the different sites were carried out in the laboratory, it was found that high lead toxicity was present in the soil at 0.3 km from the plant site. At this site it was noted that the first flowering of the plants was delayed and it was considered that the high lead toxicity may have masked any mutagenic effect. In addition to Pb, high concentrations of Cd, Cu, and Zn were present in the soil.

The *Tradescantia* stamen hair system achieves its greatest sensitivity with chronic exposures comparable to those occurring in populated areas near industrial pollution sources. Thus, the stamen hair system is highly adaptable for monitoring ambient air pollution for mutagenicity in urban and industrial sites. As a result, a minivan trailer has been equipped with growth chambers to provide a constant controlled environment for control plants (using filters) and for exposing plants concurrently to the surrounding atmosphere (Schairer *et al.*, 1978a,b, 1979). Many urban and industrial sites have been monitored for mutagenic air-borne pollutants in the United States. Specific sites monitored have been given (Schairer *et al.*, 1978a; Hughes *et al.*, 1980). It has been observed that atmospheric contaminants from petroleum and chemical processing plants increased the number of mutant cells from 14.5 to 71% above control levels; an increase in pink mutations as high as 30.9% was obtained during July, 1976, at Elizabeth, New Jersey, in a region of high-density automotive traffic. Contaminants from smelters, automotive combustion, and photochemical reactions produced increases from 0 to 8% over control levels.

The stamen hair system has been used also to monitor the air around a nuclear power plant in Japan (Ichikawa, 1981).

D. Chromosome Aberrations in Meiosis

In a number of species, meiotic chromosome abnormalities in pollen mother cells can be studied as indicators of DNA damage by atmospheric pollutants.

Somatic cells of *Tradescantia* have a complement of 12 large chromosomes that are well suited for detailed cytological analyses. This small number of large chromosomes makes *Tradescantia* excellent material for the scoring of both somatic and meiotic chromosome abnormalities (Evans, 1962).

E. Micronucleus System

Chromosomes of early meiotic prophase of *Tradescantia* are extremely sensitive to mutagens and are subject to breakage with the subsequent production of acentric fragments which can end up as micronuclei in the quartet (tetrad) stage. Ma (1979) has developed a bioassay for atmospheric mutagens in an assay which he refers to as "*Tradescantia* Micronuclei-in-Tetrad Test" or "Trad-MCN Test." Exposure of inflorescences with buds in early prophase to gaseous clastogenic agents will cause chromosome aberrations. The number of micronuclei in the tetrads can be scored 24–30 hours after treatment.

The mutagenic efficiency of ethylene dibromide for induction of MCN in tetrads (6-hour treatment, range 0.001–0.002 MCN/tetrad/ppm-hour) is about 36 times that for induction of pink mutations in stamen hairs (Ma *et al.*, 1978). Later experiments have confirmed the efficiency of this assay. When inflorescences were exposed to 20 and 40 R of X-rays, an average of 22.8 and 66.6 MCN/100 tetrads, respectively, were observed. Liquid EMS at 50 and 100 mM, absorbed through the stem, induced 13.2 and 15.2 MCN/100 tetrads, respectively, while gaseous EMS (1000 ppm) induced 17.4 MCN/tetrads. At 0.2 mM liquid sodium azide (NaN$_3$) induced 10.1 MCN/tetrads, while 136 ppm of gaseous hydrazoic acid (HN$_3$ fumes released from NaN$_3$ solution in acidic media) induced 21.2 MCN/100 tetrads. The control yielded around 5 MCN/100 tetrads (Ma, 1979).

Ma *et al.* (1980) have used the *Tradescantia* MCN/tetrad test for *in situ* monitoring of air pollutants at 11 sites which included industrial complexes, public parking garages, and truck and bus stops. Monitoring was conducted by exposing the flower buds of plant cuttings to the atmosphere of the site for 1–6 hours. Six sites (unspecified) yielded positive results and five sites negative results.

Recently, T.-H. Ma (personal communication) and colleagues in China (Ma *et al.*, 1981) have selected and cloned a plant of *Tradescantia reflexa* Raf. that is tetraploid ($2n = 24$). In a micronucleus assay, Ma found that X-rayed florets of *T. reflexa* are up to four times more sensitive than *T. paludosa*. Furthermore, treatments with sodium azide in concentrations used by Nilan *et al.* (1973) in barley showed that there was about a 10-fold increase in mutagen sensitivity of *T. reflexa* over that of barley.

F. Microspore Assay

When pollen is grown in culture, the pollen tubes grow partially on the surface halfway embedded in the solidified medium. This provides the ideal conditions for gaseous agents to enter the cells directly from the atmosphere (Ma *et al.*, 1973). The six chromosomes of the mitotic generative nucleus usually line up end-to-end within the pollen tube, thus facilitating the analysis of chromosome aberrations (Ma, 1967; Rushton, 1969).

Ma et al. (1973) report that pollen tube culture is extremely sensitive to gaseous impurities in the air, including tobacco smoke, perfume or antiseptic compounds. A sensitive indicator is pollen tube growth and pollen germination. At 0.1 ppm or higher, SO_2 inhibits pollen germination, tube growth, or mitotic activity (Ma and Khan, 1976).

III. POLLEN SYSTEMS FOR THE DETECTION OF ATMOSPHERIC MUTAGENS

Pollen systems show considerable promise for the monitoring of atmospheric mutagens because they provide a highly sensitive detection system. Such systems are unique among higher eukaryotes in terms of low cost and high genetic resolution—several hundred-thousand pollen grains with their own haploid genotypes can be scored—paralleling that of prokaryotes. However, at present, two of the most developed systems are in commercial crops, namely, Zea mays and Hordeum vulgare. Therefore, these plants would have to be grown, which may be impossible, or transported to the site in the case of screening or monitoring for mutagenicity. On the other hand, such plant systems are excellent for in situ screening or monitoring in the case of aerial spraying of herbicides or other chemicals under agriculture practices.

Most of the apparently useful heritable traits that can be identified in pollen test systems of various species are proteins which may be stained (Lindgren and Lindgren, 1972; Plewa, 1978; Freeling, 1978), or electrophoretically (Schwartz and Osterman, 1976; Freeling, 1978), or immunologically (Schwartz, 1972) identified.

A. The *waxy* Locus

The wx locus assay in maize (Zea mays) can be used to measure forward mutation frequency (Wx → wx), reverse mutation frequency (wx → Wx), and the frequency of intragenic recombination. Alleles of the waxy locus of barley (Hordeum vulgare; Rosichan et al., 1979), as well as specific base substitution, frameshift, and deletion mutant lines (Nilan et al., 1981) are being developed for screening and monitoring mutagens. The waxy (wx) allele is recessive to starchy (Wx).

When subjected to an iodine solution, kernels carrying the *Wx* allele stain blue-black; *wx/wx* kernels stain red. The starch type of pollen grain is controlled by the genetic constitution of that pollen grain, not by that of the parental sporophyte. Hence, a genetic reversion of *wx* to *Wx* can be detected by scoring for pollen grains from plants that are homozygous for *wx*.

The procedure used by Plewa for detecting the mutagenic effects of pesticides with the *waxy* locus in *Zea mays* has been described in detail (Plewa, 1978; Grant *et al.*, 1981).

In an *in situ* test for mutagenicity at three locations (0.3, 1.7, and 11.4 km) from a lead smelter plant, Lower *et al.* (1978) used two *Zea* systems, the *waxy* locus and a Neuffer strain of *Zea mays*. The latter strain detects somatic mutations in leaf tissue. The strain is heterozygous at the yellow locus (*Y/y*) and the mutant event is scored as a yellow streak (*y/y*) in an otherwise green (*Y/y*)leaf tissue. At the *waxy* locus, there was a nonlinear increase in mutation frequency with distance; the 1.7-km site had a slightly greater mutation frequency than the 0.3-km site with the mutation frequency at both sites being highly significant. In the case of the Neuffer strain, environmental damage to the leaves did not allow scoring of the leaf mutations. However, the frequency of pollen abortion of both strains was significantly greater at the 0.3- and 1.7-km sites than at the 11.4-km site.

In a laboratory experiment to determine if ethylene oxide would cause chromosome aberrations, Fabergé (1955) treated pollen carrying genes *I Sh Bz Wx* in a rotating glass tube through which a mixture of 1 part ethylene oxide to 20 parts air by volume flowed at 50 ml/minute for 2 or 4 minutes. He found a 4-minute exposure reduced seed set drastically; a 2 minute, slightly. Roughly 1% of the kernels had aberrations in the short arm of chromosome 9, about 0.1 of the rate that could be achieved at the same sterility level by radiation. The effects of ethylene oxide seemed quite similar to those of X-rays, the same kinds of chromosome aberration were produced in about the same proportions.

Seedlings of *Zea mays* of the genotype $A_1A_2C^1Wx$ were fumigated in growth chambers with hydrogen fluoride at a concentration of 3 µg/m^3 for 4, 6, 8, or 10 days. This was below the threshold of visible injury. Meiotic stages were studied for chromosome anomalies (Mohamed, 1969, 1970). A positive correlation was found between the frequency of chromosomal aberration and duration of treatment. Asynapsis, translocations, inversions, bridges, and fragments were observed in addition to physiological effects which were believed to account for chromosome stickiness.

Lindgren and Lindgren (1972) reported that ethylene oxide had a detectable mutagenic effect in the *waxy* assay in barley at levels that were regarded as acceptable for occupational exposure. Tobacco smoke, car exhausts, and city air were tested in this assay but no strong mutagenic effect was found.

B. The *yg-2* Locus

The *yg-2* allele is a chlorophyll deficiency that is detected and scored in seedling leaves of *Zea mays* (Conger and Carabia, 1977). At present, kernels heterozygous *Yg-2/yg-2* are treated with a putative mutagen and *yg-2* sectors on the fourth and fifth leaves are counted in seedlings of 3 to 4 weeks of age. Tests to determine the efficiency of this system under monitoring conditions are required.

C. The *Adh* Locus

The alcohol dehydrogenase (ADH) system in *Zea mays* detects mutants that result in the absence of or a drastic reduction in enzyme activity. Pollen shed from plants which are heterozygous for the *Adh* locus (Adh^+/Adh^-) may be detected cytochemically; the ADH^- grains stain yellow and are translucent whereas the ADH^+ grains stain blue and are opaque (Freeling, 1978). For monitoring, Adh^- mutant plants may be used in which case the number of blue grains—phenotypic mutants—can be determined. A typical maize plant sheds 10^7 pollen grains which permits the quantitation of ADH^+ gametophytes at frequencies below 10^{-5}. Between 1 and 3% of EMS-treated seeds produce ears with sectors of kernels carrying a mutation at the *Adh* locus (Schwartz, 1981).

D. Chromosome Nondisjunction

Weber (1981) is exploring the use of pollen test systems to detect nondisjunction in *Zea mays*. Each member of a quartet of haploid microspores produced by meiosis contains a single chromosome 6 which carries the only nucleolar organizing region in the maize genome. Thus each member of a normal quartet contains one nucleolus. If nondisjunction of chromosome 6 takes place at either meiotic division, this may be readily detected as an abnormal distribution of nucleoli in the quartets by staining and cytological examination. The feasibility of this system in monitoring is still to be evaluated.

E. Pollen Viability

Pollen viability per se may be used as an indicator of mutagenic events. Because pollen viability is dependent on a large number of loci, it provides an extremely sensitive indicator of mutagenesis. Mulcahy (1981) is developing the use of plants shedding pollen in quartets, characteristic of some 40 families, such as the Ericaceae, rather than single pollen grains. An estimate of mutagenic events is provided by determining the number of quartets which contain two, instead of four, viable grains.

F. Self-Incompatibility

Gametophytic self-incompatibility assay systems have been developed in *Oenothera, Nicotiana,* and *Petunia* (Nettancourt, 1977); Mulcahy and Johnson (1978) have described a method for the construction of a largely autonomous incompatibility system which would provide continuous monitoring of the environment. All these systems are based on the principle that pollen cannot fertilize a plant bearing the same S allele. For example, pollen carrying an S_1 allele cannot fertilize a plant homozygous for S_1, but can effect 50% fertilization in plants with the genotype S_1S_2, and can completely fertilize plants not containing the S_1 allele. Fruit- and seed-set are used as criteria of compatibility (Pandey, 1964). Since the incompatible style acts as a sieve excluding all but mutant pollen, scoring is limited to the small number of pollinations involving mutant pollen grains.

IV. OTHER HIGHER-PLANT MUTAGEN ASSAY SYSTEMS

See Table II for a listing of mutagen assay systems.

A. *Allium cepa*

The common onion, *Allium cepa* ($2n = 16$), is an excellent species for assaying for chromosome aberrations (Grant, 1982). The "*Allium* test" is the classical test for studying the effects of chemicals on plant chromosomes (Levan, 1938, 1949).

Engle and Gabelman (1966) studied the inheritance to ozone damage in the onion by taking leaf disks, 1 cm^2, from parental and hybrid strains and floating these on the surface of a water reservoir in a microchamber. With the stomata open, ozone was pumped through the microchamber at 0.3 ppm. Plants were also placed in airtight

TABLE II

Studies Using Plant Test Systems Other than *Tradescantia*

Agent	Species	Stage treated	Target	Range	References
Ozone	*Allium cepa*	First leaves of young plants	Stomata closure	0.3 ppm until stomatal closure	Engle and Gabelman, 1966
Nitrous oxide	*Allium cepa*	Sprouting bulbs	C-mitosis	6 atm for 6 hours	Ferguson *et al.*, 1950
Nitrous acid	*Arabidopsis thaliana*	Developing buds	Sterility as unfertilized ovules; embryonic lethals	100 pphm (hundred million) for 4 and 8 hours	Janakiraman and Harney, 1975
Nitrous oxide	*Avena sativa*	Tillers	Induction of polyploids	4 atm for 24 hours	Dvorak and Harvey, 1973
Nitrous oxide	*Crepis capillaris*, *Phalaris* species	Plants when youngest flowers completed flowering	Induction of polyploids	10 atm for 6–12 hours	Ostergren, 1954, 1957
Nitrous oxide	*Hordeum vulgare*	Tillers	Doubling haploids	21.1, 31.6, and 42.2 × 10³ kg/m³ atm for 12 and 24 hours	Subrahmanyam and Kasha, 1975
Nitrous oxide	*Hordeum vulgare*	Tillers	Induction of polyploids	4 atm for 24 hours	Dvorak *et al.*, 1973
Nitrous acid	*Hordeum vulgare*	Embryos and tillers in culture medium	Induction of polyploidy in embryos	4 atm for 24 hours	Dvorak *et al.*, 1973
Hydrogen fluoride	*Lycopersicon esculentum*	Plants with developing buds and young leaves	Mitotic and meiotic chromosome aberrations	3 μg/m³ for 4, 6, 8, 10, and 12 days	Mohamed *et al.*, 1966; Mohamed 1968, 1969
Hydrogen fluoride	*Lycopersicon esculentum*	Seeds	Meiotic chromosome aberrations	3 μg/m³ for 4 days	Mohamed, 1968

Agent	Species	Material	Effect	Treatment	Reference
Nitrous oxide	Melandrium species	Young plants	Induction of polyploids	2–5 atm for 5–16 hours	Nygren, 1955, 1957
Hydrogen fluoride	Phaseolus vulgaris	Continuous from seed to seed maturity	Effect on progeny	0.58, 2.1, 9.1, and 10.5 µg F/m³, continuous treatment	Pack, 1970
Nitrous oxide	Solanum tuberosum × S. phureja and Datura stramonium	Whole plants and cut flowering branches	Haploid mitotic nuclei for C-metaphases	4, 6, or 9 atm for 2 hours	Montezuma-de-Carvalho, 1967
Nitrous oxide	Trifolium species	Excised heads	Induction of polyploids	6 bars atm for 24 hours	Taylor et al., 1976
Nitrous acid	Trifolium pratense	Excised stems bearing inflorescences	Induction of polyploidy	6 bars atm pressure for 24 hours	Taylor et al., 1976
Nitrous acid	Triticum aestivum	Embryos and tillers in culture medium	Induction of polyploidy in embryos	4 atm for 24 hours	Dvorak et al., 1973
Nitrous oxide	Triticum vulgare	Seedlings	Induction of polyploids	4 atm for 24 hours	Dvorak et al., 1973
Ozone	Vicia faba	Seeds	Somatic chromosome aberrations	0.4 weight percent ozone for 15, 30, and 60 minutes	Fetner, 1958
Ozone	Vicia faba	Inflorescences with developing buds	Meiotic chromosome aberrations	100 pphm (hundred million)–200 pphm for 4 or 8 hours	Janakiraman and Harney, 1976
Ethylene oxide (EO)	Zea mays	Pollen	Chromosome aberrations in kernels	1 EO:20 air at 50 ml/minute for 2 or 4 minutes	Fabergé, 1955
Hydrogen fluoride	Zea mays	Plants with developing buds	Meiotic chromosome aberrations	3 µg/m³ for 4, 6, 8, 10, and 12 days	Mohamed et al., 1966; Mohamed, 1969, 1970

polyethylene chambers and ozone at 0.4 ppm was circulated for 1.5 hours. Resistance to ozone damage was found to be dominantly controlled, possibly by a single gene. In resistant plants, the stomata close, whereas in sensitive plants ozone passes through the stomata injuring the susceptible plants in a way similar to the type of damage produced in the field by naturally occurring ozone. The feasibility of such a system for the detection of atmospheric mutagens remains to be elucidated.

Ferguson *et al.* (1950) induced partial C-mitosis in *Allium cepa* by treating sprouting bulbs with nitrous oxide at a pressure of 6 atm for 16 hours. He failed to achieve complete endoreduplication since increased pressure caused mitotic inhibition.

B. *Arabidopsis thaliana*

This crucifer has excellent features for testing the effects of mutagens (Rédei, 1974). It is a diploid ($2n = 10$), self-fertilized species with a short life-cycle (28 days). A single plant may produce more than 50,000 seeds. Rédei and Acedo (1976) have described procedures for using this species in mutagenesis testing.

Janakiraman and Harney (1975) have tested for mutagenicity of ozone by fumigating plants of *Arabidopsis* in the bud stage in Plexiglas chambers with a concentration of 100 pphm for 4 and 8 hours. Increases in embryonic lethals were observed after the 4- and 8-hour treatments of 19.12 and 37.75%, respectively. Seed germination decreased from 71 (control) to 50 and 30%, respectively. The effects were found only in the treated generation. Ozone lowered fertility but did not cause any morphological mutagenic effects.

C. *Glycine max*

Vig (1975) has developed a strain of soybeans in which somatic mutational events are observed as twin spots in the leaves of heterozygous light green plants, or as light green spots on yellow or dark green leaves of homozygous plants. Other plants which use this feature for mutagenic testing have been listed by Vig (1978). In an *in situ* test for mutagenesis using this soybean strain, Lower *et al.* (1978) found conditions at a site 0.3 km distant from a lead smelter plant too toxic and the plants died.

D. *Hordeum vulgare*

Barley (2n = 14) has long been used for mutagenicity testing. Considerable knowledge has been accumulated on both the cytology and the genetics of this species (Nilan, 1974). In a study designed to double the chromosome number of haploid barley plants, tillers were treated with nitrous oxide at pressures of 21.1×10^3, 31.6×10^3, and 42.2×10^3 kg/m^2 (30, 48, and 60 psi, respectively) for 12 and 24 hours, commencing 18, 24, or 30 hours after pollination with the objective of affecting the first mitotic divisions of the fertilized embryo (Subrahmanyam and Kasha, 1975). The treatment which was most effective produced 0.5% of the seedlings with the double number of chromosomes, whereas 17% of the florets pollinated in the control resulted in seedlings. In contrast, the proportion of doubled tillers per plant was 10.8% from colchicine treatment, and 61.6% in colchicine plus dimethyl sulfoxide.

In a similar study, Dvorak *et al.* (1973) treated tillers of barley with 4 atm of N_2O for 24 hours, commencing the treatments 4, 24, and 48 hours after pollination. The authors also treated wheat seedlings with 4 atm of N_2O for 24 hours. Treatments of 4 to 24 hours after pollination were most effective in barley and 24 to 48 hours in wheat; 75% of the barley and 54% of the wheat embryos were polyploid. In barley 5% of the treated embryos were aneuploid, whereas 42% of the wheat embryos were aneuploid. A high frequency of aneuploids were also recovered in *Avena sativa* after nitrous oxide treatment in the same manner (Dvorak and Harvey, 1973).

E. *Lycopersicon esculentum*

The tomato is a particularly favorable experimental plant for the study of induced mutations since a large number of seedling mutants are known enabling one to score mutations in the seedling stage without growing plants to maturity (Rick and Butler, 1956). In addition, chromosome aberrations may be scored in both mitotic and meiotic cells (Grant and Harney, 1960).

Mohamed (1968, 1969; see also Mohamed *et al.*, 1966) fumigated tomato plants with hydrogen fluoride at 3 μg/m^3 for 4, 6, 10, and 12 days. Alternatively, he fumigated seeds at 3 μg/m^3 for 4 days (Mohamed, 1968). He studied the effect on mitotic chromosomes in young leaf tips and on meiosis in developing microsporocytes in flower buds. The plants showed no visible injury except for the 12-day exposures. A positive correlation was found between the frequency of chromosomal aberrations and duration of treatment. The response was more pronounced in the meiotic chromosome analyses. Mitotic chromosome aberrations in leaf cells were mainly anaphase bridges with or without fragments, or fragments alone. Anaphase bridges and fragments were found also in both meiotic divisions. It was considered that the production of bridges was a physiological effect of the hydrogen fluoride causing the chromosomes to become sticky, break, and form dicentric chromosomes. The results of these studies suggested that hydrogen fluoride may be a mutagenic agent.

F. *Vicia faba*

Broad beans have long been used for cytological studies (Kihlman, 1975). The species possesses six pairs of large chromosomes, one pair being more than twice the length of the others and possessing a large satellite on the short arm. As a result, the karyotype is very favorable for the assessment of clastogenic potential of putative mutagenic chemicals.

Fetner (1958) observed chromosome bridges in root tip cells after seeds of *Vicia faba* were exposed to 0.4 weight percent ozone. Treatments of 15, 30, and 60 minutes duration induced chromosome aberrations at a frequency of 10, 29, and 42%, respectively. Thus, there was a positive correlation between the frequency of chromosomal aberrations and duration of treatment.

Janakiraman and Harney (1976) treated *Vicia faba* plants with ozone at a concentration of 100 and 200 pphm for 4 or 8 hours when the first florets were in bud formation. Another group of plants was fumigated at a concentration of 50 pphm for 24 hours, allowed 24 hours recovery, and fumigated again. Meiosis was studied for chromosome abnormalities in buds collected 24 hours after treatment at the MI, AI-TI, and AII-TII stages. Only the dose of 200 pphm caused chromosome damage which included bridges, fragments, and stickiness. Significant chromosomal damage appeared 24 hours after fumigation in MI and AI-TI but not in later stages.

Pack (1970) studied the F_1, F_2, and F_3 generations of tendergreen bean plants (*Phaseolus vulgaris*) that were fumigated with hydrogen fluoride under continuous exposure at concentrations of 0.58, 2.1, 9.1, and 10.5 $\mu g/m^3$ from seeding to the time the plants were mature and set seed. The progeny were grown in a fluoride-free atmosphere. No effects were found among the progeny grown at 0.58 $\mu g/m^3$. The F_1 progeny of the treated plants were less vigorous as shown by later emergence, smaller primary leaves, and slower stem growth. The primary leaves were severely stunted and distorted. Abnormal trifoliate leaves were present in both F_1 and F_2 generations. Considerably fewer

F_3 generation plants had abnormal trifoliate leaves. These occurred most frequently among plants whose parents were abnormal.

G. Miscellaneous Species

Nygren (1955) observed polyploidy in species of *Melandrium* after 4- to 48-hour treatments with nitrous oxide at pressures between 2 and 10 atm. All plants treated for longer than 16 hours set no seeds. Twelve triploids and seven hexaploids were produced. In a later study, using *Melandrium album*, Nygren (1957) produced triploid, tetraploid, and hexaploid plants from nitrous oxide treatment. Seven hours at 5 atm and 16 hours at 2 atm pressure gave the best results.

When the youngest flowers of *Phalaris canariensis* ($2n = 12$) and *P. paradoxa* ($2n = 14$) had just completed their flowering, Ostergren (1957) treated the plants with nitrous oxide at 10 atm pressure for 6–12 hours. Seed was collected and progeny examined for polyploids. Four tetraploids of *P. canariensis* and two aneuploids of *P. paradoxa* were recovered. Ostergren (1954) had previously obtained polyploids of *Crepis capillaris* by this same method. Ostergren suggested that nitrous oxide inactivates the spindle, thus resulting in the production of polyploids. Montezuma-de-Carvalho (1967) cites several studies using nitrous oxide to induce polyploidy. He treated whole plants of *Datura* and cut flowering branches of an interspecific cross *Solanum tuberosum* × *S. phureja* to 4, 6, or 9 atm of nitrous oxide for 2 hours. He obtained C-mitotic effects in pollen tube mitoses. Bhatia and Sybenga (1965) have shown that nitric oxide enhances the effect of X-rays.

Taylor *et al.* (1976) treated excised heads of diploid red clover (*Trifolium pratense*) with nitrous acid at 6 bars of atmospheric pressure 24 hours after crossing for a period of 24 hours. Of the plants which flowered 71% were tetraploid. This same technique produced 49 and 79% tetraploids in *T. alpestre* and *T. rubens*, respectively. Nitrous oxide was considered far more successful for the production of tetraploids than colchicine.

V. THE USE OF *IN SITU* WEED COMMUNITIES FOR THE DETECTION OF ATMOSPHERIC MUTAGENS

Weeds exist both in open situations resulting from disturbances by man, and in closed communities, the product of secondary succession after abandonment of intensely used lands. Although these weedy habitats may be well defined, changes do arise as a result of chemical insults to the environment, such as herbicide clearance of roadsides and rights-of-way. Several studies have shown that the vegetational composition of an area can be radically altered after herbicide treatment and that, as a result of such treatments, weeds may develop resistance through mutation (Grant, 1972; Pinthus et al., 1972).

The use of weeds to detect symptoms of plant injury and altered growth habit from environmental pollutants is well known (Feder, 1978). However, the use of weeds to detect altered reproductive patterns through the study of chromosome irregularities and pollen abortion has only recently been considered as a means of detecting environmental mutagens (Tomkins and Grant, 1976).

A study to determine weed species that might be suitable for the detection of environmental mutagens and grow naturally in the vicinity of an industrial plant has been carried out at the Macdonald Campus of McGill University. This study shows the usefulness of this approach for the detection of mutagenic atmospheric pollutants, and, therefore, will be outlined in some detail.

Weed species were studied in an area along a transmission line right-of-way running through an abandoned agriculture field (Tomkins and Grant, 1974). Plots 4 × 4 m each, separated from adjacent plots by a 1-m border strip, were sprayed with one of seven commercial herbicides. From almost 56,000 individual plants identified, more than 50 species were represented. Of these, 12 species representing more than 1.0% of the total number of individuals were studied for somatic and meiotic chromosome aberrations from flower buds collected at different sampling periods (June, July, August, September) (Tomkins and Grant, 1976). Mitotic anaphase aberrations were scored as well as meiotic chromosome aberrations at metaphase I and anaphase I. Aberrations included acentric fragments, anaphase bridges, lagging chromosomes, multipolar anaphases, sticky bridges, univalents, and trivalents. Flower buds from five of the 12 species collected at different sampling dates proved to be particularly useful for monitoring populations for clastogenic (chromosome breaking) effects. These included one annual, *Ambrosia artemisiifolia* L., two biennials, *Melilotus alba* Desr. and *Pastinaca sativa* L., and two perennials, *Solidago canadensis* L. and *Vicia cracca* L.

The average spontaneous mitotic chromosome aberration rate determined in the 12 weed species growing under natural conditions was about 0.4%, a level similar to that observed in several populations of newborn humans (Hamerton et al., 1972). Significant increases in the frequency of aberrant cells were observed in the treated populations after exposure to some herbicides. The most frequently observed abnormalities were fragmentation of the chromosomes, lagging chromosomes, multiple spindles, and bridges. The distribution of aberrant types varied with the different herbicides and provided some insight into the interaction of these chemicals with chromosome structure and the spindle apparatus.

From the frequency of chromosome aberrations found in sampling plants at different intervals (months) after spraying had occurred, an indication of mutagenic damage may be obtained. As may be seen from Table III, the frequency of chromosome aberrations in *Pastinaca sativa* was the highest in June and July (2.5 and 2.6%, respectively) but declined by August and September (1.1 and 0.8%, respectively). This recovery was similar to the results found in human leukocyte cultures from occupationally exposed agricultural workers who had a higher incidence of chromosome abnormalities during midseason exposure to a combination of herbicides than off-season (Yoder *et al.,* 1973).

The use of wild plants *in situ* around urban and industrial complexes for the monitoring of mutagenic air pollutants has not been carried out to date, although the fern *Osmunda regalis* has been successfully used as an *in situ* bioassay for mutagens in an aquatic ecosystem downstream from a paper processing plant (Klekowski, 1978a,b). From our studies (Tomkins and Grant, 1976), it is apparent that chromosomes can be sensitive indicators of clastogenic effects of environ-

TABLE III

Somatic Chromosome Aberration Rates at
Sampling Time for *Pastinaca sativa*

Herbicide	Month	Aberrations (%)
Picloram	July	1.4
	August	1.9
Simazine	June	2.5
	August	1.1
Diuron	June	1.3
	July	2.6
	August	1.4
	September	0.8
Control	June	0.0
	July	0.0
	August	0.7
	September	1.6

mental airborne mutagens. This type of study needs to be carried out under actual application, that is, for weeds in the vicinity of an industrial complex with control groups some distance from the experimental monitoring site.

VI. DISCUSSION AND CONCLUSIONS

Higher plants can play a major role in the detection and monitoring of mutagenic airborne pollutants. There are many advantages in utilizing plant systems and these have been reviewed (Nilan and Vig, 1976). Like mammals, higher plants possess a genome organization and cell heredity similar to that of humans; consequently, the usual difficulties in extrapolating the potential genetic hazards to man are less than those associated with microorganism-derived mutagenicity data. It has been found that a good correlation exists where mutagenicity data from higher plant assays have been compared with data from nonmammalian and mammalian assays (Grant, 1978, 1982; Grant et al., 1981; Constantin and Owens, 1982). Thus, information derived from higher plants may provide data as useful as that from microorganisms on the mutagenic effects of air pollutants.

At the present time, only the *Tradescantia* genetic systems have received the extensive field studies that have proved this species excellent and highly recommendable for *in situ* testing and monitoring of atmospheric mutagenic pollutants.

Several pollen assay systems, especially the *waxy* locus in *Hordeum vulgare* and *Zea mays*, show considerable promise and may prove equally valuable for monitoring airborne mutagens, but require more extensive trials under *in situ* conditions similar to that which has been carried out by Lower et al. (1978). Identical arguments apply to other higher-plant mutagen assay systems, such as somatic mosaicism in *Glycine max*.

It is also clear from the review that many higher plants can be used for testing for mutagenic gaseous compounds under laboratory conditions. *Allium cepa* and *Vicia faba* are classical species used to test for chromosome aberrations (Kihlman, 1975; Grant, 1978, 1982). *Arabidopsia thaliana* and *Lycopersicon esculentum* (as well as barley and corn) are highly suitable for assessing genetic damage from pollutants (Grant et al., 1981).

Higher plants have long served as detectors of pollution in the environment; for example, ozone-sensitive populations of tobacco, *Petunia*, and tomato are being used to monitor for phytochemical oxidants (Feder, 1978). However, the use of *in situ* weed communities is only currently being seriously considered for the detection and monitoring of airborne mutagens. Although it is possible to transport and place *in situ* plants at high-risk sites to serve as monitors of mutagens (Schairer *et al.*, 1979), the selection of species growing naturally in the environment may prove highly useful for such purposes (Tomkins and Grant, 1976; Klekowski, 1978a). Further research is required for the development of such naturally occurring *in situ* systems for the detection and monitoring of environmental mutagens.

ACKNOWLEDGMENTS

Financial support from the National Sciences and Engineering Research Council of Canada for studies in genetic toxicity of environmental chemicals is gratefully acknowledged.

REFERENCES

Ames, B. N. (1979). Identifying environmental chemicals causing mutations and cancer. *Science* **204**, 587–593.

Anonymous (1975). Environmental mutagenic hazards. *Science* **187**, 503–514.

Bhatia, C. R., and Sybenga, J. (1965). Effect of nitric oxide on X-ray sensitivity of *Crotalaria intermedia* seeds. *Mutat. Res.* **2**, 332–338.

Cataldo, D. A., and Wildung, R. E. (1978). Soil and plant factors influencing the accumulation of heavy metals by plants. *Environ. Health Perspect.* **27**, 149–159.

Conger, A. D., and Carabia, J. V. (1977). Mutagenic effectiveness and efficiency of sodium azide versus ethyl methanesulfonate in maize: Induction of somatic mutations at the yg_2 locus by treatment of seeds differing in metabolic state and cell population. *Mutat. Res.* **46**, 285–296.

Constantin, M. J., and Owens, E. T. (1982). Plant genetic and cytogenetic assays: Introduction and perspectives. *Mutat. Res.* (in press).

de Serres, F. J. (1978). Introduction: Utilization of higher plant systems as monitors of environmental mutagens. *Environ. Health Perspect.* **27**, 3–6.

Dvorak, J., and Harvey, B. L. (1973). Production of aneuploids in *Avena sativa* L. by nitrous oxide. *Can. J. Genet. Cytol.* **15**, 649–651.

Dvorak, J., Harvey, B. L., and Coulman, B. E. (1973). The use of nitrous oxide for producing eupolyploids and aneuploids in wheat and barley. *Can. J. Genet. Cytol.* **15**, 205–214.

Emmerling-Thompson, M., and Nawrocky, M. M. (1980). Genetic basis for using *Tradescantia* clone 4430 as an environmental monitor of mutagens. *J. Hered.* **71**, 261–265.

Engle, R. L., and Gabelman, W. H. (1966). Inheritance and mechanisms for resistance to ozone damage in onion, *Allium cepa* L. *Proc. Am. Soc. Hortic. Sci.* **89**, 423–430.

Evans, H. J. (1962). Chromosome aberrations induced by ionizing radiation. *Int. Rev. Cytol.* **13**, 221–231.

Fabergé, A. C. (1955). Types of chromosome aberrations induced by ethylene oxide in maize. *Genetics* **40**, 571.

Feder, W. A. (1978). Plants as bioassay systems for monitoring atmospheric pollutants. *Environ. Health Perspect.* **27**, 139–147.

Ferguson, J., Hawkins, S. W., and Doxey, D. (1950). C-mitotic action of some simple gases. *Nature (London)* **165**, 1020–1022.

Fetner, R. H. (1958). Chromosome breakage in *Vicia faba* by ozone. *Nature (London)* **181**, 504–505.

Fishbein, L. (1976). Atmospheric mutagens. *Chem. Mutagens* **4**, 219–319.

Freeling, M. (1978). Maize *Adhl* as a monitor of environmental mutagens. *Environ. Health Perspect.* **27**, 91–97.

Grant, W. F. (1972). Pesticides—Subtle promoters of evolution. *Symp. Biol. Hung.* **12**, 43–50.

Grant, W. F. (1978). Chromosome aberrations in plants as a monitoring system. *Environ. Health Perspect.* **27**, 37–43.

Grant, W. F. (1982). Chromosome aberration assays in *Allium. Mutat. Res.* (in press).

Grant, W. F., and Harney, P. M. (1960). Cytogenetic effects of maleic hydrazide treatment of tomato seed. *Can. J. Genet. Cytol.* **2**, 162–174.

Grant, W. F., Zinov'eva-Stahevitch, A. E., and Zura, K. D. (1981). Plant genetic test systems for the detection of chemical mutagens. *In* "Short-Term Tests for Chemical Carcinogens." (H. F. Stich and R. H. C. San, eds.), pp. 200–216. Springer-Verlag, New York.

Hamerton, J. L., Ray, M., Abbott, J., Williamson, C., and Ducasse, G. C. (1972). Chromosome studies in a neonatal population. *Can. Med. Assoc. J.* **106**, 776–779.

Heath, C. W., Jr. (1978). Environmental pollutants and the epidemiology of cancer. *Environ. Health Perspect.* **27**, 7–10.

Hughes, T. J., Pellizzari, E., Little, L., Sparacino, C., and Kolber, A. (1980). Ambient air pollutants: Collection, chemical characterization and mutagenicity testing. *Mutat. Res.* **76**, 51–83.

Ichikawa, S. (1981). *In situ* monitoring with *Tradescantia* around nuclear power plants. *Environ. Health Perspect.* **37**, 145–164.

Janakiraman, R., and Harney, P. M. (1975). Effect of ozone on *Arabidopsis. Arabidopsis Inf. Serv.* **12**, 4–5.

Janakiraman, R., and Harney, P. M. (1976). Effects of ozone on meiotic chromosomes of *Vicia faba. Can. J. Genet. Cytol.* **18**, 727–730.

Kihlman, B. A. (1975). Root tips of *Vicia faba* for the study of the induction of chromosomal aberrations. *Mutat. Res.* **31**, 401–412.

Klekowski, E. J., Jr. (1978a). Screening aquatic ecosystems for mutagens with fern bioassays. *Environ. Health Perspect.* **27**, 99–102.

Klekowski, E. J., Jr. (1978b). Detection of mutational damage in fern populations: An *in situ* bioassay for mutagens in aquatic ecosystems. *Chem. Mutagens* **5**, 79–99.

Kostoff, D. (1931). Heteroploidy in *Nicotiana tabacum* and *Solanum melongena* caused by fumigation with nicotine sulfate. *Bull Soc. Bot. Bulgar.* **4**, 87–92. *Biol. Abstr.* **8**, 10 (1934).

Levan, A. (1938). The effect of colchicine on root mitosis of *Allium. Hereditas* **24**, 471–486.

Levan, A. (1949). The influence on chromosomes and mitosis of chemicals, as studied by the *Allium* test. *Proc. Int. Congr. Genet.*, 8th, 1948 pp. 325–337.

Lindgren, D., and Lindgren, K. (1972). Investigation of environmental mutagens by the waxy method. *Environ. Mutagen Soc. Newsl.* **6**, 22.

Lower, W. R., Rose, P. S.,and Drobney, V. K. (1978). *In situ* mutagenic and other effects associated with lead smelting. *Mutat. Res.* **54,** 83–93.

Ma, T.-H. (1967). Thin-layer lactose agar for pollen-tube culture of *Tradescantia* to enhance planar distribution of chromosomes. *Stain Technol.* **42,** 285–291.

Ma, T.-H. (1979). Micronuclei induced by X-rays and chemical mutagens in meiotic pollen mother cells of *Tradescantia*. A promising mutagen test system. *Mutat. Res.* **64,** 307–313.

Ma, T.-H., and Khan, S. H. (1976). Pollen mitosis and pollen tube growth inhibition by SO_2 in cultured pollen tubes of *Tradescantia*. *Environ. Res.* **12,** 144–149.

Ma, T.-H., Isbandi, D., Khan, S. H., and Tseng, Y.-S. (1973). Low level of SO_2 enhanced chromatid aberrations in *Tradescantia* pollen tubes and seasonal variation of the aberration rates. *Mutat. Res.* **21,** 93–100.

Ma, T.-H., Sparrow, A. H., Schairer, L. A., and Nauman, A. F. (1978). Effect of 1,2-dibromoethane (DBE) on meiotic chromosomes of *Tradescantia*. *Mutat. Res.* **58,** 251–258.

Ma, T.-H., Anderson, V. A., and Ahmed, I. (1980). *In situ* monitoring of air pollutants and screening of chemical mutagens using *Tradescantia*-micronucleus bioassay. *Environ. Mutagen.* **2,** 287.

Ma, T.-H., Fang, T., Ho, J., Chen, D., Zhou, R., Lin, G., Dai, J., and Li, J. (1981). Hypersensitivity of *Tradescantia reflexa* meiotic chromsomes to mutagens. *Mutat. Res.* (in press).

Maugh, T. H., II (1978). Chemicals: How many are there? *Science* **199,** 162.

Mericle, L. W., and Mericle, R. P. (1967). Genetic nature of somatic mutations for flower color in *Tradescantia*, clone 02. *Radiat. Bot.* **7,** 449–464.

Mohamed, A. H. (1968). Cytogenetic effects of hydrogen fluoride treatment in tomato plants. *J. Air. Pollut. Control. Assoc.* **18,** 395–398.

Mohamed, A. H. (1969). Cytogenetic effects of hydrogen fluoride on plants. *Fluoride* **2,** 76–84.

Mohamed, A. H. (1970). Chromosomal changes in maize induced by hydrogen fluoride gas. *Can. J. Genet. Cytol.* **12,** 614–620.

Mohamed, A. H., Smith, J. D., and Applegate, H. G. (1966). Cytological effects of hydrogen fluoride on tomato chromosomes. *Can. J. Genet. Cytol.* **8,** 575–583.

Montezuma-de-Carvalho, J. (1967). The effect of N_2O on pollen tube mitosis in styles and its potential significance for inducing haploidy in potato. *Euphytica* **16,** 190–198.

Mulcahy, D. L. (1981). Pollen in tetrads as indicators of environmental mutagenesis. *Environ. Health Perspect.* **37,** 91–94.

Mulcahy, D. L., and Johnson, C. M. (1978). Self-incompatibility systems as bioassays for mutagens. *Environ. Health Perspect.* **27,** 85–90.

Nauman, C. H., and Grant, W. F. (1982). The mutagenicity of ethyl methanesulfonate, mitomycin C, and vinyl chloride in the *Tradescantia* test systems. "NIEHS Comparative Mutagenesis Conference Papers." Plenum Press, New York. (in press).

Nauman, C. H., Villalobos-Pietrini, R., and Sautkulis, R. C. (1974). Response of a mutable clone of *Tradescantia* to chemical mutagens and to ionizing radiation. *Mutat. Res.* **26,** 444.

Nauman, C. H., Sparrow, A. H., Schairer, L. A., and Sautkulis, R. C. (1975). Influences of temperature, ionizing radiation, and chemical mutagens on somatic mutation rate in *Tradescantia*. *Mutat. Res.* **31,** 318–319.

Nauman, C. H., Sparrow, A. H., and Schairer, L. A. (1976). Comparative effects of ionizing radiation and two gaseous chemical mutagens on somatic mutation induction in one mutable and two non-mutable clones of *Tradescantia*. *Mutat. Res.* **38**, 53–70.

Nauman, C. H., Klotz, P. J., and Schairer, L. A. (1979). Uptake of tritrated 1,2-dibromoethane by *Tradescantia* floral tissues: Relation to induced mutation frequency in stamen hair cells. *Environ. Exp. Bot.* **19**, 201–215.

Nettancourt, D., de. (1977). "Incompatibility in Angiosperms." Springer-Verlag, Berlin and New York.

Nilan, R. A. (1974). Barley (*Hordeum vulgare*) *In* "Handbook of Genetics: Plants, Plant Viruses and Protists." Vol. 2. (R. C. King, ed.), pp. 93–110. Plenum Press, New York.

Nilan, R. A. (1978). Potential of plant genetic systems for monitoring and screening mutagens. *Environ. Health Perspect.* **27**, 181–186.

Nilan, R. A., and Vig, B. K. (1976). Plant test systems for detection of chemical mutagens. Vol. 4 *In* "Chemical Mutagens." (A. Hollaender, ed.), pp. 143–170. Plenum Press, New York.

Nilan, R. A., Sideris, E. G., Kleinhofs, A., Sander, C., and Konzak, C. F. (1973). Azide— A potent mutagen. *Mutat. Res.* **17**, 142–144.

Nilan, R. A., Rosichan, J. L., Arenaz, P., and Hodgdon, A. L. (1981). Pollen genetic markers for detection of mutagens in the environment. *Environ. Health Perspect.* **37**, 19–25.

Nygren, A. (1955). Polyploids in *Melandrium* produced by nitrous oxide. *Hereditas* **41**, 287–290.

Nygren, A. (1957). Studies in polyploid *Melandrium album* by nitrous oxide, N_2O. *Ann. R. Agric. Coll. Sweden* **23**, 393–404.

Odum, W. E., and Drifmeyer, J. E. (1978). Sorption of pollutants by plant detritus: A review. *Environ. Health Perspect.* **27**, 133–137.

Ostergren, G. (1954). Polyploids and aneuploids of *Crepis capillaris* produced by treatment with nitrous oxide. *Genetica* **27**, 54–64.

Ostergren, G. (1957). Production of polyploids and aneuploids of *Phalaris* by means of nitrous oxide. *Hereditas* **43**, 512–516.

Pack, M. R. (1970). Effects of hydrogen fluoride on bean reproduction. *J. Air. Pollut. Control Assoc.* 70–134. *Annu. Meet. St. Louis* **21**(3), 133–137.

Pandey, K. K. (1964). Elements of the S-gene complex. I. The S_{FI} alleles in *Nicotiana*. *Genet. Res.* **5**, 397–409.

Pinthus, M. H., Eshel, Y., and Schchori, Y. (1972). Field and vegetable crop mutants with increased resistance to herbicides. *Science* **177**, 715–716.

Plewa, M. J. (1978). Activation of chemicals into mutagens by green plants: A preliminary discussion. *Environ. Health Perspect.* **27**, 45–50.

Rédei, G. P. (1974). *Arabidopsis thaliana*. *In* "Handbook of Genetics: Plants, Plant Viruses and Protists" (R. C. King, ed.), Vol. 2, pp. 151–180. Plenum, New York.

Rédei, G. P., and Acedo, G. (1976). Biochemical mutants in higher plants. *In* "Cell Genetics in Higher Plants" (D. Dudits, G. L. Farkas, and P. Maliga, ed.), p. 39. Akademiai Kaido, Budapest.

Rick, C. M., and Butler, L. (1956). Cytogenetics of the tomato. *Adv. Genet.* **8**, 267–382.

Rosichan, J., Arenaz, P., and Nilan, R. A. (1979). A high resolution plant mutagen monitoring system. *Genetics* **92**, s108.

Rushton, P. S. (1969). The effects of 5-fluorodeoxyuridine in radiation-induced chromatid aberrations in *Tradescantia* microspores. *Radiat. Res.* **38**, 404–413.

Schairer, L. A., Sparrow, A. H., and Sautkulis, R. C. (1973). Somatic mutation rates induced by 0.43 MeV neutrons, X-rays, and ethyl methane sulfonate (EMS) in *Tradescantia. Radiat. Res.* **55**, 599–600.

Schairer, L. A., Van't Hof, J., Hayes, C. G., Burton, R. M., and de Serres, F. J. (1978a). Exploratory monitoring of air pollutants for mutagenicity activity with the *Tradescantia* stamen hair system. *Environ. Health Perspect.* **27**, 51–60.

Schairer, L. A., Sparrow, A. H., and Tempel, N. R. (1978b). Mobile monitoring vehicle designed to assess the mutagenicity of ambient air in high pollution areas. *Mutat. Res.* **53**, 111–112.

Schairer, L. A., Klug, E. E., Pond, V., Sautkulis, R. C., and Tempel, N. R. (1979). Mutagenicity monitoring of ambient air at selected sites in the U.S. using *Tradescantia* in a mobile laboratory. *Environ. Mutagen.* **2**, 187–188.

Schwartz, D. (1972). A method of high resolution immunoelectrophoresis for alcohol dehydrogenase isozymes. *J. Chromatogr.* **67**, 385–388.

Schwartz, D. (1981). *Adh* locus in maize for detection of mutagens in the environment. *Environ. Health Perspect.* **37**, 75–77.

Schwartz, D., and Osterman, J. (1976). A pollen selection system for alcohol-dehydrogenase-negative mutants in plants. *Genetics* **83**, 63–65.

Smith, H. H., and Lotfy, T. A. (1954). Comparative effects of certain chemicals on *Tradescantia* chromosomes as observed at pollen tube mitosis. *Am. J. Bot.* **41**, 589–593.

Sparrow, A. H., and Schairer, L. A. (1971). Mutational response in *Tradescantia* after accidental exposure to a chemical mutagen. *Environ. Mutagen. Soc. Newsl.* **5**, 16–19.

Sparrow, A. H., and Schairer, L. A. (1974). Mutagen response of *Tradescantia* to treatment with X-rays, EMS, DBE, ozone, SO_2, N_2O and several insecticides. *Mutat. Res.* **26**, 445.

Sparrow, A. H., and Schairer, L. A. (1976). Response of somatic mutation frequency in *Tradescantia* to exposure time and concentration of gaseous mutagens. *Mutat. Res.* **38**, 405–406.

Sparrow, A. H., Schairer, L. A., and Villalobos-Pietrini, R. (1974). Comparison of somatic mutation rates induced in *Tradescantia* by chemical and physical mutagens. *Mutat. Res.* **26**, 265–276.

Stich, H. F., and San, R. H. C. (1980). International workshop on short-term tests for chemical carcinogens. *Genet. Soc. Can. Bull.* **11**, 25–28.

Subrahmanyam, N. C., and Kasha, K. J. (1975). Chromosome doubling of barley haploids by nitrous oxide and colchicine treatments. *Can. J. Genet. Cytol.* **17**, 573–583.

Taylor, N. L., Anderson, M. L., Quesenberry, K. H., and Watson, L. (1976). Doubling the chromosome number of *Trifolium* species using nitrous oxide. *Crop Sci.* **16**, 516–518.

Tomkins, D. J., and Grant, W. F. (1974). Differential response of 14 weed species to seven herbicides in two plant communities. *Can. J. Bot.* **52**, 525–533.

Tomkins, D. J., and Grant, W. F. (1976). Monitoring natural vegetation for herbicide-induced chromosomal aberrations. *Mutat. Res.* **36**, 73–84.

Underbrink, A. G., Schairer, L. A., and Sparrow, A. H. (1973). *Tradescantia* stamen hairs: A radiobiological test system applicable to chemical mutagenesis. *Chem. Mutagens* **3**, 171–207.

Vig, B. K. (1975). Soybean (*Glycine max*): A new test system for study of genetic parameters as affected by environmental mutagens. *Mutat. Res.* **31**, 49–56.

Vig, B. K. (1978). Somatic mosaicism in plants with special reference to somatic crossing over. *Environ. Health Perspect.* **27**, 27–36.

Villalobos-Pietrini, R., Sparrow, A. H., Schairer, L. A., and Sparrow, R. C. (1974). Variation in somatic mutation rates induced by X-rays, DBE and EMS in several *Tradescantia* species and hybrids. *Radiat. Res.* **59,** 153.

Weber, D. F. (1981). Maize pollen grains as a screen for non-disjunction. *Environ. Health Perspect.* **37,** 79–84.

Yoder, J., Watson, M., and Benson, W. W. (1973). Lymphocyte chromosome analysis of agricultural workers during extensive occupational exposure to pesticides. *Mutat. Res.* **21,** 335–340.

Chapter 16

Fishes as Biological Detectors of the Effects of Genotoxic Agents

A. D. Kligerman

I. INTRODUCTION

A. Overview

In the past man has considered the aquatic environment as a dumping ground to rid the terrestrial environment of biological and technological wastes. The immense volume of water present on earth (approximately 1.34×10^6 km³; Hynes, 1972) was thought to render most toxic compounds innocuous through dilution. However, in the last few decades it has become apparent that disposal of toxicants into aquatic ecosystems can lead to their accumulation both in the sediments (Sonstegard, 1977) and upper levels of the food chain (Stathan et al., 1976).

435

In the United States, the Council of Environmental Quality estimates that 72% of all water basins are contaminated by industrial pollution (Stara et al., 1980). Low concentrations of more than 300 biorefractories have been identified in the drinking water. Many of these are known or suspected carcinogens (Kraybill, 1977). The problem is not confined to the United States. Pesticides, polychlorinated biphenyls, and polycyclic aromatic hydrocarbons reach significant levels in many of the world's water systems and animals living in these waters (Kraybill, 1976). Thus, man is exposed throughout his lifetime to low levels of toxicants present in both the water and many of the aquatic food species consumed. The public health hazards these may pose is as yet unknown.

The science of aquatic toxicology has developed in an attempt to determine what effects xenobiotics have on the structure and function of aquatic ecosystems. Fish species are often used in such studies because they are ecologically and economically important parts of aquatic systems, are sensitive to adverse environmental conditions, and can be studied relatively easily in a laboratory. As early as the 1930s, experiments were being performed to establish a cause and effect relationship between chemical contamination and acute toxicity in fish populations (Macek, 1980). In the 1940s and 1950s, a large amount of data was gathered showing the vast differences in species susceptibility to toxicants. By the mid-1960s through the early 1970s, subacute and chronic toxicity studies were in wide use examining such end points as growth, hatchability, and reproductive fitness (Macek, 1980).

While aquatic toxicology was becoming a respected discipline for scientific endeavor, the field of genetic toxicology was experiencing a period of exponential growth brought on primarily by studies supporting the somatic mutation theory of cancer (Ames et al., 1973; Miller and Miller, 1975; McCann et al., 1975). With the heightened concern and awareness of the problems caused by environmental pollution coupled with the development of new assays for the studying of genotoxic effects, a great deal of effort, time, and money has been devoted to trying to understand better the relationship between chemical insult to the DNA and end points such as carcinogenicity, mutagenicity, and teratogenicity. Thus, it is only logical that genetic toxicologists have become interested in aquatic organisms, and aquatic toxicologists have begun to investigate genotoxic phenomena. However, the melding of these two fields into what may be called "aquatic genetic toxicology" has been difficult and has only very recently showed signs of progress. The main problems have been that techniques, species, and training in one discipline are either unknown, unusable, or unfamiliar to workers in the other discipline.

B. Historical Perspective

Contrary to what might be expected when viewing the limited number of publications dealing with aquatic genetic toxicology, fish species were used in some of the first cytogenetic investigations. In 1887, Schwarz viewed colored *Schleifen* in dividing cells from fertilized eggs of *Salmo trutta* (*fario*), which he identified as similar to the chromosomes studied by Flemming (Svärdson, 1945). A quarter of a century later, Oppermann (1913) found that radiation could affect the course of cleavage in fish eggs and lead to abnormalities in the chromatin. Solberg (1938) discovered that the X-ray sensitivity of *Fundulus* embryos varied with the mitotic activity at the time of irradiation. However, for the most part genetic studies with fishes were concerned only with karyotype description, cytotaxonomy, and gene identification and mapping.

This descriptive phase of endeavor changed to a more experimentally oriented approach in the late 1950s and early 1960s when scientists in the Soviet Union began to investigate the effects of radiation on fish reproductive physiology and development (for review, see Kligerman, 1979a, 1980). Most of these studies reported an increase in anaphase and telophase bridges, multipolar mitoses, and chromosome fragments after fish embryos, sperm, or ova were X-irradiated or exposed to waterborne radionuclides. Tsoi (1970, 1974) and coworkers (1975) were among the first to investigate the effects of chemical mutagens on the genetic material of fishes. They used dimethyl sulfate, nitrosomethyl urea, nitrosoethyl urea, and 4-bisdiazoacetyl butane to induce gynogenesis in a diversity of fish species. They also reported the occurrence of chromosome abnormalities (e.g., anaphase and telophase bridges, chromosome fragments, multipolar mitoses) in the developing gastrulas following chemical exposure.

In one of the first studies undertaken to analyze chromosome damage from fishes taken from a polluted environment, Longwell (1976) reported that mackerel (*Scombus scombus*) collected from the surface waters of the New York Bight displayed chromosome bridges, chromosome "stickiness," and abnormal spindle orientation.

Unfortunately, most of the aforementioned studies are only of qualitative or historical importance. Some are plagued by unrealistically high control levels of damage due to inadequate methodologies while others lack adequate controls. Because of unsatisfactory methodologies and the use of inappropriate species for cytogenetic investigations, quantitative estimates of induced damage with reproducible dose–response curves are absent from most of these initial studies in aquatic genetic toxicology (Kligerman, 1979a, 1980).

More recently, the rapid advances that have been made in the field of genetic toxicology have led to improved methods for studying the genotoxic effects of waterborne chemicals using aquatic organisms. Two basic approaches to aquatic genetic toxicology are currently in use: (1) long-term protocols involving epidemiology and laboratory carcinogenicity bioassays; and (2) short-term protocols making use of dominant lethal, specific locus, and cytogenetic assays.

Both long- and short-term methods for using fishes to detect the effects of aquatic genotoxicants are reviewed below, and the advantages and disadvantages of the different techniques discussed. Emphasis will be placed on the use of short-term cytogenetic methods for detecting the effects of waterborne genotoxic agents.

II. METHODOLOGIES FOR AQUATIC GENETIC TOXICOLOGY

A. Epidemiology

In theory epidemiological investigations offer the most realistic and conclusive means to determine if a water system carries carcinogenic agents. However, in practice good epidemiological studies are difficult to carry out and often give equivocal results.

Brown et al. (1973, 1979), studied watersheds that were classified as highly, moderately, or essentially pollution-free and found that there was a direct relationship between the level of pollution and the percentage of fishes with oncogenic diseases. Fishes from the polluted Fox River system in Illinois had greater than a fivefold higher incidence of tumors than species from the unpolluted Lake of the Woods system in Canada. Similarly, Sonstegard (1977) reported that a survey of the Great Lakes conducted between 1973 and 1976 showed high incidences of gonadal tumors in carp (*Cyprinus carpio*) and goldfish (*Cassarius auratus*) and papillomas in white suckers (*Catostomus commersomi*). A review of the museum collections taken prior to 1952 from the same area revealed an absence of gonadal tumors in the carp and goldfish. He associated the prevalence of gonadal tumor with the increased industrial pollution of the Great Lakes. However, the papillomas of the white suckers appear to be related to endogenous C-type viruses. Sonstegard hypothesizes that chemical carcinogen contamination of the waters and sediments acts as a "triggering mechanism" which causes the production of papillomas by the viruses. He proposes that papilloma and C-virus incidence in white suckers would be "an ideal indicator for carcinogen detection."

Stich et al. (1976) believe that monitoring the skin papilloma frequencies of flatfishes (Pleuronectidae and Bothidae) might prove to be a useful indicator of carcinogenic contamination of marine and estuarine environments. They report that there appears to be a link between the papilloma incidence and the discharge of industrial and biological waste into the waters; however, they do caution that more work needs to be done on the epidemiology of such tumors (Stich and Acton, 1976). This is emphasized by Mearns and Sherwood (1977) who found that the occurrences of skin papillomas on Dover soles (*Microstomas pacificus*) and lip papillomas on white croakers (*Genyonemus limeatus*) were not related to waste water discharge. Similar findings by Oishi et al. (1976) led them to conclude that a parasitic etiology for epidermal papillomas of flatfishes should not be overlooked.

Epidemiological investigations of the incidences of neoplastic diseases in fishes may yet prove valuable when a better data base is obtained. At present such studies appear to be capable at best of pointing out only the obvious areas of high-level carcinogenic contamination. Furthermore, these types of investigations are time consuming and labor intensive, and do not pinpoint the agent or agents responsible. Viral and parasitic factors may confound the problem, and adequate control populations are difficult if not impossible to obtain. Thus, epidemiology may warn of significant environmental deterioration, but it does not appear to be the most appropriate or sensitive means to monitor the aquatic environment for the presence of genotoxic chemicals.

B. Carcinogenicity Bioassays

The epidemic of hepatocellular carcinomas that affected hatchery raised rainbow trout (*Salmo gairdneri*) led to the discovery of the highly potent carcinogen aflatoxin B_1 in moldy fish feed (reviewed by Sinnhuber et al., 1977). This discovery initiated research into the induction of neoplasia in fishes (Ishikawa and Takayama, 1979) and led to the use of fishes in bioassay programs analogous to those used by the National Cancer Institute for carcinogenicity testing with small rodents. Already, fishes have proved valuable test organisms as studies have shown that the Shasta strain of rainbow trout is far more sensitive to the carcinogenic effects of aflatoxin B_1 than any other animal as yet tested (Sinnhuber et al., 1977).

Stanton (1965) was the first to show the utility of using small aquarium fishes for carcinogenicity studies. He induced hepatomas and cholangiomas in Zebra fish (Brachydanio rerio) by exposing them to diethylnitrosamine. Subsequently many investigators have demonstrated the high sensitivity of small aquarium fishes to the induction of tumors by a variety of carcinogens such as benzo[a]pyrene, dimethylnitrosamine, and o-aminoazotoluene (see review by Matsushima and Sugimura, 1976). Besides sensitivity, other advantages offered by the use of small aquarium fishes in carcinogen bioassay programs include their resistance to some of the toxic effects of carcinogens, the rapidity with which they develop tumors, and the ease with which large numbers of animals can be maintained in the laboratory (Matsushima and Sugimura, 1976).

Hendricks et al. (1980) have developed an innovative system for the detection of hepatocarcinogens. They treat embryos of rainbow trout (S. gairdneri) for brief periods of time (usually 30 minutes) and raise the fish under standard conditions. After 1 year the fishes are examined for the presence of hepatocellular carcinomas. Though the system is still in the testing stage, it appears to be quite sensitive for detecting the tumorigenic effects of aflatoxin B_1, sterigmatocystin, and N-methyl-N'-nitro-N-nitrosoguanidine.

Hart and Setlow (1975) have been experimenting with a system to study the mechanisms involved in the induction of tumors. They treat isolated cells from embryos of the gynogenetic, clonal Amazon mollie (Poecilia formosa) with carcinogenic agents and inject the treated cells into the abdominal cavity of isogenic recipients. After 6 to 9 months the fish are examined for the presence of tumors. Thus far, cells treated with UV light and N-acetoxyacetylaminofluorene have caused the production of thyroid tumors in recipient animals.

Some of these systems may prove applicable for the detection of environmental carcinogens. Chromatographic methods could be used to determine the type and concentration of chemicals present in the water samples, and the carcinogenic potential of these compounds could be tested individually or in specific mixtures. A more feasible approach would be to concentrate organics present in the water samples using techniques such as reverse osmosis (Kopfler et al., 1977). These concentrates could then be tested directly.

The advantages of these experimental carcinogenicity systems compared to epidemiological studies are that they are sensitive, the species examined do not have to be endogenous to the water system under investigation, and conditions of exposures can be rigorously controlled. The major disadvantage of all the aforementioned methods is the time (usually 6 months to 1 year) required to determine if a chemical or mixture is carcinogenic.

C. *In Vitro* Short-Term Methods

Because the number of chemicals that may have to be analyzed for carcinogenicity is extensive, there has been a trend toward the development of short-term genetic tests for identifying genotoxic agents. These are based on the premise that both mutagenic and carcinogenic events are initiated through changes in the genome. Thus, by screening for chemicals that can interact with DNA, one is searching, in essence, for both carcinogenic and mutagenic potential.

In general short-term assays are of two types: *in vivo* or *in vitro*. Since most genotoxic compounds must be metabolized to exert their genotoxic effects, and cells grown in culture retain very limited metabolic capabilities, *in vitro* systems usually contain a microsomal or feeder-cell component to supply metabolic activation. One area of research where fishes may be useful is as a source of microsomes in *in vitro* assays. Such an approach has been used by Stott and Sinnhuber (1978). They showed that aflatoxin B_1 could be activated by the postmitochondrial fraction of rainbow trout (*S. gairdneri*) liver and produce mutations in strain TA1538 of *Salmonella*. Kurelec *et al.* (1979) used the postmitochondrial fraction from carp and mullet (*Mugil cephalus*) liver to determine mixed-function oxidase-inducing potential of seawater extracts from areas of differing hydrocarbon contamination. These microsomes were then used in an Ames-type *Salmonella* test to assess the mutagenicity of seawater extracts from the sampled areas. The extracts from polluted water had much higher benzo[a]pyrene monooxidase-inducing potential and produced significantly more mutations in the *Salmonella* test than extracts from the "clean" zone. The authors suggest that the monitoring of mixed-function oxidase activity in fish liver may be useful as an indicator of xenobiotic contamination. Another approach might be to use fish hepatocytes in coculture with fibroblasts. The cocultures could be treated with concentrated extracts from water systems, and cytogenetic damage could be scored in the fibroblasts. Fibroblasts from the fish

Ameca splendens have already been used to study radiation- and chemically induced chromosome breakage and sister chromatid exchange (Woodhead, 1976; Barker and Rackham, 1979). However, this *in vitro* system lacks metabolic activation.

Aside from the use of fish microsomal preparations, it is difficult to see the value of piscine *in vitro* systems for detecting waterborne genotoxicants. Standard Ames-type tests among others should suffice. The value of using fish tissues in *in vitro* experimentation should be to gain insight into piscine metabolism (Elcombe and Lech, 1979), chromosome structure and sensitivity (Woodhead, 1976; Barker and Rackham, 1979), and DNA repair (Stich and Acton, 1976).

D. *In Vivo* Short-Term Methods

In vivo systems offer a realistic approach to the study of the genotoxic potential of water samples. By using intact animals, the processes of dilution, excretion, concentration, and metabolism of xenobiotics are accounted for inherently by the animal itself. The use of fishes in *in vivo* short-term tests for waterborne genotoxicants seems a logical choice. To this end, studies have been undertaken to investigate the production of dominant lethals, point mutations, and cytogenetic damage in the genomes of fishes.

1. *Dominant Lethal Studies*

Generally, dominant lethal studies involve the exposure of males to the test agent followed by the mating of these to virgin females. When fishes are used, fry or embryos are examined to determine the number of living, dead, and/or abnormal progeny produced. Results are compared to controls, and a reduction in the number of viable, normal embryos coupled with an increase in dead embryos is evidence for the production of dominant lethal mutations. The majority of these lethals are believed to result from chromosome breakage and elimination through the breakage–fusion–breakage cycle (Generoso, 1973; Binkert and Schmid, 1977; Bürki and Sheridan, 1978).

Egami and Hyodo-Taguchi (1973) X-irradiated male medakas (*Oryzias latipes*) with 2000–16,000 R and demonstrated a dose-dependent increase in dominant lethals in the resulting embryos and fry. Hemsworth and Wardhaugh (1978) and Mathews *et al.* (1978) have shown that fishes can be used to detect chemically induced dominant lethal mutations. The former group injected male *Tilapia mossambica* with either methylmethane sulfonate (50 or 70 mg/kg) or dimethylmyleran

(0.8 or 4 mg/kg) and produced a dose-related increase in "embryopathies incompatible with survival." Mathews et al. (1978) exposed male guppies (*Poecilia reticulata*) by injection with or immersion in 0.1, 0.2, or 0.4 mg triethylenemelamine/kg. These treatments resulted in a dose-dependent increase in the percentages of dead embryos with a corresponding reduction in brood size.

Assaying for dominant lethals can be done quite efficiently if the number of fertilization events can be scored accurately (i.e., the dead embryos are not reabsorbed by ovoviviparous females). Although dominant lethal studies usually are performed under tightly controlled laboratory conditions, it does seem feasible to develop protocols for environmental monitoring. Males of a selected species could be "caged" and exposed in the water system under study. After an appropriate exposure these could be brought back into the laboratory and mated to unexposed females. Alternatively, water samples could be obtained, the chemical constituents concentrated, and the males exposed in the laboratory. However, the main disadvantages of dominant lethal studies for use in the detection of waterborne genotoxicants are that these tests are not sensitive to low dose effects and will generally only detect agents that are capable of causing chromosome breakage.

2. Specific-Locus Assays

Specific-locus tests are another type of relatively short-term genetic assay that may prove useful in aquatic research. In these tests genetic changes expressed as alterations in phenotypic characters such as spinal curvature, color, eye formation, aggressiveness, number of vertebrae, sex ratio, and scale characteristics are examined in the progeny after treatment of the P_1 generation or fertilized eggs. Fishes are ideal subjects for specific-locus tests because of their high fecundity and rapid developmental period.

A number of studies have shown that phenotypic changes can be caused by irradiation (see review by Schröder 1973, 1979), but few studies have been undertaken in which the type of genetic damage or the frequency of specific-locus mutations could be determined accurately. Promising work has been done by Schröder (1969) and Purdom and Woodhead (1973) using the guppy (*P. reticulata*). This species was chosen for study because of its short generation time (60–100 days) and the availability of pure breeding strains with both autosomal and sex-linked markers (Schröder, 1979). Schröder (1969) found that X-irradiating both parents leads to mutations affecting spinal curvature and pigmentation. He concluded that the X-ray-induced mutation rate for the guppy is 2.5×10^{-7} mutations/R/locus, making it less X-ray sensitive than the mouse (*Mus musculus*) but more sensitive than *Drosophilia melanogaster*. Similarly, Purdom and Woodhead (1973) observed three color mutations in 853 F_1 guppies after irradiating parental males. However, they concluded that no valid comparison could be made with mice due to the differences between the two species in the types of gene complexes involved.

Another approach was taken by Dubinin *et al.* (1975). They analyzed electrophoretic variants of 11 polymorphic proteins in clones of unisex goldfish (*Carassius auratus* gibelio) to investigate the mutational effects of water containing strontium-90 and cesium-137. They found a dose-related increase in diversity and number of variants brought about by induced mutations and/or selection pressure.

Only a few papers have been published that have dealt with the chemical induction of point mutations in fishes. Tsoi (1971) and co-workers (1976) have characterized the genetics that determine scale coverings in the carp (*Cyprinus carpio*). They treated sperm with either 0.005% dimethyl sulfate or 0.005% nitrosoethyl urea and found that the frequency of scale mutations in the progeny increased 100- and 500-fold respectively.

In one of the few *in situ* environmental studies done to date, Blaylock and Frank (1980) investigated the fecundity and viability of mosquito fish (*Gambusia affinis*) that inhabited a radioactive settling basin at Oak Ridge, Tennessee. They found that chronically irradiated fish produced larger broods than control populations, but the embryos from the irradiated populations had lower viability and higher frequencies of phenotypic abnormalities. The authors surmised that the latter effects were due to increased frequencies of deleterious and recessive lethal genes in the gene pool.

Specific-locus assays have the advantage of demonstrating actual heritable gene damage in organisms. However, they are often time consuming and cumbersome. Large numbers of animals are needed in order to obtain the statistically adequate sample sizes required to detect small increases in mutation frequencies. Furthermore, due to the lack of well characterized gene loci in fishes, the choice of species for such studies is limited. The use of small aquarium species such as the guppy coupled with relatively simple multiloci screening procedure like electrophoretic protein analyses might reduce some of the problems associated with specific locus procedures and make them more practical for use in the detection of the effects of waterborne genotoxicants.

3. Cytogenetic Studies

Cytogenetic methods offer a relatively fast and simple means for detecting genotoxic effects. However, most fishes are not well suited for cytogenetic investigations because they have large numbers ($2n \geq 46$) of small chromosomes (≤ 5 μm) (Fig. 1A, B, and C) and generally display low mitotic activity. These attributes have discouraged the use of fishes in cytogenetic studies.

Prior to 1975 a number of attempts had been made to analyze the effects of genotoxic agents on the chromosomes of fishes. These investigations made use of a chromosome "squash technique" for scoring cytogenetic damage at the anaphase–telophase stage of cell division. With this methodology dividing cells are scored for the presence of chromosome bridges, fragments, "lagging chromosomes," and multipolar mitoses. This technique avoids the problem of analyzing numerous small chromosomes for breakage (clastogenicity), but it also has serious drawbacks. Due to the unavailability of methods to accumulate cells at the anaphase–telophase stage of the cell cycle, only very rapidly dividing cells such as those found in embryos and fry can be used. In addition the specific type of damage (e.g., chromatid, chromosome, aneuploidy) is often difficult to quantify, and the magnitude of an induced response seems to be quite susceptible to artifactual increase caused by poor cell viability (Pechkurenkov, 1973) or

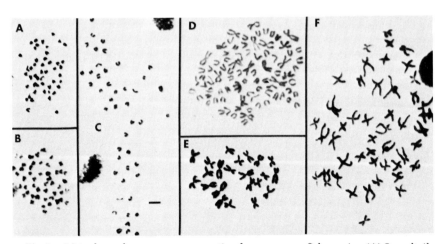

Fig. 1. Metaphase chromosome preparation from common fish species. (A) Swordtail (*Xiphophorus helleri*) $2n = 48$. (b) Fathead minnow (*Pimephales promelas*) $2n = 50$. (C) Sailfin mollie (*Poecilia latipinna*) $2n = 46$. (D) Brook trout (*Salvelinus fontinalis*) $2n = 84$. (E) Central mudminnow (*Umbra limi*) $2n = 22$. (F) Bronze catfish (*Corydoras aeneus*) $2n = 46$ (Scale = 5 μm).

the "squash technique" itself. This is shown by the unusually high control levels of damage seen in many studies (Kligerman, 1979a). Although it is not impossible to use this methodology to produce good dose–response curves with low control levels of damage (see Suyama et al., 1980), the technique is not recommended for clastogenicity studies (Evans, 1976).

Within the past decade investigators have begun to adopt mammalian cytogenetic methods of metaphase chromosome analysis for the study of clastogenicity in fishes. The examination of metaphase chromosomes offers two major advantages: (1) relatively large numbers of dividing cells can be accumulated at metaphase with spindle poisons such as colchicine or its derivatives; and (2) chromosome morphology is defined clearly at metaphase which leads to more accurate scoring of chromosome damage. As shown schematically in Fig. 2, the methodology is straightforward and applicable to both chemical and physical agents.

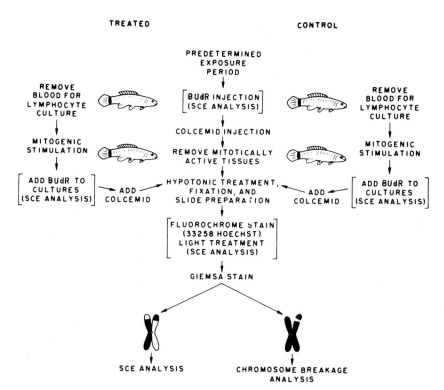

Fig. 2. Schematic representation of methodologies used to study chromosome breakage and sister chromatid exchange in fishes exposed to genotoxic agents.

Kligerman *et al.* (1975) used metaphase chromosome analysis to examine the effects of X-irradiation on the chromosomes of the central mudminnow (*Umbra limi*). They found that an exposure of 325 R of X-rays produced chromatid breaks and gaps in approximately 30% of the G_2-exposed cells. Mong and Berra (1979) expanded on this work and derived an X-irradiation dose–response curve for the mudminnow.

Kligerman and Bloom (1975) and Sugatt (1978) observed chemically induced chromosome breakage with fishes. The former group saw a slight but significant increase in chromosome breakage in mudminnows exposed to extremely high concentrations (500–3000 ppm) of maleic hydrazide. The latter investigators found a dose-dependent increase in chromosome breakage in eastern mudminnows (*U. pygmaea*) that swam in water containing 4–20 µg Trenimon/liter. Prein *et al.* (1978) and Sugatt (1978) looked at the clastogenic effects of polluted river water on the chromosomes of the eastern mudminnow. Amazingly, 28% of the metaphases from fish exposed to Rhine River water for 11 days showed chromosome damage (Prein *et al.*, 1978). However, Sugatt (1978) found only control levels (approximately 1.0%) in the same species "caged" in the heavily polluted Lek and Maas Rivers for from 1 to 4 weeks.

The importance of these studies is not in showing that radiation or chemicals can break chromosomes of fishes, but in demonstrating that fishes can be used to study clastogenicity in both the laboratory and field. Furthermore, these investigators departed from precedent by using species that were well suited for cytogenetic work instead of fishes that were of commercial importance.

Kligerman *et al.* (1975) and Kligerman (1979a, 1980) proposed a model system approach to the study of chromosome damage in fishes. This involves the selection of a representative fish species for use in studies of chromosome damage, much the way the mouse (*Mus musculus*), the Chinese hamster (*Cricetulus griseus*), and muntjac (*Muntiacus muntjac*) cells are used for mammalian cytogenetic research. Although organisms chosen for such studies should be hardy, easy to handle, and sensitive to the effects to be measured (Table I), for cytogenetic research they should have a suitable karyotype consisting of small numbers of large chromosomes. Fishes of the genus *Umbra* meet the first 4 criteria in Table I as well as having excellent karyotypes for clastogenicity studies. They possess 22 moderate to large meta- and submetacentric chromosomes (Fig. 1E). Other fish species with exceptional karyotypes are listed in Table II. Species with large numbers of very large chromosomes (Fig. 1F) such as *Coryadoras anaeus* ($2n = 46$) may also prove useful, but species of Salmonids should be avoided because of their complex and unstable karyotypes (Fig. 1D).

TABLE I

Criteria for Choosing an *in Vivo* Cytogenetics
Model Organism[a]

Species	
1.	Should possess tissues that yield adequate numbers of well-spread metaphases
2.	Should be able to withstand experimental conditions
3.	Should be easy to obtain and maintain in the laboratory
4.	Should be relatively small in size
5.	Should have a satisfactory karyotype

[a] Modified from Kligerman, 1980.

Studies of chromosome breakage are not the only means to use cytogenetics for detecting genotoxic agents. While clastogenicity is a sensitive indicator of radiation-induced genetic damage, many genotoxic chemicals do not produce significant levels of chromosome breakage until high levels of exposure are reached.

Perry and Evans (1975) using the 5-bromodeoxyuridine (BUdR) methodology for visualizing sister chromatid exchange (SCE) (Zak-

TABLE II

Fish Species with Karyotypes Suitable for
Clastogenic Studies[a]

Species	Diploid chromosome number
Ameca splendens	$2n = 26$
Aphyosemion celiae	$2n = 20$
Aphyosemion christyi	$2n = 18$
Aphyosemion franzwerneri	$2n = 22$
Apteronotus albifrons	$2n = 24$
Characodon lateralis	$2n = 24$
Galaxias maculatus	$2n = 22$
Nothobranchius rachowi	$2n = 16$
Sphaericthys osphoromoides	$2n = 16$
Umbra limi	$2n = 22$
Umbra pygmaea	$2n = 22$

[a] Modified from Kligerman, 1980.

harov and Egolina, 1972; Latt, 1973) found that many mutagenic carcinogens induced significant elevations in SCE frequencies at dose
levels sometimes orders of magnitude lower than those causing significant increases in chromosome breakage. Furthermore, Latt et al.
(1979) state that although the exact meaning and mechanisms of SCE
are not known, there have been no convincing reports of chemicals
that induce high levels of SCE that are in turn not mutagenic or carcinogenic in some test system. Thus, incorporation of a SCE methodology into an aquatic model system for the detection of genotoxic
agents should improve its sensitivity.

This methodology entails the examination of metaphase chromosomes for switches in label (SCE) after the cells have replicated twice
in the presence of BUdR. Sister chromatid exchange can be induced
either prior to BUdR exposure or after BUdR administration up until
the final replication period before the metaphases are examined. In
addition, the protocol is quite compatible with standard clastogenicity
methods (Fig. 2).

Kligerman and Bloom (1976) showed that injection of 500 μg BUdR/
gm into central mudminnows permitted the observation of SCEs in
metaphases of the gills, kidneys, and intestines. In a later study,
Kligerman (1979b) found that ip injections of the carcinogens, methyl
methane sulfonate, and cyclophosphamide, or the addition of the dye,
neutral red, to the aquarium water resulted in a dose-dependent increase in SCE (Figs. 3 and 4A, B, and C). Only at the highest doses
examined was chromosome breakage evident (Fig. 4D). These studies
proved that fishes could be used to detect low levels of both direct-
and indirect-acting genotoxicants administered directly to the animal
or added to its aquatic environment.

Recently, other researchers have utilized this procedure to study the
effects of genotoxic agents on fishes. Bishop and Valentine (1982)
verified and extended Kligerman's SCE work with the central mudminnow by producing dose-response curves for the carcinogens methylmethane sulfonate, cyclophosphamide, dimethylnitrosamine, and N-
methyl-N'-nitro-N-nitrosoguanidine. Alink et al. (1980) applied the
BUdR SCE methodology to the eastern mudminnow (U. pygmaea) and
found significant increases in SCE frequencies in the fish exposed to
Rhine River water van der Hoeven et al. (1980) compared the SCE-
inducing potential of ethylmethane sulfonate in U. pygmaea and No-
tobranchius rachowi and concluded that both species were equally
sensitive to this alkylating agent.

The in vitro technique of lymphocyte culture may have application
in in vivo environmental monitoring programs (Kligerman, 1979a).

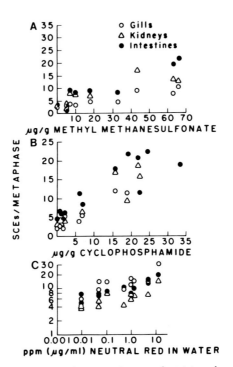

Fig. 3. Effects of mutagen administration on the sister chromatid exchange frequencies of the gills, intestines, and kidneys of central mudminnows. Fishes were injected with methylmethane sulfonate (A) or cyclophosphamide (B), or exposed to neutral red dye (C) in the aquarium water. (Modified from Kligerman, 1979b.)

Maddock and Kelly (1980) have cultured lymphocytes from the oyster toadfish (*Opsanus tau*) in the presence of BUdR and ethylmethane sulfonate. As expected, they observed elevated SCE levels. However, using this technique, fishes could be exposed *in situ,* and blood removed and cultured either with or without BUdR to score for either SCE or chromosome breakage. The use of ubiquitous fishes like the toadfish could prove valuable in a program designed to detect the presence of genotoxic contamination in aquatic environments. Moreover, because blood can be removed nonlethally from some of the larger fish species, each animal can serve as its own control; that is, blood can be removed and cultured before and after treatment.

Thus, these initial investigations show that cytogenetic techniques hold significant though underutilized potential for inclusion in aquatic toxicology programs. The methods are relatively simple and quick, readily adaptable to suitable fish species, and sensitive to the effects

Fig. 4. Metaphases from central mudminnows showing sister chromatid exchange (A, B, C) and chromosome breakage (D). (A) Control showing 4 SCEs (arrows). (B) Renal metaphase with 10 SCEs from a fish exposed to 7 μg/gm methylmethane sulfonate (MMS). Large arrows point to chromosome with multiple SCEs. (C) Intestinal metaphase with over 30 SCEs from a fish injected with 66 μg/gm MMS. Arrows point to some of the many multiple SCEs. (D) Complex aberration (arrow) from a fish injected with 103 μg/gm MMS. (Scale = 5 μm.) (From Kligerman, 1980.)

of many genotoxicants. Furthermore, both clastogenicity and SCE could be analyzed in the same organism (Fig. 2) adding to the resolving power of the system.

III. SUMMARY AND CONCLUSIONS

In the preceding pages much of the work in the field of aquatic genetic toxicology has been reviewed. Emphasis has been focused on *in vivo* methods that make use of fishes to detect the effects of genotoxicants. This is because fishes are sensitive indicator organisms, economically important human food sources, higher-order predators capable of concentrating xenobiotics, and a major part of most aquatic communities. They are thus ideally suited for use in *in vivo* investigations of waterborne genotoxic agents.

During the last decade there has been renewed interest in using fishes for genetic and cytogenetic research. Many types of long- and short-term systems have been developed; each with advantages and disadvantages. This chapter espouses the point of view that cytogenetic methods are presently the most sensitive and efficient means to use fishes to detect the effects of genotoxicants. While other methods are quite valuable, especially for investigating basic problems such as metabolism, neoplastic initiation and promotion, and mechanisms of chemical–genome interactions, *in vivo* cytogenetic assays require smaller sample sizes and can be rigorously controlled and carried out in shorter periods of time. With cytogenetic studies, the manifestation of the genotoxicant's interaction with the chromatin can be observed directly, and the entire genome, not just one or two loci, is subject to scrutiny.

One can envision a system similar to that depicted in Fig. 2. Fishes with appropriate karyotypes could be "caged" in water systems under investigation or exposed in the laboratory to test chemicals, concentrates from water systems, or other environmental agents. Mitotically active tissues could then be removed, metaphases scored for cytogenetic damage, and results compared to controls.

At present the main priorities for research should be the improvement and validation of existing systems along with the compilation of a data base from which meaningful comparisons among organisms and test systems can be made. These should involve both laboratory and field studies, and adequate positive and negative controls must be utilized to determine the sensitivity and selectivity of the systems under study. Finally, investigators must feel free to abandon impractical methodologies or species. Previously, aquatic toxicologists have been partial to the fathead minnow (*Pimephales promelas*), the bluegill (*Lepomis macrochirus*), and certain Salmonid species. However, these organisms may not be suitable for genetic toxicological research. Novel methods and species must continue to be the subjects of research in the field of aquatic genetic toxicology in order that satisfactory systems can be developed for the detection of the effects of waterborne genotoxicants.

ACKNOWLEDGMENTS

The author wishes to thank Dr. D. B. Couch, Ms. I. Utterback and Mr. J. L. Wilmer for critical review of the manuscript and Ms. L. Smith, J. Quate, and G. Erexson for help in preparing the manuscript.

REFERENCES

Alink, G. M., Frederix-Wolters, E. M. H., van der Gaag, M. A., van de Kerkhoff, J. F. J., and Poels, C. L. M. (1980). Induction of sister-chromatid exchanges in fish exposed to Rhine water. *Mutat. Res.* **78**, 369–374.

Ames, B. N., Durston, W. E., Yamasaki, E., and Lee, F. D. (1973). Carcinogens are mutagens: A simple test system combining liver homogenates for activation and bacteria for detection. *Proc. Natl. Acad. Sci. U.S.A.* **70**, 2281–2285.

Barker, C. J., and Rackham, B. D. (1979). The induction of sister chromatid exchange in cultured fish cells (*Ameca splendens*) by carcinogenic mutagens. *Mutat. Res.* **68**, 381–387.

Binkert, F., and Schmid, W. (1977). Pre-implantation embryos of Chinese hamster. II. Incidence and type of karyotype anomalies after treatment of the paternal post-meiotic germ cells with an alkylating mutagen. *Mutat. Res.* **46**, 77–86.

Bishop, W. E., and Valentine, L. C. (1982). Use of the central mudminnow (*Umbra limi*) in the development and evaluation of a sister chromatid exchange test for detecting mutagens *in vivo*. *Proc. 5th ASTM symposium on Aquatic Toxicology 1981* (in press).

Blaylock, B. G., and Frank, M. L. (1980). Effects of chronic low-level irradiation on *Gambusia affinis*. In "Radiation Effects on Aquatic Organisms" (N. Egami, ed.), pp. 81–90. Japan Sci. Soc. Press, Tokyo.

Brown, E. R., Hazdra, J. J., Keith, L., Greenspan, I., Kwapinski, J. B. G., and Beamer, P. (1973). Frequency of fish tumors found in a polluted watershed as compared to nonpolluted Canadian waters. *Cancer Res.* **33**, 189–198.

Brown, E. R., Koch, E., Sinclair, T. F., Spitzer, R., and Callaghan, O. (1979). Water pollution and diseases in fish (an epizootiologic survey). *J. Environ. Pathol. Toxicol.* **2**, 917–925.

Bürki, K., and Sheridan, W. (1978). Expression of TEM-induced damage to postmeiotic stages of spermatogenesis of the mouse during early embryogenesis. *Mutat. Res.* **52**, 107–115.

Dubinin, N. P., Altukhov, Y. P., Salmenkova, E. A., Milishnikov, A. N., and Novikova, T. A. (1975). Analysis of monomorphic markers of genes in populations as a method of evaluating the mutagenicity of the environment. *Dokl. Biol. Sci. (Engl. Transl.)* **225**, 527–530.

Egami, N., and Hyodo-Taguchi, Y. (1973). Dominant lethal mutation rates in the fish, *Oryzias latipes*, irradiated at various stage of gametogenesis. In "Genetics and Mutagenesis of Fish" (J. H. Schröder, ed.), pp. 75–81. Springer-Verlag, Berlin and New York.

Elcombe, C. R., and Lech, J. J. (1979). Induction and characterization of hemoprotein(s) P-450 and monooxygenation in rainbow trout (*Salmo gairdneri*). *Toxicol. Appl. Pharmacol.* **49**, 437–450.

Evans, H. J. (1976). Cytological methods for detecting chemical mutagens. *Chem. Mutagens* **4**, 1–29.

Generoso, W. M. (1973). Evaluation of chromosome aberration effects of chemicals on mouse germ cells. *Environ. Health Perspect.* **6**, 13–22.

Hart, R. W., and Setlow, R. B. (1975). Direct evidence that pyrimidine dimers in DNA result in neoplastic transformation. In "Molecular Mechanisms for Repair of DNA, Part B" (P. C. Hanawalt and R. B. Setlow, eds.), pp. 719–724. Plenum, New York.

Hemsworth, B. N., and Wardhaugh, A. A. (1978). The induction of dominant lethal mutations in *Tilapia mossambica* by alkane sulfonic esters. *Mutat. Res.* **58**, 263–268.

Hendricks, J. D., Wales, J. H., Sinnhuber, R. O., Nixon, J. E., Loveland, P. M., and Scanlan, R. A. (1980). Rainbow trout (*Salmo gairdneri*) embryos: A sensitive animal model for experimental carcinogenesis. *Fed. Proc. Fed. Am. Soc. Exp. Biol.* **39**, 3222–3229.

Hynes, H. B. N. (1972), "The Ecology of Rung Waters." Univ. of Toronto Press, Toronto.

Ishikawa, T., and Takayama, S. (1979). Importance of hepatic neoplasms in lower vertebrate animals as a tool in cancer research. *J. Toxicol. Environ. Health* **5**, 537–550.

Kligerman, A. D. (1979a). Cytogenetic methods for the detection of radiation-induced chromosome damage in aquatic organisms. In "Technical Report Series No. 190: Methodology for Assessing Impacts of Radioactivity on Aquatic Ecosystems," pp. 349–367. IAEA, Vienna.

Kligerman, A. D. (1979b). Induction of sister chromatid exchange in the central mudminnow following *in vivo* exposure to mutagenic agents. *Mutat. Res.* **64**, 205–217.

Kligerman, A. D. (1980). The use of aquatic organisms to detect mutagens that cause cytogenetic damage. In "Radiation Effects on Aquatic Organisms" (N. Egami, ed.), pp. 241–252. Japan Sci. Soc. Press, Tokyo.

Kligerman, A. D., and Bloom, S. E. (1975). A cytogenetics model for the study of chromosome aberrations in fishes. *Mutat. Res.* **31**, 334–335a.

Kligerman, A. D., and Bloom, S. E. (1976). Sister chromatid differentiation and exchanges in adult mudminnows (*Umbra limi*) after *in vivo* exposure to 5-bromodeoxyuridine. *Chromosoma* **56**, 101–109.

Kligerman, A. D., Bloom, S. E., and Howell, W. M. (1975). *Umbra limi*: A model for the study of chromosome aberrations in fishes. *Mutat. Res.* **31**, 225–233.

Kopfler, F. C., Coleman, W. E., Melton, R. G., and Tardiff, R. G. (1977). Extraction and identification of organic micropollutants: Reverse osmosis method. *Ann. N.Y. Acad. Sci.* **298**, 20–30.

Kraybill, H. F. (1976). Distribution of chemical carcinogens in aquatic environments. *Prog. Exp. Tumor Res.* **20**, 3–34.

Kraybill, H. F. (1977). Global distribution of carcinogenic pollutants in water. *Ann. N.Y. Acad. Sci.* **298**, 80–89.

Kurelec, B., Matijasevic, Z., Rijavec, M., Alacevic, M., Britvic, S., Müller, W. E. G., and Zahn, R. K. (1979). Induction of benzo[a]pyrene monooxygenase in fish and the *Salmonella* test as a tool for detecting mutagenic/carcinogenic xenobiotics in the aquatic environment. *Bull. Environ. Contam. Toxicol.* **21**, 799–807.

Latt, S. A. (1973). Microfluorometric detection of deoxyribonucleic acid replication in human metaphase chromosomes. *Proc. Nat. Acad. Sci. U.S.A.* **70**, 3395–3399.

Latt, S. A., Schreck, R. R., Loveday, K. S., and Shuler, C. F. (1979). *In vitro* and *in vivo* analysis of sister chromatid exchange. *Pharmacol. Rev.* **30**, 501–535.

Longwell, A. C. (1976), "Chromosome Mutagenesis in Developing Mackeral Eggs Sampled from the New York Bight." NOAA Technical Memorandum ERL MESA-7, Marine Ecosystems Analysis Program Office, Boulder, Colorado.

Macek, K. J. (1980). Aquatic toxicology: Fact or fiction? *Environ. Health Perspect.* **34**, 159–163.

Maddock, M. B., and Kelly, J. J. (1980). A sister chromatid exchange assay for detecting genetic damage to marine fish exposed to mutagens and carcinogens. In "Water Chlorination Environmental Impact and Health Effects" (R. L. Jolley, W. A. Brungs, R. B. Cumming, and V. A. Jacobs, eds.), Vol. 3, pp. 835–844. Ann Arbor Sci. Publ., Ann Arbor, Michigan.

Matsushima, T., and Sugimura, T. (1976). Experimental carcinogenesis in small aquarium fishes. *Prog. Exp. Tumor Res.* **20**, 367–379.

Mathews, J. G., Favor, J. B., and Crenshaw, J. W. (1978). Dominant lethal effects of triethylenemelamine in the guppy *Poecilia reticulata*. *Mutat. Res.* **54**, 149–157.

McCann, J., Choi, E., Yamasaki, E., and Ames, B. N. (1975). Detection of carcinogens as mutagens in the *Salmonella*/microsome test: Assay of 300 chemicals. *Proc. Nat. Acad. Sci. U.S.A.* **72**, 5135–5139.

Mearns, A. J., and Sherwood, M. J. (1977). Distribution of neoplasms and other diseases in marine fishes relative to the discharge of waste water. *Ann. N.Y. Acad. Sci.* **298**, 210–224.

Miller, J. A., and Miller, E. C. (1975). Chemical and radiation carcinogenesis in man and experimental animals. *Radiat. Res. Proc. Int. Congr.*, 5th, 1974 pp. 158–168.

Mong, S. J., and Berra, T. M. (1979). The effects of increasing dosages of x-irradiation on the chromosomes of the central mudminnow, *Umbra limi* (Kirkland) (Salmomiformes:Umbridae). *J. Fish Biol.* **14**, 523–527.

Oishi, K., Yamazaki, F., and Harada, T. (1976). Epidermal papillomas of flatfish in the coastal waters of Hokkaido, Japan. *J. Fish. Res. Board Can.* **33**, 2011–2017.

Oppermann, K. (1913). Die Entwicklung von Forelleneiern nach Befruchtung mit Radiumbestrahltgen Samenfäden. *Arch. Mikrosk. Anat.* **83**, 307–323.

Pechkurenkov, V. L. (1973). Appearance of chromosome aberrations in larvae of the loach (*Misgurnus fossilis* L.) developing in solutions of strontium-90 and yttrium-90 of various activities. *Sov. Genet. (Engl. Transl.)* **6**, 1323–1332.

Perry, P., and Evans, H. J. (1975). Cytological detection of mutagen-carcinogen exposure by sister chromatid exchange. *Nature (London)* **258**, 121–125.

Prein, A. E., Thie, G. M., Alink, G. M., Koeman, J. H., and Poels, C. L. M. (1978). Cytogenetic changes in fish exposed to water of the river Rhine. *Sci. Total Environ.* **9**, 287–291.

Purdom, C. E., and Woodhead, D. S. (1973). Radiation damage in fish. In "Genetics and Mutagenesis of Fish" (J. H. Schröder, ed.), pp. 67–73. Springer-Verlag, Berlin and New York.

Schröder, J. H. (1969). X-ray-induced mutations in the poecillid fish, *Lebistes reticulatus* Peters. *Mutat. Res.* **7**, 75–90.

Schröder, J. H. (1973). Teleosts as a tool in mutation research. In "Genetics and Mutagenesis of Fish" (J. H. Schröder, ed.), pp. 91–99. Springer-Verlag, Berlin and New York.

Schröder, J. H. (1979). Methods for screening radiation-induced mutations in fish. In "Methodology for Assessing Impacts of Radioactivity on Aquatic Ecosystems," Technical Report Series 190, pp. 371–402. IAEA, Vienna.

Sinnhuber, R. O., Hendricks, J. D., Wales, J. H., and Putnam, G. B. (1977). Neoplasms in rainbow trout, a sensitive animal model for environmental carcinogenesis. *Ann. N.Y. Acad. Sci.* **298**, 389–408.

Solberg, A. N. (1938). The susceptibility of *Fundulus heteroclitus* embryos to x-radiation. *J. Exp. Zool.* **78**, 441–465.

Sonstegard, R. A. (1977). Environmental carcinogenesis studies in fishes of the Great Lakes of North America. *Ann. N.Y. Acad. Sci.* **298**, 261–269.

Stanton, M. F. (1965). Diethylnitrosamine-induced hepatic degeneration and neoplasia in the aquarium fish, *Branchydanio rerio*. *J. Natl. Cancer Inst.* **34**, 117–130.

Stara, J. F., Kello, D., and Durkin, P. (1980). Human health hazards associated with chemical contamination of aquatic environment. *Environ. Health Perspect.* **34**, 145–158.

Statham, C. N., Melancon, M. J., Jr., and Lech, J. J. (1976). Bioconcentration of xenobiotics in trout bile: A proposed monitoring aid for some waterborne chemicals. *Science* **193**, 680–681.

Stich, H. F., and Acton, A. B. (1976). The possible use of fish tumors in monitoring for carcinogens in the marine environment. *Prog. Exp. Tumor Res.* **20**, 44–54.

Stich, H. F., Acton, A. B., and Forrester, C. R. (1976). Fish tumors and sublethal effects of pollutants. *J. Fish. Res. Board Can.* **33**, 1993–2001.

Stott, W. T., and Sinnhuber, R. O. (1978). Trout hepatic enzyme activation of aflatoxin B_1 in a mutagen assay system and the inhibitory effect of PCBs. *Bull. Environ. Contam. Toxicol.* **19**, 35–41.

Sugatt, R. H. (1978). Chromosome aberrations in the eastern mudminnow (*Umbra pygmaea*) exposed *in vivo* to Trenimon or river water. Report MD-N&E 78/3 of Central Laboratory TNO, Delft, Netherlands.

Suyama, I., Etoh, H., and Ichikawa, R. (1980). Effects of ionizing radiation on the development of *Limanda* eggs. In "Radiation Effects on Aquatic Organisms" (N. Egami, ed.), pp. 205–207. Japan Sci. Soc. Press, Tokyo.

Svärdson, G. (1945). Chromosome studies on Salmonidae. Reports from the Swedish State Institute of Fresh-Water Fishery Research, No. 23, pp. 1–151, Drottningholm.

Tsoi, R. M. (1970). Effect of nitrosomethyl urea and dimethyl sulfate on sperm of rainbow trout (*Salmo irideus* Gibb.) and Peled (*Coregonus peled* Gmel.). *Dokl. Biol. Sci. (Engl. Trans.)* **189**, 849–851.

Tsoi, R. M. (1971). Effect of methyl sulfate on mutation frequency of genes S and n in the carp (*Cyprinus carpio* L.). *Dokl. Biol. Sci. (Engl. Trans.)* **197**, 197–200.

Tsoi, R. M. (1974). Chemical gynogenesis in *Salmo irideus* and *Coregonus peled*. *Soviet Genet. (Engl. Trans.)* **8**, 275–277.

Tsoi, R. M., Men'shova, A. I., and Golodov, Y. F. (1975). Specificity of the influence of chemical mutagens on spermatozoids of *Cyprinus carpio* L. *Soviet Genet. (Engl. Trans.)* **10**, 190–193.

Tsoi, R. M., Men'shova, A. I., and Golodov, Y. F. (1976). Frequency of spontaneous and induced mutations in genes determining carp scales. *Soviet Genet. (Engl. Trans.)* **10**, 1368–1370.

van der Hoeven, J. C. M., Bruggeman, I. M., Alink, G. M., and Koeman, J. H. (1980). *Nothobranchius rachowi*, a new experimental animal for cytogenetic studies. *Mutat. Res.* **85**, 240–241a.

Woodhead, D. S. (1976). Influence of acute irradiation on induction of chromosome aberrations in cultured cells of the fish *Ameca splendens*. In "Biological and Environmental Effects of Low-Level Radiation," (M. Lewis, ed.), Vol. 1, pp. 67–76, IAEA, Vienna.

Zakharov, A. F., and Egolina, N. A. (1972). Differential spiralization along mammalian mitotic chromosomes. I. BUdR-revealed differentiation in Chinese hamster chromosomes. *Chromosoma* **38**, 341–365.

Index

2-Amino-9*H*-pyrido(2,3-*b*)indole, 77
3-Aminotriazole, 11
Amitrole, 40, 41, 45
Anchorage dependence, loss of, 229
Anesthetic gases, 342, 343
Aneuploidy, 2, 5, 423
Antabuse, 204
Anthracene, 105, 124
Antigens, 149
Antimutagens, 135, 137
Antioxidants, 84
Antipain, 185
Antitumor, 2
Apple brandy, 83
Apple juice, 128
Apricots, 128
Aquatic ecosystem, 427
Aquatic environment, 435
Aquatic genetic toxicology, 436, 438
Arabidopsis thaliana, 422
Arabinose, 128
Aramite, 48
Arginine, 26
L-Arginine, 130
Aromatic amines, 146
Arsenic compounds, 38, 39, 44, 341
  cytogenetic monitoring of, 257
Aryl hydrocarbon hydroxylase (AHH) in
  lymphocytes, 202
Aryl hydroxylamines, 146
Asbestos, 39, 41, 44
Aspartic acid, 341
Assay
  for chromosome breakage, 179
  comparison of cytogenetic and other,
    190
  cytogenic, 171
  for micronuclei, 180
  for sister chromatid exchanges, 179
  for transformation, comparison of, 153
  for unscheduled DNA synthesis, 181
  validated for transformation, 151
Assessment of risk, 36
Astragalin, 80
Asynapsis, 417
Ataxia telangiectasia, 189, 369
Atmospheric mutagens, 407, 408, 409,
  416
Auramine, 38, 40
  dye mixture, 44

manufacture of, 39
pure, 45
technical grade, 11
8-Azaadenine, 220
8-Azaguanine resistance, *see* 6-Thioguan-
  ine resistance, 220, 308, 328
Azaserine, 48
Azathioprine, 40, 45

B

3t3/Balb, 182
Balb/C-3t3, 151
Bananas, 128
Banding techniques, 244
Battery, 144
Beef hamburger, 126
Benz(*a*)anthracene, 48, 54, 55, 59, 124
Benzanthracene, 5
Benzene, 5, 12, 38, 39, 41, 44, 58, 59
Benzidine, 39, 44, 188
Benzo(*b*)fluoranthene, 48, 124
Benzo(*a*)pyrene, 11, 22, 48, 55, 58, 59,
  105, 124, 126, 188
Benzo(*a*)pyrenediolepoxide, 147
3-*OH*-Benzo(*a*)pyrene glucuronyltransfer-
  ase, 182
Benzo(*a*)pyrene monooxygenase, 182
Benzyl violet 4b, 48
Beryllium
  certain beryllium compounds, 40, 45
  oxide, 48
  phosphate, 48
  sulfate, 48
Betel nuts, 123
BHK-TK, 182
Bioassays, 90
Biological detectors, 435
Bis(2-chloroethyl)methylamine, 5
N,N-Bis(2-chloroethyl)-2-naphthylamine,
  39, 44
Bis(chloromethyl)ether, 39, 44
Black pepper, 126
Black tea, 82
Blackfan-diamond, 189
Bleomycin, 183
Bloom's syndrome, 183, 189, 368, 370
Bone marrow micronucleus, 9
Boot and shoe manufacture and repair,
  38, 39

CELL BIOLOGY: A Series of Monographs

EDITORS

D. E. BUETOW

Department of Physiology
and Biophysics
University of Illinois
Urbana, Illinois

I. L. CAMERON

Department of Anatomy
University of Texas
Health Science Center at San Antonio
San Antonio, Texas

G. M. PADILLA

Department of Physiology
Duke University Medical Center
Durham, North Carolina

A. M. ZIMMERMAN

Department of Zoology
University of Toronto
Toronto, Ontario, Canada

Stuart Coward (editor). DEVELOPMENTAL REGULATION: Aspects of Cell Differentiation, 1973

I. L. Cameron and J. R. Jeter, Jr. (editors). ACIDIC PROTEINS OF THE NUCLEUS, 1974

Govindjee (editor). BIOENERGETICS OF PHOTOSYNTHESIS, 1975

James R. Jeter, Jr., Ivan L. Cameron, George M. Padilla, and Arthur M. Zimmerman (editors). CELL CYCLE REGULATION, 1978

Gary L. Whitson (editor). NUCLEAR–CYTOPLASMIC INTERACTIONS IN THE CELL CYCLE, 1980

Danton H. O'Day and Paul A. Horgen (editors). SEXUAL INTERACTIONS IN EUKARYOTIC MICROBES, 1981

Ivan L. Cameron and Thomas B. Pool (editors). THE TRANSFORMED CELL, 1981

Arthur M. Zimmerman and Arthur Forer (editors). MITOSIS/CYTOKINESIS, 1981

Ian R. Brown (editor). MOLECULAR APPROACHES TO NEUROBIOLOGY, 1982

Henry C. Aldrich and John W. Daniel (editors). CELL BIOLOGY OF *PHYSARUM* AND *DIDYMIUM*, Volume I: Organisms, Nucleus, and Cell Cycle, 1982

John A. Heddle (editor). MUTAGENICITY: New Horizons in Genetic Toxicology, 1982

Potu N. Rao, Robert T. Johnson, and Karl Sperling (editors). PREMATURE CHROMOSOME CONDENSATION: Application in Basic, Clinical, and Mutation Research, 1982

George M. Padilla and Kenneth S. McCarty, Sr. (editors). GENETIC EXPRESSION IN THE CELL CYCLE, 1982

In preparation

Henry C. Aldrich and John W. Daniel (editors). CELL BIOLOGY OF *PHYSARUM* AND *DIDYMIUM*, Volume II: Differentiation, Metabolism, and Methodology, 1982.

David S. McDevitt (editor). CELL BIOLOGY OF THE EYE, 1982

Govindjee (editor). PHOTOSYNTHESIS, Volume I: Energy Conversion by Plants and Bacteria, 1982; Volume II: Development, Carbon Metabolism, and Plant Productivity, 1982

P. Michael Conn (editor) CELLULAR REGULATION OF SECRETION AND RELEASE, 1982